Starch-Based Polymeric Materials and Nanocomposites

Chemistry, Processing, and Applications

Edited by
Jasim Ahmed
Brijesh K. Tiwari
Syed H. Imam
M.A. Rao

CRC Press
Taylor & Francis Group
Boca Raton   London   New York

CRC Press is an imprint of the
Taylor & Francis Group, an **informa** business

CRC Press
Taylor & Francis Group
6000 Broken Sound Parkway NW, Suite 300
Boca Raton, FL 33487-2742

First issued in paperback 2016

Version Date: 20120224

ISBN 13: 978-1-138-19862-3 (pbk)
ISBN 13: 978-1-4398-5116-6 (hbk)

Visit the Taylor & Francis Web site at
http://www.taylorandfrancis.com

and the CRC Press Web site at
http://www.crcpress.com

Contents

Preface

Biodegradable polymers from renewable resources have attracted much attention in recent years. Starch has been considered as one of the most promising candidates among biopolymers mainly because of its attractive combination of availability and price. Starch is unique in its application and versatility. In addition to its contribution to processing industries, research on starch is currently more focused on the development of biodegradable packaging materials and the synthesis of starch derivatives, bionanoparticles, and bionanocomposites. In order to extend applications of starch as primary packaging materials, current research has focused on starch blends, especially starch/polyester (e.g., polylactides [PLA], polycaprolactone [PCL], and polybutylene succinate [PBS]) biodegradable blends. However, starch/polyester blends are not compatible for blending and, therefore, a reactive compatibilization process known as reactive extrusion is used to synthesize starch/polyester blends at high starch levels (>20 wt%). The process can also be expanded to plasticized starch and nanocomposites by incorporating nanoparticles in the blend. Such nanocomposites exhibit improved mechanical strength, solvent and UV resistance, gas barrier properties, thermal stability, and flame retardancy as compared to conventional composites. Rheological and mechanical properties during plasticization and nanocomposite formation provide a better understanding of structural modification and degree of dispersions of nanoparticles into biopolymers.

The starch industry is one the largest manufacturing industries, and the significant contribution of food professionals to the industry is well recognized. A professional food/starch scientist should be aware of the latest developments in starch chemistry, rheology, starch derivatives, starch-based nanocomposites, and their applications. *Starch-Based Polymeric Materials and Nanocomposites: Chemistry, Processing and Applications* has been written primarily to fulfill those expectations and is intended for students undertaking graduate courses in food science, food technology, biotechnology, human nutrition, and related areas. The book has particular relevance to students' understanding of starch from basics to applications as well as for their future professions. It should also be of interest to graduates of other disciplines who are working in starch-related industries.

The chapters in this book have been contributed by leading experts who have both academic and professional credentials. We are thankful to our contributors for their wonderful cooperation and, finally, we are indebted to our families for their continued support throughout the project. It is worth mentioning that there was an excellent coordination among editors that resulted in delivering the book on time. Our special thanks to Steve Zollo, Senior Editor of Taylor & Francis, for his constant encouragement and excellent communication.

Editors

Dr. Jasim Ahmed has been associated with food processing teaching, research, and industry in India, Canada, and the Middle East for about two decades. He is mostly involved in research areas on food product development, food rheology and structure, novel food processing, and the thermal behavior of foods. His current research focus is on bionanocomposites and application to food packaging. Dr. Ahmed recently moved to the Kuwait Institute for Scientific Research, Kuwait as research scientist to develop the Food Processing Program. He has published 92 peer-reviewed research papers and 21 book chapters. Dr. Ahmed is coeditor of four books including *Novel Food Processing: Effects on Rheological and Functional Properties* (CRC Press Publication) and *Handbook of Food Process Design* (Wiley-Blackwell). Dr. Ahmed is one of the editors of the *International Journal of Food Properties*.

Dr. Brijesh K. Tiwari is a lecturer in food engineering at Manchester Metropolitan University (MMU), United Kingdom, and a senior consultant at Manchester Food Research Centre, United Kingdom. Prior to joining MMU, he was a lecturer in biosystems engineering at University College Dublin, Ireland. Dr. Tiwari served the Indian Institute of Crop Processing Technology as a research scientist (2004–2006), where he was actively involved in developing a research portfolio in grain processing. He was also responsible for numerous consultancy assignments, technology transfer projects, and the management of a number of industry-focused research projects. His main research accomplishments are in the areas of novel food processing and preservation technologies, grain processing, and mathematical modeling of food processes. To date he has contributed about 50 peer-reviewed Science Citation Index–listed journal publications and 20 chapters published in books by leading publishers; he has also presented over 30 refereed conference papers at key international research conferences. Dr. Tiwari is a coeditor of four books: *Pulse Foods: Quality, Technology, and Nutraceutical Application* (Elsevier); *Ozone in Food Processing* (Wiley-Blackwell, in development); *Novel Thermal and Non-Thermal Technologies for Fluid Foods* (Elsevier); and the *Handbook of Phytochemicals: Sources, Stability and Extraction* (Wiley-Blackwell). He is also the book series editor for IFST Food Processing Technology.

Dr. Syed H. Imam is a senior research chemist at the United States Department of Agriculture, Albany. He conducts research to enhance the utilization of agriculturally derived renewable polymers for nonfood applications as an alternative to petroleum chemicals and provides fundamental knowledge in the development of polymer-based materials of tailored properties. With over 25 years of research experience, Dr. Imam is internationally recognized for his accomplishments and contributions in the field. He has authored or coauthored over 130 publications in peer-reviewed professional journals, delivered over 180 invited talks at national and international meetings, and has numerous patents to his credit. He has served on the ASTM-D20 Committee on Biodegradable Polymers and currently serves as a U.S. representative on the UNIDO–ICS Expert Panel on Biodegradable Polymers and Materials. He is a member of the editorial boards of the *Journal of Polymers and the Environment* and the *International Journal of Biochemistry and Biotechnology* and serves as an associate editor of the *Journal of Bioenergy and Biomaterials*. He is a chief editor of the ACS book titled *Biopolymers: Natures Advanced Materials* published in 1998 by the Oxford University Press. He is also a coeditor of another book titled *Produce Degradation: Reaction Pathways and Prevention* published in 2005 by CRC Press.

Dr. M.A. Rao is an emeritus professor at Cornell University, Geneva, New York. His main research interest is in rheological properties of foods: principles, measurement, and applications. He is a fellow of the Institute of Food Technologists, the Association of Food Scientists and Technologists

(India), and the International Academy of Food Science and Technology. He was conferred the Lifetime Achievement Award by the International Association of Engineering and Food in 2011, the Scott Blair Award for his contributions to food rheology by the American Association of Cereal Chemists in 2000, and an honorary membership by the Brazilian Association of Rheology in 2011. He has also been chosen as a distinguished food engineer by the Food Process Engineering Institute of the American Society of Agricultural Engineers in 2003. Dr. Rao was awarded Fulbright–Hays senior scholarships to Brazil, 1980–1981, and to Portugal, 1988–1989. He has contributed about 250 journal papers and book chapters on food rheology and food engineering. He is the editor of the book *Rheology of Fluid and Semisolid Foods: Principles and Applications*, second edition, Springer, New York, 2007, and a coeditor of *Engineering Properties of Foods*, third edition, Taylor & Francis, Boca Raton, Florida, 2005. He is also a scientific editor of the Food Engineering and Physical Properties section of the *Journal of Food Science*.

Contributors

Mohamed A. Abdelwahab
Bioproduct Chemistry and Engineering
  Research
Western Regional Research Center
Agricultural Research Service
United States Department of Agriculture
Albany, California
and
Department of chemistry and Industrial
  Chemistry
University of Pisa
Pisa, Italy
and
University of Tanta
Tanta, Egypt

Jasim Ahmed
Food and Nutrition Program
Kuwait Institute for Scientific Research
Safat, Kuwait

Debbie P. Anderson
Bioproducts and Bioprocesses National Science
  Program
Agriculture and Agri-Food Canada
Saskatoon, Saskatchewan, Canada

G. Astray
Department of Physical Chemistry
University of Vigo at Ourense
Ourense, Spain

Charles Brennan
Department of Wine, Food and Molecular
  Biosciences
Lincoln University
Lincoln Christchurch, New Zealand

David Bressler
Department of Agricultural, Food and
  Nutritional Science
University of Alberta
Edmonton, Alberta, Canada

Peter R. Chang
Bioproducts and Bioprocesses National Science
  Program
Agriculture and Agri-Food Canada
and
Department of Chemical and Biological
  Engineering
University of Saskatchewan
Saskatoon, Saskatchewan, Canada

Bor-Sen Chiou
Bioproduct Chemistry and Engineering
  Research
Western Regional Research Center
Agricultural Research Service
United States Department of Agriculture
Albany, California

A. Cid
Department of Physical Chemistry
University of Vigo at Ourense
Ourense, Spain

Anitha R. Dudhani
Department of Pharmacy Practice
Monash Institute of Pharmaceutical Sciences
Monash University
and
School of Biomedical and Health Sciences
Victoria University
and
Healthline Pharmacy
Melbourne, Victoria, Australia

Gregory M. Glenn
Bioproduct Chemistry and Engineering
  Research
Western Regional Research Center
Agricultural Research Service
United States Department of Agriculture
Albany, California

C. González-Barreiro
Department of Analytical and Food Chemistry
University of Vigo at Ourense
Ourense, Spain

Sherald H. Gordon
Plant Polymer Research Unit
National Center for Agricultural Utilization
  Research
Agricultural Research Service
United States Department of Agriculture
Peoria, Illinois

Mahesh Gupta
School of Food Science and Environmental
  Health
Dublin Institute of Technology
Dublin, Ireland

R.E. Harry-O'kuru
Bio-Oils Research Unit
National Center for Agricultural Utilization
  Research
Agricultural Research Service
United States Department of Agriculture
Peoria, Illinois

Kevin Holtman
Bioproduct Chemistry and Engineering
  Research
Western Regional Research Center
Agricultural Research Service
United States Department of Agriculture
Albany, California

Ratnajothi Hoover
Department of Biochemistry
Memorial University of Newfoundland
St. John's, Newfoundland, Canada

Syed H. Imam
Bioproduct Chemistry and Engineering
  Research
Western Regional Research Center
Agricultural Research Service
United States Department of Agriculture
Albany, California

Jihong Li
Department of Agricultural, Food and
  Nutritional Science
University of Alberta
Edmonton, Alberta, Canada

Hung-Ju Liao
Department of Food Science
National Chiayi University
Chiayi City, Taiwan

Sathya B. Kalambur
Frito Lay
Plano, Texas
and
Department of Food Sciences
Cornell University
Ithaca, New York

Xiaofei Ma
Department of Chemistry
School of Science
Tianjin University
Tianjin, China

A.V. Machado
Institute of Polymers and Composites/I3N
University of Minho
Guimarães, Portugal

J.A. Manso
Department of Physical Chemistry
University of Vigo at Ourense
Ourense, Spain

J.C. Mejuto
Department of Physical Chemistry
University of Vigo at Ourense
Ourense, Spain

J. Morales
Department of Physical Chemistry
University of Vigo at Ourense
Ourense, Spain

I. Moura
Institute of Polymers and Composites/I3N
University of Minho
Guimarães, Portugal

William J. Orts
Bioproduct Chemistry and Engineering
  Research
Western Regional Research Center
Agricultural Research Service
United States Department of Agriculture
Albany, California

M.A. Rao
Department of Food Science
Cornell University
Geneva, New York

R. Rial-Otero
Department of Analytical and Food
  Chemistry
University of Vigo at Ourense
Ourense, Spain

Randal L. Shogren
National Center for Agricultural Utilization
  Research
Agricultural Research Service
United States Department of Agriculture
Peoria, Illinois

J. Simal-Gándara
Department of Analytical and Food
  Chemistry
University of Vigo at Ourense
Ourense, Spain

Preeti Singh
Department of Material Development
Fraunhofer Institute for Process Engineering
  and Packaging
and
Food Packaging Technology
Technical University of Munich
Freising, Germany

Jirarat Tattiyakul
Department of Food Technology
Chulalongkorn University
Bangkok, Thailand

Brijesh K. Tiwari
School of Food and Consumer Technology
Manchester Metropolitan University
Manchester, United Kingdom

Thava Vasanthan
Department of Agricultural, Food and
  Nutritional Science
University of Alberta
Edmonton, Alberta, Canada

Ali Abas Wani
Department of Process Development for Plant
  Raw Materials
Fraunhofer Institute for Process Engineering
  and Packaging
and
Food Packaging Technology
Technical University of Munich
Freising, Germany

Tina G. Williams
Bioproduct Chemistry and Engineering
  Research
Western Regional Research Center
Agricultural Research Service
United States Department of Agriculture
Albany, California

Delilah F. Wood
Bioproduct Chemistry and Engineering
  Research
Western Regional Research Center
Agricultural Research Service
United States Department of Agriculture
Albany, California

Shan-Jing Yao
Department of Chemical and Biological
  Engineering
Zhejiang University
Hangzhou, China

Jiugao Yu
Department of Chemistry
School of Science
Tianjin University
Tianjin, China

Jun Zhao
Department of Chemical and Biological
 Engineering
Zhejiang University
Hangzhou, China

and

Department of Bioengineering and
 Biotechnology
College of Chemical Engineering
Huaqiao University
Xiamen, China

CHAPTER 1

Introduction
Starch as Biopolymer and Nanocomposite

Jasim Ahmed, Brijesh K. Tiwari, Syed H. Imam, and M.A. Rao

CONTENT

Among the natural polymers, starch is of interest. Starch is the second most abundant natural polymer after cellulose which is derived from renewable resources such as corn, wheat, potato, tapioca, and legumes. Starch occurs as granules in plant tissue, from which it can easily be recovered in large quantities. Due to its unique physical and chemical characteristics, starch is increasingly viewed as a potential sustainable alternative to many petroleum-based polymers. Many efforts have been made to develop starch-based polymers for reducing environmental impact and in searching more applications. Particularly, two features in a starch molecule make it a unique and versatile material. First, starch is naturally biodegradable and degrades into sugars and organic acids that serve as feedstock for manufacturing many industrial chemicals, thermoplastics, and biofuels. Second, due to its chemical structure, the starch polymer is amenable to a variety of chemical and enzymatic modifications that permits creating new and novel functionalities in starch polymer.

Starch is not truly thermoplastic in nature, but it can be converted into a continuous polymeric entangled phase by mixing with enough aqueous or nonaqueous plasticizer (polyols such as glycerol) (Angellier et al. 2006). Despite their improved mechanical properties, plasticized starch (PS)–based films cannot meet all the requirements of packaging applications. Especially, these materials remain water sensitive and, therefore, lose their barrier properties upon hydration (Gaudin et al. 2000; Dole et al. 2004). Furthermore, retrogradation and crystallization of the mobile starch chains significantly affect thermomechanical properties. The mechanical weaknesses of these materials for packaging application can be, however, improved by incorporation of a synthetic polymer or an inorganic/organic reinforcing material, including nanoparticles (Curvelo et al. 2001; Wilhelm et al. 2003; Sorrentino et al. 2007). The combinations of PS matrix with nanofillers such as layer silicates, carbon nanotubes, carbon black (CB), and cellulose and starch nanocrystals are termed as bionanocomposites. The layer silicates are generally modified as inorganic nanofillers, giving intercalation or exfoliation compounds, thereby resulting in improved mechanical properties in PS matrix (Ma et al. 2007; Chivrac et al. 2008).

Polymer/biopolymer nanocomposites have emerged as a new class of materials and attracted considerable interest and investment in research and development around the world. This happened due to their much improved mechanical, thermal, electrical, and optical properties as compared to their macro- and microcounterparts. Nanoparticles can be three-dimensional spherical and polyhedral nanoparticles, two-dimensional nanofibers, or one-dimensional disk-like nanoparticles. These nanoparticles offer enormous advantages over traditional macro- or microparticles (e.g., talc, glass, carbon fibers) due to their higher surface area and aspect ratio, improved adhesion between nanoparticle and polymer, and lower amount of loading to achieve equivalent properties (Zeng et al. 2005).

New bionanocomposites are impacting diverse areas, in particular, biomedical science. Nanocomposites based on biopolymers (polylactide, polycaprolactone, starch grafting) have been extensively used in biomedical engineering and drug delivery system. A powerful approach for controlling drug delivery is to incorporate the drug into biodegradable polymeric nano- or microparticles, which can achieve a controlled and sustained fashion through the drug diffusion and/or the polymeric carrier degradation. Nanocomposites can also be intended to be used as a carrier of antimicrobials, drug delivery, and additives. Recent studies have demonstrated their ability to stabilize the additives and efficiently control their diffusion into the food system. This control can be especially important for long-term storage of food or for imparting specific desirable characteristics, such as flavor, to a food system.

Information on starch as biopolymers and nanocomposites and their potential applications are scattered in the literature. In this book, an attempt is made to compile the latest developments in starch chemistry, technology, starch-derived products, and bionanocomposites. This book also includes current efforts and focuses on key research challenges in the emerging usage of starch for potential conventional and nonconventional applications.

The contributions culminating in the present form of the book are briefly outlined in this opening chapter so that readers can obtain an overview of the book and its highlights. There are 18 chapters in total, covering basics of starch to plasticization, starch-derived cyclodextrin (CD) and its functionality, rheological properties of starch blend and nanocomposites, and starch-based advanced materials and nanocomposites. It is worth mentioning that information on starch-based nanocomposites is limited and is still in academic research levels; however, the latest development in the field has been covered in this book.

Starch chemistry, structure, property, and susceptibility to various chemical, physical, and enzymatic modifications offer a high technological value which contributes enormously to both the food and nonfood industries. The presence of many hydroxyl groups on starch permits easy alteration of its properties through chemical derivatization. Acetate esters and carboxymethyl and hydroxypropyl ethers exemplify starch derivatives. The chemistry, microstructure, processing, and enzymatic degradation of starch are elegantly covered in Chapter 2. Chapter 3 deals with the importance and role of starch as gelling agent. Details of starch gelatinization and hydrogels have been discussed. Starch-based hydrogels play a significant role as a delivery vehicle in biomedical and pharmaceutical applications due to their hydrophilicity, biocompatibility, and biodegradability. Starch-based hydrogels are used as drug carriers for sustained release of drugs to specific site by swelling, diffusion, and/or hydrolytic degradation.

Granular starch has been converted into a thermoplastic material by incorporating suitable plasticizers at high temperature and shear. The crystalline structure of the starch is destroyed during processing by exposing starch hydroxyl groups form hydrogen bonds with water or plasticizer molecules. Further, starch is transformed into a moldable thermoplastic. After plasticization, starch is ready for use as an injection, extrusion, or blow-molding material. Details of plasticization and role of plasticizers are described in Chapter 4.

Thermal analysis measures the gelatinization and glass transition properties of starch polymers. Differential scanning calorimetry (DSC) is mostly used to detect these phase changes. The DSC peak for starch gelatinization widens as the amount of water is reduced. Rheological characteristics of starch dispersions can be ascribed in terms of the intrinsic viscosity–molecular weight relationships when the starch product can be dissolved in water. Chapter 5 discusses various rheological techniques applied to starch-related products, and furthermore, the structural and rheological characteristics of starch dispersions are elucidated.

Reactive extrusion (REX) has been used as an attractive method to prepare new materials based on biodegradable polymer. Polymerization and chemical reactions associated with polymers are carried out in situ, while processing is in progress. This method has been successful as a route for polymerization, chemical modification, blending, and filler dispersion. There are two chapters on REX: Chapter 6 deals with mostly polymeric aspects, and Chapter 7 focuses on starch and other biopolymers.

CDs are natural cyclic oligosaccharides formed by the enzymatic degradation of starch or starch derivates, composed of glucose units, using CD glycosyltransferase. The CDs family comprise several cyclic oligosaccharides, among which α-(**1**), β-(**2**), and γ-(**3**) CDs are well known and most commonly used, containing six, seven, and eight D-glucopyranose units, respectively. Chapter 8 ascribes functional and technological interests of CDs and their industrial applications. CD-based supramole and CD-based polymers have been covered in Chapter 9.

Petroleum-based aliphatic polyesters are expensive, which restricts their extensive application. Blending of the petroleum-based aliphatic polyesters with starch is an effective way to reduce the cost of the biodegradable materials. Packaging materials based on polymers that are derived from renewable sources may be a solution to the problems mentioned. Usually, the components to blend with starch are aliphatic polyesters, polyvinyl alcohol (PVA), polylactide (PLA), and other biopolymers. The resultant blends can be better processed via extrusion or film blowing and have mechanical and/or barrier properties superior to starch alone, and addition of compatibilizers to lower the interfacial energy and increase miscibility of two incompatible phases (starch and synthetic biodegradable polymers), leading to a stable blend with improved characteristics. Chapters 10 through 12 discuss potential of starch in development of food packaging, edible packaging, and feedstock for bioproducts. Starch is also being used as advanced material for industrial and consumer products (Chapter 14).

Chemometric method based on absorbance ratios from Fourier transform infrared (FTIR) spectra has been used to analyze multicomponent biodegradable plastics. The method uses the Beer-Lambert law to directly compute individual component concentrations and weight losses before and after biodegradation of composite plastics. The method employs a new concept of absorbance ratios, the ratio matrix or **R**-matrix, that not only renders a multivariate system of Beer–Lambert law equations amenable to solution but also affords the unknown absorbance coefficients from as few as two sample mixtures without external calibration. Chapter 13 describes the theoretical basis and derivation of the mathematical model for multicomponent systems and its solution algorithm.

Nanometer-sized biofillers from renewable resources show unique advantages over traditional inorganic nanoparticles by virtue of their biodegradability and biocompatibility. Currently, in practical research and applications, the category of biomass-based nanofillers includes mainly the rodlike whiskers of cellulose and chitin and the platelet-like nanocrystals of starch. Recent progress in biodegradable starch-based hybrids (where starch serves as a co-continuous phase with other biodegradable natural polymers) and nanomaterials (where starch serves as a matrix) is therefore focused in Chapter 15. Incorporation of nanoparticles on mechanical and thermal properties of starch and PS is elaborately discussed in Chapter 16. Development of biopolymeric nanocomposites (BNCs) and nanoparticles (BNPs) from biodegradable materials has received significant attention over the past few years. BNPs are expected to address some specific issues in the drug delivery system including (1) improvement in solubility, in vivo absorption, and mucosal permeability of the drug due to their size and surface characters (Singh and Lillard 2009) and (2) increased stability and controlled drug release. Details are discussed in Chapter 17.

The exponential growth of starch-based polymeric materials has generated an interest in the sustainability of starch-based new materials. The sustainable development involves the simultaneous pursuit of economic prosperity, environmental protection, and social equity. This means that the starch and its blends with their applications to packaging and nanocomposites have to expand their responsibility by including the environmental and social dimensions to avert the involved risks. Life cycle assessment (LCA) is used as a measurement tool to assess "products" or "services" for their environmental impact. Furthermore, LCA can be useful to evaluate and predict all the environmental impacts associated with starch and starch-based products at each stage of the product's life. The stages will start from the extraction of resources to the ultimate disposal of the products being marketed and the waste disposal. Chapter 18 focuses on LCA of starch-based polymeric materials for packaging and allied applications.

REFERENCES

Angellier, H., Molina-Boisseau, S., Dole, P., and Dufresne, A. 2006. Thermoplastic starch-waxy maize starch nanocrystals nanocomposites. *Biomacromolecules*, 7, 531–539.

Chivrac, F., Pollet, E., Schmutz, M., and Avérous, L. 2008. New approach to elaborate exfoliated starch-based nano-biocomposites. *Biomacromolecules*, 9, 896–900.

Curvelo, A. A. S., Carvalho, A. J. F., and Agnelli, J. A. M. 2001. Thermoplastic starch–cellulosic fibers composites: Preliminary results. *Carbohydrate Polymers*, 45,183–188.

Dole, P., Joly, C., Espuche, E., Alric, I., and Gontard, N. 2004. Gas transport properties of starch based films, *Carbohydrate Polymer*, 58, 335–343.

Gaudin, S., Lourdin, D., Forssell, P. M., and Colonna, P. 2000. Antiplasticisation and oxygen permeability of starch–sorbitol films, *Carbohydrate Polymer*, 43, 33–37.

Ma, X. F., Yu, J. G., and Wang, N. 2007. Production of thermoplastic starch/MMT-sorbitol NCs by dual-melt extrusion processing. *Macromolecular Materials and Engineering*, 292, 723–728.

Singh, R. and Lillard, J. W. Jr. 2009. Nanoparticle-based targeted drug delivery. *Experimental and Molecular Pathology*, 86, 215–223.

Sorrentino, A., Gorrasia, G., and Vittoria, V. 2007. Potential perspectives of bio-nanocomposites for food packaging applications. *Trends in Food Science & Technology*, 18, 84–95.

Wilhelm, H. M., Sierakowski, M. R., Souza, G. P., and Wypych, F. 2003. Starch films reinforced with mineral clay. *Carbohydrate Polymers*, 52, 101–110.

Zeng, Q. H., Yu, A. B., Lu, G. Q. (Max), and Paul, D. R. 2005. Clay-based polymer nanocomposites: Research and commercial development. *Journal of Nanoscience and Nanotechnology*, 5, 1574–1592.

CHAPTER 2

Starch
Chemistry, Microstructure, Processing, and Enzymatic Degradation

Syed H. Imam, Delilah F. Wood, Mohamed A. Abdelwahab, Bor-Sen Chiou,
Tina G. Williams, Gregory M. Glenn, and William J. Orts

CONTENTS

2.1 INTRODUCTION

Starch, a glucose polymer, is one of the most abundant natural polysaccharides synthesized by plants. Corn, potato, wheat, rice, cassava (tapioca), and sorghum are the major botanical sources for most commercial starches. Table 2.1 lists the major crop sources of starch worldwide. In its native state, starch is produced in granular form in plant cells mostly located in fruits, grains, and roots/tubers which is recovered during commercial processing and refining (Eksteen et al. 2003; Rausch and Belyea 2006). The United States is the largest producer of starch in the world and contributes about half of the commercial starch produced globally. Roughly, 6.5 billion pounds of cornstarch and specialty corn chemicals are produced every year in the United States (www.faostat. fao.org). Many streams of industrially useful chemicals and by-products are also produced during the commercial processing of corn (Kim et al. 2008). These include protein, corn fiber, C5 and C6

5

Table 2.1   Major Crop Sources of Starch in the World

Plant Sources of Starch	Regions of the World Where Produced
Corn/maize	North America, South America, Europe, Asia
Potato	United States, Europe, Asia (sweet potato)
Rice	Europe, United States, Asia
Wheat	Europe, United States, South America, Asia
Cassava	Asia, Africa, South America, United States, Europe,
Tapioca	Asia, Africa, Europe
Sorghum	Asia, Africa, North America, South America, Europe
Banana	Africa, South America, Asia, Caribbean
Sago palm	Mostly Southeast Asia

sugars, gums, corn oil, gluten, and sweeteners which are used in a wide array of food and nonfood consumer products (Johnson and May 2003).

Morphologically, starch consists of mostly spherical granules having diameters ranging between 1 and 150 μm depending on its botanical origin. Chemically, starch is composed of two distinct molecules: amylose and amylopectin. Starch polymers are inherently biodegradable in nature by a wide variety of enzymes and microbes. Advancements in the production of novel and efficient enzymes and fermentation technology have led to the development of many industrially useful specialty chemicals from starch (Peters 2006). Particularly, enzymatic conversions of starch polymers have been exploited to obtain many commercially important and useful chemicals such as fuel ethanol, high fructose corn syrup (HFCS), lactic acid, citric acid, 1,3-propanediol, pullulan, and dextrins (Paster et al. 2003). Other notable industrial chemicals derived from starch derived from corn and sorghum include sorbitol, polylactide, ethyl lactate, succinic acid, levulinic acid, and itaconic acid (Paster et al. 2003; Werpy and Petersen 2004). These chemicals are used in a wide range of industrial products such as polymers, composites, solvents, coatings, inks, detergents, pharmaceuticals, laminates, paints, and cosmetics (Paster et al. 2003).

Starch polymers are also amenable to a variety of chemical modifications which include derivatization, substitution, grafting, and cross-linking reactions to produce starches with novel properties. Starch polymers have been processed, either alone or in conjunction with other polymers, plasticizers, and additives to produce plastics, foams, adhesives, super absorbents, and time-release drug delivery systems (BeMiller and Whistler 2009; Galicia-García et al. 2011; Werpy and Petersen 2004). Companies all over the world such as National Starch, Cargill, Cargill-Dow, ADM, DuPont, Tate & Lyle, Penford, CPC International, Global Starch, National Starch and Chemicals, CPL, Sweco Europe, Novamont, Agrana Beteiligung-AG, and Zhucheng RS Starch Co. Ltd. are leading the way in producing starch-based chemicals for commercial use. In Europe, Cargill, Syral (part of Tereos), Roquette Frères, Danisco, and Kerry and National Starch are the main producers of starch chemicals. Table 2.2 includes a list of major commercial producers of starch-based products in the United States.

Remarkable progress has been made in the past decade with respect to food processing technologies (BeMiller and Whistler 2009; Lamikanra et al. 2005). Compounding/blending and extrusion processing of starch is quite attractive for the food industry because it offers versatility, high productivity, low cost, and energy efficiency. Nevertheless, the processing of starch is quite challenging. Particularly, extrusion processing of starch under high-temperature, high-pressure, and low-moisture conditions causes starch to undergo complex structural transformation and phase transitions, which impacts starch properties (Galicia-García et al. 2011; Grassia and D'Amore 2009). But several strategies have emerged to overcome these challenges. Successful extrusion processing of starch has found diverse applications in both food and nonfood industries.

Plasticization of starch and its compatibility with other synthetic, bioderived, and renewable polymers have provided an opportunity to create scores of bioplastic composites that are environmentally

Table 2.2   The Major U.S. Producers of Starch-Based Commercial Chemicals and Products

Company	Product(s)
Archer Daniels Midland Company (ADM)	Ethanol, citric acid, sorbitol, high fructose corn syrup (HFCS)
Cargill, Inc.	Citric acid, sorbitol
Cargill-Dow	PLA, ethyl lactate
Midwest Grain Products, Inc. (MGPI)	Ethanol
DuPont	1,3-Propanediol
Tate & Lyle	Citric acid, starch, ethanol
Williams Bio-Energy	Ethanol
National Starch	Starch, specialty starches
Penford, Inc.	Specialty starches

biodegradable or compostable (Fakirov and Bhattacharyya 2007; Grassia and D'Amore 2009; Janssen and Moscicki 2009; Long 2009; Mohanty et al. 2005). Such materials are ideal for single-use consumer packaging. Biodegradable packaging, foams, shopping bags, and agricultural mulch films are the best known examples of such products currently available to consumers in many countries.

This chapter will focus mostly on starch chemistry, microstructure, processing, and starch-hydrolyzing enzymes. Certain starch applications relevant to the scope of this chapter will also be mentioned.

2.2 STARCH CHEMISTRY AND PROPERTIES

Starch chemistry and properties are only briefly discussed providing the most pertinent and relevant information. For detailed information on this subject, readers are referred elsewhere (BeMiller and Whistler 2009; Galicia-García et al. 2011; Grassia and D'Amore 2009). Starch is essentially a condensation polymer of glucose molecules connected by acetal linkages. Glucose sugars in starch are the principal food molecules of the cell. All sugars contain hydroxyl groups and either an aldehyde ($_H$>C=O) or a ketone (>C=O) group (Figure 2.1). The hydroxyl group of one sugar can combine with the aldehyde or ketone group of a second sugar with the elimination of water to form a disaccharide. The addition of more monosaccharides in this manner yields very large polysaccharide molecules containing thousands of monosaccharide units (residues).

Starch polymers are composed of two major components: amylose and amylopectin. Amylose is composed mostly of linear α-D-(1→4)-glucan units (Figure 2.2), whereas amylopectin is a highly branched α-D-(1→4)-glucan with α-D-(1→6) linkages at the branch points (Figure 2.3). The linear amylose molecules comprise about 30% of common cornstarch and have molecular weights of 200,000–700,000, while the branched amylopectin molecules have molecular weights as high as 100–200 million (Robyt 1998).

Numerous factors influence the properties and function of starch. Starch granules are hydrophilic since each starch monomer unit contains three free hydroxyl groups. Consequently, starch moisture content changes as relative humidity (RH) changes. Cornstarch granules retain about 6% moisture at 0% RH but contain 20% moisture when in equilibrium with about 80% RH.

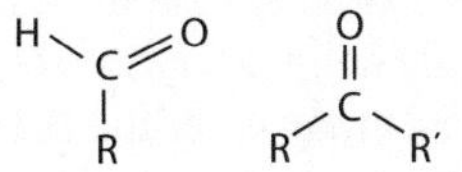

$$H\diagdown C\!\!=\!\!O \qquad \underset{\underset{R'}{\overset{\|}{C}}}{\overset{O}{}} $$

$$\underset{R}{|}$$

$$\text{(A)} \qquad\qquad \text{(B)}$$

Figure 2.1   Chemical structure of aldehyde (A) and ketone (B) groups.

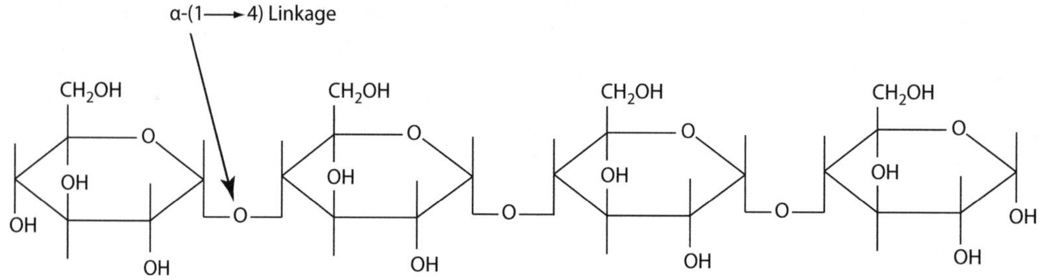

α-(1⟶4) Linkage

Figure 2.2 Linear amylose chain.

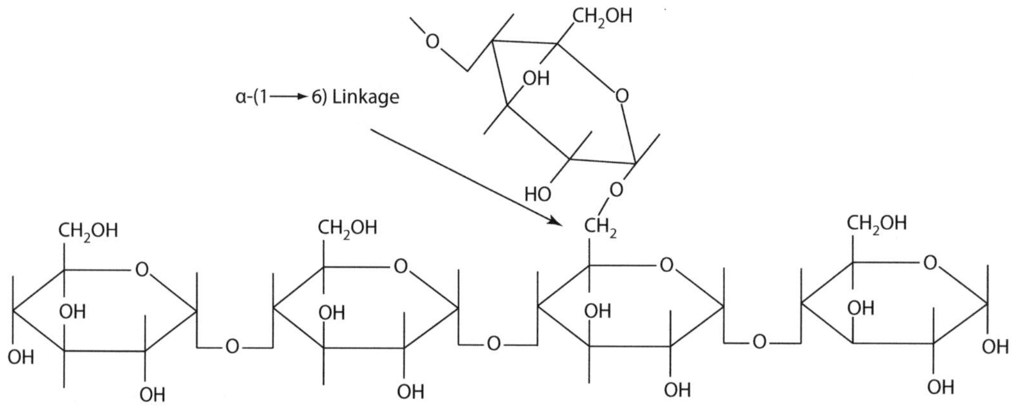

α-(1⟶6) Linkage

Figure 2.3 Branched amylopectin molecule.

Starch granules are thermally stable when heated in an open atmosphere to about 250°C. Above that temperature, the starch molecules begin to decompose. Dry granules absorb moisture when immersed in water but retain their basic structure due to their crystallinity and hydrogen bonding within the granules. Native granular starch contains crystalline areas within the amylopectin (branched) component, but the linear amylose component is largely amorphous and can be mostly extracted in cold water. The granular structure is ruptured by heating in water or treating with aqueous solutions of reagents that disrupt crystalline areas and hydrogen bonding within the granules. The constituent molecules are completely soluble in water at 130°C–150°C and at lower temperatures in alkaline solutions. Starch granules that have been ruptured in aqueous media are commonly referred as gelatinized or destructurized starches (Otey and Westhoff 1982). The temperature at which starch granules are completely gelatinized is known as the gelatinization temperature, which varies depending on the botanical source of the starch. Application of high pressure and shear to starch granules permits disruption of the organized structure at lower water content than is possible at atmospheric pressure.

Starch solutions are unstable at low temperatures. On standing in dilute solutions, the linear amylose component crystallizes. Many branches of amylopectin may also crystallize. Rapid cooling of concentrated starch dispersions creates stiff gels, which crystallize more slowly. Amylose, and to a lesser degree the outer branches of amylopectin, can assume helical conformations which have a hydrophobic core. Each turn of the helix comprises of about six monomer units. Iodine, fatty acids, lipids, alcohols, and other materials may enter the core of the helix to form stable complexes with starch. Small amounts of crystalline amylose–lipid V-type complexes are usually found in starches such as corn and wheat, which contain free fatty acids and phospholipids (Imam et al. 1993; Shogren et al. 1991, 1992).

Starch molecules readily depolymerize into glucose monomer units when heated in acidic solutions or when treated with a variety of amylolytic enzymes. They are generally stable under alkaline conditions at moderate temperatures. When heated with amines under alkaline conditions, starch undergoes complex Maillard reactions to form brown-colored products with caramel-like odors. Such treated starch biodegrades fairly quickly.

Starch chemistry has provided opportunities to produce an array of useful chemicals by modifying starch both chemically and enzymatically. The presence of many hydroxyl groups on starch permits easy alteration of its properties through chemical derivatization. Acetate esters and carboxymethyl and hydroxypropyl ethers exemplify starch derivatives. Most commercial starch products have a low degree of substitution levels designed to alter their solution properties or adhesion to paper. The primary users of starch in the United States are the paper, fermentation, and food industries. Research in our laboratory (Agüeros et al. 2011) yielded several useful starch derivatives of commercial significance. Some of these modifications include creation of 2-O-butyl starch, starch palmitate, and starch-ethyl acrylic acid ester (Figure 2.4).

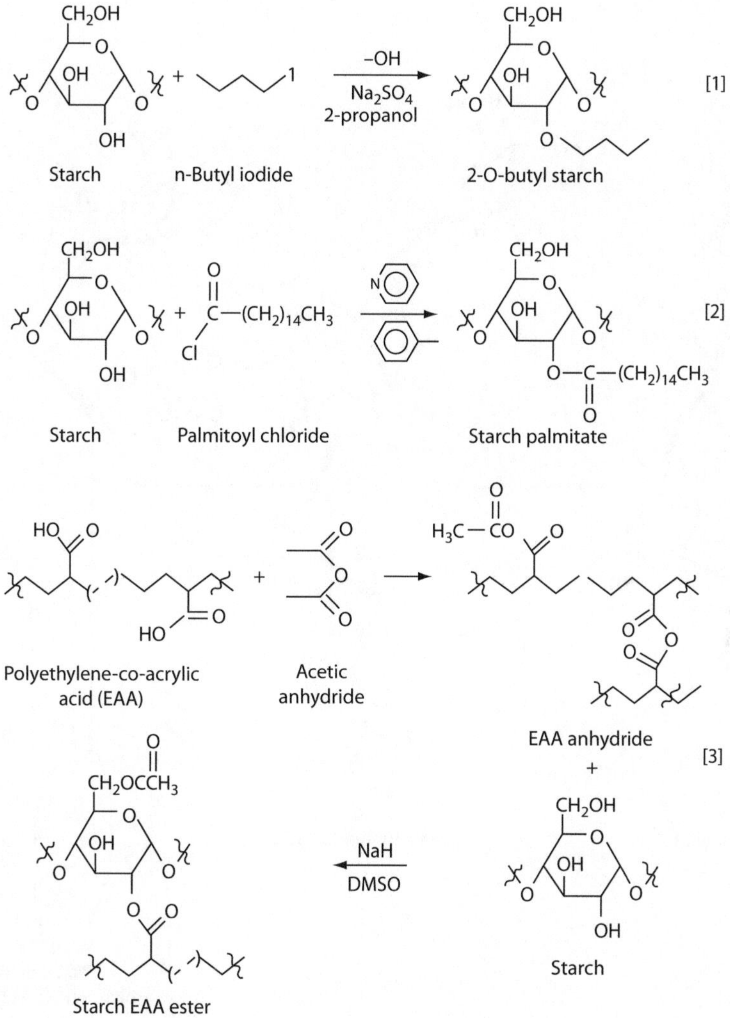

Figure 2.4 Derivatization of starch polymer. (From Imam, S.H. and Harry-O'Kuru, R.E., *Appl. Environ. Microbiol.*, 57, 1128, 1991. Printed with copyright permission.)

Starch may also be modified by grafting synthetic polymer branches onto the starch backbone. The subject of starch graft (starch-g) copolymers has been reviewed providing an overview of the numerous monomers and initiating systems (Jasberg et al. 1992; Jyothi 2010; Kalichevsky et al. 1993; Kampeerapappun et al. 2007; Khanna et al. 2011; Koenig and Huang 1995; Laza-Knoerr et al. 2010; Lee and McCarthy 2009; Lee et al. 2001; Xie et al. 2006; Zhang and Zhou 2007). Graft copolymers containing thermoplastic branches are prepared by generating free radicals on starch and allowing these free radicals to initiate polymerization of vinyl or acrylic monomers (Chen et al. 2004). The ceric ion initiates grafting of unsaturated monomer onto starch. Usually, free radical–initiated grafted branches are infrequently spaced along the starch backbone and have high molecular weights. Free radicals are generated on the starch backbone by chemical means like ceric salts or by high-energy irradiation produced by cobalt-60. The chemical reaction of grafting is presented in Figure 2.5. Starch-g-PMA at PMA levels exceeding 60% is conveniently prepared by ceric-initiated graft polymerization. The starch-g-PMA was extruded to produce packaging peanuts and was among the first product generated from cornstarch that was successfully commercialized. Photographs of the packaging peanuts produced via single-screw and twin-screw extruders are shown in Figure 2.6.

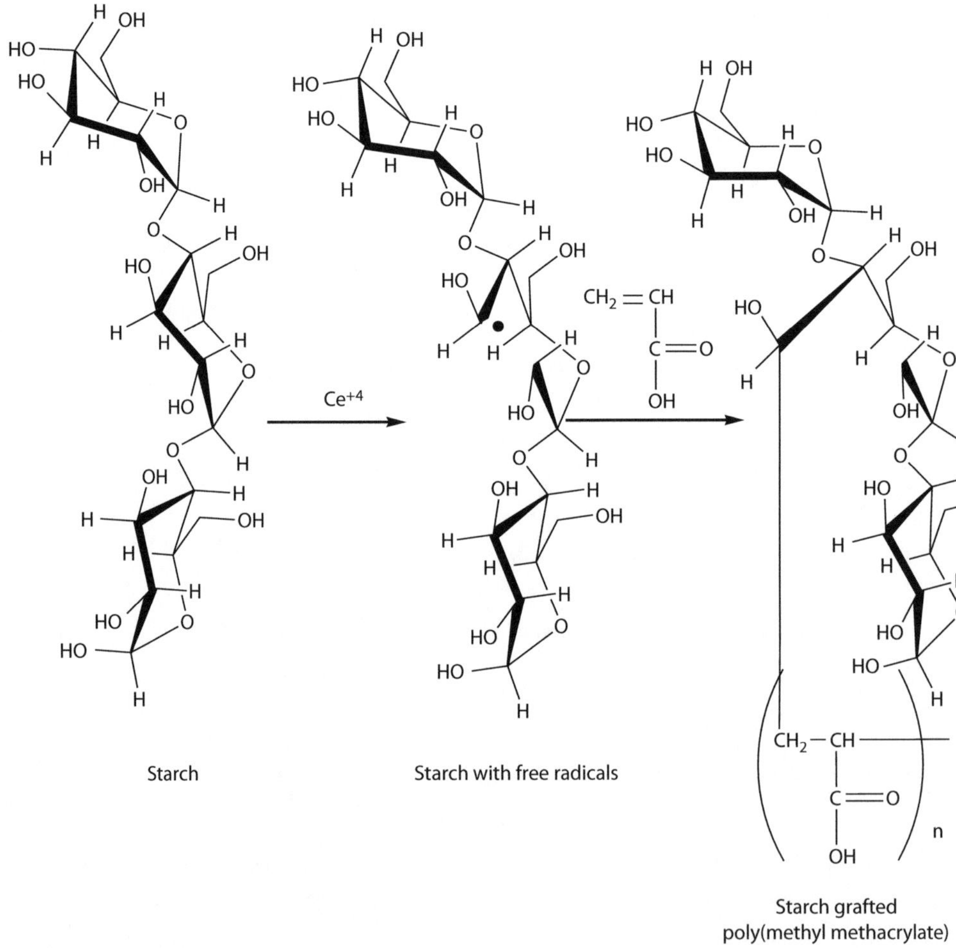

Figure 2.5 Chemical reaction showing the starch grafting.

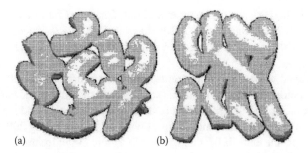

(a)   (b)

Figure 2.6 Photographs of starch graft poly(methyl acrylate) loose-fill foams produced by (a) twin-screw and (b) single-screw extrusion. (From Chen, L. et al., *Biomacromolecules*, 5(1), 238, 2004. Printed with copyright permission.)

Glucose syrups are widely used as a sweetener and structural agent in many foods. Glucose syrup is a solution of pure glucose in water, but in addition to glucose, also contains maltose and smaller maltodextrins. To increase the sweetness, glucose syrups are treated with the enzyme glucose isomerase to produce HFCS (Eliason 2004). The HFCS is the preferred sweetener used by the beverage industry worldwide. While the beverage industry in the United States principally relies on cornstarch as their feedstock to obtain HFCS, other starches are also used in other countries. Dextrins or dextrinized starches are yet another example of a starch hydrolysis product that is commercially produced. Industrial dextrins are a mixture of the degradation product from both amylose and amylopectin molecules that are produced by using enzymes, acid, or chemicals. Both dextrins and HFCS are completely digested in the small intestine and provide the same amount of energy as glucose. Cyclodextrins are fascinating and novel structures (Figure 2.7) and are a special class of circular dextrins produced by enzymes from several bacteria. While cyclodextrins are rarely used in foods, they have many industrial uses (Del Valle 2004), particularly in the pharmaceutical industry, where they are used for encapsulation of bioactive compounds and other drugs

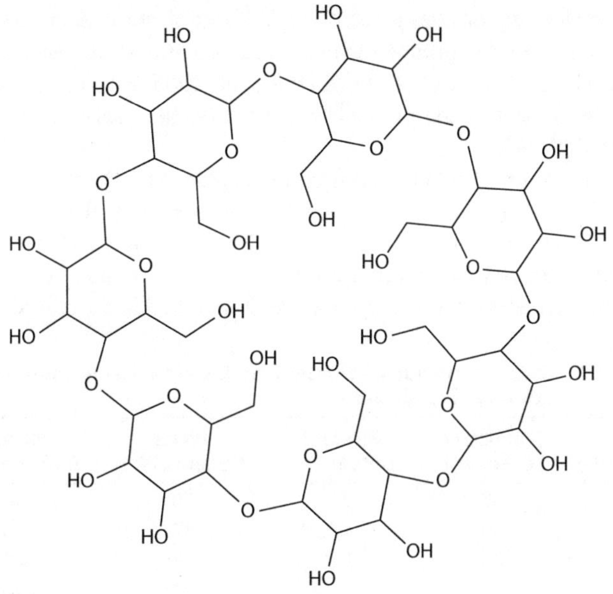

Figure 2.7 β-Cyclodextrin.

for delivery and controlled release (Agüeros et al. 2011; Chen and Jiang 2011; Chen and Liu 2010; Chernykh and Brichkin 2010; Davis et al. 2010; Hu et al. 2009; Laza-Knoerr et al. 2010; Loftsson and Brewster 2010; van de Manakker et al. 2009; Otero-Espinar et al. 2010; Patil et al. 2010; Sasaki and Akiyoshi 2010).

Starch polymer aged at constant temperature and moisture levels results in starch embrittlement. Differential scanning calorimetric (DSC) studies have shown that the embrittlement is due to structural relaxation of starch chains, leading to decreases in enthalpy and free volume with time. This type of aging is typical of most amorphous polymers (Amy and Jörg 2009; Chen et al. 2009b; Dobreva et al. 2010; Etienne et al. 2007; Grassia and D'Amore 2009; Imam et al. 1995; Jasberg et al. 1992; Kierkels et al. 2008; Koenig and Huang 1995; Lee and McCarthy 2009; Tsuji and Tsuruno 2010; Xie et al. 2007). The rates of aging vary with polymer structure, but the reason for such variation is not fully understood at present. Gelatinized starch also tends to swell in water leading to its hydrolytic degradation.

Recent advances in the field of genetic engineering have increased our knowledge and understanding of metabolic pathways in starch synthesis (Rahman et al. 2000). Starches from different botanical origin have different biosynthetic mechanisms and may exhibit distinct molecular structure and characteristics as well as diversity in shape, size, composition, and other macroscale constituents. Thus, the ultimate processing and properties of starch are linked to starch genetics as well as to various levels found in granule structure, macromolecular structure, and crystalline macrostructures.

The glass transition temperature (T_g) of dry amorphous starch is experimentally inaccessible owing to the thermal degradation of starch polymers at elevated temperatures. It is estimated that the T_g of the dry starch is in the range of 240°C–250°C (Poutanen and Forssell 1996). Native starch is a nonplasticized material because of the intra- and intermolecular hydrogen bonds between the hydroxyl groups of starch molecules. During the thermoplastic process, in the presence of a plasticizer, a semicrystalline granule of starch is transformed into a homogeneous material with hydrogen bond cleavage between starch molecules, leading to loss of crystallinity.

The physical properties of the thermoplastic starch are greatly influenced by the amount of plasticizer present. In most literature for thermoplastic starch, polyols are the preferred plasticizers, of which glycerol is the major one. Besides glycerol, glycol, xylitol, sorbitol, sugars, and ethanolamine, chemical agents containing amide groups such as urea, formamide, and acetamide have also been used to plasticize starch (Da Róz et al. 2006; Huang et al. 2005; Rodriguez-Gonzalez et al. 2004; Yang et al. 2006). The impact of the plastification level on the glass transition of thermoplastic starch is presented in Table 2.3.

The mechanical properties of a low- and high-molecular-mass thermoplastic starch at various moisture levels (5%–30%) indicated that stress–strain properties of the material were dependent on the water content. Materials containing less than 9% water were glassy with an elastic modulus between 400 and 1000 MPa. Starches from a variety of sources extruded with the plasticizer glycerol showed that above certain glycerol levels, a lower T_g resulted in decreased modulus and tensile

Table 2.3   Glass Transition of Thermoplastic Starch Using Different Levels of Plasticizer

% Starch	Plasticizer Level, Wt%	Glycerol Content, Wt%	Water Content, Wt%	Glass Transition, °C
74	26	10	16	43
70	30	18	12	8
67	33	24	9	−7
65	35	35	0	−20

Source: From Avérous, L. and Fringant, C., *Polym. Eng. Sci.*, 41(5), 727, 2001. Printed with copyright permission.

strengths and increased elongations. The T_gs for pea, wheat, potato, and waxy maize starches were 75°C, 143°C, 152°C, and 158°C, respectively (Van Soest et al. 1996).

The type and amount of plasticizer also impact mechanical, thermal, and water absorption properties of melt-processed starch (Da Róz et al. 2006). In general, monohydroxy alcohols and high-molecular-weight glycols failed to plasticize starch, whereas shorter glycols were highly effective. In this regard, Thunwall et al. (2006) investigated the mechanical and melt flow properties of two thermoplastic potato starch materials with different amylose contents and found that after conditioning at 53% RH and 23°C, the glycerol-plasticized sheets with a high amylose content were stronger and stiffer than the normal thermoplastic starch with an amylose content typical for common potato starch. The tensile modulus at 53% RH was about 160 MPa for the high-amylose material and about 120 MPa for the plasticized native potato starch. The strain at break was about 50% for both materials (Thuwall et al. 2006).

2.3 STARCH MORPHOLOGY AND MICROSTRUCTURE

Starch is a storage product found in most plant cells in the form of granules or solid particles comprised of molecules of both amylose and amylopectin. Other minor components associated with the granules include small polymers, lipids, proteins, and phosphate ester groups that have a huge impact on the functionality of starch pastes and gels (Jane 2009). The granules vary in size depending on their botanical source, and they exhibit variable morphology. Scanning electron microscopic images of several starches are shown in Figure 2.8. Granules exhibit highly diverse morphologies that are discussed later in this section.

Starch is formed in amyloplasts, unpigmented plastids which are a type of leucoplast (Esau 1965). Figure 2.9 is a scanning electron micrograph showing numerous starch granules within plant cells (Figure 2.9a) and remnants of amyloplasts and cytoplasm surrounding the starch granules (Figure 2.9b). The starch granule initiates at a nucleation point, known as the hilum, and biosynthesis starts in the amyloplast (Jane 2009). Examination of starch granules via a combination of techniques such as x-ray diffraction, atomic force microscopy (AFM), scanning electron microscopy (SEM), and transmission electron microscopy (TEM) revealed a unique organizational structure. In their native form, starch granules contain a highly structured internal order where there is a central hilum surrounded by concentric growth rings. Amylose and amylopectin molecules are deposited and organized as alternating semicrystalline and amorphous layers forming growth rings (Figure 2.10). The semicrystalline layer consists of ordered regions composed of double helices formed by short amylopectin branches, most of which are further ordered into crystalline structures known as the crystalline lamellae. Amorphous regions of the semicrystalline layers and the amorphous layers are composed of amylose and nonordered amylopectin branches (Ball and Morell 2003). Figure 2.10a shows a scanning electron micrograph of a cross-section of wheat starch granule that had been exposed to a mild α-amylase treatment in order to reveal the ring structure. Figure 2.10b shows a transmission electron micrograph of cross section of waxy maize starch following acid treatment (Yamaguchi et al. 1979).

The growth rings, containing different amorphous–crystalline ratios (Wellner et al. 2011), are about 100–400 μm thick. Pea starch has been shown to consist of alternating layers of crystalline and less crystalline regions (Parker et al. 2008). The alternating layer idea had its origins long before the advent of modern microscopes or chemical techniques (Gallant et al. 1997) and was coined the "blocklet concept" (Badenhuizen 1937). However, later the "blocklet concept" was rejected and thought to be a "purely mechanical process" having nothing to do with the inherent unit structure of starch but rather due to the differences in swelling that the various treatments caused (Badenhuizen 1956). The "extended cluster model" was proposed for waxy maize (Yamaguchi et al. 1979). The cluster model was also mentioned as a possibility for the organization of rice starch

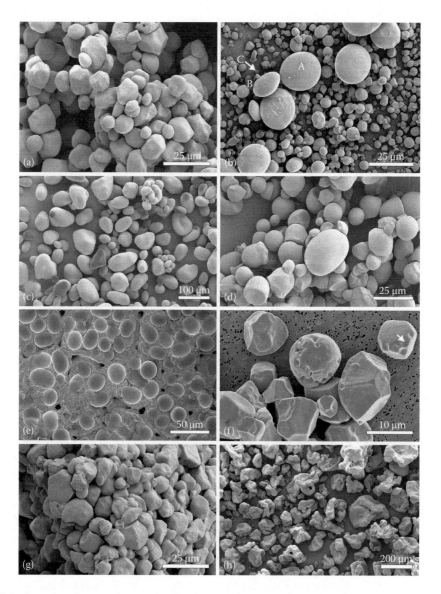

Figure 2.8 Scanning electron micrographs showing granular morphology of starches derived from native corn (a), wheat (b), potato (c), tapioca (d), red bean (e), rice (f), waxy corn (g), and Cargill pregel (h).

(Ohtani et al. 2000). More recently, the blocklet concept seems to be the accepted model for starch structure (Wellner et al. 2011).

These models were developed after acid, alkali, or enzymatic treatment to dissolve the less resistant parts of the starch granule in order to visualize what remained. The components of native starch granules have no inherent density differences that may be discerned by TEM, and native starch granules are not effectively fixed during preparation for TEM. The unfixed starch granules retain their ability to absorb water. Thin sections are typically collected in a water trough prior to being placed on a grid for TEM observation. Thus, the sectioned starch granules take up water from the trough and fold creating dense areas or bands (Gallant and Guilbot 1971) when viewed in the TEM. Some of the granules may also pull away from the surrounding resin creating holes or spaces in the sections (Bechtel and Wilson 2003). Thus, dissolving portions of the starch granules allows

Figure 2.9 Tapioca starch in the cassava tuber showing numerous starch granules in a plant cell (a) and remnants of amyloplasts and cytoplasm surrounding starch granules (b).

better penetration of solvents used for TEM. Better preservation of the remaining structure coupled with the resulting density differences allows the visualization of the parts of the remaining starch granule. Therefore, removing various components by selective erosion is a useful alternative for interpreting the structure (Gallant et al. 1973). This phenomenon is illustrated in Figure 2.10b which is waxy maize starch following extensive acid treatment. A low-molecular-weight, highly crystalline residue remains and is visible in the electron micrograph (Yamaguchi et al. 1979).

The crystallinity of starch is 15%–45% and is responsible for the characteristic maltese cross pattern of birefringence of starch granules when viewed through crossed polarizers in a light microscope (Figure 2.11). The maltese cross is centered on the hilum of the starch granule. The crystallinity, thus the birefringence, disappears in damaged or gelatinized starch. The growth rings and crystallinity give starch some of its basic properties determined by x-ray crystallography as consisting of an orderly, radial arrangement of material (Alsberg 1938). There are three x-ray diffraction patterns in native starch: A, B, and C. Cereals generally contain starch with the A diffraction pattern. The B pattern starches are synthesized at low T and high humidity; hence, most tubers fall into this category. Starches having the B pattern may be converted irreversibly to A-type starches. Legumes generally have the C pattern, although there are exceptions (Gallant et al. 1992). Following gelatinization, amyloses are prone to retrograde, or recrystallize, forming the V pattern. These x-ray diffraction patterns are due to specific conformations of the helical structures of starch (Zobel 1988).

Simple starch granules form in a single amyloplast and may take on different shapes including spherical, polyhedral, ovoid, ellipsoid, or lenticular, as depicted in Figure 2.8. Some granules have

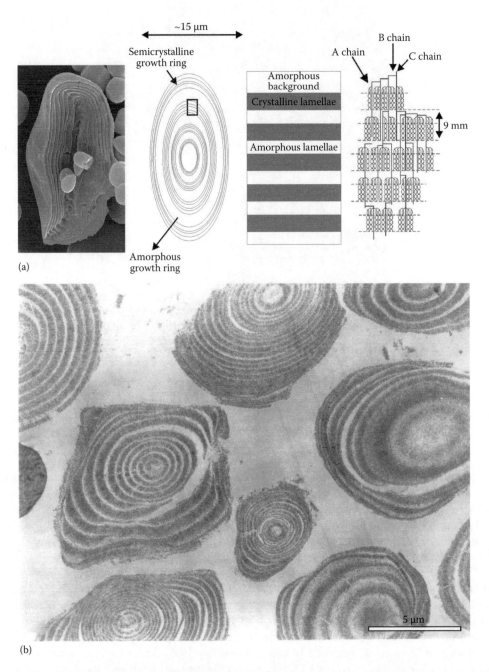

Figure 2.10 (a) Schematic view of the structure of a starch granule, with alternating amorphous and semi-crystalline regions consisting of growth rings with SEM of the granule cross section. (From Ball, S.G. and Morell, M.K., *Annu. Rev. Plant Biol.*, 54, 207, 2003. Printed with copyright permission.) (b) Transmission electron micrograph of acid-treated waxy maize starch. Note difference in thickness of growth rings. (Reprinted from Electron microscopic observations of waxy maize starch, 69(2), Yamaguchi, M., Kainuma, K., and French, D., *J. Ultrastruct. Res.*, 249, 1979. Copyright 1979, with permission from Elsevier.)

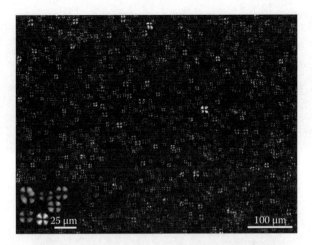

Figure 2.11 Starch viewed through crossed polarizers in a light microscope showing the characteristic maltese cross patterns. High-amylose corn and bean (inset) starches.

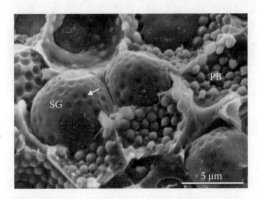

Figure 2.12 Scanning electron micrograph of *Tripsacum dactyloides* L. central endosperm showing spherical starch granules with regular concavities caused by protein bodies. SG, starch granule; PB, protein bodies; arrow, concavity due to protein body which was fractured away.

a slight bend in the middle and are kidney shaped. Simple starch granules may also be deformed by surrounding structures, such as protein bodies (Figure 2.12) in a cell, which are responsible for some of the depressions on the surfaces of starch granules. These depressions were also recognized to be due to protein bodies in cornstarch granules (Fitt and Snyder 1984). Multiple starch granules forming in a single amyloplast develop into compound starch granules (Figure 2.13b). The individual granules in compound starch granules are typically polyhedral since they develop in close proximity to granules in the same amyloplast. Starch granules from compound granules are also difficult to separate (Jane 2009). According to Pérez et al. (2009), starch granules range in size from about 0.1 to 200 μm (Gallant et al. 1992; Pérez et al. 2009); more commonly reported, however, are the values between 1 and 150 μm (Moss 1976) or <1 to 100 μm (Jane 2009).

Among the most studied starches are those of wheat. Two types of wheat starch granule are well accepted: A, the larger, lenticular granules; and B, the smaller spherical granules (Evers 1969). The two types of starch granules are not only very different in size and shape, developmental initiation also occurs at different times (Baruch et al. 1979). The A-type starch granules are initiated as soon as 4 days after flowering and continue growing until they reach their maximum diameters of >16 μm

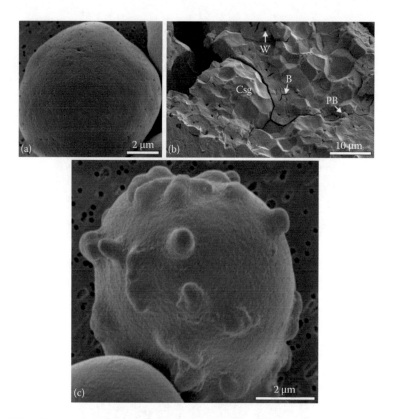

Figure 2.13 SEMs of starch granule showing pin holes (pitted surface) (a), cavities visible in fractured waxy rice starch granules (b), and high-amylose rice starch granule with surface protrusions (c).

at 19 days after flowering or later, whereas the B-type granules are initiated at about 10 days after flowering (Bechtel et al. 1990) and range from 5.3 to 16 µm in diameter. The A and B starch granules have virtually the same densities, but the A granules are more susceptible to enzymatic degradation (Meredith 1981) probably because of their increased surface area compared to the B granules. Early workers suggested that there might be a third type of starch granule in wheat, and the third type was demonstrated by using mathematical models (Baruch et al. 1983). Bechtel and coworkers described the third type of starch granule, designated C, as initiating at about 21 days after flowering and having diameters <5.3 µm at maturity. Thus, the A, B, and C starch granules in wheat are initiated at different developmental times and are of three size distributions in the mature wheat kernel (Bechtel and Wilson 2003; Bechtel et al. 1990).

Cornstarch is another very important commercial starch and has three major classes of native starches. Normal cornstarch has polyhedral and mostly spherical or ovoid starch granules ranging from 5 to 16 µm in diameter (Figure 2.8). Waxy cornstarch (i.e., high in amylopectin) has more irregularly shaped granules with a slightly larger range in sizes from roughly 4 to 17 µm in diameter. High-amylose cornstarch has very irregular starch granules; some are elongated, others have various large protrusions from their surfaces (Figure 2.13c), while others are regular spheres. The budding or deformity of cornstarch granules increases with increasing amylose content (Fitt and Snyder 1984). Cornstarch granules from other types of plants, such as sorghum and millet, contain pores or channels such as those found in high-amylose and normal cornstarches.

The pores or channels have been shown to extend into the granules (Fannon et al. 1992, 1993; Huber and Bemiller 1997). The pores were first described as pin holes in cornstarch (Figure 2.13a) by Hall and Sayre (1970). Not surprisingly, the pores are lined with proteins (Benmoussa et al. 2010) which are presumably enzymes that have bored the holes into the granules. Pores were also found in wheat starch granules grown under heat stress, conditions that accelerated the maturity of the wheat kernels. The pitting in wheat starch granules occurred along the equatorial groove (Hurkman and Wood 2011), likely due to enzymatic degradation as suggested by Jane (2009). Pores, although few, and larger surface deformations were also seen in rice starch granules.

Pores or cavities also occur regularly in waxy rice starch granules in the centers and between starch granules. The cavities in waxy rice starch, however, are air spaces and probably reduce the milling strength of waxy rice (Ibáñez et al. 2007). The cavities are only visible in rice starch granules that have been fractured (Figure 2.13b) so will not be visible in scanning electron micrographs that only depict surfaces. The cavities do not occur in rice starch with normal amylose contents.

Pores or cavities have not been reported in potato starch granules. Potato starch granules are very large ellipsoidal granules with apparently smooth surfaces (Figure 2.8). However, surfaces of potato starch granules viewed by low-voltage SEM and AFM show that potato starch granules have surface protrusions. Although smaller, wheat starch and waxy cornstarch surfaces also have surface protrusions. High-amylose cornstarch can have many large surface protrusions (Figure 2.13c). Baldwin et al. (1997) suggested that the protrusions on potato and wheat starch surfaces are the amylopectin side chain cluster ends occurring at the granule surface. They rule out the possibility of the protrusions being protein because the potato starch granule has less associated protein than does wheat starch, and yet, the protrusions are larger on potato starch than on wheat granule surfaces. Research on starch granular morphology, microstructure, rheology, spherulite formation, and internal structure and phase transition has been the subject of many recent investigations utilizing state-of-the-art techniques. Specifically, microstructural characterization, impact of amylose/amylopectin ratios, and other chemical, physical, and processing parameters on granular microstructure have been studied in detail (Chen et al. 2006, 2009b, 2011b; Galicia-García et al. 2011; Ma et al. 2011; Vu Dang et al. 2009). Particularly, Chen et al. (2009a) have reported on the morphologies and microstructures of cornstarches with different amylose–amylopectin ratios and showed that waxy and maize starches have clear internal cavities and channels, and sharp growth ring structures that were clearly observed in low-amylose starches (waxy and maize) after acid hydrolysis. Gelatinization of all starches started at the hilum and areas adjacent to the channels, spreading rapidly to the periphery.

Taro is a tropical plant from which starch is extracted from the corm (70%–80% starch). Individual granules range 1–5 µm in diameter (Aboubakar et al. 2008; Jane et al. 1992). The starch granules of taro are mostly polyhedral. Also tropical, cassava starch (or tapioca) has unique thickening properties and is high in purity and low in cost and produces viscous, clear pastes (Chen et al. 2011a). Tapioca starch is isolated from the cassava also known as Manihot (*Manihot esculenta*) tuber and is native to South America. Hall and Sayre (1970) described some of the cassava starch granules using SEM as having a spongy appearance and that the spongy appearance was likely due to the isolation technique used for starch. However, the same type of sponginess appears in rice starch, and this is due to the development of the starch granules in close proximity to protein bodies (shown in *Tripsacum dactyloides* starch in Figure 2.12). Tapioca was enzymatically hydrolyzed and studied by SEM by Chen and coworkers (Chen et al. 2011a). The hydrolyzation or modification of starches improves their physical properties by increasing the surface areas of the starch granules.

2.4 STARCH-HYDROLYZING ENZYMES

Starch-degrading enzymes are the principal commercial enzymes used by the food and beverage industry worldwide. Much of our knowledge of starch-degrading enzymes is based on the efforts and advancements made by the food industry in seeking and developing novel and efficient enzyme systems (Aiyer 2005; Sun et al. 2010). More recently, the fuel ethanol industry has joined forces to seek efficient starch-hydrolyzing enzymes to improve the efficiency of converting corn-starch into ethanol (Rattanachomsri et al. 2009).

Considerable literature is available indicating that under appropriate conditions, the starch polymer is susceptible to biodegradation, photo degradation, or chemical degradation processes. Particularly, biodegradation of starch and starch-based materials has been studied in detail (Imam and Snell 1988; Imam et al. 1991; Swanson et al. 1993). The most efficient and rapid starch degradation usually requires a combination of degradation mechanisms and is mediated by microorganisms and their hydrolytic enzymes. Native starch is unquestionably biodegradable because it is readily metabolized by microorganisms such as bacteria, fungi, algae, and yeasts. Additionally, biosynthetic enzymes inherently present and stored within the cells and tissues of plants participate in the starch degradation process. Efforts are continuing globally, to create genetically engineered microbes to develop improved and more effective commercial enzymes that are critical for food and the alcohol fuel industries (Eksteen et al. 2003; Houghton-Larsen and Pedersen 2003; Lee et al. 2001). Future progress of the whole biofuel industry is dependent on the availability of cheaper and more efficient biocatalysts. In the last decade, great progress has been made in the understanding of the function of starch-degrading enzymes. Combining mutational analysis, multiple sequence alignment, and three-dimensional structures of enzyme–substrate analogue complexes formed the basis for developing amylolytic and related enzymes through protein engineering (Svensson et al. 1999). Studies have also been focused on other important and interesting applications of many starch-hydrolyzing enzymes. For example, enzyme systems have been developed to modify starches in order to create starches with novel functionalities and to derive useful products of commercial potential (Aiyer 2005; Ba et al. 2010; Blazek and Copeland 2010; Morán et al. 2011).

Mechanisms for starch biodegradation are numerous and depend on the type of hydrolytic enzymes present (Aiyer 2005; Smith et al. 2005). Enzymes are extremely efficient and specific biocatalysts produced by plants, animals, and microbes, are known for their catalytic efficiency and regio- and stereoselectivity. These biocatalysts are involved in the (1) biosynthesis of polysaccharides to assemble complex polymeric structures, (2) polymer modifications to render useful functionalities, and (3) biodegradation process (recycling) of natural polymers. A comprehensive monograph on the roles of enzymes in food science has been published (Whitaker 1994).

2.4.1 Starch Depolymerases

The enzyme specificity and the availability of multiple enzyme reactive sites on the starch polymer are two characteristics that are critically important in dictating efficient starch degradation. Amylases act on starch polymers to hydrolyze α-1,4 and α-1,6 glucosidic bonds. These bonds are cleaved via the exo- or endo-splitting mode of hydrolysis. A list of common starch-hydrolyzing enzymes and their catalytic activity is provided in Table 2.4. In complex branched polymers like

Table 2.4 Starch-Hydrolyzing Amylases from Microbial Sources

Enzyme	Activity
Endo-acting α-amylases	Hydrolyze α-1,4 bonds and bypass α-1,6 linkages
Exo-acting β-amylases	Hydrolyze α-1,4 bonds and cannot bypass α-1,6 linkages. Produce maltose as a major end product
Exo-acting glucoamylases (amyloglucosidase)	Hydrolyze α-1,4 and α-1,6 linkages
Pullulanase and other debranching amylases	Hydrolyze only α-1,6 linkages
α-Glucosidases	Hydrolyze preferentially α-1,4 linkages in short-chain oligosaccharides produced by the action of other enzymes on amylose and amylopectin
Cyclodextrin producing amylases	Hydrolyze starch to a series of nonreducing cyclic D-glucosyl polymers called cyclodextrins or dextrins

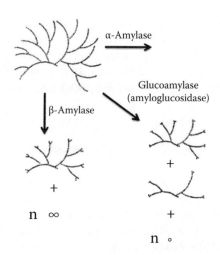

α-Amylase

β-Amylase

Glucoamylase
(amyloglucosidase)

+

+

n ∞

+

n ∘

α-Amylase hydrolyzes amylopectin to
oligosaccharides containing two to
six monomers.

Glucoamylase hydrolyzes
preferentially α-1,4 linkages from
nonreducing ends releasing glucose.

β-Amylase hydrolyzes
each chain to produce maltose
until it reaches an α-1,6-linkage.

Figure 2.14 Major starch degrading enzymes and their mode of action.

starch, branching creates a large number of nonreducing terminal residues, which require a combination of both exo- and endo-splitting enzymes to achieve a complete polymer hydrolysis. For example, β-amylase (1,4-α-D-glucan maltohydrolase, EC 3.2.1.2), an exo-splitting enzyme acts only on residues at the nonreducing terminus until it reaches α-1→6 linkage, whereas α-amylase (1,4-α-D-glucan glucanohydrolase, EC 3.2.1.1) is an endo-splitting enzyme that can randomly cleave α-1→4 glycosidic linkages throughout the linear chain. Glucoamylase (1,4-α-D-glucan glucohydrolase, EC 3.2.1.3 or 1,4-α-glucosidase) can act on α-1→4 bonds but can also hydrolyze α-1→6 linkages, albeit at a much slower rate. The enzymatic breakdown of the highly branched amylopectin polymer by α- and β-amylases, and glucoamylases is also depicted in Figure 2.14. A comprehensive review of pathways of starch degradation has recently been published (Smith et al. 2005).

A study of six starches of different botanical origin was carried out utilizing a combination of techniques including in situ time-resolved small-angle neutron scattering (SANS) complemented by the analysis of native and digested material by x-ray diffraction, DSC, small-angle x-ray scattering, and SEM to follow changes in starch granule nanostructure during enzymatic digestion. Starches varied in their digestibility and displayed structural differences in the course of enzymatic digestion. As evidenced by differing morphologies of enzymatic attack among varieties, the existence of granular pores and channels and physical penetrability of the amorphous growth ring affected the accessibility of the enzyme to the substrate. The combined effects of the granule microstructure and the nanostructure of the growth rings influence the opportunity of the enzyme to access its substrate; as a consequence, these structures determine the enzymatic digestibility of granular starches more than the absolute physical densities of the amorphous growth rings and amorphous and crystalline regions of the semicrystalline growth rings (Blazek and Gilbert 2010).

Understanding amylases is important to agriculture. Loss of viscosity, which is critical for texture and taste of juicy fruits, is related to the type of amylase used. In starchy fruits and vegetables, the loss of starch polymer viscosity is much more rapid with α-amylase; however, both β-amylase and glucoamylase are slow acting and would have a lesser impact on the polymer viscosity. It has been observed that the oxidation and etherification of starch significantly alter the extent of starch degradation by commercial amylases without impacting the rate of degradation (Wolf et al. 1999). The development of genetically modified variety of fruits and vegetables with starch that is resistant to enzymatic breakdown can be an important part in the diet of diabetic patients. In this regard, preliminary work in our laboratory on the enzymatic modification of granular starch (pH 5.5 from 0 to 8 h) yielded granules that were morphologically coherent with an intact crystalline structure, but suffered a 5%–6% loss of the amorphous region and a shift in the glass transition (T_g) value of the starch polymer. However, the most dramatic effect of this treatment was on starch degradation behavior. With time, as the amorphous portion of the starch polymer became more solubilized, crystalline domains in the starch were increasingly resistant to degradation. Such modified starches offer potentially interesting opportunities in the design of food products where delayed carbohydrate degradation is desired.

2.5 EXTRUSION PROCESSING OF STARCH AND STARCH-BASED BIOPLASTICS

In granular form, starch has been used as filler in polymers, but it can also be plasticized and processed using classical plastic processing technologies. These include extrusion, thermoforming, injection molding, compression molding, film casting, extrusion blowing, jet cooking, etc. The main issue for starch is its hydrophilic nature that limits its use in high-moisture environments (Brouillet-Fourmann et al. 2002). Compared to conventional polymers, starch processing is complex and difficult to control. For example, during processing, starch is susceptible to phase transition, exhibits high viscosity and rapid moisture loss, and retrogrades. However, using proper formulation and suitable processing conditions, many of these challenges can be overcome. Such strategies include the use of (1) appropriate plasticizers; (2) lubricants; (3) modified starch in which the hydroxyls have been replaced with ester and ether groups such as in carboxymethyl and hydroxypropyl starches; (4) blending starch with a hydrophobic polymer like polylactic acid (PLA), polycaprolactone (PCL), or cellulose in the presence of an appropriate compatibilizer (often a starch-graft copolymer grafted with the hydrophobic polymer); (5) copolymers of starch-graft-hydrophobic polymer, such as starch-graft PLA, starch-graft PCL, etc.; and (6) blending of starch with nanoclays to form starch nanocomposites.

2.5.1 Sheet/Film Extrusion

Twin-screw extrusion is the simple and well-established technique for producing sheets or films from starch-based polymers (Dean et al. 2007). However, a two-stage sheet/film extrusion-processing technique has also been used (Fishman et al. 2006; Matzinos et al. 2002). Though more time-consuming, this two-stage extrusion technique offers much more stable starch sheets/films because the high-pressure capacity of the single-screw extruder overcomes the high viscosity and poor processability of starch-based materials. To successfully produce sheets or films by extrusion, raw starches are normally blended with other additives and plasticizers to enhance their processability and to improve the properties of final products. For this purpose, the most commonly used plasticizers are water and/or glycerol (Delville et al. 2002; Fishman et al. 2000, 2004, 2006; Thuwall et al. 2006), but various other additives have also been evaluated. Urea has been reported to improve the gelatinization of starch at low-moisture levels, thus allowing direct extrusion of a uniform film from a semidry blend. Thuwall et al. (2006) used a fluoroelastomer lubricant to reduce the tendency of a material to stick to and, therefore, clog the die, and used dextrin to lower the viscosity by virtue of its low molecular mass. Stearic acid and poly(ethylene glycol) have been used to improve the rheological behavior of starch blends (Walenta et al. 2001a,b). Products obtained by blending poly(vinyl alcohol) (PVOH) with starch also exhibit excellent mechanical properties (Fishman et al. 2000).

A multilayered coextrusion technique has been used for preparing starch sheets or films (Dole et al. 2005; Martin et al. 2001; Wang et al. 2000), where thermoplastic starch is laminated with appropriate biodegradable polymers to improve the mechanical properties, water resistance, and gas-barrier properties of the film. These products have shown potential for applications such as food packaging and disposable product manufacture. Three-layer coextrusion is most often practiced, in which a coextrusion line consists of two single-screw extruders (one for the inner starch layer and the other for the outer polymer layers), a feedblock, a coat-hanger-type sheet die, and a three-roll calendaring system (Martin et al. 2001). Biodegradable polyesters such as PCL (Wang et al. 2000), PLA (Fang et al. 2005) and polyesteramide, poly(butylene succinate adipate), and poly(hydroxybutyrate-co-valerate) are often used for the outer layer (Martin et al. 2001).

2.5.2 Foaming (Expanded Starch) Extrusion

Foaming extrusion has mainly been used to produce loose-fill packaging materials, mimicking the process quite similar to that used for producing expanded snack foods (Ilo et al. 1999). While single-screw extrusion (Miladinov and Hanna 2001) has been used for this purpose, twin-screw extrusion (Guan et al. 2005; Nabar et al. 2006; Xu et al. 2005) is most widely used and preferable because of high throughput, longer residence time, high shear, and more flexible temperature control.

In the development of starch foams with physical and mechanical properties equivalent to those of traditional expanded polystyrene foam, efforts have been to use modified cornstarch in conjunction with other biodegradable polymers such as PLA (Guan et al. 2005), natural fibers (Guan et al. 2004), poly(ethylene vinyl alcohol), polystyrene maleic anhydride (Cha et al. 2001), and poly(hydroxy amino ether) (Nabar et al. 2006). Some postextrusion treatments may also be beneficial to achieve desirable physical and mechanical properties in the final starch product (Chen et al. 2004). Achieving ideal processing temperatures during extrusion is critical, as fluctuation in processing temperatures could significantly impact the quality of loose-fill products. Low processing temperatures may affect the foam density, whereas high temperatures could result in excessive moisture loss during the expansion process.

2.5.3 Injection Molding

The high viscosity and poor flow properties of starch-based materials present difficulties during injection molding. Additionally, lack of reliable parameters makes it more difficult to predict and design the optimum processing conditions. For example, because all starch-based formulations contain water, which evaporates during the heating process, it is impossible to measure the melting flow index using a conventional facility. In this regard, efforts have been made to gain a better understanding of the injection-molding behavior of starch-based materials. Stepto (2003) and Onteniente et al. (2000) used a given mold (shot) volume, screw speed, and temperature profile, and then measured the variation of refill times for materials to feed in front of a screw at different screw-rotation speeds and under different applied back pressure. In this way, the shear rate of the screw was determined by its rotational speed, and the back pressure defined the reverse pressure drop along the metering zone of the screw. Due to the poor mechanical properties and instability of a pure raw-starch material, it is frequently reinforced with fibers (Avérous and Boquillon 2004) and/or blended with other synthetic polymers to improve processing (Mani and Bhattacharya 1998). Compression molding has been intensively investigated for processing starch-based plastics, particularly for producing foamed containers, and generally involves starch gelatinization, expanding, and drying. Apart from gelatinization agents, mold-releasing agents such as magnesium stearate and stearic acid are often used in formulations to prevent the starch from sticking to the mold (Glenn and Orts 2001; Glenn et al. 2007). Explosion puffing is the oldest technique used to create starch-based foams from starch feedstock with low-moisture content. Explosion puffing can produce low-density starch-based foams within several seconds; however, starch-based foamed products are susceptible to moisture and heat. Baked foams made from potato amylopectin, PVOH, aspen fiber, and monostearyl citrate appeared to have adequate flexibility and water resistance to function as clamshell-type hot sandwich containers (Shogren et al. 2002).

2.5.4 Casting

The technique of casting starch-based film has been widely reported (Cyras et al. 2006; Kampeerapappun et al. 2007; Lee et al. 2007; Mali et al. 2005; Shen and Lai 2006) and typically includes solution preparation, gelatinization, casting, and drying. For casting, typically, starch and plasticizers are mixed together and transferred to Brabender-type viscograph where the solution is heated to 95°C and maintained for 10 min while mixing. A higher temperature is used with a plasticizer like glycerol. The gelatinized suspensions are then poured onto a flat Teflon or acrylic plate and left to dry for 24 h in an oven at 40°C–75°C to a constant weight. The final thickness of the cast films is normally 0.02–0.10 mm, which is controlled by calculating the quantity of starch suspension poured onto the plate. Plasticizers not only play an important role in processing starch but also improve mechanical properties of starch-based films. Water is an excellent plasticizer for starch; however, at low humidity, starch plastics are brittle, and at high humidity, starch plastics are soft. Glycerol and sorbitol are the two most widely used plasticizers to help films be less brittle and more flexible. Other agents such as sugars, sucrose, glucose, xylose, fructose, urea, and various glycols have also been evaluated (Kalichevsky et al. 1993). Different plasticizers have regularly been used in combination with water in order to approach the conditions suitable for gelatinization. The coating technology has also been used for improving the mechanical properties and moisture sensitivity in starch films (Bangyekan et al. 2006).

The reactive extrusion (REX) technology has witnessed a tremendous growth in the past few years. In this process, an extruder is used as a continuous reactor to carry out chemical and/or structural modifications of polymers within the extruder. The technology was originally developed in the 1980s primarily for the modification of synthetic polymers, but has evolved quickly and is now applied in various areas such as polymerization, grafting, and cross-linking of starch-based thermoplastic blends (Moad 1999; Xanthos 1992). The most popular REX systems for starch modification include twin-screw extruders, although single-screw extruders were once used. Most REX systems prepare starch-based products with higher efficiency and better product properties compared to traditional batch processes (Liu et al. 2009). In addition to functionalization, the REX technology allows the control of polymer rheology (melt flow, melt strength). REX technology is gaining increasing popularity and competes with diluent-free operations with respect to efficiency and economics.

The introduction of starch into plastics signifies a major milestone in the birth and development of bioplastics. In 1973, Griffin first proposed the addition of 6%–7% granular starch into polyethylene films (Griffin 1973). Around same time, Otey and coworkers developed film composites based on plasticized starch compounded with a synthetic polymer (Otey et al. 1977) and also developed a process for compounding and melt extrusion of thermoplastic starch into blown films at 5%–10% moisture content (Otey and Westhoff 1982). These extruded films were proposed for applications such as agricultural mulch and disposable packaging, where deterioration of physical properties as the starch biodegraded was deemed advantageous. When tested, such films biodegraded in a laboratory where films were exposed to amylolytic microbes and/or starch-degrading amylases and also when exposed to environments such as compost (Imam et al. 1998), river (Imam et al. 1992), sludge (Imam et al. 1995), and coastal waters (Imam et al. 1999). Further examination suggested that while a starch-polyethylene matrix substantially deteriorated, only starch in the films biodegraded leaving behind mostly chewed-up or weakened polyethylene matrix which formed the nonbiodegradable part of the film. Although there are many advantages and sound reasoning for developing and marketing of starch-plastic products, numerous challenges remain. Products that satisfy the real needs of the public must be developed. Performance must meet consumer expectations, and costs must be competitive to feedstock currently used for the same applications. Study of the influence of compounding variables on morphology, physical properties, and biodegradability can provide a basis for tailoring properties of starch plastics to fit specific applications. Many opportunities and challenges accrue from the molecular structure and chemistry of starch.

The last few decades have seen an explosive growth in bioplastics research and development owing mostly to the success in processing of biopolymers using conventional polymer-processing techniques such as extrusion, blow molding, thermoforming, calendaring, and injection molding. Particularly, many plastic blends containing starch in combination with other biopolymers such as proteins, celluloses, oils, and chitin/chitosan have been developed. The appearance of many bioderived polymers like PLA, polyhydroxybutyrate (PHB), thermoplastic starch, and biodegradable synthetic polymers, such as PVOH and ε-PCL, has inspired the field of bioplastics with array of innovative and functional single-use products that are also biodegradable and exhibit superior performance. Such products include packaging, delivery systems, encapsulants, coatings, specialty plastics, and films. Most biopolymers show excellent compatibility with a wide variety of plasticizers, fillers, and other additives. Starch has also been expanded into foams under heat and pressure using water as a foaming agent. The USDA-ARS and EK Industries, Inc. have jointly developed starch-based disposable plates, cups, and bowls that are now commercially available in stores in the United States. One recent development with respect to biodegradable plastics is the use of oxidative additives in conjunction with polyethylene to develop "oxobiodegradable" plastics. Several studies have shown that oxidative additives expedited the polymer chain scission under heat and sunlight reducing the molecular weight of the polymer to an extent that was readily attacked and mineralized by soil fungi (Corti et al. 2010).

2.6 CONCLUSION

Starch is recognized as one of the most abundant and important commodities containing value-added attributes for a vast number of industrial applications. Its chemistry, structure, property, and susceptibility to various chemical, physical, and enzymatic modifications offer a high technological value which contributes enormously to both the food and nonfood industries. Recent advances in processing technologies for starch and genetic manipulation to create new starch varieties having useful functionalities had a profound impact as many new products have been developed around starch as a feedstock for foods, chemicals, plastics, and fuels. Recent progress in developing effective and efficient enzyme systems for starch biocatalysis points to a considerable opportunity and scope for exciting industrial materials and the future of starch science and technology.

REFERENCES

Aboubakar, Y.N. Njintang, J. Scher, and C.M.F. Mbofung. 2008. Physicochemical, thermal properties and microstructure of six varieties of taro (*Colocasia esculenta* L. Schott) flours and starches. *Journal of Food Engineering* 86(2):294–305.

Agüeros, M., S. Espuelas, I. Esparza, P. Calleja, I. Peñuelas, G. Ponchel, and J.M. Irache. 2011. Cyclodextrin-poly(anhydride) nanoparticles as new vehicles for oral drug delivery. *Expert Opinion on Drug Delivery* 8(6):721–734.

Aiyer, P.V. 2005. Amylases and their applications. *African Journal of Biotechnology* 4(13):1525–1529.

Alsberg, C.L. 1938. Structure of the starch granule. *Plant Physiology* 13:295–330.

Amy, Y.H.L. and R. Jörg. 2009. Physical aging and structural relaxation in polymer nanocomposites. *Journal of Polymer Science, Part B: Polymer Physics* 47(18):1789–1798.

Avérous, L. and N. Boquillon. 2004. Biocomposites based on plasticized starch: Thermal and mechanical behaviours. *Carbohydrate Polymers* 56(2):111–122.

Avérous, L. and C. Fringant. 2001. Association between plasticized starch and polyesters: Processing and performances of injected biodegradable systems. *Polymer Engineering & Science* 41(5):727–734.

Ba, K., J. Destain, and P. Thonart. 2010. Hydrolysis of starch by sorghum malt for maltodextrin production. *Biotechnology, Agronomy and Society and Environment* 14(2):537.

Badenhuizen, N.P. 1937. Die Struktur des Stärkekorns. *Protoplasma* 28(1):293–326.

Badenhuizen, N.P. 1956. The structure of the starch granule. *Protoplasma* 45(3):315–326.

Baldwin, P.M., M.C. Davies, and C.D. Melia. 1997. Starch granule surface imaging using low-voltage scanning electron microscopy and atomic force microscopy. *International Journal of Biological Macromolecules* 21(1–2):103–107.

Ball, S.G. and M.K. Morell. 2003. From bacterial glycogen to starch: Understanding the biogenesis of the plant starch granule. *Annual Review of Plant Biology* 54:207–233.

Bangyekan, C., D. Aht-Ong, and K. Srikulkit. 2006. Preparation and properties evaluation of chitosan-coated cassava starch films. *Carbohydrate Polymers* 63(1):61–71.

Baruch, D.W., L.D. Jenkins, H.N. Dengate, and P. Meredith. 1983. Nonlinear model of wheat starch granule distribution at several stages of development. *Cereal Chemistry* 60:32–35.

Baruch, D.W., P. Meredith, L.D. Jenkins, and L.D. Simmons. 1979. Starch granules of developing wheat kernels. *Cereal Chemistry* 56:554–558.

Bechtel, D.B. and J.D. Wilson. 2003. Amyloplast formation and starch granule development in hard red winter wheat. *Cereal Chemistry* 80(2):175–183.

Bechtel, D.B., I. Zayas, L. Kaleikau, and Y. Pomeranz. 1990. Size-distribution of wheat starch granules during endosperm development. *Cereal Chemistry* 67:59–63.

BeMiller, J.N. and R.L. Whistler. 2009. *Starch: Chemistry and Technology*. New York: Academic Press.

Benmoussa, M., B.R. Hamaker, C.P. Huang, D.M. Sherman, C.F. Weil, and J.N. BeMiller. 2010. Elucidation of maize endosperm starch granule channel proteins and evidence for plastoskeletal structures in maize endosperm amyloplasts. *Journal of Cereal Science* 52(1):22–29.

Blazek, J. and L. Copeland. 2010. Amylolysis of wheat starches. I. Digestion kinetics of starches with varying functional properties. *Journal of Cereal Science* 51(3):265–270.

Blazek, J. and E.P. Gilbert. 2010. Effect of enzymatic hydrolysis on native starch granule structure. *Biomacromolecules* 11(12):3275–3289.

Brouillet-Fourmann, S., C. Carrot, C. Lacabanne, N. Mignard, and V. Samouillan. 2002. Evolution of interactions between water and native corn starch as a function of moisture content. *Journal of Applied Polymer Science* 86(11):2860–2865.

Cha, J.Y., D.S. Chung, P.A. Seib, R.A. Flores, and M.A. Hanna. 2001. Physical properties of starch-based foams as affected by extrusion temperature and moisture content. *Industrial Crops and Products* 14(1):23–30.

Chen, L., S.H. Gordon, and S.H. Imam. 2004. Starch graft poly(methyl acrylate) loose-fill foam: Preparation, properties and degradation. *Biomacromolecules* 5(1):238–244.

Chen, Y., S. Huang, Z. Tang, X. Chen, and Z. Zhang. 2011a. Structural changes of cassava starch granules hydrolyzed by a mixture of α-amylase and glucoamylase. *Carbohydrate Polymers* 85(1):272–275.

Chen, G. and M. Jiang. 2011. Cyclodextrin-based inclusion complexation bridging supramolecular chemistry and macromolecular self-assembly. *Chemical Society Reviews* 40(5):2254–2266.

Chen, Y. and Y. Liu. 2010. Cyclodextrin-based bioactive supramolecular assemblies. *Chemical Society Reviews* 39(2):495–505.

Chen, P., L. Yu, L. Chen, and X. Li. 2006. Morphology and microstructure of maize starches with different amylose/amylopectin content. *Starch—Stärke* 58(12):611–615.

Chen, P., L. Yu, G.P. Simon, X. Liu, K. Dean, and L. Chen. 2011b. Internal structures and phase-transitions of starch granules during gelatinization. *Carbohydrate Polymers* 83(4):1975–1983.

Chen, P., L. Yu, G. Simon, E. Petinakis, K. Dean, and L. Chen. 2009a. Morphologies and microstructures of cornstarches with different amylose-amylopectin ratios studied by confocal laser scanning microscope. *Journal of Cereal Science* 50(2):241–247.

Chen, S., J. Zhang, and J. Su. 2009b. Effect of damp-heat aging on the properties of ethylene-vinyl acetate copolymer and ethylene-acrylic acid copolymer blends. *Journal of Applied Polymer Science* 114(5):3110–3117.

Chernykh, E.V. and S.B. Brichkin. 2010. Supramolecular complexes based on cyclodextrins. *High Energy Chemistry* 44(2):83–100.

Corti, A., S. Muniyasamy, M. Vitali, S.H. Imam, and E. Chiellini. 2010. Oxidation and biodegradation of polyethylene films containing pro-oxidant additives: Synergistic effects of sunlight exposure, thermal aging and fungal biodegradation. *Polymer Degradation and Stability* 95(6):1106–1114.

Cyras, V.P., M.C.T. Zenklusen, and A. Vazquez. 2006. Relationship between structure and properties of modified potato starch biodegradable films. *Journal of Applied Polymer Science* 101(6):4313–4319.

Da Róz, A.L., A.J.F. Carvalho, A. Gandini, and A.A.S. Curvelo. 2006. The effect of plasticizers on thermoplastic starch compositions obtained by melt processing. *Carbohydrate Polymers* 63(3):417–424.

Davis, J.T., O. Okunola, and R. Quesada. 2010. Recent advances in the transmembrane transport of anions. *Chemical Society Reviews* 39(10):3843–3862.

Dean, K., L. Yu, and D.Y. Wu. 2007. Preparation and characterization of melt-extruded thermoplastic starch/clay nanocomposites. *Composites Science and Technology* 67(3–4):413–421.

Del Valle, E.M.M. 2004. Cyclodextrins and their uses: A review. *Process Biochemistry* 39(9):1033–1046.

Delville, J., C. Joly, P. Dole, and C. Bliard. 2002. Solid state photocrosslinked starch based films: A new family of homogeneous modified starches. *Carbohydrate Polymers* 49(1):71–81.

Dobreva, T., R. Benavente, J.M. Pereña, E. Pérez, M. Avella, M. García, and G. Bogoeva-Gaceva. 2010. Effect of different thermal treatments on the mechanical performance of poly(L-lactic acid) based eco-composites. *Journal of Applied Polymer Science* 116(2):1088–1098.

Dole, P., L. Avérous, C. Joly, G. Della Valle, and C. Bliard. 2005. Evaluation of starch-PE multilayers: Processing and properties. *Polymer Engineering and Science* 45(2):217–224.

Eksteen, J.M., A.J.C. Steyn, P. van Rensburg, R.R. Cordero Otero, and I.S. Pretorius. 2003. Cloning and characterization of a second α-amylase gene (LKA2) from *Lipomyces kononenkoae* IGC4052B and its expression in *Saccharomyces cerevisiae*. *Yeast* 20(1):69–78.

Eliason, A.-C. 2004. *Starch in Food: Structure, Function and Applications*. Boca Raton, FL: CRC Press.

Esau, K. 1965. *Plant Anatomy*, 2nd edn. New York: John Wiley & Sons, Inc.

Etienne, S., N. Hazeg, E. Duval, A. Mermet, A. Wypych, and L. David. 2007. Physical aging and molecular mobility of amorphous polymers. *Journal of Non-Crystalline Solids* 353(41–43):3871–3878.

Evers, A.D. 1969. Scanning electron microscopy of wheat starch I. Entire granules. *Starch—Stärke* 21(4):96–99.

Fakirov, S. and D. Bhattacharyya. 2007. *Handbook of Engineering Biopolymers: Homopolymers, Blends and Composites*. Cincinnati, OH: Hanser Gardner.

Fang, J.M., P.A. Fowler, C. Escrig, R. Gonzalez, J.A. Costa, and L. Chamudis. 2005. Development of biodegradable laminate films derived from naturally occurring carbohydrate polymers. *Carbohydrate Polymers* 60(1):39–42.

Fannon, J.E., R.J. Hauber, and J.N. BeMiller. 1992. Surface pores of starch granules. *Cereal Chemistry* 69:284–288.

Fannon, J.E., J.M. Shull, and J.N. BeMiller. 1993. Interior channels of starch granules. *Cereal Chemistry* 70:611–613.

Fishman, M.L., D.R. Coffin, R.P. Konstance, and C.I. Onwulata. 2000. Extrusion of pectin/starch blends plasticized with glycerol. *Carbohydrate Polymers* 41(4):317–325.

Fishman, M.L., D.R. Coffin, C.I. Onwulata, and R.P. Konstance. 2004. Extrusion of pectin and glycerol with various combinations of orange albedo and starch. *Carbohydrate Polymers* 57(4):401–413.

Fishman, M.L., D.R. Coffin, C.I. Onwulata, and J.L. Willett. 2006. Two stage extrusion of plasticized pectin/poly(vinyl alcohol) blends. *Carbohydrate Polymers* 65(4):421–429.

Fitt, L.E. and E.M. Snyder. 1984. Photomicrographs of starches. In *Starch: Chemistry and Technology*, 2nd edn., R.L. Whistler, J.N. Bemiller, and E.F. Paschall (eds.). San Francisco, CA: Academic Press, Inc.

Galicia-García, T., F. Martínez-Bustos, O. Jiménez-Arevalo, A.B. Martínez, R. Ibarra-Gómez, M. Gaytán-Martínez, and M. Mendoza-Duarte. 2011. Thermal and microstructural characterization of biodegradable films prepared by extrusion-calendering process. *Carbohydrate Polymers* 83(2):354–361.

Gallant, D.J., B. Bouchet, and P.M. Baldwin. 1997. Microscopy of starch: Evidence of a new level of granule organization. *Carbohydrate Polymers* 32(3–4):177–191.

Gallant, D.J., B. Bouchet, A. Buléon, and S. Pérez. 1992. Physical characteristics of starch granules and susceptibility to enzymatic degradation. *European Journal of Clinical Nutrition* 46(Suppl 2):S3–S16.

Gallant, D., A. Derrien, A. Aumaitre, and A. Guilbot. 1973. Dégradation in vitro de l'amidon par le suc pancréatique. Etude par microscopie électronique à transmission et à balayage. *Starch—Stärke* 25(2):56–64.

Gallant, D. and A. Guilbot. 1971. Artefacts au cours de la préparation de coupes de grains d'amidon. Etude par microscopie photonique et électronique. *Starch—Stärke* 23(7):244–250.

Glenn, G.M., A.K. Klamczynski, K.M. Holtman, J. Shey, B.S. Chiou, J. Berrios, D. Wood, W.J. Orts, and S.H. Imam. 2007. Heat expanded starch-based compositions. *Journal of Agricultural and Food Chemistry* 55(10):3936–3943.

Glenn, G.M. and W.J. Orts. 2001. Properties of starch-based foam formed by compression/explosion processing. *Industrial Crops and Products* 13(2):135–143.

Grassia, L. and A. D'Amore. 2009. On the interplay between viscoelasticity and structural relaxation in glassy amorphous polymers. *Journal of Polymer Science, Part B: Polymer Physics* 47(7):724–739.

Griffin, G.J.L. 1973. Biodegradable fillers in thermoplastics. *ACS Organic Coatings and Plastics Chemistry Division Preprints* 33(2 Meet):88–96.

Guan, J., K.M. Eskridge, and M.A. Hanna. 2005. Acetylated starch-polylactic acid loose-fill packaging materials. *Industrial Crops and Products* 22(2):109–123.

Guan, J., Q. Fang, and M.A. Hanna. 2004. Macrostructural characteristics of starch acetate extruded with natural fibers. *Transactions of the American Society of Agricultural Engineers* 47(1):205–212.

Hall, D.M. and J.G. Sayre. 1970. Internal architecture of potato and canna starch part I: Crushing studies. *Textile Research Journal* 40(2):147–157.

Houghton-Larsen, J. and P.A. Pedersen. 2003. Cloning and characterisation of a glucoamylase gene (GlaM) from the dimorphic zygomycete *Mucor circinelloides*. *Applied Microbiology and Biotechnology* 62(2–3):210–217.

Hu, J., Z. Tao, and S. Cao. 2009. Unique structure and property of cyclodextrins and their application in transdermal drug delivery. *Methods and Findings in Experimental and Clinical Pharmacology* 31(7): 449–456.

Huang, M., J. Yu, and X. Ma. 2005. Ethanolamine as a novel plasticiser for thermoplastic starch. *Polymer Degradation and Stability* 90(3):501–507.

Huber, K.C. and J.N. Bemiller. 1997. Visualization of channels and cavities of corn and sorghum starch granules. *Cereal Chemistry* 74(5):537–541.

Hurkman, W.J. and D.F. Wood. 2011. High temperature during grain fill alters the morphology of protein and starch deposits in the starchy endosperm cells of developing wheat (*Triticum aestivum* L.) grain. *Journal of Agricultural and Food Chemistry* 59(9):4938–4946.

Ibáñez, A.M., D.F. Wood, W.H. Yokoyama, I.M. Park, M.A. Tinoco, C.A. Hudson, K.S. Mckenzie, and C.F. Shoemaker. 2007. Viscoelastic properties of waxy and nonwaxy rice flours, their fat and protein-free starch, and the microstructure of their cooked kernels. *Journal of Agricultural and Food Chemistry* 55:6761–6771.

Ilo, S., Y. Liu, and E. Berghofer. 1999. Extrusion cooking of rice flour and amaranth blends. *LWT—Food Science and Technology* 32(2):79–88.

Imam, S.H., A. Burgess-Cassler, G.L. Cote, S.H. Gordon, and F.L. Baker. 1991. A study of cornstarch granule digestion by an unusually high molecular weight α-amylase secreted by *Lactobacillus amylovorus*. *Current Microbiology* 22(6):365–370.

Imam, S.H., L. Chen, S.H. Gorden, R.L. Shogren, D. Weisleder, and R.V. Greene. 1998. Biodegradation of injection molded starch-poly(3-hydroxybutyrate-co-3-hydroxyvalerate) blends in a natural compost environment. *Journal of Environmental Polymer Degradation* 6(2):91–98.

Imam, S.H., S.H. Gordon, R.L. Shogren, and R.V. Greene. 1995. Biodegradation of starch-poly (β-hydroxybutyrate-co-valerate) composites in municipal activated sludge. *Journal of Environmental Polymer Degradation* 3(4):205–213.

Imam, S.H., S.H. Gordon, R.L. Shogren, T.R. Tosteson, N.S. Govind, and R.V. Greene. 1999. Degradation of starch-poly(β-hydroxybutyrate-Co-β-hydroxyvalerate) bioplastic in tropical coastal waters. *Applied and Environmental Microbiology* 65(2):431–437.

Imam, S.H., S.H. Gordon, A.R. Thompson, R.E. Harry-O'kuru, and R.V. Greene. 1993. The use of CP/MAS 13C-NMR for evaluating starch degradation in injection molded starch-plastic composites. *Biotechnology Techniques* 7(11):791–794.

Imam, S.H., J.M. Gould, S.H. Gordon, M.P. Kinney, A.M. Ramsey, and T.R. Tosteson. 1992. Fate of starch-containing plastic films exposed in aquatic habitats. *Current Microbiology* 25(1):1–8.

Imam, S.H. and R.E. Harry-O'Kuru. 1991. Adhesive interactions between amylolytic bacterium Lactobacillus amylovorus and insoluble cornstarch granules. *Appl. Environ. Microbiol.* 57:1128–1133.

Imam, S.H. and W.J. Snell. 1988. The *Chlamydomonas* cell wall degrading enzyme, lysine, acts on two substrates within the framework of the wall. *The Journal of Cell Biology* 106:2211–2221.

Jane, J.-L. 2009. Structural features of starch granules II. In *Starch: Chemistry and Technology*, R.L. Whistler and J.N. BeMiller (eds.). New York: Elsevier.

Jane, J., L. Shen, J. Chen, S. Lim, T. Kasemsuwan, and W.K. Nip. 1992. Physical and chemical studies of taro starches and flours. *Cereal Chemistry* 69:528–535.

Janssen, L.P.B.M. and L. Moscicki. 2009. *Thermoplastic Starch: Green Material for Various Industries.* Weinheim, Germany: Wiley-VCH.

Jasberg, B.K., C.L. Swanson, R.L. Shogren, and W.M. Doane. 1992. Effect of moisture on injection molded starch-EAA-HDPE composites. *Journal of Polymeric Material* 9:163–170.

Johnson, L.A. and J.B. May. 2003. Wet milling: The basis for corn biorefineries. In *Corn: Chemistry and Technology*, P.J. White and L.A. Johnson (eds.). St. Paul, MN: American Association of Cereal Chemists.

Jyothi, A.N. 2010. Starch graft copolymers: Novel applications in industry. *Composite Interfaces* 17(2–3):165–174.

Kalichevsky, M.T., E.M. Jaroszkiewicz, and J.M.V. Blanshard. 1993. A study of the glass transition of amylopectin-sugar mixtures. *Polymer* 34(2):346–358.

Kampeerapappun, P., D. Aht-ong, D. Pentrakoon, and K. Srikulkit. 2007. Preparation of cassava starch/montmorillonite composite film. *Carbohydrate Polymers* 67(2):155–163.

Khanna, N.D., I. Kaur, and A. Kumar. 2011. Starch-grafted polypropylene: Synthesis and characterization. *Journal of Applied Polymer Science* 119(1):602–612.

Kierkels, J.T.A., C.L. Dona, T.A. Tervoort, and L.E. Govaert. 2008. Kinetics of re-embrittlement of (anti) plasticized glassy polymers after mechanical rejuvenation. *Journal of Polymer Science, Part B: Polymer Physics* 46(2):134–147.

Kim, Y., N.S. Mosier, R. Hendrickson, T. Ezeji, H. Blaschek, B. Dien, M. Cotta, B. Dale, and M.R. Ladisch. 2008. Composition of corn dry-grind ethanol by-products: DDGS, wet cake, and thin stillage. *Bioresource Technology* 99(12):5165–5176.

Koenig, M.F. and S.J. Huang. 1995. Biodegradable blends and composites of polycaprolactone and starch derivatives. *Polymer* 36(9):1877–1882.

Lamikanra, O., S.H. Imam, and D. Ukuku. 2005. *Produce Degradation: Pathways and Prevention.* Boca Raton, FL: Taylor & Francis.

Laza-Knoerr, A.L., R. Gref, and P. Couvreur. 2010. Cyclodextrins for drug delivery. *Journal of Drug Targeting* 18(9):645–656.

Lee, J. and S. McCarthy. 2009. Biodegradable poly(lactic acid) blends with chemically modified polyhydroxyoctanoate through chain extension. *Journal of Polymers and the Environment* 17(4):240–247.

Lee, J.S., K.D. Wittchen, C. Stahl, J. Strey, and F. Meinhardt. 2001. Cloning, expression, and carbon catabolite repression of the bamM gene encoding β-amylase of *Bacillus megaterium* DSM319. *Applied Microbiology and Biotechnology* 56(1–2):205–211.

Lee, W.J., Y.N. Youn, Y.H. Yun, and S.D. Yoon. 2007. Physical properties of chemically modified starch(RS4)/PVA blend films—Part 1. *Journal of Polymers and the Environment* 15(1):35–42.

Liu, H., F. Xie, L. Yu, L. Chen, and L. Li. 2009. Thermal processing of starch-based polymers. *Progress in Polymer Science (Oxford)* 34(12):1348–1368.

Loftsson, T. and M.E. Brewster. 2010. Pharmaceutical applications of cyclodextrins: Basic science and product development. *Journal of Pharmacy and Pharmacology* 62(11):1607–1621.

Long, Y. 2009. *Biodegradable Polymer Blends and Composites from Renewable Resources.* Hoboken, NJ: John Wiley & Sons, Inc.

Ma, U.V.L., J.D. Floros, and G.R. Ziegler. 2011. Effect of starch fractions on spherulite formation and microstructure. *Carbohydrate Polymers* 83(4):1757–1765.

Mali, S., L.S. Sakanaka, F. Yamashita, and M.V.E. Grossmann. 2005. Water sorption and mechanical properties of cassava starch films and their relation to plasticizing effect. *Carbohydrate Polymers* 60(3):283–289.

van de Manakker, F., T. Vermonden, C.F. van Nostrum, and W.E. Hennink. 2009. Cyclodextrin-based polymeric materials: Synthesis, properties, and pharmaceutical/biomedical applications. *Biomacromolecules* 10(12):3157–3175.

Mani, R. and M. Bhattacharya. 1998. Properties of injection moulded starch/synthetic polymer blends—IV. Thermal and morphological properties. *European Polymer Journal* 34(10):1477–1487.

Martin, O., E. Schwach, L. Avérous, and Y. Couturier. 2001. Properties of biodegradable multilayer films based on plasticized wheat starch. *Starch—Stärke* 53(8):372–380.

Matzinos, P., V. Tserki, A. Kontoyiannis, and C. Panayiotou. 2002. Processing and characterization of starch/
    polycaprolactone products. *Polymer Degradation and Stability* 77(1):17–24.
Meredith, P. 1981. Large and small starch granules in wheat—Are they really different? *Starch—Stärke*
    33(2):40–44.
Miladinov, V.D. and M.A. Hanna. 2001. Temperatures and ethanol effects on the properties of extruded modi-
    fied starch. *Industrial Crops and Products* 13(1):21–28.
Moad, G. 1999. Synthesis of polyolefin graft copolymers by reactive extrusion. *Progress in Polymer Science
    (Oxford)* 24(1):81–142.
Mohanty, A.K., M. Misra, and L.T. Drzal. 2005. *Natural Fibers, Biopolymers and Biocomposites*. Boca Raton,
    FL: Taylor & Francis.
Morán, J.I., V.P. Cyras, A. Vázquez, and M.L. Foresti. 2011. Characterization of chemically modified potato
    starch films through enzymatic degradation. *Journal of Polymers and the Environment* 19(1):217–224.
Moss, G.E. 1976. The microscopy of starch. In *Examination and Analysis of Starch and Starch Products*, J.A.
    Radley (ed.). London, U.K.: Applied Science Publishers Ltd.
Nabar, Y., R. Narayan, and M. Schindler. 2006. Twin-screw extrusion production and characterization of starch
    foam products for use in cushioning and insulation applications. *Polymer Engineering and Science*
    46(4):438–451.
Ohtani, T., T. Yoshino, T. Ushiki, S. Hagiwara, and T. Maekawa. 2000. Structure of rice starch granules
    in nanometre scale as revealed by atomic force microscopy. *Journal of Electron Microscopy* 49(3):
    487–489.
Onteniente, J.-P., B. Abbès, and L.H. Safa. 2000. Fully biodegradable lubricated thermoplastic wheat starch:
    Mechanical and rheological properties of an injection grade. *Starch—Stärke* 52(4):112–117.
Otero-Espinar, F.J., J.J. Torres-Labandeira, C. Alvarez-Lorenzo, and J. Blanco-Méndez. 2010. Cyclodextrins in
    drug delivery systems. *Journal of Drug Delivery Science and Technology* 20(4):289–301.
Otey, F.H. and R.P. Westhoff. 1982. Biodegradable starch-based blown films. US Patent 4337181.
Otey, F.H., R.P. Westhoff, and C.R. Russell. 1977. Biodegradable films from starch and ethylene-acrylic acid
    copolymer. *Industrial & Engineering Chemistry Product Research and Development* 16(4):305–308.
Parker, M.L., A.R. Kirby, and V.J. Morris. 2008. In situ imaging of pea starch in seeds. *Food Biophysics*
    3(1):66–76.
Paster, M., J.L. Pelligrino, and T.M. Carole. 2003. *Industrial Bioproducts: Today and Tomorrow*. Washington,
    DC: Department of Energy, Office of Energy Efficiency and Renewable Energy/Office of the Biomass
    Program.
Patil, J.S., D.V. Kadam, S.C. Marapur, and M.V. Kamalapur. 2010. Inclusion complex system; a novel technique
    to improve the solubility and bioavailability of poorly soluble drugs: A review. *International Journal of
    Pharmaceutical Sciences Review and Research* 2(2):29–34.
Pérez, S., P. Baldwin, and D.J. Gallant. 2009. Structural features of starch granules I. In *Starch: Chemistry and
    Technology*, R.L. Whistler and J.N. BeMiller (eds.). New York: Elsevier.
Peters, D. 2006. Carbohydrates as fermentation raw materials. *Kohlenhydrate als fermentationsrohstoff*
    78(3):229–238.
Poutanen, K. and P. Forssell. 1996. Modification of starch properties with plasticizers. *Trends in Polymer
    Science* 4:128–132.
Rahman, S., Z. Li, I. Batey, M.P. Cochrane, R. Appels, and M. Morell. 2000. Genetic alteration of starch func-
    tionality in wheat. *Journal of Cereal Science* 31(1):91–110.
Rattanachomsri, U., S. Tanapongpipat, L. Eurwilaichitr, and V. Champreda. 2009. Simultaneous non-thermal
    saccharification of cassava pulp by multi-enzyme activity and ethanol fermentation by *Candida tropica-
    lis*. *Journal of Bioscience and Bioengineering* 107(5):488–493.
Rausch, K.D. and R.L. Belyea. 2006. The future of coproducts from corn processing. *Applied Biochemistry and
    Biotechnology* 128(1):47–86.
Robyt, J.F. 1998. *Essentials of Carbohydrate Chemistry*. New York: Springer-Verlag.
Rodriguez-Gonzalez, F.J., B.A. Ramsay, and B.D. Favis. 2004. Rheological and thermal properties of thermo-
    plastic starch with high glycerol content. *Carbohydrate Polymers* 58(2):139–147.
Sasaki, Y. and K. Akiyoshi. 2010. Development of an artificial chaperone system based on cyclodextrin. *Current
    Pharmaceutical Biotechnology* 11(3):300–305.
Shen, Y.D. and X.J. Lai. 2006. Structure and mechanical properties of polyethylene glycol modified poly (lactic
    acid)/thermoplastic starch blend films. *Xiandai Huagong/Modern Chemical Industry* 26(5):35–37+39.

Shogren, R.L., R.V. Greene, and Y.V. Wu. 1991. Complexes of starch polysaccharides and poly(ethylene co-acrylic acid). Structure and stability in solution. *Journal of Applied Polymer Science* 42(6):1701–1709.

Shogren, R.L., J.W. Lawton, and K.F. Tiefenbacher. 2002. Baked starch foams: Starch modifications and additives improve process parameters, structure and properties. *Industrial Crops and Products* 16(1):69–79.

Shogren, R.I., A.R. Thompson, F.C. Felker, R.E. Harry-O'Kuru, S.H. Gordon, R.V. Greene, and J.M. Gould. 1992. Polymer compatibility and biodegradation of starch-poly(ethylene-co-acrylic acid)-polyethylene blends. *Journal of Applied Polymer Science* 44(11):1971–1978.

Smith, A.M., S.C. Zeeman, and S.M. Smith. 2005. Starch degradation. *Annual Review of Plant Biology* 56:73–97.

Stepto, R.F.T. 2003. The processing of starch as a thermoplastic. *Macromolecular Symposia* 201:203–212.

Sun, H., P. Zhao, X. Ge, Y. Xia, Z. Hao, J. Liu, and M. Peng. 2010. Recent advances in microbial raw starch degrading enzymes. *Applied Biochemistry and Biotechnology* 160(4):988–1003.

Svensson, B., K.S. Bak-Jensen, M.T. Jensen, J. Sauer, T.E. Gottschalk, and K.W. Rodenburg. 1999. Studies on structure, function, and protein engineering of starch-degrading enzymes. *Journal of Applied Glycoscience* 46(1):49–63.

Swanson, C.L., R.L. Shogren, G.F. Fanta, and S.H. Imam. 1993. Starch-plastic materials—Preparation, physical properties, and biodegradability. *Journal of Environmental Polymer Degradation* 1(2):155–166.

Thunwall, M., A. Boldizar, and M. Rigdahl. 2006. Compression molding and tensile properties of thermoplastic potato starch materials. *Biomacromolecules* 7(3):981–986.

Thuwall, M., A. Boldizar, and M. Rigdahl. 2006. Extrusion processing of high amylose potato starch materials. *Carbohydrate Polymers* 65(4):441–446.

Tsuji, H. and T. Tsuruno. 2010. Accelerated hydrolytic degradation of Poly(L-lactide)/Poly(D-lactide) stereocomplex up to late stage. *Polymer Degradation and Stability* 95(4):477–484.

Van Soest, J.J.G., S.H.D. Hulleman, D. De Wit, and J.F.G. Vliegenthart. 1996. Changes in the mechanical properties of thermoplastic potato starch in relation with changes in B-type crystallinity. *Carbohydrate Polymers* 29(3):225–232.

Vu Dang, H., C. Loisel, A. Desrumaux, and J.L. Doublier. 2009. Rheology and microstructure of cross-linked waxy maize starch/whey protein suspensions. *Food Hydrocolloids* 23(7):1678–1686.

Walenta, E., H.P. Fink, P. Weigel, and J. Ganster. 2001a. Structure-property relationships in extruded starch, 1: Supermolecular structure of pea amylose and extruded pea amylose. *Macromolecular Materials and Engineering* 286(8):456–461.

Walenta, E., H.P. Fink, P. Weigel, J. Ganster, and E. Schaaf. 2001b. Structure-property relationships of extruded starch, 2: Extrusion products from native starch. *Macromolecular Materials and Engineering* 286(8):462–471.

Wang, L., R.L. Shogren, and C. Carriere. 2000. Preparation and properties of thermoplastic starch-polyester laminate sheets by coextrusion. *Polymer Engineering and Science* 40(2):499–506.

Wellner, N., D.M.R. Georget, M.L. Parker, and V.J. Morris. 2011. In situ Raman microscopy of starch granule structures in wild type and ae mutant maize kernels. *Starch—Stärke* 63(3):128–138.

Werpy, T. and G. Petersen. 2004. *Top Value-Added Chemicals from Biomass, Vol 1. Results of Screening for Potential Candidates from Sugars and Synthesis Gas*. Washington, DC: U.S. Department of Energy, Office of Biomass Program.

Whitaker, J.R. 1994. *Principles of Enzymology for the Food Sciences*, 2nd edn. New York: Marcel Dekker, Inc.

Wolf, B.W., L.L. Bauer, and G.C. Fahey, Jr. 1999. Effects of chemical modification on in vitro rate and extent of food starch digestion: An attempt to discover a slowly digested starch. *Journal of Agricultural and Food Chemistry* 47(10):4178–4183.

Xanthos, M. 1992. *Reactive Extrusion: Principles and Practice*. New York: Hanser Publishers.

Xie, F., L. Yu, H. Liu, and L. Chen. 2006. Starch modification using reactive extrusion. *Starch—Stärke* 58(3–4):131–139.

Xie, Y., D. Yu, W. Wan, X. Guo, and S. Li. 2007. Effects of heating, tensile, and high pressure treatments on aggregate structure of a low density polyethylene. *Polymer—Plastics Technology and Engineering* 46(4):377–383.

Xu, Y.X., Y. Dzenis, and M.A. Hanna. 2005. Water solubility, thermal characteristics and biodegradability of extruded starch acetate foams. *Industrial Crops and Products* 21(3):361–368.

Yamaguchi, M., K. Kainuma, and D. French. 1979. Electron microscopic observations of waxy maize starch. *Journal of Ultrastructure Research* 69(2):249–261.

Yang, J.H., J.G. Yu, and X.F. Ma. 2006. Study on the properties of ethylenebisformamide and sorbitol plasticized corn starch (ESPTPS). *Carbohydrate Polymers* 66(1):110–116.
Zhang, B. and Y.Y. Zhou. 2007. Development of initiators in graft copolymerization of acrylic monomers
   onto starch. *Gaofenzi Cailiao Kexue Yu Gongcheng/Polymeric Materials Science and Engineering*
   23(2):36–40.
Zobel, H.F. 1988. Starch crystal transformations and their industrial importance. *Starch—Stärke* 40(1):1–7.

CHAPTER 3

Starch as Gelling Agent

Thava Vasanthan, Jihong Li, David Bressler, and Ratnajothi Hoover

CONTENTS

3.1 INTRODUCTION

Starch, as the second most abundant carbohydrate in nature, is a biodegradable and renewable biopolymer reserved by plants in all major agricultural crops (cereals, pulses, tubers, yams, roots, etc.) and is simply isolated through various extraction processes. It is a major ingredient in food products contributing to nutrition, texture, appearance, sensory characteristics, and functional properties in most food systems and also as an economically favorable source of polymer for other industrial and energy end uses. As a hydrocolloid gelling agent, starch in its native state can be directly used as an ingredient with its gelling properties and other functionalities for food processing. Furthermore, it is largely modified physically, chemically, and enzymatically to improve its functional properties for desired applications.

A gel is a form of matter intermediate between solid and liquid with mechanical rigidity, which consists of polymer molecules cross-linked to form tangled and interconnected molecular network immersed in a liquid medium. A gel can be defined by its structure and rheological behavior. From the point of view of structure, a gel is a system consisting of molecules, chains, particles, etc., which are partially connected to each other in a fluid medium by cross-links to the macroscopic dimensions (Nishinari 2009). From the point of view of rheology, a gel is a viscoelastic system characterized by the presence of a plateau region of storage modulus (G′) and the low tan δ (<0.1) at an angular frequency range from 10^{-3} to 10^{-2} rad/s, which is accessible by many commercially available rheometers (Nishinari 2009). Gels are commonly classified based on their source (natural and synthetic), constitution (polymer and molecular), cross-linking manner (chemical and physical), solvent (organogel and hydrogel), and drying process (aerogel, cryogel, and xerogel). As the most popular type of gels, hydrogel is a colloidal gel in which water is the dispersion medium. The three-dimensional network of a hydrogel can be formed physically by association of hydrophilic polymers through hydrogen bonding, molecular entanglements, hydrophobic interaction, and cation-mediated cross-linking and/or chemically by covalent cross-linking of polymer chains, which possesses high-water-holding capacity and mechanical stability. The phenomenon of water sorption by a hydrogel depends mechanistically on the diffusion of water molecules into the hydrogel matrix and subsequent relaxation of macromolecular chains of the hydrogel (Zhang et al. 2005a). Starch hydrogel is considered as a complex composite, in which an interpenetrating amylose gel matrix is filled with swollen granules (Morris 1990; Ring and Stainsby 1982). The physical properties of starch hydrogels depend on the rheology of the amylose matrix, the rigidity of the amylopectin granules, the volume fraction, and the shape of granules as well as the filler–matrix interaction. All these factors depend on the source of starch, processing conditions, and product formulation (Oakenfull 1987).

Starch-based hydrogel is a water swollen network and chemical gel of copolymers, which is polymerized with multifunctional agents, showing great water absorbent ability with advantages of non-toxicity, biocompatibility, biodegradability, and antibacterial activity. It is classified as a unique category of polymer material for widespread applications in the fields of biomedical, pharmaceutical, agriculture, biology, and environmental remediation and protection (Zhang et al. 2005b). This chapter starts with the introduction of the basic knowledge of starch structure and properties for better understanding of starch gelling behavior and then overviews the general preparation of gelling starches, starch-based hydrogels, and their applications in food and nonfood industries.

3.2 STARCH SOURCE, COMPOSITION, AND STRUCTURE

3.2.1 Source

Starch is commercially isolated in pure form from a variety of sources. Over the years, maize has been the world's major source of starch, followed by potato, cassava, and wheat. Minor amounts of starch from rice, sorghum, sago, pea, and amaranth are also produced commercially. The United States has the largest starch industry in the world starch production, almost entirely depending upon maize. Starches from maize, wheat, and potato share the starch industry in the European Union. Cassava starch is produced mainly in South East Asia. Starches are also commercially produced in Asia from rice, sorghum, sweet potato, arrowroot, mung bean, and sago. Generally, cereal grains, roots and tubers, and legumes contain 40%–90%, 65%–85%, and 30%–70% of starch on dry basis, respectively (Guilbot and Mercier 1985).

3.2.2 Composition and Type

Starch consists of two glucosidic macromolecular polymers: amylose and amylopectin, which together account for 98%–99% of the dry weight (Tester et al. 2004). The ratio of amylose to amylopectin varies depending on the starch origin. Starch is generally grouped into waxy, normal/regular, and high-amylose types when amylose constitutes approximately 0%–15%, 20%–35%, and higher than 40% of total starch, respectively (Tester et al. 2004). Waxy starches are available in maize, barley, wheat, potato, sorghum, and rice mutants, while high-amylose starches are mainly developed from maize and barley. In legume starches, amylose content varies from 12% to 88% with a general range of 30%–40% (Hoover et al. 2010). Wrinkled pea starch and mutant maize starches especially from *amylose-extender* and *sugary* maize contain the intermediate materials, which are branched polyglucans with molecular size intermediate to amylose and amylopectin (Kasemsuwan et al. 1995; Klucinec and Thompson 1998; Perera et al. 2001; Takeda and Preiss 1993; Wang et al. 1993b). In addition, starch also contains small quantities of proteins (surface and integral) and lipids (free and bound) as well as a trace amount of minerals.

3.2.3 Granular Structure

Starch is synthesized in plants as semicrystalline granules, which are densely packed (density of about 1.5 g/cm^3) with varying polymorphic types and degree of crystallinity. The structural complexity and heterogeneity in starch granules at different levels of structural scales both within a single granule and the natural variation inherent among granule populations make starch one of the most complex materials in nature. Different levels of structural scales are present within a single starch granule: (1) whole granule varying in shape and size (in μm scale), (2) alternative semicrystalline and amorphous growth rings (~100s nm), (3) amylopectin superhelix (~10s nm), (4) crystalline and amorphous lamellae (~9 nm), and (5) amylose and amylopectin molecules (~Å) (Gidley 2001; Tester and Debon 2000).

3.2.3.1 Granule Morphology

Starch from different botanical sources exhibits its inherently characteristic granule morphology specific to each species. Starch granules vary in shape (e.g., oval, spherical, polygonal, lenticular, disk/plate shaped, or irregular in various starches), size (diameter ranges of 0.5–110 μm), size distribution (e.g., mono-, bi-, or trimodal distribution), and surface features (e.g., smooth or rough surfaces, distribution of pores) (Hoover 2001; Jane et al. 1994). Generally, starch granules from tubers and roots are relatively larger in size (e.g., canna and potato) than those from cereals. Among cereal starches, large round disk-shaped granules are found in wheat, barley, and rye starches with a size of 15–36 μm in diameter and 6–10 μm in thickness, while the very small spherical granules with a diameter of 2–3 μm are present in small amounts (Jane et al. 1994). Legume starch granules are generally oval in shape with diameter of 2–70 μm (Hoover et al. 2010). Oat and rice starches are a mixture of irregularly shaped single and compound granules (a cluster of several single granules with a relatively small size of 2–15 μm) (Hoseney 1986). Compound granules are also found in smooth pea (Bertoft et al. 1993), wrinkled pea (Colonna et al. 1982), tapioca (Hoover 2001), and barley (Li et al. 2001). Maize starch granules are irregularly shaped with a number of polyhedral faces in the size of 5–15 μm. However, amylomaize starches also contain elongated and filamentous granules (Jane 2009; Jiang et al. 2010). Most of the cereal starches, including wheat, barley, rye, and sorghum, show a distinct bimodal granule size distribution, whereas other starches (maize, legumes, tubers, and roots) give monomodal distribution (Buléon et al. 1998). The large and small granules are referred to as A- and B-granules, which are commonly distinguished at 10 μm in diameter. The number proportion of A-granules is always lower than those of B-granules (each amyloplast of endosperm contains one large granule and a variable number of small granules) but represents the major proportion of the starch mass (Parker 1985). Trimodal distributions of granule sizes have also been reported in wheat and barley starches (Bechtel et al. 1990; Makela et al. 1982; Raeker et al. 1998).

Although most starch granules when viewed under scanning electron microscope appear relatively smooth with no evidence of pores, cracks, or fissures (Hoover 2001; Hoover and Sosulski 1991; Jane et al. 1994), studies on high-resolution imaging of granule surface using atomic force microscopy (Baldwin et al. 1997, 1998) revealed that potato and wheat starches possess a rough granule appearance with many protrusions (10–300 nm) on the granule surface, which is believed to be the ends of the starch polymers within the crystalline amylopectin side-chain clusters at the granule surface (Baldwin et al. 1998). Baldwin (2001) postulated that the region near the granule surface may hinder granule hydration, and diffusion of enzymes and chemical reagents into the granule interior.

Surface pores and internal channels are found in maize, wheat, barley, rye, sorghum, and millet starch granules (Fannon et al. 1992, 1993; Huber and BeMiller 1997; Kim and Huber 2008; Li et al. 2003; Naguleswaran et al. 2011). These surface pores are exterior openings that extend from internal channels, which penetrate into the granule interior radially. Pores are randomly distributed on the granule surface, whereas channels vary in depth and dimension, most extending into the central region of the granule, and appear to facilitate the transfer of chemical reagents into the granule matrix (Huber and BeMiller 2000, 2001; Kim and Huber 2008). Surface pores and internal channels are true architectural features of starch granules, potentially increasing the granule surface area available for chemical and enzymatic reactions (Huber and BeMiller 2000; Kim and Huber 2008; Sujka and Jamroz 2007). Starches with high pore and channel density show higher susceptibility to enzymes than those with low density (Benmoussa et al. 2006; Fannon et al. 1992). Kim and Huber (2008) and Naguleswaran et al. (2011) postulated that the presence of minor components, such as proteins and lipids on granule surface and within internal channels, may restrict the pathway for the diffusion of chemicals and enzymes into the starch matrix.

3.2.3.2 Macrostructure of Granule

Within a single starch granule, starch molecules are packed into more or less concentric layers, namely, semicrystalline growth rings and amorphous growth rings (French 1984; Jenkins et al. 1994) with a thickness of 120–400 nm (French 1984). The semicrystalline growth rings alternate with the amorphous growth rings. Each semicrystalline growth ring contains periodic crystalline and amorphous lamellae, while amorphous growth rings contain mostly amylose (Cameron and Donald 1992). The alternate growth rings, in which starch molecules are periodically deposited by increasing diameter from the hilum toward the surface of granule, represent two distinct regions of the granule with high and low refractive indices, densities, crystallinities, and resistances to acid and enzymatic hydrolysis (French 1984). The width of the growth rings becomes progressively thinner toward the exterior of the starch granule. The number and size of the growth rings depend on the botanical origin of the starch and the amount of carbohydrates available for biosynthesis. Gallant et al. (1997) proposed that growth rings are composed of spherical structures, namely, blocklets, stacked on top of each other and having sizes between 20 and 500 nm in diameter depending on the starch source and the location in the granule. Semicrystalline rings contain large blocklets of about 100 nm in diameter, compared to 20–50 nm blocklets in the less organized amorphous rings. A-crystalline-type starches, mostly from cereals, have smaller blocklets (80–120 nm), whereas potato and other B- and C-type starches show much larger blocklets (200–500 nm) at the granule surface to a depth of about 10 μm (Gallant et al. 1992, 1997).

Within a starch granule, both amylose and amylopectin molecules orientate perpendicularly to the surface of the granule (Blanshard 1987; French 1984; Waigh et al. 1997). Amylose is interspersed in amorphous regions of the granule, whereas amylopectin molecules are arranged into clusters with short chains (Hizukuri 1986; Robin et al. 1974). Neighboring short amylopectin chains form double helices and are further packed into crystallites. Crystallites fall within crystalline lamellae, while amylopectin branch points are located in the amorphous lamellae. The stacks of crystalline and amorphous lamellae form semicrystalline growth rings alternating with completely amorphous growth rings of a similar size. The nonstarch minor components, such as protein and phosphorus, bound to starch molecules may stabilize the double-helical cross-linking of glucan chains, especially in the outer region of the granule (Debet and Gidley 2007), playing an important role in maintaining the structural stability of the granule (Baldwin 2001; Naguleswaran et al. 2011).

3.2.4 Molecular Structure of Amylose

Amylose is an essentially linear α-1,4-D-glucan chain with a low degree of α-1,6-linked branch points. The molecular weight and degree of polymerization (DP) of amylose generally range from 10^5 to 10^6 and 700 to 5000 anhydroglucose units, respectively (Biliaderis 1991; Hizukuri 1996). Amylose from cereal starches has a much smaller average molecular size than those from tuber and root starches. The molecular weight distribution of amylose is monomodal, but amyloses from tuber and root starches show narrower distribution with less small molecules than those from cereal starches (Hizukuri 1996; Hizukuri and Takagi 1984). About 25%–55% of the total amylose (mole basis) in starch granules is branched with 4–18 branch points per molecule and branch chain length (CL) of 4 to over 100 (Hizukuri et al. 1981, 1997). Overall, amylose molecules contain 0.27%–0.68% branching points with an average CL of 100–550 anhydroglucose units, and large amylose molecules have more branched linkages than small amylose molecules (Hizukuri et al. 1997). The presence of branched chains (predominant DP 10–30) does not significantly influence the general properties of amylose. In general, branched amylose molecules from various sources differ with respect to molecular size, side CL, and the number of chains. Research has shown that the molecular size of amylose decreases with the increase of amylose content in maize and barley starches (Shi et al. 1998; Yasushi et al. 2002; Yoshimoto et al. 2000).

The location of amylose in starch granules has been shown to be influenced by amylose content (Atkin et al. 1999). In starches with low amylose content (e.g., potato), amylose mainly localizes in amorphous growth rings alternating with semicrystalline growth rings, whereas with high amylose content (e.g., amylomaize), amylose is encapsulated by an amylopectin surface but separated with an amylopectin center, suggesting that an increase in amylose content causes more separation of amylopectin molecules. Within a starch granule, amylose molecules are possibly located in (1) amorphous growth rings, (2) intercrystalline amorphous lamellae (including amylopectin branching zone), (3) between crystallites/blocklets within crystalline lamellae, and (4) radial channels and central cavities (Gallant et al. 1997). Thus, amylose may be randomly interspersed as individual molecules in both the amorphous and crystalline regions of the granule or cocrystallized with amylopectin molecules (Atkin et al. 1999; Jane et al. 1992; Jenkins and Donald 1995; Kasemsuwan and Jane 1994). Chemical surface gelatinization of potato and corn starches has indicated that amylose is more concentrated at the periphery of the granule than in the inner part of the granule (Jane and Shen 1993; Pan and Jane 2000), with smaller amylose chains predominating near the surface of the granule (Jane and Shen 1993).

Amylose in neutral, aqueous solution tends to be slightly helical due to the natural twist present in the chair conformation of the glucose units. It behaves as a random, flexible coil, which consists of extended helical segments (helical regions stabilized by intermolecular hydrogen bonds) interspaced by nonhelical segments (Appelqvist and Debet 1997) with a typical hydrodynamic radius (a measure of the volume occupied by the polymer in solution depending on the molecule weight) of 7–22 nm (Ring et al. 1985).

3.2.5 Molecular Structure of Amylopectin

3.2.5.1 Molecular Size and Chain Length

Amylopectin is the major component of most starches and is generally characterized as a cluster of glucan branch chains with polymodal CL distribution (Hizukuri 1996) and a nonrandom nature of branching (Thompson 2000). The highly branched macromolecule is composed of thousands of linear α-1,4-D-glucan unit chains and 4%–5.5% of α-1,6-glucosidic branch points (Hizukuri et al. 1997). The average CL is 18–25 anhydroglucose units. However, these unit chains are linked in a unique way to make up an average molecular weight of about 10^7–10^9 (Hizukuri 1996; Yasushi et al. 2002; Yoshimoto et al. 2000) with a hydrodynamic radius of 21–75 nm (Bello-Pérez et al. 1996). The DP by number (DP_n) of amylopectin molecule is typically in the range of 9,600–15,900 with the presence of three major molecular species at 13,400–26,500, 4,400–8,400, and 700–1,200 (Takeda et al. 2003). Amylopectin in high-amylose starches has a lower degree of branching, higher average CL, and a large proportion of long chains than those in waxy and normal starches (Cheetham and Tao 1997; Salomonsson and Sundberg 1994; Shi et al. 1998; Takeda et al. 1993a,b).

3.2.5.2 Cluster Structure

According to the cluster model of amylopectin molecular structure (Hizukuri 1986; Robin et al. 1974), the amylopectin molecule consists of three types of unit chains, referred to as A-, B-, and C-chains. A-chains are unbranched and linked to B-chains through α-1,6-bonds at their reducing end group. B-chains are linked to other B-chains or C-chain in the same manner, carrying either A-chains or B-chains. C-chain is the only reducing end group carrying numerous B-chains. Unit chains are arranged in groups/clusters, where A- and B1-chains are located within individual clusters, whereas B2- to B4-chains connect 2–4 clusters. The average CLs of debranched amylopectin molecule are 12–16, 20–24, 42–48, 69–75, and 101–119 for A- and B1- to B4-chains, respectively (Hizukuri et al. 1997). The A- and B1-chains account for 80%–90% (mole basis) of total chains, which represent outer short chains in a single molecule, and B2-chains 10%, B3-chains 1%–3%, and B4-chains 0.1%–0.6% (Hizukuri 1986). The ratio of A- to B-chains is 0.8–1.5 to 1 depending on the genetic origin (Bertoft et al. 2008; Manners 1989). The CL of C-chain is in the range of 10–130 with a majority at DP 40 (Hanashiro et al. 2002). The shortest CL in amylopectin is 6, and the short chains at DP 6–17 are characteristic for amylopectin as a "fingerprint" (Koizumi et al. 1991). Chains at DP 6–8 that constitute a subgroup of A-chains are termed "fingerprint A-chains" (Bertoft 2004; Bertoft et al. 2008). The short chains with DP < 10, which do not readily form double helices, are considered to introduce structural defects into the crystalline lamellae of the granules and thereby interfere with the organization of the crystals (Noda et al. 2009).

3.2.5.3 Chain Length Distribution

The branch-CL distribution of amylopectin is related to the crystalline polymorphs (A-, B-, and C-types in x-ray diffraction pattern), and its profile is characteristic for each starch even in the same polymorphic type (Hanashiro et al. 1996; Hizukuri 1985; Jane et al. 1999). Potato and other B-type starches contain less short (A- and B1-) chains and more long B-chains (B2–B4) than A-type cereal starches, with a mole ratio of 5 for potato and 8–10 for cereal starch amylopectins (Hizukuri 1996). In mutant maize starches, A-type crystalline amylopectin contains larger clusters, more short branch chains, and more branching points than B-type crystalline amylopectin (Gérard et al. 2000). A-type starches have shorter average CL than B-type starches. Also, A-type starches have relatively higher proportions of short chains (DP 6–12) than B-type starches, and C-type starches have intermediate amounts. In addition, amylopectins from amylomaize starches have relatively longer average CL and higher proportions of long chains (DP ≥ 37) compared to those from waxy and normal maize starches (Jane and Chen 1992; Jane et al. 1999; Shi et al. 1998). Branch-CL distribution of amylopectin influences starch physicochemical properties such as gelatinization, pasting, retrogradation, and acid/enzyme hydrolysis (Franco et al. 2002; Jane and Chen 1992; Jane et al. 1999; McPherson and Jane 1999; Shi et al. 1998; Shi and Seib 1992, 1995).

3.2.5.4 Double Helices, Crystallites, and Crystallinity

Within starch granules, about 40%–50% of chains on a weight basis exist as double helices with approximately half of these helical chains present in the form of crystallites (Gidley 2001; Gidley and Bociek 1985). Double helices are formed between the outer branch chains (A- and B1-chains) of amylopectin (French 1972). Two neighboring short chains fit together compactly, with the hydrophobic parts of the opposed glucose units in close contact inside the structure, and the hydroxyl groups at the outside of the double helix resulting in strong interchain hydrogen bonding (helical order or short-range order). French and Murphy (1977) proposed the first detailed computer model for the starch double helix with no intrachain hydrogen bonds. The stability of the helix is attained by interchain hydrogen bonding between hydroxyl groups at positions C2 and C6 and from van der Waals forces. The helical core is highly hydrophobic and compact so that there is no room for water or any other molecule to reside in it. Close packing of neighboring double helices (crystalline order or long-range order) via direct or indirect (through a bridge of water molecules) hydrogen bonding forms granule crystallites (Imberty et al. 1988, 1991; Imberty and Perez 1988; Wu and Sarko 1978a,b).

Crystallites, which are roughly spherical in shape with a diameter of about 10 nm in wheat starch and somewhat smaller in potato starch (Muhrbeck 1991), may occur either between adjacent branch chains in the same amylopectin branch cluster or between adjacent clusters in three dimensions (Oates 1997). The unit cell structures of crystallites are classified as A- and B-type (Imberty and Perez 1988; Imberty et al. 1988, 1991). The double helices are identical for both A- and B-types, but their packing manner (helix geometry) and water content differ. The left-handed, parallel-stranded double helices are packed in hexagonal arrays and stabilized by a network of hydrogen bonds (Gidley and Bociek 1985; Imberty and Perez 1988; Imberty et al. 1988).

Degree of crystallinity of starch granules ranges from 15% to 45% by weight depending on starch origin, hydration, and method used (Zobel 1988). The degree of crystallinity is inversely proportional to the amylose content and average branch CL in maize starches (Cheetham and Tao 1998). The double-helical content determined by nuclear magnetic resonance (NMR) (corresponds to crystalline and noncrystalline domains) is higher than the crystalline order determined by x-ray diffraction (which corresponds only to the crystalline order of the double helices), suggesting that a portion of the double helices exist individually and not further ordered to form crystalline order within the starch granule (Cooke and Gidley 1992; Gidley and Bociek 1985).

3.2.5.5 Polymorphic Patterns

The crystalline types within starch granules from various botanical sources are shown as one of three well-defined x-ray diffraction patterns, classified as A-, B-, and C-type polymorphs. The A-type occurs in most cereal starches (e.g., normal maize, rice, wheat, barley, and oat) as well as in some root and tuber starches (e.g., taro, some sweet potatoes, tapioca, and iris) (Cheetham and Tao 1998; Hizukuri 1996; Hoover 2001). The B-type occurs in tuber and root starches (e.g., potato, lily, canna, and tulip) and high-amylose (>40%) cereal starches (e.g., amylomaize, high-amylose barley, and high-amylose rice) (Cheetham and Tao 1998; Hizukuri 1996; Hoover 2001). The A- and B-types are believed to be independent, whereas the C-type, which is commonly observed in legume starches, is a mixture of A- and B-type crystallites in varying proportions (Gernat et al. 1990; Hizukuri 1996; Sarko and Wu 1978). Mutant maize starches containing A- and B-type polymorphs in various ratios (Gérard et al. 2001) and wheat and rye starches containing low levels of the B-type polymorph (up to 10%) (Gernat et al. 1993) have been reported. Cheetham and Tao (1998) have shown that a transition of crystalline types in maize starches from A- to B-types via C-type occurs at approximately 40% amylose.

Except for the differences in their crystallite arrangements and water content, the amylopectin CL (Gidley and Cooke 1991; Hizukuri 1996; Pfannemuller 1987) and branching pattern (Jane et al. 1997) also differ with each other. Relatively shorter branch chains lead to A-type, whereas longer branch chains show B-type, and intermediate branch chains are associated with C-type polymorph (Hizukuri 1996; Hizukuri et al. 1983). The difference in the average CL between A- and B-types can be as little as one glucose unit (Hizukuri 1996; Hizukuri et al. 1983). These are probably due to mixtures of A- and B-type crystallites either within individual granules or as mixtures of A- and B-type granules (Lineback 1984). In C-type pea starches, Bogracheva et al. (1998) found that B-type was present in the granule interior, while A-type was located in the periphery of the granule. Synthesis of both polymorphs from linear α-glucans (Gidley and Bulpin 1987; Pfannemuller 1987) showed that at least 10 glucose units of a linear chain are required for the crystallization of double helices, but that a CL of more than DP 50 does not form crystallites. Also, shorter CL, higher crystallization temperature, greater polysaccharide concentration, and slower crystallization favor formation of the A-type polymorph (Gidley and Bulpin 1987).

3.2.5.6 Lamellae and Superhelices

Kassenbeck (1975, 1978) and Yamaguchi et al. (1979) observed rippled fibrillar structures with a periodicity of 6–8 nm in the growth rings of maize and wheat starch granules using electron microscopy. After extensive acid hydrolysis, the residual starch granules (acid-resistant amylodextrins) showed lamellar structures of 5 nm thickness, which are stacked together side-by-side and oriented tangentially to the growth rings and to the granule surface (Yamaguchi et al. 1979). These lamellar structures were believed to be the crystalline region of the ripples alternating with amorphous lamellae. The sizes of crystalline lamellae and the combined period (repeat distance) of crystalline and amorphous lamellae corresponded to the linear CL and a single amylopectin cluster, respectively, as proposed by French (1972) and Robin et al. (1974). By optical diffraction analysis of electron micrographs, Oostergetel and van Bruggen (1989) found that the repeat distance agreed with the value determined by small-angle x-ray diffraction at approximately 10 nm (at an angle of approximately 25° along the molecular axis). Jenkins et al. (1993) confirmed that the lamellae spacing is maintained across starch sources with A-, B-, and C-type x-ray diffraction patterns and is consistent at 9 nm, despite differing lengths of the amylopectin side chains formed among the species (Hizukuri 1986) and variations in the size of crystalline lamellae in starches with various amylose contents (Jenkins and Donald 1995). Oostergetel and van Bruggen (1993) modeled the crystalline lamellae as a continuous, left-hand-twisted, superhelical structure. Each superhelix is interlocked to neighboring superhelices to give a tetragonal array, which is filled with amorphous materials. Gallant et al. (1997) further modified this model in which the spherical blocklets range from 20 to 500 nm in size.

3.3 STARCH CHARACTERISTICS AND GEL FORMATION

Starch in its native state is insoluble in cold water. In order to achieve its functional attributes in starch-containing products, starch has to be cooked in almost all food and other industrial applications. Starch gels formed upon heating in excess water, cooling, and storage involve several phase transitions such as gelatinization, pasting, gelation, and retrogradation.

3.3.1 Gelatinization

Gelatinization is a major step in the starch gelling process. It occurs when starch is heated in the presence of excess water (the volume fraction of water $v > 0.7$) (Donovan 1979) to its gelatinization temperature (typically in the range of 60°C–70°C for most starches depending on starch source) (Hermansson and Svegmark 1996). Gelatinization is an irreversible order–disorder phase transition, which involves granule hydration and swelling, uptake of heat, loss of birefringence, crystallite melting, uncoiling/dissociation of double helices (in crystalline regions and those not located with ordered crystallites), and leaching of amylose and amylopectin (French 1984; Tester 1997). Of the various methods, which have been used for the characterization of starch gelatinization, such as polarized light microscopy, x-ray diffraction, differential scanning calorimetry (DSC), NMR spectroscopy, enzymatic digestibility, viscoamylography, and small-angle light scattering, DSC has been more widely used for evaluating gelatinization parameters (T_o [onset temperature], T_p [peak temperature], T_c [conclusion temperature], and ΔH [enthalpy change]) (Tester and Debon 2000; Zobel 1984). According to Jenkins and Donald (1998) and Donovan (1979), gelatinization in excess water is primarily a swelling-driven process. The swelling within the amorphous regions destabilizes the amylopectin crystallites within the crystalline lamellae via ripping apart the edges of crystallites (smaller crystallites are destroyed first), leading to a disruption of the crystallinity of starch granule. This process occurs rapidly for an individual crystallite, but over a limited temperature range for a single granule (1°C–2°C) and a wide range (10°C–15°C) for a whole population of granules with endothermic ΔH values in the range of 10–20 J/g (Eliasson and Gudmundsson 2006; French 1984; Liu and Lelievre 1993). However, at intermediate or lower water content, there is insufficient water present for hydration and gelatinization of all crystallites resulting in simply melting of remaining crystallites at high temperature and consequently increasing the gelatinization temperature range.

Gelatinization is primarily a property of amylopectin (Tester and Morrison 1990a,b), in which gelatinization temperature reflects crystalline perfection (Tester 1997) and gelatinization enthalpy is a measure of the overall crystallinity of amylopectin (i.e., the quality and quantity of crystallites) mainly reflecting the loss of double-helical order (Cooke and Gidley 1992). Gelatinization and swelling properties of starch are mainly controlled by the molecular structure of amylopectin (branch CL, the degree and pattern of branching, molecular weight, and polydispersity), starch composition (amylose to amylopectin ratio, lipid-complexed amylose chains, and phosphorous content), and granule architecture (crystalline to amorphous ratio) (Tester 1997). Other factors also influence starch gelatinization. Increasing the heating rate increases endotherm temperature (Donovan 1979; Shiotsubo and Takahashi 1984). The ratio of starch and water is used to control the degree of gelatinization during extrusion process (Fishman et al. 2000). The presence of sugars, polyhydric alcohols, and salts may increase gelatinization temperature with invariant ΔH (Evans and Haisman 1982). Addition of sugar affects starch gelatinization by delaying starch granule swelling, amylose leaching, and granule fragmentation. The sugar molecules possibly interact with starch granule amorphous regions and increase energy required to melt them, and/or stabilize the crystalline regions of the starch and thus immobilize water molecules, hindering starch gelatinization (Beleia et al. 1996). Heat-moisture treatment with low moisture levels (<35%) at a temperature above the glass transition temperature (T_g) but below the gelatinization temperature (84°C–120°C) for a certain time period (15 min–16 h) causes an increase in gelatinization temperature, broadening of the gelatinization temperature range, and a decrease in enthalpies with a transformation of B- or C-type polymorphs to more stable A-type polymorph (Hoover 2010; Jacobs and Delcour 1998). Annealing at water content above 40% and below the gelatinization temperature for a certain period of time increases gelatinization temperature and enthalpy, narrowing the gelatinization temperature range (Jacobs and Delcour 1998; Jayakody and Hoover 2008). Heat-moisture treatment and annealing may be associated with crystallite growth or perfection of already existing crystallites, rearrangement of double helices, partial melting and realignment of polymer chains, formation of crystalline amylose–lipid complexes, interaction between amylose chains or between amylose and amylopectin, and reorientation of the crystallites within the amorphous matrix (Jacobs and Delcour 1998). Partial acid hydrolysis increases gelatinization temperature and gelatinization temperature range with varied enthalpy depending on starch source and hydrolysis time (Hoover 2000). Chemical modification methods such as hydroxypropylation and acetylation also decrease gelatinization parameters and enthalpy (Hoover et al. 1988; Kim and Eliasson 1993; White et al. 1989).

3.3.1.1 Gelatinization and Loss of Granular Order

During gelatinization, various forms of order (orientational, lamellar, crystalline, and double helical) are lost over a broadly similar temperature range (Parker and Ring 2001). Loss of birefringence (measured by polarized light microscopy) occurs in a narrower temperature range while crystalline order (measured by x-ray diffraction) and molecular order (measured by NMR spectroscopy) are lost concurrently with the endotherm change (Cooke and Gidley 1992; Gidley and Cooke 1991; Liu and Lelievre 1993; Liu et al. 1991). The melting disruption of crystallinity starts from the area around the hilum and spreads along the granule, accompanied by swelling of disrupted areas (Bogracheva et al. 1998; French 1984; Garcia et al. 1997). Jenkins and Donald (1998) and Donald et al. (1997) have shown by a combination of x-ray and small-angle neutron scattering studies that initial swelling following rapid uptake of water in the amorphous regions imposes a stress upon the double helices within the amylopectin crystallites. Consequently, the amylopectin crystallites are disrupted, but do not completely correspond to the gelatinization endotherm, even at excess water content (most but not all crystallinity loss occurs throughout the gelatinization endotherm).

The lamellar order remains unchanged (the crystalline lamellae do not expand radially) as long as the crystallites can still be identified and rather beyond the end of the endotherm revealed by DSC.

3.3.1.2 Gelatinization and Amylose/Amylopectin Ratio

Gelatinization temperature (T_p) increases, and endotherm enthalpy (ΔH) gradually decreases with increasing amylose content in maize starches with a range of 1%–70% amylose (Matveev et al. 2001) and in pea starches with a range of 36%–86% amylose content (Czuchajowska et al. 1998). High-amylose maize starches (57%–90%) have broader gelatinization range (T_c–T_o) up to 58.8°C compared to waxy and normal maize starches (Jane et al. 1999; Shi et al. 1998). Waxy starches from wheat, barley, maize, and potato show higher gelatinization temperatures and enthalpy than their normal counterparts. The high gelatinization temperature and enthalpy are closely related with a high degree of crystallinity caused by high amylopectin content (Cheetham and Tao 1998; Fujita et al. 1998; Gernat et al. 1993; Matveev et al. 2001). Therefore, waxy starches require higher gelatinization temperature and more energy for gelatinization compared to nonwaxy starches. Gelatinization properties of starches are also influenced by amylose molecular size, amylopectin CL (Cheetham and Tao 1997), crystalline (polymorph) type, and the ratio of A- and B-types and distribution within C-type starch granules (Cheetham and Tao 1998; Gérard et al. 2001; Matveev et al. 2001), cocrystallinity of a part of amylose with the short amylopectin chains (forming double helices and crystalline lamellae) (Jenkins and Donald 1995), and arrangement of part of amylose chains between the crystalline lamellae (acting as defects and does not contribute to the melting enthalpy of starches) (Matveev et al. 2001).

3.3.1.3 Gelatinization and Branch Chain Length of Amylopectin

A larger proportion of long branch chains of amylopectin in starches (e.g., amylomaize and *ae* waxy maize) have been shown to display a higher gelatinization temperature and enthalpy (due to large and more ordered double-helical crystallites, which require higher temperature and energy to uncoil and dissociate) (Franco et al. 2002; Jane et al. 1999; McPherson and Jane 1999; Sanders et al. 1990; Shi et al. 1998; Yuan et al. 1993). Potato starch has higher proportion of long branch chains than cereal starches but contains high phosphate monoester derivatives, which contribute to low gelatinization temperature (Jane et al. 1999; McPherson and Jane 1999). Starches with short average branch CL (e.g., waxy rice) and a large proportion of short branch CL (DP 6–12) as well as those with relatively short second peak CL (DP 41–43) (e.g., wheat and barley) contribute to low gelatinization temperature and enthalpy (Jane et al. 1999). High-amylose barley starch does not show significantly higher gelatinization temperature and enthalpy than waxy and normal barley starches, corresponding to their similar amylopectin branch-chain distribution and small amount of long branch chains (Song and Jane 2000; Yasushi et al. 2002; Yoshimoto et al. 2000).

3.3.1.4 Gelatinization and Crystalline Type

A-polymorph starches (most cereals), which contain shorter average branch CL and large portion of short chains with densely packed cluster structure, show lower transition temperatures and enthalpy change than B-polymorph starches (amylomaize and most tubers and roots), which contain longer average branch CL and high proportion of long chains with loosely packed clusters (Fredriksson et al. 1998; Jane et al. 1999; McPherson and Jane 1999). In a study of C-polymorph pea starches using DSC, x-ray diffraction, NMR, and polarized light microscopy with excess salt solution, Bogracheva et al. (1998) found that the B-polymorph is present in the granule interior while the A-polymorph is located at the periphery. A wide gelatinization temperature with two endothermic peaks in C-polymorph starches corresponds to the disruption of B-polymorph and A-polymorph.

3.3.1.5 *Swelling and Solubilization*

Swelling and subsequent solubilization of amylose and amylopectin are the most important structural changes during and after gelatinization of starch granules. The irreversible swelling starts at a temperature corresponding to T_o in DSC measurements and continues to a much higher temperature than T_c (Tester and Morrison 1990a,b). During granule swelling, macromolecules leach out of granules resulting in a parallel increase in starch solubility. In general, starches from tubers and roots show a rapid single-stage swelling and solubilization pattern (Hoover 2001). Of these, potato starch has the highest swelling power due to a high phosphate ester content of amylopectin (repulsion between phosphate groups on adjacent chains will increase hydration by weakening the extent of bonding with the crystalline domain) (Garlliard and Bowler 1987; Swinkels 1985a). Most legume starches also exhibit a restricted single-stage swelling and low solubility pattern (Hoover and Sosulski 1991). In contrast to tuber, root, and legume starches, normal cereal starches show a two-stage swelling and solubilization pattern (Langton and Hermansson 1989; Leach et al. 1959). Amylose-free waxy starches swell much faster without the formation of typical swollen granules (Hermansson and Svegmark 1996) and show unrestricted swelling and great solubility (Wang et al. 1993a). But amylose-containing waxy starches have lower solubility than normal and high-amylose starches (Lorenz 1995). A starch granule could swell 100-fold its original volume in potato and up to 30-fold in cereal starches (Hermansson and Svegmark 1996), but cooking at high temperatures (above 95°C) or the influence of shear forces breaks down the swollen granules. Swelling is a property of amylopectin, which is regulated by the crystallinity of the starch before full gelatinization, and amylose acts as a diluent of amylopectin (Tester and Karkalas 1996; Tester and Morrison 1990a,b). Amylose is believed to act as a restraint on swelling and maintains the integrity of swollen granules, and cereal starch granules do not show complete swelling until amylose has been leached out of the granule (Hermansson and Svegmark 1996). In cereal starches, lipid-complexed amylose restrains swelling (Morrison et al. 1993b).

In addition to amylose/amylopectin ratio, molecular weight, branch CL and distribution of amylopectin, and degree of crystallinity (Sasaki and Matsuki 1998; Wang et al. 1993a), physical (e.g., pregelatinization, shearing, heat-moisture treatment, annealing, and radiation) and chemical modifications (e.g., oxidation, acid thinning, substitution and cross-linking, dextrinization) also influence swelling and solubilization properties (Chiu and Solarek 2009; Mason 2009).

Amylose leaches out of granules in the temperature range of 50°C–70°C (Doublier 1981; Svegmark and Hermansson 1991; Tester and Morrison 1990a,b). In the first stage of swelling, amylose leaching is limited, and an amylose-rich phase is formed in the center of the granule (Langton and Hermansson 1989). Once amylose–lipid complexes dissolve at the second stage, amylose leaching is enhanced, and amylopectin-rich granules start to deform and lose their original shape (Doublier 1981; Eliasson 1986; Ghiasi et al. 1982; Zeleznak and Hoseney 1987). The solubilized molecules increase in molecular weight and become more branched with increasing temperature (Doublier 1987; Ellis and Ring 1985; Prentice et al. 1992). Further heating (above 95°C) increases solubility substantially and enhances degradation of the granules, and separation into amylose phase continues where amylopectin fragments are dispersed (Langton and Hermansson 1989; Virtanen et al. 1993). The residue of swollen granules (ghosts) contains mainly amylopectin without any crystalline order, and amylose content decreases to 16% in pea starch, 8.3% in wheat, and 8.0% in maize starches (Ellis et al. 1988). Amylose–lipid complexes dissociate into lipid-free amylose and dispersed lipids (Morrison et al. 1993a,b; Shamekh et al. 1999). True solubilization of all the starch substance does not occur unless the paste is cooked at 100°C–160°C (in an autoclave or in a commercial steam-jet cooker) to give a low-viscosity starch macromolecular solution (Swinkels 1985b) or by dissolution of starch in dimethyl sulfoxide (DMSO) (French 1984).

3.3.2 Pasting

The term pasting is used to describe the phenomenon following gelatinization in the dissolution of starch (Atwell et al. 1988). Gelatinization induces substantial rheological behavior of starch suspension as a result of structural changes at both molecular and granular levels. During gelatinization, starch granules swell with progressive hydration, but the more tightly bound crystalline lamellae of amylopectin remain intact, holding the granules together. As the granules continue to expand, more water is imbibed, and swollen granules occupy the most space. The movement of swollen granules is restricted, and viscosity increases rapidly. When maximum swelling is reached, granules start to rupture with further heating and shearing, which induces a viscosity reduction. Depending on the starch source, type, and concentration, heating and shearing result in a dispersed phase (swollen granules or granule remnants) embedded in a continuous solution (rich in amylose) phase.

The Brabender Visco-Amylography is the most common method to evaluate starch pasting properties. The viscosity measured in Brabender Units (BU) or torque (g cm) is recorded as a function of temperature and time during a programmed heating-holding-cooling cycle at a specified rotational speed (rpm/min). The starch paste is subjected to both thermal and mechanical treatment to reveal the paste stability during heating (breakdown) and the consistency (retrogradation tendency) during cooling (setback) (Shuey and Tipples 1980). Pasting temperature (where a perceptible increase in viscosity occurs and thus is always higher than gelatinization temperature) and peak viscosity (reflects the ability of starch granules to swell freely before their physical breakdown at a given concentration) are also recorded. However, most recently, the Rapid Visco-Analyzer (RVA), which has a high viscosity correlation with Brabender Visco-Amylograph, is used to rapidly characterize starch pasting with a small sample requirement (Deffenbaugh and Walker 1989; Thiewes and Steeneken 1997).

Starches with higher amylose and lipid/phospholipid contents have higher pasting temperature, lower peak viscosity and breakdown viscosity (shear thinning), and higher setback viscosity, whereas starches with more amylopectin long branch chains (DP ≥ 37) display lower pasting temperature and higher peak and breakdown viscosities (Franco et al. 2002; Jane et al. 1999; Zeng et al. 1997). Also, the long-CL amylopectin and intermediate molecular size amylose produce the synergistic effect on the viscosity of starch pastes (Jane and Chen 1992). Both significant negative correlations between amylose content and pasting parameters (Wootton et al. 1998; Zeng et al. 1997) and significant positive correlations between long branch CL and pasting temperature as well as peak viscosity (Franco et al. 2002) are found in wheat starches. In general, waxy cereal starches have lower pasting temperatures, higher peak viscosity, and lower setback viscosity than normal cereal starches. Increase in pasting temperature, reduction of peak viscosity, and high resistance to shear thinning are more pronounced in normal wheat and barley starches than in normal maize and rice starches resulted from a larger amount of lipid-complexed amylose (Jane et al. 1999). Among tuber starches, potato starch has a very high peak viscosity as a result of the high phosphate monoester content and long branch chains (Jane et al. 1996, 1999; McPherson and Jane 1999). Tuber and root starches have lower pasting value than normal cereal starches due to absence of lipids and phospholipids (Hoover 2001; Jane et al. 1999; Kasemsuwan and Jane 1996; Lim et al. 1994). In contrast to most waxy and normal cereal starches, waxy potato starch displays a lower peak viscosity than normal potato starch due to free swelling and rapid dispersion under shearing as well as a lower proportion of long branch chains and high phosphate monoester (Jane et al. 1999; McPherson and Jane 1999). Legume starches show restricted swelling, the absence of peak viscosity, constant or increasing hot paste viscosity, a high setback, and a constant cold paste viscosity during cold holding (Hoover and Sosulski 1991; Hoover et al. 2010; Ratnayake et al. 2002).

3.3.3 Gelation and Retrogradation

Gelation is the characteristic process that gelatinized starch molecules start to reassociate and form a three-dimensional network of hydrogel upon cooling mainly due to the hydrogen bonding between hydroxyl groups of starch chains. After gelatinization and upon cooling, especially during storage, retrogradation occurs as results of reassociation and recrystallization of starch molecules (formation of double helices and crystallites), mainly from amylose, which further modify the network structure of the gel, thus influences the texture and shelf life of starch-gelled products.

During retrogradation, amylose crystallization occurs quickly to form a gel network by nucleation (formation of double helices of 40–70 glucose units between the ends of amylose molecules, favoring chain elongation) and propagation (packing of double-helical regions by chain folding), contributing to the initial development of gel firmness (Jane and Robyt 1984; Leloup et al. 1992; Miles et al. 1985; Ring et al. 1987). However, amylopectin recrystallization occurs more slowly (continuing over a period of several days or weeks) by association of the outmost short branches, contributing to the subsequent slow increase in gel firmness (Ring et al. 1987). While recrystallized amylopectin melts in the temperature range of 40°C–100°C, amylose crystallites do so only at much higher temperature (120°C–170°C) (Eerlingen et al. 1994; Sievert and Pomeranz 1989).

Retrogradation of starch has mainly been investigated by x-ray diffraction, DSC, spectroscopic methods (e.g., NMR spectroscopy, Fourier transform mid-infrared [FTIR] spectroscopy, and Raman spectroscopy), and rheological techniques (e.g., uniaxial compression and texture profile analysis, dynamic oscillatory rheometry, creep compliance and recovery, and stress relaxation) (Abd Karim et al. 2000; Eliasson and Gudmundsson 2006). The retrograded starch, which shows a B-type x-ray diffraction pattern, contains both crystalline and amorphous regions (Zobel 1988). The transition temperatures and retrogradation enthalpy of retrograded starch are usually 10°C–26°C lower and 60%–80% smaller, respectively. However, temperature ranges $(T_c$–$T_o)$ are wider than those for gelatinization of native starch (Baker and Rayas-Duarte 1998; White et al. 1989; Yuan et al. 1993) due to the weaker crystallinity and a varying crystalline stability in retrograded starches (Cooke and Gidley 1992; Fredriksson et al. 1998; Sasaki et al. 2000; Shi and Seib 1992). The retrogradation tendency of starches from different origins varies greatly. The rate and extent of retrogradation of starches are mainly influenced by starch composition and structure, starch concentration, storage condition, pH, sugars, lipids, and surfactants (Eliasson and Gudmundsson 2006; Fredriksson et al. 1998; Kalichevsky et al. 1990; Lai et al. 2000; Shi and Seib 1992). The rate and extent of retrogradation has been shown to increase with increase in amylose content in high-amylose starches (75%–90%) (due to a synergistic interaction between amylose and amylopectin) (Gudmundsson and Eliasson 1990; Russell et al. 1987). Amylopectin from cereal starches has been shown to retrograde to a lesser extent than those from pea, potato, and canna starches. This was attributed to shorter average CL in cereal amylopectins (Kalichevsky et al. 1990; Orford et al. 1987). A large amount of short chains over 15 glucose units and an increased ratio of A-chains to B-chains promote retrogradation, whereas very short chains (6–9 glucose units) retard retrogradation of starch gels (Kalichevsky et al. 1990; Shi and Seib 1992). Observations of two-stage retrogradation behavior in rice amylopectins (Lai et al. 2000) and multistage retrogradation behavior in waxy maize and potato starches (Bulkin et al. 1987; Vansoest et al. 1994) appear to involve complicated crystallization mechanisms caused by amylose and amylopectin molecules. In gelled starch, sugars tend to stabilize the gel structure by inhibiting chain association, and the inhibiting effect follows the order of sucrose > glucose > fructose (Prokopowich and Biliaderis 1995).

3.3.4 Gelation of Amylose

Amylose is stable in aqueous solution only at temperatures higher than 60°C–70°C (Hermansson and Svegmark 1996). Upon cooling, development of turbidity and formation of a precipitate or a gel occurs, due to chain aggregation. The extent of chain aggregation is influenced by the concentration and the molecular size of amylose (Clark et al. 1989). Due to the dynamic nature (nonequilibrium) of amylose structure in aqueous solution, a phase separation of amylose into polymer-rich regions interpenetrated by polymer-deficient regions on gelation occurs initially, which makes the gel turbid and opaque (Miles et al. 1985). In the polymer-rich regions of the gel, locally concentrated amylose leads to interchain association in the form of double helices (short-range ordering) followed by crystallization (long-range of ordering) (Ianson et al. 1988; Miles et al. 1985). Gidley (1989) proposed that amylose gelation is dominated by the formation and subsequent aggregation of B-type double helices and that amylose gels contain rigid double-helical junction zones interconnected by more mobile amorphous single-chain segments.

The minimum concentration of amylose for gelation is found to be around 1.0% (C_o, the critical concentration for gelation), which is independent of amylose molecular size (Clark et al. 1989; Doublier and Choplin 1989). Increasing amylose concentration causes faster gelation and a higher increase in turbidity. At a certain concentration of amylose, development of turbidity with short-chain amylose (DP<300) is more rapid and precedes gelation, but with long-chain amylose (DP<1100), gelation precedes the turbidity increase (Clark et al. 1989). Generally, the longer CL, high amylose concentration, and faster cooling rates favor gel formation. Gidley and Bulpin (1989) studied the influence of amylose CL on the aggregation and gelation behavior of aqueous solutions of short linear (0.5%–5.0%) amyloses. They found that the physical form (precipitate or gel), the kinetics of aggregation, and the variation of gel strength with concentration all are dependent on amylose CL. Aggregation was found to be most rapid for CLs of ~100 residues for initial aggregation rates in 0.1% solution. Amyloses having CLs of <100 residues were found to precipitate from aqueous solution. However, both precipitation and gelation occur for CLs of 250–660 residues, whereas for longer chains (>1100 residues), gelation predominates over precipitation.

Leloup et al. (1992) showed that amylose gels (2%–8% w/v) display a macroporous structure (mesh size 100–1000 nm) containing filaments 20 ± 10 nm wide with both amorphous (18%–33%) and crystalline (67%–82% of the gel) fractions of the polysaccharide. The filaments are formed from the association of segments of amylose chains (DP 26–31), which are oriented obliquely to the filament axis. The amorphous fraction is easily degraded by acid and consists of dangling chains (DP 6–30) located in the macroporous structure. The crystalline fraction consists of chains that are poorly associated with a "B"-type crystalline array. A structural model for amylose gels proposed consists of a crystalline fraction embedded in an amorphous matrix, where double helices of the "B"-type crystallites are linked to others by loops of amorphous amylose segments dangling in the gel pores (Leloup et al. 1992).

3.3.5 Gelation of Amylopectin

Gelation of amylopectin slowly occurs at high concentration (>10%) and low temperature (<5°C) (Ring et al. 1987). In contrast to amylose gelation, where segments (~DP 50) of entire backbone are involved, amylopectin gelation is a much slow process due to its high molecular size and high degree of branching, in which only short branch chains are involved (~DP 15) (Kalichevsky et al. 1990; Ring et al. 1987). The branch CL of amylopectin greatly affects the rate of gelation and retrogradation of amylopectin gel. Longer average branch chains have a tendency for faster gelation. For example, gels form faster for amylopectin from legume and tuber starches than those from cereal starches (the shorter CL of amylopectin in cereal starches than in legume and tuber starches) (Kalichevsky et al. 1990; Orford et al. 1987). Based on the observation of the strong influence of temperature on the rate of storage modulus (G′) development in 40% (w/w) amylopectin gels, Biliaderis and Zawistowski (1990) suggested that gelation of amylopectin follows nucleation kinetics in a manner typical of polymer crystallization in the presence of a diluent. Storage of gel at low temperature causes a slow crystallization and turbidity increase by partial crystalline phase separation. Amylopectin gel is thermoreversible and melts at temperature below 100°C (Ring et al. 1987).

3.3.6 Gelation of Starch

Amylose and amylopectin are thermodynamically incompatible in aqueous solution that promotes phase separation of the two polymers (Kalichevsky and Ring 1987) into amylose-rich phase and amylopectin-rich phase. One phase is the continuous phase where the other is dispersed, depending on the amylose/amylopectin ratio. When the ratio of amylose to amylopectin reaches a certain value (r is about 0.43 for a concentration of 8% gel), namely, the phase inversion point, the continuous phase becomes the discontinuous dispersed filler or vice versa (Leloup et al. 1991). In cooked starch-water system, starch gelation is such a complex system that is governed by starch concentration/the amount of available water, the ratio of amylose to amylopectin, and the molecular structures of amylose and amylopectin (Klucinec and Thompson 2002).

Upon cooling, gelatinized starch suspensions turn into a turbid viscoelastic paste at low concentrations or into an opaque elastic gel at concentrations exceeding 6% (Morris 1990; Ring 1985). In such a paste or gel matrix, the structure of swollen starch granule and granule remnants (rich in amylopectin) predominantly affect the rheological behavior of starch during heating, especially in a high-swelling or highly concentrated close-packed system (in which the swollen granules bind practically all of the available water and the volume of the phase outside the granules is small), whereas leached amylose reinforces the gel structure during cooling (Hermansson and Svegmark 1996; Lii et al. 1996; Tsai et al. 1997). If the amylose-rich phase is continuous, gelation of starch on cooling results in a strong gel formation (Hermansson and Svegmark 1996). The short-term development of gel structure and crystallinity in starch gels is dominated by irreversible gelation at <100°C and crystallization within amylose matrix, whereas long-term increases in the modulus of starch gels are linked to a reversible amylopectin crystallization within the granules on storage. The crystallization of amylopectin within swollen granules, possibly includes the cocrystallization between amylose and amylopectin, results in an increase in the rigidity of the granules and thus enhances their reinforcement of the amylose matrix (Miles et al. 1985). Tuber and root starches have high swelling capacities but form gels much more slowly than cereal starches, which may be due to the lower amylose content, higher molecular weight, and insufficient release of amylose from potato starch (Hermansson and Svegmark 1996).

In food systems, swollen granules are often required, and starch is often modified through cross-linking to enhance the integrity of swollen granules. However, a macromolecular solution is often preferred in most nonfood applications (Hermansson and Svegmark 1996). Molecularly dissolved starch solution can be obtained by a heating starch suspension up to 120°C–180°C (Aberle et al. 1994). When starch at moderate concentration in water is heated above 170°C followed by rapid cooling, spherulites with double-helical B-type crystallinity are formed (Ma et al. 2011; Nordmark and Ziegler 2002a,b; Ziegler et al. 2003), resulting in the formation of spreadable particle gels, namely, superheated starch, with spherulite morphology and a cream-like texture (Steeneken and Woortman 2009). The superheated starch is a slightly to moderately degraded starch with instant gel formation ability in cold water and is potentially used as a fat mimetic, and it can be produced by jet cooking followed by spray drying (Steeneken and Woortman 2009). The formation of spherulites varies with starch source and is affected by the ratio of linear to branched molecular chains, the degree of branching, and the CL and CL distribution of amylopectin (Ma et al. 2011).

3.4 CONTROL OF GEL FORMATION

Formation of a gel and gel properties are not only controlled by starch composition and structure of granule and molecules but also are greatly affected by starch source, starch modification, preparation conditions (e.g., temperature, heating rate and time, mechanical treatment), presence of other components (e.g., proteins, lipids, hydrocolloids, sugar, salt, and acids), and storage (e.g., temperature, time, freezing and thawing cycles) (Biliaderis 2009). Thus, by selection of starch through breeding program, improvement of gelling performance through physical and chemical modification and addition of other ingredients or additives, optimization of preparation process, and control of storage conditions, formation of a starch gel can be controlled to meet various application requirements.

3.4.1 Starch Selection

Selection of starch is prerequisite and has a direct impact on successful formulation and the quality of the finished gelling products. Gelling properties of starch are diverse with starch sources. With advances of agronomy, new starches become available from new sources. Besides maximum utilization of conventional starch source, discovery of new starch source from nonconventional crops and biosynthesis of novel starches are important for meeting industry needs. Although potato, maize, wheat, cassava, rice, sorghum, arrowroot, and sago starches are commercially used in industries, starches from many tropical tuber crops and pulses have not been exploited for industrial applications partially due to extraction cost and limited information on the structure and properties of these lesser known starches (Chibbar et al. 2010; Hoover et al. 2010; Moorthy 2002). However, these starches show many structural features and unique physicochemical properties, which are not perceived/ possessed by common cereal starches. Utilization of selected tuber and pulse starches as a new source of gelling agents could bring different functional characteristics to starch-based gelling products in specific industrial applications.

With advantages of conventional hybrid breeding program and genetic engineering, development of novel starches from new genotypes and mutants of crops with altered granule composition and structure becomes possible. Recently, newly developed waxy and high-amylose wheat starches have been shown to possess unique structures and characteristics that are believed to enhance the quality of food products (Hung et al. 2006). Other structural alterations could potentially improve gel formation and viscosity stability with significantly economic value and reduction of the needs for in vitro starch modification (Davis et al. 2003). For example, increase of phytoglycogen (a highly branched, water-soluble polysaccharide) in mutants of maize and rice to level of higher than 30% could produce a polysaccharide with reduced viscosity, gel formation and retrogradation rate, and increased water-holding capacity (Johnson et al. 2001). Development of phosphorylated cereal starches may increase pasting peak viscosity, decrease setback viscosity, and reduce gel hardness and stickiness (Davis et al. 2003). Partially waxy starch and starch with intermediate level of amylose from wheat lines have higher degree of crystallinity, gelatinization temperatures, and enthalpy value than does wild wheat starch (Demeke et al. 1999). Alteration of the ratio of large to small granules in starch, creation of starches with altered CL and branch pattern, and introduction of β-linkages in starch chains may also result in changes of gelatinization properties, improvement of gel formation, and viscosity stability (Davis et al. 2003; Johnson et al. 2001).

3.4.2 Modification of Starch

Although native starches can be used as gelling agents in many gelling products depending on their source and availability, native starches from various plant species have limited physicochemical properties, which are not suitable for relatively specific end applications. For the majority of industrial utilizations, physically and chemically modified starches are commonly used, providing controlled viscosity, improved tolerance to rigorous processing conditions (e.g., temperature, pH, and shear), desirable texture, and prolonged stability.

3.4.2.1 Physical Modification

The gelling properties of starch can be improved by physical modification, such as pregelatinization in water or solvents by drum drying, spray drying, and extrusion cooking. Pregelatinized starches (drum or spray dried) and cold-water-swelling starches (heated in aqueous alcohol solution at high temperature or alcoholic alkali solution at 25°C–35°C) thicken instantly without heat and form gels more slowly. Cold-water-swelling starches have smooth texture similar to cook-up starch and more processing tolerance than conventional pregelatinized starches. Extruded starches have reduced viscosity, lower cold-water-swelling capacity, and greater solubility than drum dried starches due to degradation of amylose and amylopectin by high temperature and shear stress (Mason 2009). Retrograded amylose is also added to food products to provide a smooth mouth feel, low water affinity, and health benefits. Heat-moisture treatment and annealing cause a physical modification of starch without starch gelatinization and damage of granular integrity by interactions between amylose chains, amylose–amylopectin chains, and/or amylose lipid (Hoover 2010; Jayakody and Hoover 2008).

Radiation treatments, including gamma rays, electron beam, x-rays, ultraviolet (UV) light, and microwave with different wavelength, dose, and exposure time, lead to a variety of property changes, such as degradation of starch molecules and/or destruction of crystalline ordering, resulting in increasing gelatinization temperature, improving stability, reducing viscosity and gel strength, decreasing swelling capacity, and increasing solubility (Bhat and Karim 2009).

3.4.2.2 Chemical Modification

Currently, in order to achieve desired properties for industrial products, starch is typically modified by conversion (e.g., acid thinning and dextrinization), cross-linking, and substitution. Conversion process by acid hydrolysis and dextrinization reduces the viscosity of native starch, increases water solubility, controls gel strength, and modifies stability of starch, thus allowing uses of starch at high concentration level, such as soft candy in confection industry (Light 1990). Acid thinning of starch is conducted by adding hydrochloric acid or sulfuric acid to a starch slurry heated to just below its gelatinization temperature. Acid-thinned starch has a lower hot peak viscosity when gelatinized and imparts high gel strength on cooling and stable water retention when used at high solids concentration. Dextrinization, as a roasting process of starch in the presence of acid, is commonly carried out in a roasting kiln or fluidized bed reactor by random hydrolysis. The dextrinized starch has low viscosity and good clarity. Cross-linking of starch is to build ester links between the hydroxyl groups on starch by mixing multifunctional reagents, such as adipic–acetic mixed anhydride, phosphorous oxychloride, or sodium trimetaphosphate with a starch slurry under alkaline conditions to produce starch adipates or starch phosphates. Cross-linking increases gelatinization temperature and hot peak viscosity and increases resistance to overcooking and processing conditions, giving a short paste texture. But high level of cross-linking reduces granule swelling and decreases viscosity. Substitution by acetylation with acetic anhydride and by hydroxypropylation with propylene oxide reduces starch gelatinization temperature, inhibits retrogradation, increases freeze–thaw stability, hinders syneresis, and thus improves product texture and shelf life. Double modification by cross-linking and substitution is also useful for both degradation prevention and process tolerance.

3.4.2.3 Enzymatic Modification

Industrial modification of starch by enzyme is based on the use of amylases such as α-amylase, pullulanase, and glucoamylase to produce maltodextrins. Maltodextrins with low dextrose equivalent (DE) value (DE<5) are soluble in cold water and have low viscosity in solution, but at high concentrations

(>20% w/w), they form gels (Marchal et al. 1999). The low-DE maltodextrins (DE 1–10) have been particularly used as fat mimics (Roller 1996). However, maltodextrins are mixtures of glucose to long polymeric (linear and branched) chains, and their molecular weight distribution may be more important in determining the gelation properties of maltodextrins. Amylomaltase (4-α-glucanotransferase) is a disproportionating enzyme, which possesses both hydrolyzing and glucan-transferring activities (able to attack an α-1,4-glucosidic bond and transfer part of the glucan donor molecule to another glucan acceptor by forming a new α-1,4-glucosidic linkage), and has been used to modify starches for preparation of strong and stiff thermoreversible gels (Lee et al. 2008; van der Maarel et al. 2005). The amylomaltase-modified starches have a potential as a plant-derived gelatin replacer.

3.4.3 Optimization of Process Conditions

Starch pastes/gels can have a very different structure when gelatinized, ranging from a swollen particle suspension, giving a nonhomogeneous suspension, to a macromolecular dispersion, largely depending on the starch sources used and process conditions.

Depending on the end use of the product, starch should be properly cooked. For applications requiring maximum viscosity (highly swollen granules with minimum loss of granule integrity), proper cooking of starch provides excellent viscosity development, texture characteristics (such as good clarity, heavy-bodied short texture), and good stability. Undercooking (granules are slightly swollen and highly intact) results in low viscosity, poor clarity, starchy taste, and poor stability, while overcooking (granules are ruptured with a large amount of fragments) results in a cohesive texture and a much lower viscosity. Proper cooking is normally achieved by choosing appropriately cross-linked starch, proper equipment, cooking time and temperature, and control of pH and shear (e.g., mixing, pumping, homogenization, extrusion, steam injection, jet cooking). Understanding of unit operations and the conditions of the starch at each point throughout the process also helps to optimize starch gelling performance.

3.4.4 Interaction of Starch with Sugars, Hydrocolloids, Lipids, Acids, and Salts

Food system is a complex mixture of various ingredients. Any ingredient interaction with starch that involves forming a complex, binding, or coating can impede starch hydration and cause an impact on viscosity development, performance, and texture of hydrogels.

Sugar and hydrocolloids greatly affect starch cooking by increasing the gelatinization temperature and by decreasing maximum viscosity (Christianson 1982; Mason 2009; Mitolo 2006). The presence of these ingredients may hinder granule hydration and amylose leaching, thereby reducing gel strength. In such cases, either starch is cooked prior to the addition of ingredients or a properly modified starch is chosen for gelatinization. The effect of sugars on starch gelatinization and swelling is caused by stabilization of the amorphous region through a cross-linking reaction between sugars and starch chains (Gunaratne et al. 2007). The ability to stabilize the gel structure and to reduce starch retrogradation has been shown to follow the order: sucrose > glucose ≥ fructose (Katsuta et al. 1992a). Among monosaccharides (xylose, ribose, glucose, fructose), hexoses have been shown to be more than pentoses in reducing retrogradation (Katsuta et al. 1992b). The ability of saccharides to stabilize gel structure is closely related to the stereochemical conformation of saccharide molecules (i.e., the mean value of the number of equatorial OH groups in the molecule) and the hydration of saccharides (i.e., the dynamic hydration number of the molecule) (Katsuta et al. 1992b,c). The ability to stabilize gels increases with an increase in the mean number of equatorial OH groups and the dynamic hydration number of the saccharide molecules. Hydrocolloids, such as alginates, xanthan gum, guar gum, and cellulose derivatives, modify the gelatinization, pasting, swelling, and gelation properties of starch, depending on the hydrocolloid structure and concentration (Appelqvist and Debet 1997; Christianson 1982; Rosell et al. 2011). For example, guar and xanthan gums induce the greatest effect on the pasting properties of rice starch, whereas hydroxypropyl methylcellulose only causes a change in the viscosity during cooling. Starch–hydrocolloid pastes form weaker gels compared to those of the starch alone. In general, starch interacts with hydrocolloids synergistically to produce more desirable texture of gelling products, but phase separation occurs under certain conditions, such as heating–cooling cycles (Appelqvist and Debet 1997; Rosell et al. 2011).

It is well-known that amylose forms complexes (V-type) with polar lipids, such as mono-glycerides, fatty acids, and lipid-containing surfactants and emulsifiers. Complex formation is greatly influenced by lipid type, aliphatic CL, degree of unsaturation, amylose CL, ratio of amylose to lipid, and formation method (Eliasson and Gudmundsson 2006; Putseys et al. 2010b). Amylose–lipid complex increases gelatinization temperature, restricts swelling, hinders amylose leaching, and decreases granule disruption and viscosity development (Putseys et al. 2010b). During heating and cooling, the presence of lipids raises a competition between amylose–lipid complex formation and amylose crystallization, eliminating the contribution of amylose to crystallization and cocrystallization with amylopectin, and also raises the possibility of amylopectin–lipid complex formation, interfering with amylopectin crystallization, resulting in retrogradation inhibition (Eliasson and Gudmundsson 2006; Putseys et al. 2010b). In order to meet the process tolerance requirement and provide the desired texture in high-fat-containing foods, such as salad dressings and cream-based sauces, highly cross-linked and moderately substituted waxy starches in combination with an amylose-containing starch are often chosen (Mitolo 2006). Recently, addition of amylose–lipid complexes to starch-containing system has been used to improve starch gel characteristics and resistant starch formation by controlled release of lipid and short-chain amylose at their dissociation temperature and recomplex during cooling in starch hydrogels (Gelders et al. 2006; Putseys et al. 2010a). Short-chain amylose–lipid complexes stabilized starch paste in high temperature and high shear conditions more than longer amylose complexes do (Gelders et al. 2006).

In many starches, the effect of pH on gel strength is marginal. However, in potato starch, maximum gel strength occurs at pH 8.5 and is markedly reduced on the addition of salts due to electrostatic interactions between phosphate groups and added cations, which blocks the normal phosphate-to-phosphate cross-linking (Muhrbeck and Eliasson 1987). The presence of salts affected starch gelatinization, swelling, rheology, gelation rate, gel strength, and retrogradation, depending on the type of salt and the concentration (Ahmad and Williams 1999). The effect of salts follows the Hofmeister series with the anions having a greater influence than the cations on polymer–solvent interactions, whereas the cations interact with the hydroxyl groups of starch molecules to form complexes that disrupt chain aggregation (Ahmad and Williams 1999; Katsuta 1998). Addition of organic acids in salad dressings and confections aids in hydrogen bond disruption, causing rapid swelling and weakened gel strength.

3.5 PRODUCTION OF STARCH-BASED HYDROGELS

Starch-based hydrogels are usually prepared by radiation and chemical modification to enhance hydrophilicity of native starch and to take advantage of their nontoxicity, biocompatibility, biodegradability, and antibacterial activity for broad applications in the fields of biomedical, pharmaceutical, agriculture, biology, and environmental remediation and protection (Zhang et al. 2005b). Gamma rays, electron beams, x-rays, or UV light have been shown to induce cross-linking points by radiation, whereas chemical modification involves either direct cross-linking with multifunctional cross-linking agents or copolymerization–cross-linking reaction between starch and other monomers using polymerizable cross-linking agent. The properties of these hydrogels depend on types of starch and cross-linker used, monomer/starch ratio and concentration, initiation and polymerization method, etc., which could be largely varied (Athawale and Lele 2001; Castel et al. 1990; Zhang et al. 2005b).

3.5.1 Cross-Linking

A great number of cross-linking agents can be used for the preparation of cross-linked starch gels. Cross-linking of carboxymethyl starch and phosphorylated starch produces starch hydrogels with a considerable swelling capability and stability. Starch hydrogels produced by esterification of carboxymethyl starch using different carboxylic acids, such as malic, tartaric, citric, malonic, succinic, glutaric, and adipic acid, have variable gel stabilities and viscoelastic properties (Seidel et al. 2001), whereas hydrogels produced by etherification of carboxymethyl starch using dichloroacetic acid in the presence of monochloroacetic acid exhibit greater gel stability than esterified hydrogels, which are more suitable for ultrasonic medical examinations (Seidel et al. 2004). Carboxymethylation of mung bean starch with monochloroacetic acid in methanol yields a carboxymethyl starch with high viscosity, good stability, and compatibility with a model drug, which has a great potential for topical pharmaceutical gel preparation (Kittipongpatana et al. 2009).

Starch phosphate hydrogels with a swelling capacity of up to 185 g water/g dry gel can be prepared by cross-linking low-substituted (with degree of substitution of phosphate groups 0.04–0.65) monostarch monophosphates (phosphorylation of starch with sodium dihydrogen phosphate and/or disodium hydrogen phosphate under dry conditions at 150°C for 3 h in acidic media) with di- and tricarboxylic acids in a semidry process at 150°C for 3 h (Passauer et al. 2009). These hydrogels can be used to improve soil quality. The type of starch, concentrations of the cross-linking agent, and the length of the carbon chain between the ends of carboxylic acid groups of a polyfunctional carboxylic acid have been shown to influence the swelling capacity of hydrogels (Passauer et al. 2009).

3.5.2 Graft Copolymerization

Graft copolymerization of vinyl monomers onto starch and starch derivatives is an efficient approach to achieve starch-based superabsorbent hydrogels, which can be achieved either by simultaneous graft copolymerization as one step or by functionalization of starch with reactive double bonds first and then cross-linking by polymerization in water as two steps (Zhang et al. 2005b). For simultaneous graft polymerization, an initiator is necessary to start cross-linking via cross-linking agent. A variety of monomers, mostly acrylic monomers, have been grafted onto native and modified starches, using various cross-linking agents and different initiators. Acrylic acid (AA), acrylamide (AM), methyl acrylate (MA), ethyl acrylate (EA), butyl acrylate (BA), hydroxypropyl methacrylate (HPMA), ethylene glycol dimethacrylate (EGDMA), N-isopropylacrylamide (NIPA), and vinyl pyrrolidone (VP) have been used as copolymerization monomers with cross-linking agents, such as N,N′-methylene bisacrylamide (MBA) and EGDMA. Initiators used to start cross-linking contain potassium persulfate (KPS), ammonium persulfate (APS), N,N,N′,N′-tetramethylenediamine (TEMED), ceric ammonium nitrate (CAN), benzoyl peroxide (BPO), 4-dimethylaminobenzyl alcohol (DMOH), and so on. Native starch, starch fractions (amylose and amylopectin), soluble starch, pregelatinized starch, modified starch, and starch blends with other polymers have been used as substrates (Zhang et al. 2005b).

Copolymerization of starch and polyacrylonitrile (PAN) in sodium hydroxide solution (under conditions of alkalization at 80°C for 30 min, NaOH 10 wt%, PAN/starch weight ratio of 1.5, alkaline hydrolysis at 90°C for 90 min, and postneutralization at pH 7) yields a pH-sensitive starch–poly(sodium acrylate-co-AM) superabsorbent hydrogels with water swelling capacity higher than 500 g/g (Sadeghi and Hosseinzadeh 2008). This starch-based hydrogel displays a pH-responsive swelling–deswelling behavior in acidic and basic solutions, contributing to its suitability as a potential candidate for controlled delivery of drug and bioactive agent. Graft copolymerization of polymethacrylic acid (PMAA) onto carboxymethyl starch carried out at 70°C using bisacrylamide as a cross-linking agent and persulfate as an initiator also produces a pH-sensitive starch-based hydrogel (Saboktakin et al. 2009). The swelling and hydrolytic behavior of the hydrogels depends on the content of methacrylic acid groups, which makes the hydrogels pH dependent. Poly(acrylonitrile)-starch hydrogel produced by polymerization of acrylonitrile with starch in the presence of ceric ammonium nitrate (CAN) as initiator (starch gelatinized at 85°C for 30 min, polymerization with monomer/starch ratio of 4.8, initiator concentration of 10 mmol/L, liquor ratio of 12.5, saponification in 0.7 N sodium hydroxide at 95°C for 180 min) has absorbencies of water, synthetic urine, and mercury ions of 920 g/g, 38 mL/g, and 1250 mg/g, respectively, exhibiting suitable application as diaper and absorbent materials to remove dyes and heavy metals from industrial effluents and other wastewater (Hashem et al. 2005, 2008). Modification of cornstarch with allyl chloride in the presence of cetyltrimethyl ammonium bromide as a catalyst and pyridine as an acid acceptor at room temperature for 24 h and followed by copolymerization with methacrylic acid (allyl starch and acrylic monomer in a ratio of 1.7:1 w/w) and a combination of methacrylic acid and AM (ratio of 3.2:1 w/w) at 50°C and 70°C with KPS as an initiator, result in a superabsorbent biodegradable hydrogel with an absorption capacity of 6500% in water and 4500% in 0.9% NaCl solution (Bhuniya et al. 2003). By cross-linking polymerization of glycidyl methacrylate (GMA)–modified cornstarch using sodium persulfate as an initiating agent, the hydrogel produced is thought to be a promising candidate for transporting and preserving acid-responsive drugs in the treatment of colon-specific diseases (Reis et al. 2008). When AM and diallyldimethylammonium chloride (DMDAAC) are introduced into sodium starch sulfate by solution polymerization, a new amphoteric superabsorbent hydrogel is synthesized with a maximum swelling capacity of 1493 g/g in water and high salt tolerance (91 g/g in 0.9 wt% saline solution) (Peng et al. 2008).

Superabsorbent hydrogels can absorb and retain a large amount of water or aqueous fluids relative to their own mass (10–1000 times of its original weight or volume) in a relatively short period of time (from less than 1 min to h), while superporous hydrogel with average pore size of 200 μm can exhibit both fast swelling (within minutes) and superabsorbent properties (Omidian et al. 2005, 2007). The swelling of superabsorbent hydrogel is particle-size-dependent water diffusion, whereas superporous hydrogel swells to the equilibrium as size independent. The fast swelling of superporous hydrogel is due to water absorption through its interconnected open porous structure by capillary force (Omidian et al. 2005). Modern superabsorbent hydrogels are prepared in a fashion similar to regular hydrogels but with a simultaneous gas blowing, in which a combination of foaming agent (e.g., sodium bicarbonate), foaming aid (e.g., glacial acetic acid and AA), and foaming stabilizer (e.g., poly(ethylene oxide)–poly(propylene oxide)–poly(ethylene oxide) triblock polymers) is responsible for forming a homogeneous porous structure (Omidian et al. 2005). In order to improve the mechanical strength and elastic properties of porous superabsorbent hydrogels, a composite agent (a pre-cross-linked water-absorbent hydrophilic polymer) or a hybrid agent (natural or synthetic water-soluble or water-dispersible polymer, such as sodium alginate, pectin, chitosan, and poly(vinyl alcohol) (PVA) that can form cross-linked structure) has been used (Omidian et al. 2005, 2007).

3.5.3 Radiation

Introduction of cross-linking in aqueous starch system by ionizing radiation (e.g., high-energy gamma rays and electron beams) and by electromagnetic radiation (e.g., UV light and microwave) is regarded as a reliable, economically viable, safe, and environmentally friendly alternative to chemical modification, and possesses several advantages over other conventional methods employed for modification and cross-linking. Radiation polymerization, being a physical process, requires minimal sample preparation, short exposure time, mild reaction condition (process at room temperature, and without or low level of usage of initiators/cross-linking agent), discharge of less waste liquid, simple process, and low cost (Bhat and Karim 2009; Shu et al. 2011; Singh et al. 2006).

Gamma ray and electron beam radiation involves the formation of radicals on the polymer chains by the hemolytic scission of C–H bonds, and also the radiolysis of water molecules generates hydroxyl radicals that abstract hydrogen atoms from the polymer side chains, resulting in the formation of macroradicals. The macroradicals on different chains recombine leading to the formation of covalent bonds and a cross-linked network structure (Zhang et al. 2005b). Gamma-ray radiation of pregelatinized starches with AA at a dose rate of 1300 rad/min at room temperature results in the formation of starch-based hydrogels suitable for use as slow-release matrices for drug delivery systems (Geresh et al. 2002). Radiation of starch–PVA blend by gamma rays and electron beams causes a grafting reaction between PVA and starch molecules along with the cross-linking of PVA molecules, in which amylose influences hydrogel formation (Zhai et al. 2002). Electron beam radiation of carboxymethyl starch (degree of substitution of 0.15) at a concentration of 20%–50% (w/w) with dose up to 10 kGy results in efficient cross-linking of starch polymers forming a biodegradable starch hydrogel, in which amylopectin cross-linking is predominant (Nagasawa et al. 2004; Yoshii et al. 2003). The starch–polyvinyl pyrrolidone hydrogels prepared by electron beam radiation have sufficient strength for wound dressing and could be a good barrier against microbes (Abd El-Mohdy and Hegazy 2008), while the starch/N-vinyl pyrrolidone hydrogels grafted with AA and synthesized by gamma radiation with dose rate of 1.2 Gy/s (Eid 2008) induce a pH-dependent hydrogel (it swells minimally at pH 1 and maximally at pH 7) for potential colon-targeted drug delivery. Ali and AlArifi (2009) reported a starch/methacrylic acid copolymer hydrogel induced by gamma-ray radiation (2.03 kGy/h) which shows similar pH sensitivity. A novel superabsorbent hydrogel synthesized by graft copolymerization of AA/AM onto starch in the presence of montmorillonite via gamma-ray radiation effectively improves the water retention capacity of soil (Luo et al. 2005).

Similar to ionizing radiation, UV light also facilitates the production of free radicals, which react with a monomer to form a grafted copolymer in the free radical–initiated copolymerization of starch. However, due to its low energy levels compared to high-energy gamma rays and electron beams, a photosensitizer (photo initiator) is needed to absorb a low-energy photon and activate polymerization by formation of free radicals (Bhat and Karim 2009). UV-induced polymerization of acryloylated starch with zwitterionic monomer 3-dimethyl(methacryloyloxyethyl) ammonium propane sulfonate (DMAPS) induces a novel salt-tolerant starch-based hydrogel, in which the anti-polyelectrolyte swelling behavior in divalent salt ($CaCl_2$) solution is more pronounced than in monovalent salt (NaCl) solution (Li and Zhang 2007).

Microwave radiation is emerging as an efficient thermal energy source and can heat reactants selectively, directly, without thermal inertia, and without the involvement of a heat exchange medium (Singh et al. 2006). Graft-copolymer hydrogels, such as starch-graft-poly(AM), starch-graft-PAN, and starch-graft-poly(ethylacrylate), can be synthesized under microwave radiation either in the absence or in the presence of low concentration of initiator (Singh and Maurya 2010; Singh et al. 2006, 2007). The synthesis could be conducted efficiently in a few minutes under atmospheric conditions. For example, a superabsorbent hydrogel was prepared from cornstarch and sodium acrylate using microwave radiation at 85–90 W in 10 min. The factors that influenced the swelling ratio of the aforementioned hydrogel followed the order: microwave power > radiation time > cross-linker concentration > initiator content (Tong et al. 2005). Under microwave radiation, the synthesis and drying of a hydrogel can be completed in one step without the need for an inert atmosphere (Shu et al. 2011).

3.6 INDUSTRIAL APPLICATIONS

Starches from different source differ significantly in their physicochemical, rheological, thermal, gelatinization, and retrogradation properties, thus in turn gelling properties. Starches with specific gelling properties are in great demand in food and nonfood industries.

3.6.1 Food Applications

Starch is the most important source of carbohydrate in human and animal nutrition and is widely used in many applications throughout the food industry, alone or in conjunction with other gelling agents, to thicken and bind food products. Starches with desirable functional properties could play a significant role in improving the quality of different food products. Native and modified starches and their derivatives are widely used in confectionery as a functional ingredient to perform one or more functions. However, one of its main applications is as a gelling agent in jelly gum and hard gum candies. Thin-boiling starches obtained by acid thinning and high-amylose corn starches are most often used as gelling agents. They not only contribute to the textural stability and shelf life but also partially replace gelatin and gum arabic in gummy-type candies, especially in vegetarian/halal/kosher products for cost saving and texture modification (Zallie 2011). Potato starch is particularly more suitable for gummy-type candies due to its clarity and elasticity over other starches after cooking. Typical usage level of starches is in the range of 9%–14% on dry solids basis for soft jelly gums and 20%–30% for hard gums (Zallie 2011). For manufacture of both soft and hard gum candy, starch is cooked using either batch process (kettle cooking) or continuous process (jet cooker and tubular heat exchanger). Kettle cooking is mainly used for cooking thin-boiling starches and dextrins at the concentration below 50% S.S. by heating to 91°C–96°C and holding for 5 min (Zallie 2011). Other ingredients including syrup, sugar, color, and flavor are then added and boiled until the desired depositing solids are reached (typically 75%–78% S.S.) (Zallie 2011). Jet cooking is used for one-step cooking of starch (thin-boiling starch, dextrin, and high-amylose starch) and other ingredients in high solids concentration at 141°C–168°C (Zallie 2011). In jelly gum production, the low hot viscosity of thin-boiling starches at a high level of starch concentration enables the starches in a sugar solution to be rapidly and efficiently cooked in an open kettle or by jet cooking. An ideal thin-boiling starch has been shown to have a fluidity number in the range of 60–75 (Zallie 2011). The higher the fluidity number, the lower the hot viscosity and the gel strength. High-amylose cornstarch with the amylose content of 50%–70% (cooked at 168°C) forms much stronger gel than those boiling starches with a reduction of setting time (from 48–72 to 24 h) (Zallie 2011). High-amylose starch is typically used in combination with thin-boiling starches for improvement of gelling properties.

3.6.2 Nonfood Applications

In agriculture, starch-based hydrogels prepared either by cross-linking or by polymerization act as superabsorbents for water retention of soils and water supply of plants and also are used as controlled release devices to improve plant growth and crop production (Athawale and Lele 2001; Kazanskii and Dubrovskii 1992; Rudzinski et al. 2002). The controlled release systems may exhibit many advantages, such as in controlled or slow release of the core active ingredient (fertilizers and pesticides), reduction of dosage, stabilization of the core active ingredient against environmental degradation (light, air, humidity, and microorganisms), reduction in environment pollution and drift, and increase in the number of target organisms (Rudzinski et al. 2002). Starch-based hydrogels, such as starch-graft-AA, are successfully used to preserve food products by coating to films for control of humidity and oxygen/carbon dioxide concentration in fruits and vegetables and by using as a cooling medium to keep food fresh (Athawale and Lele 2001). Starch hydrogels have also great feasibility for the removal of heavy metal ions from wastewater (Hashem et al. 2008).

Starch-based hydrogels play a significant role as a delivery vehicle in biomedical and pharmaceutical applications due to their hydrophilicity, biocompatibility, and biodegradability (Zhang et al. 2005b). In surgical and wound-healing dressing, starch-based hydrogel helps to maintain a relatively constant temperature on application to skin and acts as barrier for microorganisms, like in medical tapes (Athawale and Lele 2001). In drug delivery systems, drugs and bioactive components are trapped in the hydrogel matrix during polymerization or introduced through the absorption of drug solution (Kishida and Ikada 2001). Starch-based hydrogels are used as drug carriers for sustained release of drugs to specific site by swelling, diffusion, and/or hydrolytic degradation (Gupta et al. 2010). In addition, starch-based hydrogels are applied in lubrication for surgical gloves, urinary catheters, surgical drainage systems, and contact lenses. In tissue engineering and cell transplantation, starch-based hydrogels can be used for temporary or semipermanent devices, such as degradable scaffolds, temporal artificial skin substitutes, tissue barriers to prevent postoperation adhesion, and coatings for cell culture devices/vessels (Barcili 2007). Other commercial potentials may include personal care products, construction of buildings, and pet and toy industries.

REFERENCES

Abd El-Mohdy, H. L. and Hegazy, E. A. 2008. Preparation of polyvinyl pyrrolidone-based hydrogels by radiation-induced crosslinking with potential application as wound dressing. *Journal of Macromolecular Science Part A—Pure and Applied Chemistry* 45:997–1004.

Abd Karim, A., Norziah, M. H., and Seow, C. C. 2000. Methods for the study of starch retrogradation. *Food Chemistry* 71:9–36.

Aberle, T., Burchard, W., Vorwerg, W., and Radosta, S. 1994. Conformational contributions of amylose and amylopectin to the structural-properties of starches from various sources. *Starch/Stärke* 46:329–335.

Ahmad, F. B. and Williams, P. A. 1999. Effect of salts on the gelatinization and rheological properties of sago starch. *Journal of Agricultural and Food Chemistry* 47:3359–3366.

Ali, A. E. H. and AlArifi, A. 2009. Characterization and in vitro evaluation of starch based hydrogels as carriers for colon specific drug delivery systems. *Carbohydrate Polymers* 78:725–730.

Appelqvist, I. A. M. and Debet, M. R. M. 1997. Starch-biopolymer interactions—A review. *Food Reviews International* 13:163–224.

Athawale, V. D. and Lele, V. 2001. Recent trends in hydrogels based on starch-graft-acrylic acid: A review. *Starch/Stärke* 53:7–13.

Atkin, N. J., Cheng, S. L., Abeysekera, R. M., and Robards, A. W. 1999. Localisation of amylose and amylopectin in starch granules using enzyme-gold labelling. *Starch-Stärke* 51:163–172.

Atwell, W. A., Hood, L. F., Lineback, D. R., Varrianomarston, E., and Zobel, H. F. 1988. The terminology and methodology associated with basic starch phenomena. *Cereal Foods World* 33:306–311.

Baker, L. A. and Rayas-Duarte, P. 1998. Freeze-thaw stability of amaranth starch and the effects of salt and sugars. *Cereal Chemistry* 75:301–307.

Baldwin, P. M. 2001. Starch granule-associated proteins and polypeptides: A review. *Starch/Stärke* 53:475–503.

Baldwin, P. M., Adler, J., Davies, M. C., and Melia, C. D. 1998. High resolution imaging of starch granule surfaces by atomic force microscopy. *Journal of Cereal Science* 27:255–265.

Baldwin, P. M., Davies, M. C., and Melia, C. D. 1997. Starch granule surface imaging using low-voltage scanning electron microscopy and atomic force microscopy. *International Journal of Biological Macromolecules* 21:103–107.

Barcili, B. 2007. Hydrogels for tissue engineering and delivery of tissue-inducing substances. *Journal of Pharmaceutical Sciences* 96:2197–2223.

Bechtel, D. B., Zayas, I., Kaleikau, L., and Pomeranz, Y. 1990. Size-distribution of wheat-starch granules during endosperm development. *Cereal Chemistry* 67:59–63.

Beleia, A., Miller, R. A., and Hoseney, R. C. 1996. Starch gelatinization in sugar solutions. *Starch/Stärke* 48:259–262.

Bello-Pérez, L. A., Paredes-López, O., Roger, P., and Colonna, P. 1996. Molecular characterization of some amylopectins. *Cereal Chemistry* 73:12–17.

Benmoussa, M., Suhendra, B., Aboubacar, A., and Hamaker, B. R. 2006. Distinctive sorghum starch granule morphologies appear to improve raw starch digestibility. *Starch/Stärke* 58:92–99.

Bertoft, E. 2004. Lintnerization of two amylose-free starches of A- and B-crystalline types, respectively. *Starch/Stärke* 56:167–180.

Bertoft, E., Manelius, R., and Zhu, Q. 1993. Studies on the structure of pea starches. Part 1: Initial-stages in alpha-amylolysis of granular smooth pea starch. *Starch/Stärke* 45:215–220.

Bertoft, E., Piyachomkwan, K., Chatakanonda, P., and Sriroth, K. 2008. Internal unit chain composition in amylopectins. *Carbohydrate Polymers* 74:527–543.

Bhat, R. and Karim, A. A. 2009. Impact of radiation processing on starch. *Comprehensive Reviews in Food Science and Food Safety* 8:44–58.

Bhuniya, S. P., Rahman, S., Satyanand, A. J., Gharia, M. M., and Dave, A. M. 2003. Novel route to synthesis of ally starch and biodegradable hydrogel by copolymerizing allyl-modified starch with methacrylic acid and acrylamide. *Journal of Polymer Science Part A: Polymer Chemistry* 41:1650–1658.

Biliaderis, C. G. 1991. The structure and interactions of starch with food constituents. *Canadian Journal of Physiology and Pharmacology* 69:60–78.

Biliaderis, C. G. 2009. Structural transitions and related physical properties of starch. In *Starch Chemistry and Technology*, 3rd edn. B. James and W. Roy, eds. Academic Press: San Diego, CA, pp. 293–372.

Biliaderis, C. G. and Zawistowski, J. 1990. Viscoelastic behavior of aging starch gels: Effects of concentration, temperature, and starch hydrolysates on network properties. *Cereal Chemistry* 67:240–246.

Blanshard, J. M. V. 1987. Starch granule structure and function: A physicochemical approach. In *Starch: Properties and Potential*. T. Galliard, ed. John Wiley & Sons: Chichester, U.K., pp. 16–54.

Bogracheva, T. Y., Morris, V. J., Ring, S. G., and Hedley, C. L. 1998. The granular structure of C-type pea starch and its role in gelatinization. *Biopolymers* 45:323–332.

Buléon, A., Colonna, P., Planchot, V., and Ball, S. 1998. Starch granules: Structure and biosynthesis. *International Journal of Biological Macromolecules* 23:85–112.

Bulkin, B. J., Kwak, Y., and Dea, I. C. M. 1987. Retrogradation kinetics of waxy-corn and potato starches— A rapid, Raman-spectroscopic study. *Carbohydrate Research* 160:95–112.

Cameron, R. E. and Donald, A. M. 1992. A small-angle x-ray-scattering study of the annealing and gelatinization of starch. *Polymer* 33:2628–2636.

Castel, D., Ricard, A., and Audebert, R. 1990. Swelling of anionic and cationic starch-based superabsorbents in water and saline solution. *Journal of Applied Polymer Science* 39:11–29.

Cheetham, N. W. H. and Tao, L. P. 1997. The effects of amylose content on the molecular size of amylose, and on the distribution of amylopectin chain length in maize starches. *Carbohydrate Polymers* 33:251–261.

Cheetham, N. W. H. and Tao, L. P. 1998. Variation in crystalline type with amylose content in maize starch granules: An X-ray powder diffraction study. *Carbohydrate Polymers* 36:277–284.

Chibbar, R. N., Ambigaipalan, P., and Hoover, R. 2010. Molecular diversity in pulse seed starch and complex carbohydrates and its role in human nutrition and health. *Cereal Chemistry* 87:342–352.

Chiu, C.-W. and Solarek, D. 2009. Modification of starches. In *Starch Chemistry and Technology*. B. James and W. Roy, eds. Academic Press: San Diego, CA, pp. 629–655.

Christianson, D. D. 1982. Hydrocolloid interactions with starches. In *Food Carbohydrates*. D. R. Lineback and G. E. Inglett, eds. The AVI Publishing Company, Inc.: Westport, CT, pp. 399–419.

Clark, A. H., Gidley, M. J., Richardson, R. K., and Rossmurphy, S. B. 1989. Rheological studies of aqueous amylose gels: The effect of chain-length and concentration on gel modulus. *Macromolecules* 22:346–351.

Colonna, P., Buleon, A., Lemaguer, M., and Mercier, C. 1982. *Pisum sativum* and *Vicia faba* carbohydrates: Part IV—Granular structure of wrinkled pea starch. *Carbohydrate Polymers* 2:43–59.

Cooke, D. and Gidley, M. J. 1992. Loss of crystalline and molecular order during starch gelatinization: Origin of the enthalpic transition. *Carbohydrate Research* 227:103–112.

Czuchajowska, Z., Otto, T., Paszczynska, B., and Baik, B. K. 1998. Composition, thermal behavior, and gel texture of prime tailings starches from garbanzo beans and peas. *Cereal Chemistry* 75:466–472.

Davis, J. P., Supatcharee, N., Khandelwal, R. L., and Chibbar, R. N. 2003. Synthesis of novel starches in plants: Opportunities and challenges. *Starch/Stärke* 55:107–120.

Debet, M. R. and Gidley, M. J. 2007. Why do gelatinized starch granules not dissolve completely? Roles for amylose, protein, and lipid in granule "Ghost" integrity. *Journal of Agricultural and Food Chemistry* 55:4752–4760.

Deffenbaugh, L. B. and Walker, C. E. 1989. Comparison of starch pasting properties in the brabender viscoamylograph and the rapid visco-analyzer. *Cereal Chemistry* 66:493–499.

Demeke, T., Huel, P., Abdel-Aal, E. S. M., Baga, M., and Chibbar, R. N. 1999. Biochemical characterization of the wheat waxy A protein and its effect on starch properties. *Cereal Chemistry* 76:694–698.

Donald, A. M., Waigh, T. A., Jenkins, P. J., Gidley, M. J., Debet, M., and Smith, A. 1997. Internal structure of starch granules revealed by scattering studies. In *Starch: Structure and Functionality*. P. J. Frazier, A. Donald, and P. Richmond, eds. The Royal Society of Chemistry: Cambridge, U.K., pp. 173–179.

Donovan, J. W. 1979. Phase-transitions of the starch-water system. *Biopolymers* 18:263–275.

Doublier, J. L. 1981. Rheological studies on starch: Flow behavior of wheat-starch pastes. *Starch/Stärke* 33:415–420.

Doublier, J. L. 1987. A rheological comparison of wheat, maize, faba bean and smooth pea starches. *Journal of Cereal Science* 5:247–262.

Doublier, J. L. and Choplin, L. 1989. A rheological description of amylose gelation. *Carbohydrate Research* 193:215–226.

Eerlingen, R. C., Cillen, G., and Delcour, J. A. 1994. Enzyme-resistant starch. 4. Effect of endogenous lipids and added sodium dodecyl-sulfate on formation of resistant starch. *Cereal Chemistry* 71:170–177.

Eid, M. 2008. In vitro release studies of vitamin B-12 from poly N-vinyl pyrrolidone/starch hydrogels grafted with acrylic acid synthesized by gamma radiation. *Nuclear Instruments and Methods in Physics Research Section B-Beam Interactions with Materials and Atoms* 266:5020–5026.

Eliasson, A. C. 1986. Viscoelastic behavior during the gelatinization of starch. 1. Comparison of wheat, maize, potato and waxy-barley starches. *Journal of Texture Studies* 17:253–265.

Eliasson, A.-C. and Gudmundsson, M. 2006. Starch: Physicochemical and functional aspects. In *Carbohydrates in Food*, 2nd edn. A.-C. Eliasson, ed. CRC Press: Boca Raton, FL, pp. 391–469.

Ellis, H. S. and Ring, S. G. 1985. A study of some factors influencing amylose gelation. *Carbohydrate Polymers* 5:201–213.

Ellis, H. S., Ring, S. G., and Whittam, M. A. 1988. Time-dependent changes in the size and volume of gelatinized starch granules on storage. *Food Hydrocolloids* 2:321–328.

Evans, I. D. and Haisman, D. R. 1982. The effect of solutes on the gelatinization temperature-range of potato starch. *Starke* 34:224–231.

Fannon, J. E., Hauber, R. J., and Bemiller, J. N. 1992. Surface pores of starch granules. *Cereal Chemistry* 69:284–288.

Fannon, J. E., Shull, J. M., and Bemiller, J. N. 1993. Interior channels of starch granules. *Cereal Chemistry* 70:611–613.

Fishman, M. L., Coffin, D. R., Konstance, R. P., and Onwulata, C. I. 2000. Extrusion of pectin/starch blends plasticized with glycerol. *Carbohydrate Polymers* 41:317–325.

Franco, C. M. L., Wong, K. S., Yoo, S. H., and Jane, J. L. 2002. Structural and functional characteristics of selected soft wheat starches. *Cereal Chemistry* 79:243–248.

Fredriksson, H., Silverio, J., Andersson, R., Eliasson, A. C., and Aman, P. 1998. The influence of amylose and amylopectin characteristics on gelatinization and retrogradation properties of different starches. *Carbohydrate Polymers* 35:119–134.

French, D. 1972. Fine structure of starch and its relationship to the organisation of starch granules. *Journal of the Japanese Society of Starch Science* 19:8–25.

French, D. 1984. Organization of starch granules. In *Starch Chemistry and Technology*. R. L. Whistler, J. N. BeMiller, and E. F. Paschall, eds. Academic Press: Orlando, FL, pp. 183–247.

French, A. D. and Murphy, V. G. 1977. Computer modeling in the study of starch. *Cereal Foods World* 22:61–70.

Fujita, S., Yamamoto, H., Sugimoto, Y., Morita, N., and Yamamori, M. 1998. Thermal and crystalline properties of waxy wheat (*Triticum aestivum* L.) starch. *Journal of Cereal Science* 27:1–5.

Gallant, D. J., Bouchet, B., and Baldwin, P. M. 1997. Microscopy of starch: Evidence of a new level of granule organization. *Carbohydrate Polymers* 32:177–191.

Gallant, D. J., Bouchet, B., Buleon, A., and Perez, S. 1992. Physical characteristics of starch granules and susceptibility to enzymatic degradation. *European Journal of Clinical Nutrition* 46:S3–S16.

Garcia, V., Colonna, P., Bouchet, B., and Gallant, D. J. 1997. Structural changes of cassava starch granules after heating at intermediate water contents. *Starch/Stärke* 49:171–179.

Garlliard, T. and Bowler, P. 1987. Morphology and composition of starch. In *Starch: Properties and Potential*. T. Galliard, ed. John Wiley & Sons: Chichester, U.K., pp. 55–87.

Gelders, G. G., Goesaert, H., and Delcour, J. A. 2006. Amylose-lipid complexes as controlled lipid release agents during starch gelatinization and pasting. *Journal of Agricultural and Food Chemistry* 54:1493–1499.

Gérard, C., Colonna, P., Buléon, A., and Planchot, V. 2001. Amylolysis of maize mutant starches. *Journal of the Science of Food and Agriculture* 81:1281–1287.

Gérard, C., Planchot, V., Colonna, P., and Bertoft, E. 2000. Relationship between branching density and crystalline structure of A- and B-type maize mutant starches. *Carbohydrate Research* 326:130–144.

Geresh, S., Gilboa, Y., Peisahov-Korol, J., Gdalevsky, G., Voorspoels, J., Remon, J. P., and Kost, J. 2002. Preparation and characterization of bioadhesive grafted starch copolymers as platforms for controlled drug delivery. *Journal of Applied Polymer Science* 86:1157–1162.

Gernat, C., Radosta, S., Anger, H., and Damaschun, G. 1993. Crystalline parts of 3 different conformations detected in native and enzymatically degraded starches. *Starch/Stärke* 45:309–314.

Gernat, C., Radosta, S., Damaschun, G., and Schierbaum, F. 1990. Supramolecular structure of legume starches revealed by X-ray-scattering. *Starch/Stärke* 42:175–178.

Ghiasi, K., Hoseney, R. C., and Varrianomarston, E. 1982. Gelatinization of wheat-starch. 1. Excess-water systems. *Cereal Chemistry* 59:81–85.

Gidley, M. J. 1989. Molecular mechanisms underlying amylose aggregation and gelation. *Macromolecules* 22:351–358.

Gidley, M. J. 2001. Starch structure/function relationships: Achievements and challenges. In *Starch: Advances in Structure and Function*. T. L. Barsby, A. M. Donald, and P. J. Frazier, eds. The Royal Society of Chemistry: Cambridge, U.K., pp. 1–7.

Gidley, M. J. and Bociek, S. M. 1985. Molecular-organization in starches—A C-13 CP MAS NMR-study. *Journal of the American Chemical Society* 107:7040–7044.

Gidley, M. J. and Bulpin, P. V. 1987. Crystallisation of malto-oligosaccharides as models of the crystalline forms of starch: Minimum chain-length requirement for the formation of double helices. *Carbohydrate Research* 161:291–300.

Gidley, M. J. and Bulpin, P. V. 1989. Aggregation of amylose in aqueous systems: The effect of chain-length on phase-behavior and aggregation kinetics. *Macromolecules* 22:341–346.

Gidley, M. J. and Cooke, D. 1991. Aspects of molecular-organization and ultrastructure in starch granules. *Biochemical Society Transactions* 19:551–555.

Gudmundsson, M. and Eliasson, A. C. 1990. Retrogradation of amylopectin and the effects of amylose and added surfactants emulsifiers. *Carbohydrate Polymers* 13:295–315.

Guilbot, A. and Mercier, C. 1985. Starch. In *The Polysaccharides*. G. O. Aspinall, ed. Academic Press: Orlando, FL, pp. 209–282.

Gunaratne, A., Ranaweera, S., and Corke, H. 2007. Thermal, pasting, and gelling properties of wheat and potato starches in the presence of sucrose, glucose, glycerol, and hydroxypropyl beta-cyclodextrin. *Carbohydrate Polymers* 70:112–122.

Gupta, A., Kaur, K., Arora, S., and Murthy, R. S. R. 2010. Thermosensitive in-situ hydrogel: A safest way for sustained release drug delivery. *Journal of Pharmacy Research* 3:2356–2358.

Hanashiro, I., Abe, J., and Hizukuri, S. 1996. A periodic distribution of the chain length of amylopectin as revealed by high-performance anion-exchange chromatography. *Carbohydrate Research* 283:151–159.

Hanashiro, I., Tagawa, M., Shibahara, S., Iwata, K., and Takeda, Y. 2002. Examination of molar-based distribution of A, B and C chains of amylopectin by fluorescent labeling with 2-aminopyridine. *Carbohydrate Research* 337:1211–1215.

Hashem, A., Afifi, M. A., El-Alfy, E. A., and Hebeish, A. 2005. Synthesis, characterization and saponification of poly (AN)-starch composites and properties of their hydrogels. *American Journal of Applied Sciences* 2:614–621.

Hashem, A., Ahmad, F., and Fahad, R. 2008. Application of some starch hydrogels for the removal of mercury(II) ions from aqueous solutions. *Adsorption Science and Technology* 26:563–579.

Hermansson, A. M. and Svegmark, K. 1996. Developments in the understanding of starch functionality. *Trends in Food Science and Technology* 7:345–353.

Hizukuri, S. 1985. Relationship between the distribution of the chain-length of amylopectin and the crystalline-structure of starch granules. *Carbohydrate Research* 141:295–306.

Hizukuri, S. 1986. Polymodal distribution of the chain lengths of amylopectins, and its significance. *Carbohydrate Research* 147:342–347.

Hizukuri, S. 1996. Starch: Analytical aspects. In *Carbohydrates in Food*. A. C. Eliasson, ed. Marcel Dekker: New York, pp. 347–429.

Hizukuri, S., Kaneko, T., and Takeda, Y. 1983. Measurement of the chain-length of amylopectin and its relevance to the origin of crystalline polymorphism of starch granules. *Biochimica et Biophysica Acta* 760:188–191.

Hizukuri, S. and Takagi, T. 1984. Estimation of the distribution of molecular-weight for amylose by the low-angle laser-light-scattering technique combined with high-performance gel chromatography. *Carbohydrate Research* 134:1–10.

Hizukuri, S., Takeda, Y., Abe, J., Hanashiro, I., Matsunobu, G., and Kiyota, H. 1997. Analytical developments: Molecular and microstructural characterization. In *Starch: Structure and Functionality*. P. J. Frazier, P. Richmond, and A. M. Donald, eds. The Royal Society of Chemistry: Cambridge, U.K., pp. 121–128.

Hizukuri, S., Takeda, Y., Yasuda, M., and Suzuki, A. 1981. Multi-branched nature of amylose and the action of debranching enzymes. *Carbohydrate Research* 94:205–213.

Hoover, R. 2000. Acid-treated starches. *Food Reviews International* 16:369–392.

Hoover, R. 2001. Composition, molecular structure, and physicochemical properties of tuber and root starches: A review. *Carbohydrate Polymers* 45:253–267.

Hoover, R. 2010. The impact of heat-moisture treatment on molecular structures and properties of starches isolated from different botanical sources. *Critical Reviews in Food Science and Nutrition* 50:835–847.

Hoover, R., Hannouz, D., and Sosulski, F. W. 1988. Effects of hydroxypropylation on thermal-properties, starch digestibility and freeze-thaw stability of field pea (*Pisum-sativum* cv trapper) starch. *Starch/Stärke* 40:383–387.

Hoover, R., Hughes, T., Chung, H. J., and Liu, Q. 2010. Composition, molecular structure, properties, and modification of pulse starches: A review. *Food Research International* 43:399–413.

Hoover, R. and Sosulski, F. W. 1991. Composition, structure, functionality, and chemical modification of legume starches—A review. *Canadian Journal of Physiology and Pharmacology* 69:79–92.

Hoseney, R. C. 1986. Cereal starch. In *Principles of Cereal Science and Technology*. American Association of Cereal Chemists: St. Paul, MN, pp. 33–68.

Huber, K. C. and BeMiller, J. N. 1997. Visualization of channels and cavities of corn and sorghum starch granules. *Cereal Chemistry* 74:537–541.

Huber, K. C. and BeMiller, J. N. 2000. Channels of maize and sorghum starch granules. *Carbohydrate Polymers* 41:269–276.

Huber, K. C. and BeMiller, J. N. 2001. Location of sites of reaction within starch granules. *Cereal Chemistry* 78:173–180.

Hung, P. V., Maeda, T., and Morita, N. 2006. Waxy and high-amylose wheat starches and flours—Characteristics, functionality, and application. *Trends in Food Science and Technology* 17:448–456.

Ianson, K. J., Miles, M. J., Morris, V. J., Ring, S. G., and Nave, C. 1988. A study of amylose gelation using a synchrotron X-ray source. *Carbohydrate Polymers* 8:45–53.

Imberty, A., Buleon, A., Tran, V., and Perez, S. 1991. Recent advances in knowledge of starch structure. *Starch/Stärke* 43:375–384.

Imberty, A., Chanzy, H., Perez, S., Buleon, A., and Tran, V. 1988. The double-helical nature of the crystalline part of A-starch. *Journal of Molecular Biology* 201:365–378.

Imberty, A. and Perez, S. 1988. A revisit to the 3-dimensional structure of B-type starch. *Biopolymers* 27:1205–1221.

Jacobs, H. and Delcour, J. A. 1998. Hydrothermal modifications of granular starch, with retention of the granular structure: A review. *Journal of Agricultural and Food Chemistry* 46:2895–2905.

Jane, J.-L. 2009. Structural features of starch granules II. In *Starch Chemistry and Technology*, 3rd edn. J. BeMiller and R. Whistler, eds. Academic Press: San Diego, CA, pp. 193–236.

Jane, J. L. and Chen, J. F. 1992. Effect of amylose molecular-size and amylopectin branch chain-length on paste properties of starch. *Cereal Chemistry* 69:60–65.

Jane, J., Chen, Y. Y., Lee, L. F., McPherson, A. E., Wong, K. S., Radosavljevic, M., and Kasemsuwan, T. 1999. Effects of amylopectin branch chain length and amylose content on the gelatinization and pasting properties of starch. *Cereal Chemistry* 76:629–637.

Jane, J., Kasemsuwan, T., Chen, J. F., and Juliano, B. O. 1996. Phosphorus in rice and other starches. *Cereal Foods World* 41:827–832.

Jane, J. L., Kasemsuwan, T., Leas, S., Zobel, H., and Robyt, J. F. 1994. Anthology of starch granule morphology by scanning electron-microscopy. *Starch/Stärke* 46:121–129.

Jane, J. L. and Robyt, J. F. 1984. Structure studies of amylose-V complexes and retrograded amylose by action of alpha-amylases, and a new method for preparing amylodextrins. *Carbohydrate Research* 132:105–118.

Jane, J. L. and Shen, J. J. 1993. Internal structure of the potato starch granule revealed by chemical gelatinization. *Carbohydrate Research* 247:279–290.

Jane, J. L., Wong, K. S., and McPherson, A. E. 1997. Branch-structure difference in starches of A- and B-type X-ray patterns revealed by their Naegeli dextrins. *Carbohydrate Research* 300:219–227.

Jane, J., Xu, A., Radosavljevic, M., and Seib, P. A. 1992. Location of amylose in normal starch granules.1. Susceptibility of amylose and amylopectin to cross-linking reagents. *Cereal Chemistry* 69:405–409.

Jayakody, L. and Hoover, R. 2008. Effect of annealing on the molecular structure and physicochemical properties of starches from different botanical origins—A review. *Carbohydrate Polymers* 74:691–703.

Jenkins, J. P. J., Cameron, R. E., and Donald, A. M. 1993. A universal feature in the structure of starch granules from different botanical sources. *Starch/Stärke* 45:417–420.

Jenkins, P. J., Cameron, R. E., Donald, A. M., Bras, W., Derbyshire, G. E., Mant, G. R., and Ryan, A. J. 1994. In-situ simultaneous small and wide-angle X-ray-scattering: A new technique to study starch gelatinization. *Journal of Polymer Science Part B—Polymer Physics* 32:1579–1583.

Jenkins, P. J. and Donald, A. M. 1995. The influence of amylose on starch granule structure. *International Journal of Biological Macromolecules* 17:315–321.

Jenkins, P. J. and Donald, A. M. 1998. Gelatinisation of starch: A combined SAXS/WAXS/DSC and SANS study. *Carbohydrate Research* 308:133–147.

Jiang, H. X., Campbell, M., Blanco, M., and Jane, J. L. 2010. Characterization of maize amylose-extender (ae) mutant starches: Part II. Structures and properties of starch residues remaining after enzymatic hydrolysis at boiling-water temperature. *Carbohydrate Polymers* 80:1–12.

Johnson, L. A., Hardy, C. L., Baumel, C. P., and White, P. J. 2001. Identifying valuable corn quality traits for starch production. *Cereal Foods World* 46:417–423.

Kalichevsky, M. T., Orford, P. D., and Ring, S. G. 1990. The retrogradation and gelation of amylopectins from various botanical sources. *Carbohydrate Research* 198:49–55.

Kalichevsky, M. T. and Ring, S. G. 1987. Incompatibility of amylose and amylopectin in aqueous-solution. *Carbohydrate Research* 162:323–328.

Kasemsuwan, T. and Jane, J. 1994. Location of amylose in normal starch granules. 2. Locations of phosphodiester cross-linking revealed by P-31 nuclear-magnetic-resonance. *Cereal Chemistry* 71:282–287.

Kasemsuwan, T. and Jane, J. L. 1996. Quantitative method for the survey of starch phosphate derivatives and starch phospholipids by P-31 nuclear magnetic resonance spectroscopy. *Cereal Chemistry* 73:702–707.

Kasemsuwan, T., Jane, J., Schnable, P., Stinard, P., and Robertson, D. 1995. Characterization of the dominant mutant amylose-extender (Ael-5180) maize starch. *Cereal Chemistry* 72:457–464.

Kassenbeck, P. 1975. Electron-microscope contribution to study of fine-structure of wheat starch. *Starch/Stärke* 27:217–227.

Kassenbeck, P. 1978. Contribution to knowledge on distribution of amylose and amylopectin in starch granules. *Starch/Stärke* 30:40–46.

Katsuta, K. 1998. Effects of salts and saccharides on the rheological properties and pulsed NMR of rice starch during the gelatinisation and retrogradation processes. In *Gums and Stabilisers for the Food Industry 9*. P. A. Williams and G. O. Phillips, eds. Royal Society of Chemistry: Cambridge, U.K., pp. 59–68.

Katsuta, K., Miura, M., and Nishimura, A. 1992a. Kinetic treatment for rheological properties and effects of saccharides on retrogradation of rice starch gels. *Food Hydrocolloids* 6:187–198.

Katsuta, K., Nishimura, A., and Miura, M. 1992b. Effects of saccharides on stabilities of rice starch gels.1. Monosaccharides and disaccharides. *Food Hydrocolloids* 6:387–398.

Katsuta, K., Nishimura, A., and Miura, M. 1992c. Effects of saccharides on stabilities of rice starch gels. 2. Oligosaccharides. *Food Hydrocolloids* 6:399–408.

Kazanskii, K. S. and Dubrovskii, S. A. 1992. Chemistry and physics of agricultural hydrogels. *Advances in Polymer Science* 104:97–133.

Kim, H. R. and Eliasson, A. C. 1993. The influence of molar substitution on the thermal transition properties of hydroxypropyl potato starches. *Carbohydrate Polymers* 22:31–35.

Kim, H. S. and Huber, K. C. 2008. Channels within soft wheat starch A- and B-type granules. *Journal of Cereal Science* 48:159–172.

Kishida, A. and Ikada, Y. 2001. Hydrogels for biomedical and pharmaceutical applications. In *Polymeric Biomaterials, Revised and Expanded.* CRC Press: Boca Raton, FL, pp. 59–68.

Kittipongpatana, O., Burapadaja, S., and Kittipongpatana, N. 2009. Carboxymethyl mungbean starch as a new pharmaceutical gelling agent for topical preparation. *Drug Development and Industrial Pharmacy* 35:34–42.

Klucinec, J. D. and Thompson, D. B. 1998. Fractionation of high-amylose maize starches by differential alcohol precipitation and chromatography of the fractions. *Cereal Chemistry* 75:887–896.

Klucinec, J. D. and Thompson, D. B. 2002. Amylopectin nature and amylose-to-amylopectin ratio as influences on the behavior of gels of dispersed starch. *Cereal Chemistry* 79:24–35.

Koizumi, K., Fukuda, M., and Hizukuri, S. 1991. Estimation of the distributions of chain-length of amylopectins by high-performance liquid-chromatography with pulsed amperometric detection. *Journal of Chromatography* 585:233–238.

Lai, V. M. F., Lu, S., and Lii, C. 2000. Molecular characteristics influencing retrogradation kinetics of rice amylopectins. *Cereal Chemistry* 77:272–278.

Langton, M. and Hermansson, A. M. 1989. Microstructural changes in wheat-starch dispersions during heating and cooling. *Food Microstructure* 8:29–39.

Leach, H. W., McCowen, L. D., and Schoch, T. J. 1959. Structure of the starch granule. 1. Swelling and solubility patterns of various starches. *Cereal Chemistry* 36:534–544.

Lee, K. Y., Kim, Y. R., Park, K. H., and Lee, H. G. 2008. Rheological and gelation properties of rice starch modified with 4-alpha-glucanotransferase. *International Journal of Biological Macromolecules* 42:298–304.

Leloup, V. M., Colonna, P., and Buleon, A. 1991. Influence of amylose amylopectin ratio on gel properties. *Journal of Cereal Science* 13:1–13.

Leloup, V. M., Colonna, P., Ring, S. G., Roberts, K., and Wells, B. 1992. Microstructure of amylose gels. *Carbohydrate Polymers* 18:189–197.

Li, J. H., Vasanthan, T., Hoover, R., and Rossnagel, B. G. 2003. Starch from hull-less barley: Ultrastructure and distribution of granule-bound proteins. *Cereal Chemistry* 80:524–532.

Li, J. H., Vasanthan, T., Rossnagel, B., and Hoover, R. 2001. Starch from hull-less barley: I. Granule morphology, composition and amylopectin structure. *Food Chemistry* 74:395–405.

Li, J. M. and Zhang, L. M. 2007. Characteristics of novel starch-based hydrogels prepared by UV photopolymerization of acryloylated starch and a zwitterionic monomer. *Starch/Stärke* 59:418–422.

Light, J. M. 1990. Modified food starches: Why, what, where, and how. *Cereal Foods World* 35:1081–1092.

Lii, C. Y., Tsai, M. L., and Tseng, K. H. 1996. Effect of amylose content on the rheological property of rice starch. *Cereal Chemistry* 73:415–420.

Lim, S. T., Kasemsuwan, T., and Jane, J. L. 1994. Characterization of phosphorus in starch by P-31-nuclear magnetic-resonance spectroscopy. *Cereal Chemistry* 71:488–493.

Lineback, D. R. 1984. The starch granule: Organization and properties. *Bakers Digest* 58:16–21.

Liu, H. and Lelievre, J. 1993. A model of starch gelatinization linking differential scanning calorimetry and birefringence measurements. *Carbohydrate Polymers* 20:1–5.

Liu, H., Lelievre, J., and Ayoungchee, W. 1991. A study of starch gelatinization using differential scanning calorimetry, X-ray, and birefringence measurements. *Carbohydrate Research* 210:79–87.

Lorenz, K. 1995. Physicochemical characteristics and functional-properties of starch from a high beta-glucan waxy barley. *Starch/Stärke* 47:14–18.

Luo, W., Zhang, W. A., Chen, P., and Fang, Y. E. 2005. Synthesis and properties of starch grafted poly acrylamide-co-(acrylic acid)/montmorillonite nanosuperabsorbent via gamma-ray irradiation technique. *Journal of Applied Polymer Science* 96:1341–1346.

Ma, U. V. L., Floros, J. D., and Ziegler, G. R. 2011. Effect of starch fractions on spherulite formation and microstructure. *Carbohydrate Polymers* 83:1757–1765.

van der Maarel, M., Capron, I., Euverink, G. J. W., Bos, H. T., Kaper, T., Binnema, D. J., and Steeneken, P. A. M. 2005. A novel thermoreversible gelling product made by enzymatic modification of starch. *Starch/Stärke* 57:465–472.

Makela, M. J., Korpela, T., and Laakso, S. 1982. Studies of starch size and distribution in 33 barley varieties with a celloscope. *Starch/Stärke* 34:329–334.

Manners, D. J. 1989. Studies of starch size and distribution in 33 barley varieties with a celloscope. *Carbohydrate Polymers* 11:87–112.

Marchal, L. M., Beeftink, H. H., and Tramper, J. 1999. Towards a rational design of commercial maltodextrins. *Trends in Food Science and Technology* 10:345–355.

Mason, W. R. 2009. Starch use in foods. In *Starch Chemistry and Technology*. B. James and W. Roy, eds. Academic Press: San Diego, CA, pp. 745–795.

Matveev, Y. I., van Soest, J. J. G., Nieman, C., Wasserman, L. A., Protserov, V., Ezernitskaja, M., and Yuryev, V. P. 2001. The relationship between thermodynamic and structural properties of low and high amylose maize starches. *Carbohydrate Polymers* 44:151–160.

McPherson, A. E. and Jane, J. 1999. Comparison of waxy potato with other root and tuber starches. *Carbohydrate Polymers* 40:57–70.

Miles, M. J., Morris, V. J., Orford, P. D., and Ring, S. G. 1985. The roles of amylose and amylopectin in the gelation and retrogradation of starch. *Carbohydrate Research* 135:271–281.

Mitolo, J. 2006. Starch selection and interaction in foods. In *Ingredient Interactions*, 2nd edn. A. McPherson and A. G. Gaonkar, eds. CRC Press: Boca Raton, FL, pp. 139–166.

Moorthy, S. N. 2002. Physicochemical and functional properties of tropical tuber starches: A review. *Starch/Stärke* 54:559–592.

Morris, V. J. 1990. Starch gelation and retrogradation. *Trends in Food Science and Technology* 1:2–6.

Morrison, W. R., Law, R. V., and Snape, C. E. 1993a. Evidence for inclusion complexes of lipids with V-amylose in maize, rice and oat starches. *Journal of Cereal Science* 18:107–109.

Morrison, W. R., Tester, R. F., Snape, C. E., Law, R., and Gidley, M. J. 1993b. Swelling and gelatinization of cereal starches. 4. Some effects of lipid-complexed amylose and free amylose in waxy and normal barley starches. *Cereal Chemistry* 70:385–391.

Muhrbeck, P. 1991. On crystallinity in cereal and tuber starches. *Starch/Stärke* 43:347–348.

Muhrbeck, P. and Eliasson, A. C. 1987. Influence of pH and ionic-strength on the viscoelastic properties of starch gels: A comparison of potato and cassava starches. *Carbohydrate Polymers* 7:291–300.

Nagasawa, N., Yagi, T., Kume, T., and Yoshii, F. 2004. Radiation crosslinking of carboxymethyl starch. *Carbohydrate Polymers* 58:109–113.

Naguleswaran, S., Li, J., Vasanthan, T., and Bressler, D. 2011. Distribution of granule channels, protein, and phospholipid in triticale and corn starches as revealed by confocal laser scanning microscopy. *Cereal Chemistry* 88:87–94.

Nishinari, K. 2009. Some thoughts on the definition of a gel. In *Gels: Structures, Properties, and Functions— Fundamentals and Applications*. M. Tokita and K. Nishinari, eds. Springer-Verlag: Berlin, Germany, pp. 87–94.

Noda, T., Isono, N., Krivandin, A. V., Shatalova, O. V., Blaszczak, W., and Yuryev, V. P. 2009. Origin of defects in assembled supramolecular structures of sweet potato starches with different amylopectin chain-length distribution. *Carbohydrate Polymers* 76:400–409.

Nordmark, T. S. and Ziegler, G. R. 2002a. Spherulitic crystallization of gelatinized maize starch and its fractions. *Carbohydrate Polymers* 49:439–448.

Nordmark, T. S. and Ziegler, G. R. 2002b. Structural features of non-granular spherulitic maize starch. *Carbohydrate Research* 337:1467–1475.

Oakenfull, D. 1987. Gelling agents. *CRC Critical Reviews in Food Science and Nutrition* 26:1–25.

Oates, C. G. 1997. Towards an understanding of starch granule structure and hydrolysis. *Trends in Food Science and Technology* 8:375–382.

Omidian, H., Park, K., and Rocca, J. G. 2007. Recent developments in superporous hydrogels. *Journal of Pharmacy and Pharmacology* 59:317–327.

Omidian, H., Rocca, J. G., and Park, K. 2005. Advances in superporous hydrogels. *Journal of Controlled Release* 102:3–12.

Oostergetel, G. T. and van Bruggen, E. F. J. 1989. On the origin of a low-angle spacing in starch. *Starch/Stärke* 41:331–335.

Oostergetel, G. T. and van Bruggen, E. F. J. 1993. The crystalline domains in potato starch granules are arranged in a helical fashion. *Carbohydrate Polymers* 21:7–12.

Orford, P. D., Ring, S. G., Carroll, V., Miles, M. J., and Morris, V. J. 1987. The effect of concentration and botanical source on the gelation and retrogradation of starch. *Journal of the Science of Food and Agriculture* 39:169–177.

Pan, D. D. and Jane, J. L. 2000. Internal structure of normal maize starch granules revealed by chemical surface gelatinization. *Biomacromolecules* 1:126–132.

Parker, M. L. 1985. The relationship between A-type and B-type starch granules in the developing endosperm of wheat. *Journal of Cereal Science* 3:271–278.

Parker, R. and Ring, S. G. 2001. Aspects of the physical chemistry of starch. *Journal of Cereal Science* 34:1–17.

Passauer, L., Liebner, F., and Fischer, K. 2009. Starch phosphate hydrogels. Part I: Synthesis by mono-phosphorylation and cross-linking of starch. *Starch/Stärke* 61:621–627.

Peng, G., Xu, S. M., Peng, Y., Wang, J., and Zheng, L. C. 2008. A new amphoteric superabsorbent hydrogel based on sodium starch sulfate. *Bioresource Technology* 99:444–447.

Perera, C., Lu, Z., Sell, J., and Jane, J. 2001. Comparison of physicochemical properties and structures of sugary-2 cornstarch with normal and waxy cultivars. *Cereal Chemistry* 78:249–256.

Pfannemuller, B. 1987. Influence of chain-length of short monodisperse amyloses on the formation of A-type and B-type X-ray-diffraction patterns. *International Journal of Biological Macromolecules* 9:105–108.

Prentice, R. D. M., Stark, J. R., and Gidley, M. J. 1992. Granule residues and ghosts remaining after heating A-type barley-starch granules in water. *Carbohydrate Research* 227:121–130.

Prokopowich, D. J. and Biliaderis, C. G. 1995. A comparative-study of the effect of sugars on the thermal and mechanical-properties of concentrated waxy maize, wheat, potato and pea starch gels. *Food Chemistry* 52:255–262.

Putseys, J. A., Derde, L. J., Lamberts, L., Ostman, E., Bjorck, I. M., and Delcour, J. A. 2010a. Functionality of short chain amylose-lipid complexes in starch-water systems and their impact on in vitro starch degradation. *Journal of Agricultural and Food Chemistry* 58:1939–1945.

Putseys, J. A., Lamberts, L., and Delcour, J. A. 2010b. Amylose-inclusion complexes: Formation, identity and physico-chemical properties. *Journal of Cereal Science* 51:238–247.

Raeker, M. O., Gaines, C. S., Finney, P. L., and Donelson, T. 1998. Granule size distribution and chemical composition of starches from 12 soft wheat cultivars. *Cereal Chemistry* 75:721–728.

Ratnayake, W. S., Hoover, R., and Warkentin, T. 2002. Pea starch: Composition, structure and properties—A review. *Starch/Stärke* 54:217–234.

Reis, A. V., Guilherme, M. R., Moia, T. A., Mattoso, L. H. C., Muniz, E. C., and Tambourgi, E. B. 2008. Synthesis and characterization of a starch-modified hydrogel as potential carrier for drug delivery system. *Journal of Polymer Science Part A—Polymer Chemistry* 46:2567–2574.

Ring, S. G. 1985. Some studies on starch gelation. *Starch/Stärke* 37:80–83.

Ring, S. G., Colonna, P., Ianson, K. J., Kalichevsky, M. T., Miles, M. J., Morris, V. J., and Orford, P. D. 1987. The gelation and crystallization of amylopectin. *Carbohydrate Research* 162:277–293.

Ring, S. G., Lanson, K. J., and Morris, V. J. 1985. Static and dynamic light-scattering studies of amylose solutions. *Macromolecules* 18:182–188.

Ring, S. and Stainsby, G. 1982. Filler reinforcement of gels. *Progress in Food and Nutrition Science* 6:323–329.

Robin, J. P., Mercier, C., Charbonn, R., and Guilbot, A. 1974. Lintnerized starches gel: Filtration and enzymatic studies of insoluble residues from prolonged acid treatment of potato starch. *Cereal Chemistry* 51:389–406.

Roller, S. 1996. Starch-derived fat mimetics: Maltodextrins. In *Handbook of Fat Replacers*. S. Roller and S. A. Jones, eds. CRC Press: Boca Raton, FL, pp. 99–118.

Rosell, C. M., Yokoyama, W., and Shoemaker, C. 2011. Rheology of different hydrocolloids-rice starch blends. Effect of successive heating-cooling cycles. *Carbohydrate Polymers* 84:373–382.

Rudzinski, W. E., Dave, A. M., Vaishanav, U. H., Kumbar, S. G., Kulkarni, A. R., and Aminabhavi, T. M. 2002. Hydrogels as controlled release devices in agriculture. *Designed Monomers and Polymers* 5:39–65.

Russell, P. L., Gough, B. M., Greenwell, P., Fowler, A., and Munro, H. S. 1987. A study by ESCA of the surface of native and chlorine-treated wheat-starch granules: The effects of various surface treatments. *Journal of Cereal Science* 5:83–100.

Saboktakin, M. R., Maharramov, A., and Ramazanov, M. A. 2009. pH-sensitive starch hydrogels via free radical graft copolymerization, synthesis and properties. *Carbohydrate Polymers* 77:634–638.

Sadeghi, M. and Hosseinzadeh, H. 2008. Synthesis of starch-poly(sodium acrylate-co-acrylamide) superabsorbent hydrogel with salt and pH-responsiveness properties as a drug delivery system. *Journal of Bioactive and Compatible Polymers* 23:381–404.

Salomonsson, A. C. and Sundberg, B. 1994. Amylose content and chain profile of amylopectin from normal, high amylose and waxy barleys. *Starch/Stärke* 46:325–328.

Sanders, E. B., Thompson, D. B., and Boyer, C. D. 1990. Thermal-behavior during gelatinization and amylopectin fine-structure for selected maize genotypes as expressed in 4 inbred lines. *Cereal Chemistry* 67:594–602.

Sarko, A. and Wu, H. C. H. 1978. Crystal-structures of A-polymorphs, B-polymorphs and C-polymorphs of amylose and starch. *Starch/Stärke* 30:73–77.

Sasaki, T. and Matsuki, J. 1998. Effect of wheat starch structure on swelling power. *Cereal Chemistry* 75:525–529.

Sasaki, T., Yasui, T., and Matsuki, J. 2000. Effect of amylose content on gelatinization, retrogradation, and pasting properties of starches from waxy and nonwaxy wheat and their F1 seeds. *Cereal Chemistry* 77:58–63.

Seidel, C., Kulicke, W. M., Hess, C., Hartmann, B., Lechner, M. D., and Lazik, W. 2001. Influence of the cross-linking agent on the gel structure of starch derivatives. *Starch/Stärke* 53:305–310.

Seidel, C., Kulicke, W. M., Hess, C., Hartmann, B., Lechner, M. D., and Lazik, W. 2004. Synthesis and characterization of cross-linked carboxymethyl potato starch ether gels. *Starch/Stärke* 56:157–166.

Shamekh, S., Forssell, P., Suortti, T., Autio, K., and Poutanen, K. 1999. Fragmentation of oat and barley starch granules during heating. *Journal of Cereal Science* 30:173–182.

Shi, Y. C., Capitani, T., Trzasko, P., and Jeffcoat, R. 1998. Molecular structure of a low-amylopectin starch and other high-amylose maize starches. *Journal of Cereal Science* 27:289–299.

Shi, Y. C. and Seib, P. A. 1992. The structure of 4 waxy starches related to gelatinization and retrogradation. *Carbohydrate Research* 227:131–145.

Shi, Y. C. and Seib, P. A. 1995. Fine-structure of maize starches from 4 wx-containing genotypes of the W64A inbred line in relation to gelatinization and retrogradation. *Carbohydrate Polymers* 26:141–147.

Shiotsubo, T. and Takahashi, K. 1984. Study of starch gelatinization. 3. Differential thermal-analysis of potato starch gelatinization. *Agricultural and Biological Chemistry* 48:9–17.

Shu, J., Li, X. J., and Zhao, D. B. 2011. Microwave-irradiated preparation of super absorbent resin by graft copolymerization of cellulose and acrylic acid/acrylamide. In *Manufacturing Processes and Systems*, Pts 1–2. X. H. Liu, Z. Y. Jiang, and J. T. Han, eds. Trans Tech Publications Ltd.: Stafa-Zurich, Switzerland, pp. 799–802.

Shuey, W. C. and Tipples, K. H. 1980. *The Amylograph Handbook*. American Association of Cereal Chemists: St. Paul, MN.

Sievert, D. and Pomeranz, Y. 1989. Enzyme-resistant starch.1. Characterization and evaluation by enzymatic, thermoanalytical, and microscopic methods. *Cereal Chemistry* 66:342–347.

Singh, V. and Maurya, S. 2010. Microwave synthesis, characterization, and zinc uptake studies of starch-graft-poly(ethylacrylate). *International Journal of Biological Macromolecules* 47:348–355.

Singh, V., Tiwari, A., Pandey, S., and Singh, S. K. 2006. Microwave-accelerated synthesis and characterization of potato starch-g-poly(acrylamide). *Starch/Stärke* 58:536–543.

Singh, V., Tiwari, A., Pandey, S., and Singh, S. K. 2007. Peroxydisulfate initiated synthesis of potato starch-graft-poly(acrylonitrile) under microwave irradiation. *Express Polymer Letters* 1:51–58.

Song, Y. and Jane, J. 2000. Characterization of barley starches of waxy, normal, and high amylose varieties. *Carbohydrate Polymers* 41:365–377.

Steeneken, P. A. M. and Woortman, A. J. J. 2009. Superheated starch: A novel approach towards spreadable particle gels. *Food Hydrocolloids* 23:394–405.

Sujka, M. and Jamroz, J. 2007. Starch granule porosity and its changes by means of amylolysis. *International Agrophysics* 21:107–113.

Svegmark, K. and Hermansson, A. M. 1991. Distribution of amylose and amylopectin in potato starch pastes— Effects of heating and shearing. *Food Structure* 10:117–129.

Swinkels, J. J. M. 1985a. Composition and properties of commercial native starches. *Starch/Stärke* 37:1–5.

Swinkels, J. J. M. 1985b. Sources of starch, its chemistry and physics. In *Starch Conversion Technology*. G. M. A. Van Beynum and J. A. Roels, eds. Marcel Dekker: New York, pp. 15–46.

Takeda, Y. and Preiss, J. 1993. Structures of B90 (sugary) and W64A (normal) maize starches. *Carbohydrate Research* 240:265–275.

Takeda, Y., Shibahara, S., and Hanashiro, I. 2003. Examination of the structure of amylopectin molecules by fluorescent labeling. *Carbohydrate Research* 338:471–475.

Takeda, C., Takeda, Y., and Hizukuri, S. 1993a. Structure of the amylopectin fraction of amylomaize. *Carbohydrate Research* 246:273–281.

Takeda, Y., Tomooka, S., and Hizukuri, S. 1993b. Structures of branched and linear-molecules of rice amylose. *Carbohydrate Research* 246:267–272.

Tester, R. F. 1997. Starch: The polysaccharide fractions. In *Starch Structure and Functionality*. P. J. Frazier, A. M. Donald, and P. Richmond, eds. The Royal Society of Chemistry: Cambridge, U.K., pp. 163–171.

Tester, R. F. and Debon, S. J. J. 2000. Annealing of starch—A review. *International Journal of Biological Macromolecules* 27:1–12.

Tester, R. F. and Karkalas, J. 1996. Swelling and gelatinization of oat starches. *Cereal Chemistry* 73:271–277.

Tester, R. F., Karkalas, J., and Qi, X. 2004. Starch–composition, fine structure and architecture. *Journal of Cereal Science* 39:151–165.

Tester, R. F. and Morrison, W. R. 1990a. Swelling and gelatinization of cereal starches. 1. Effects of amylopectin, amylose, and lipids. *Cereal Chemistry* 67:551–557.

Tester, R. F. and Morrison, W. R. 1990b. Swelling and gelatinization of cereal starches. 2. Waxy rice starches. *Cereal Chemistry* 67:558–563.

Thiewes, H. J. and Steeneken, P. A. M. 1997. Comparison of the Brabender viskograph and the rapid visco analyser .1. Statistical evaluation of the pasting profile. *Starch/Stärke* 49:85–92.

Thompson, D. B. 2000. On the non-random nature of amylopectin branching. *Carbohydrate Polymers* 43:223–239.

Tong, Z., Peng, W., Zhiqian, Z., and Baoxiu, Z. 2005. Microwave irradiation copolymerization of superabsorbents from cornstarch and sodium acrylate. *Journal of Applied Polymer Science* 95:264–269.

Tsai, M. L., Li, C. F., and Lii, C. Y. 1997. Effects of granular structures on the pasting behaviors of starches. *Cereal Chemistry* 74:750–757.

Vansoest, J. J. G., Dewit, D., Tournois, H., and Vliegenthart, J. F. G. 1994. Retrogradation of potato starch as studied by Fourier-transform infrared-spectroscopy. *Starch/Stärke* 46:453–457.

Virtanen, T., Autio, K., Suortti, T., and Poutanen, K. 1993. Heat-induced changes in native and acid-modified oat starch pastes. *Journal of Cereal Science* 17:137–145.

Waigh, T. A., Hopkinson, I., Donald, A. M., Butler, M. F., Heidelbach, F., and Riekel, C. 1997. Analysis of the native structure of starch granules with X-ray microfocus diffraction. *Macromolecules* 30:3813–3820.

Wang, Y. J., White, P., and Pollak, L. 1993a. Physicochemical properties of starches from mutant genotypes of the Oh43 inbred line. *Cereal Chemistry* 70:199–203.

Wang, Y. J., White, P., Pollak, L., and Jane, J. 1993b. Amylopectin and intermediate materials in starches from mutant genotypes of the Oh43 inbred line. *Cereal Chemistry* 70:521–525.

White, P. J., Abbas, I. R., and Johnson, L. A. 1989. Freeze-thaw stability and refrigerated-storage retrogradation of starches. *Starch/Stärke* 41:176–180.

Wootton, M., Panozzo, J. F., and Hong, S. H. 1998. Differences in gelatinisation behaviour between starches from Australian wheat cultivars. *Starch/Stärke* 50:154–158.

Wu, H. C. H. and Sarko, A. 1978a. Packing analysis of carbohydrates and polysaccharides .9. Double-helical molecular-structure of crystalline A-amylose. *Carbohydrate Research* 61:27–40.

Wu, H. C. H. and Sarko, A. 1978b. Packing analysis of carbohydrates and polysaccharides. 8. Double-helical molecular-structure of crystalline B-amylose. *Carbohydrate Research* 61:7–25.

Yamaguchi, M., Kainuma, K., and French, D. 1979. Electron microscopic observations of waxy maize starch. *Journal of Ultrastructure Research* 69:249–261.

Yasushi, Y. B., Takenouchi, T., and Takeda, Y. 2002. Molecular structure and some physicochemical properties of waxy and low-amylose barley starches. *Carbohydrate Polymers* 47:159–167.

Yoshii, F., Zhao, L., Wach, R. A., Nagasawa, N., Mitomo, H., and Kume, T. 2003. Hydrogels of polysaccharide derivatives crosslinked with irradiation at paste-like condition. *Nuclear Instruments and Methods in Physics Research Section B-Beam Interactions with Materials and Atoms* 208:320–324.

Yoshimoto, Y., Tashiro, J., Takenouchi, T., and Takeda, Y. 2000. Molecular structure and some physicochemical properties of high-amylose barley starches. *Cereal Chemistry* 77:279–285.

Yuan, R. C., Thompson, D. B., and Boyer, C. D. 1993. Fine structure of amylopectin in relation to gelatinization and retrogradation behavior of maize starches from three wx-containing genotypes in two inbred lines. *Cereal Chemistry* 70:81–89.

Zallie, J. 2011. The role and function of specialty starches in the confection industry. http://eu.foodinnovation.com/pdfs/rolefunct.pdf (accessed on May 26, 2011).

Zeleznak, K. J. and Hoseney, R. C. 1987. The glass-transition in starch. *Cereal Chemistry* 64:121–124.

Zeng, M., Morris, C. F., Batey, I. L., and Wrigley, C. W. 1997. Sources of variation for starch gelatinization, pasting, and gelation properties in wheat. *Cereal Chemistry* 74:63–71.

Zhai, M. L., Yoshii, F., Kume, T., and Hashim, K. 2002. Syntheses of PVA/starch grafted hydrogels by irradiation. *Carbohydrate Polymers* 50:295–303.

Zhang, L. M., Wang, G. H., and Lu, H. W. 2005a. A new class of starch-based hydrogels incorporating acrylamide and vinyl pyrrolidone: Effects of reaction variables on water sorption behavior. *Journal of Bioactive and Compatible Polymers* 20:491–501.

Zhang, L. M., Yang, C., and Yan, L. 2005b. Perspectives on: Strategies to fabricate starch-based hydrogels with potential biomedical applications. *Journal of Bioactive and Compatible Polymers* 20:297–314.

Ziegler, G. R., Nordmark, T. S., and Woodling, S. E. 2003. Spherulitic crystallization of starch: Influence of botanical origin and extent of thermal treatment. *Food Hydrocolloids* 17:487–494.

Zobel, H. F. 1984. Gelatinization of starch and mechanical properties of starch pastes. In *Starch Chemistry and Technology*. R. L. Whistler, J. N. BeMiller, and E. F. Paschall, eds. Academic Press: Orlando, FL, pp. 285–309.

Zobel, H. F. 1988. Molecules to granules—A comprehensive starch review. *Starch/Stärke* 40:44–50.

CHAPTER 4

Plasticized Starch

Xiaofei Ma, Peter R. Chang, and Jiugao Yu

CONTENTS

4.1 INTRODUCTION

In view of environmental protection and sustainable use of resources, renewable biopolymers are highly sought as substitutes for traditional synthetic polymers from petroleum in applications as one-off materials. Starch is an economically important, highly functional, and renewable biopolymer present in a great variety of crops. Starch is also one of the most promising raw materials and commercially available ingredients for the production of biodegradable plastics [1]. Structurally, starch is a mixture of two main components: amylose, a linear or slightly branched (1–4)-α-D-glucan, and amylopectin, a highly branched macromolecule consisting of (1–4)-α-glucan short chains linked through α-(1–6) linkages [2]. Typically, the degree of crystallinity of natural starch granules varies from about 15%–45% [3]. The crystallinity is usually characterized by wide-angle X-ray diffraction (WXRD) analysis. Granular starch crystallinity can be mainly classified as A, B, and C types. The A type crystallinity is a closely packed arrangement with water molecules between each double helical structure, mainly found in cereal starches. The B type, usually sourced from tuber starches, is more open with more water molecules, essentially all of which are located in a central cavity surrounded by six double helices. The C pattern is a mixture of both A and B types, but also occurs naturally, for example, smooth-seeded pea starch and various bean starches. Three processing-induced single-helical amylose crystals, denoted as V_A, V_H, and E_H, were reported in the literature. Different amylose crystal structures were found after extrusion of cassava starch in the presence of fatty acids [4]. Morphological difference in structures was simply due to different arrangement of the single helices in the crystal lattice. The single helical structure, stable at low-moisture content, denoted as E_H, is shown to transform to the V_H form by increasing water content of the sample. The relative abundance of both types of structures is affected by thermal condition [4]. The V-type crystallinity is exerted by a complex of amylose and substances such as aliphatic fatty acids, emulsifiers, glycerol, butanol, and iodine [5].

Starch is a multihydroxyl polymer with three hydroxyl groups per monomer. There are so many inter- and intramolecular hydrogen bonds in starch that the natural granular starch itself does not behave like thermoplastic. Since the 1970s, granular starch has been incorporated into synthesized polymer matrices, and much effort has been made to convert natural granular starch into a thermoplastic material. At high temperature and under shear, water or plasticizers gradually swell the starch granules and disrupt some of the inter- and intramolecular hydrogen bonds among the starch molecules [6]. The crystalline structure of the starch is thus destroyed, and exposed starch hydroxyl groups form hydrogen bonds with water or plasticizer molecules. At this stage, starch is transformed into a moldable thermoplastic, known as plasticized starch (PS). After plasticization, starch is permitted for use as an injection, extrusion, or blow molding material, similar to most conventional synthetic thermoplastic polymers [7]. Commonly, there are two methods for preparing PS: melt processing and water-casting. In water-casting, a large amount of water (e.g., water/starch weight ratio = 100/5) at high temperature (90°C–95°C) transforms the granular starch into a starch paste. The water is then evaporated and a cast PS film is thus obtained. In this method, plasticizers (plasticizers/starch weight ratio = 0.3–0.4) play a marginal role in PS processing as compared with water, but they do have an important effect on the properties of the cast film. In melt processing of PS, no additional water is added and the only water present (about 10–15 wt.%) is that contained in the starch granules. Usually, PS from melt processing is also called thermoplastic starch (TPS).

4.2 EFFECT OF PLASTICIZERS ON PROPERTIES OF PS

During the plasticizing process, the plasticizers form hydrogen bonds with starch, breaking the strong interactions among the starch hydroxyl groups and causing plasticization. Prominent plasticizers in the casting process and melt process have been identified as water and the incorporated plasticizer (other than water), respectively. The properties of the plasticizer have a great influence on the properties of the PS. Plasticizers for PS can be divided into the following categories: polyols, amides, and others. Polyol plasticizers include glycerol, sorbitol [8], xylitol [9], maltitol [10], ethylene glycol, propylene glycol, butanediol, diethylene glycol [11], sucrose [12,13], and monosaccharides, that is, fructose and mannose [14]. Amide plasticizers include formamide [15], acetamide, urea [16,17], and ethylenebisformamide [18,19]. Plasticizers containing amine groups, such as ethanolamine [20], have also been used for PS. By varying the types and amounts of these plasticizers, the characteristic of PS can be tailored to specific needs.

Glycerol, a representative of the polyols, is widely used in PS because it has the advantages of low cost, nontoxicity, and a high boiling point. The hydroxyl groups of polyol plasticizers can form hydrogen bonds with the starch molecules in place of the inter- and intramolecular interactions among starch. The original crystalline structure of the starch is destroyed by the polyol plasticizers during plasticization; however, the interaction between the polyols and starch is not strong, so the starch often recrystallizes after being stored for a period of time causing the PS to become brittle [21,22]. PS with a higher glycerol content (40–60 wt.%) possesses an obvious inhibitory effect on starch recrystallization (retrogradation), but has poor tensile strength [23].

As a plasticizer, urea has been shown to prevent starch recrystallization; however, it is a high-melting solid with little internal flexibility. Urea-plasticized starch (UPS) is rigid and brittle because urea crystallizes and separates out from the PS [24]. Recrystallization of either starch or plasticizers will inhibit the application of PS. Urea and glycerol as a mixed plasticizer have not been proven to resist starch recrystallization [25].

Ma and Yu [26] analyzed the hydrogen bond interaction between plasticizer and starch in glycerol-plasticized starch (GPS) and formamide-plasticized starch (FPS). Analysis of the Fourier transform infrared (FTIR) spectra of the blends enables the hydrogen bond interactions to be identified. Using the harmonic oscillator model, reduction in the force constant (Δf) can be represented by Equation 4.1 [27]:

$$\Delta f = f_p - f_{np} = \frac{\mu \left(v_p^2 - v_{np}^2 \right)}{4\pi^2} \tag{4.1}$$

where
$\mu = m_1 m_2 / (m_1 + m_2)$ corresponds to the reduced mass of the oscillator
v is the oscillating frequency
f is the force constant
the subscripts np and p denote nonplasticized and plasticized oscillators, respectively

The reduction in force constant brought about by some interactions was directly related to the frequency (or wave number) shift of the stretching vibrations; the lower the peak frequency was, the stronger the interaction.

In the FTIR spectra of PS, the peaks characteristic of C–O stretching had shifted to a lower wave number, as compared to those of native starch. The introduction of plasticizers to the starch decreased the stretching vibration wave numbers of the C–O group, and the starch C–O group formed hydrogen bond interactions with the plasticizer (glycerol or formamide). The wave number of the C–O stretching shifted lower in FPS than in GPS with the same plasticizer content, indicating that formamide formed a stronger hydrogen bond with starch than glycerol did.

Ma and Yu [16] studied plasticization and starch recrystallization of PS plasticized by formamide, acetamide, and urea using GPS as a control. The order of their ability to form hydrogen bonds with starch was urea > formamide > acetamide > polyols, as confirmed by B3LYP chemical computation. The hydrogen-bonding interaction in 1:1 complexes formed between plasticizers (urea, formamide, acetamide, or glycerol) and starch is 14.167, 13.795, 13.698, and 12.939 kcal/mol, respectively. In order to inhibit starch recrystallization and improve the mechanical properties of PS, urea and formamide, which is a good solvent for urea, were used as a mixed plasticizer to prepare formamide and urea–plasticized starch (FUPS) [28]. The recrystallization of starch in PSs plasticized by glycerol, formamide, urea, and formamide/urea is shown in Figure 4.1. The conventional GPS stored at a relative humidity (RH) of 50% for 25 days presented V_H-style crystal peaks at 13.81° and 21.13° (Figure 4.1e) ascribed to starch recrystallization. According to van Soest [29], the V_H type is a single-helical structure "Inclusion Complex" made up of amylose and glycerol. There was no obvious starch crystal peak in the other four PSs (Figure 4.1a, b, c, and d) because urea and formamide formed stronger and more stable hydrogen bonds with starch than glycerol did, thus preventing the starch molecules from interacting and recrystallizing. The peak at 22.48° (Figure 4.1d) resulted from urea crystallization. Because formamide was a good solvent for urea, the mixture effectively inhibited the recrystallization of both PS and urea, and PS containing both formamide and urea showed no urea crystal peak. FUPS exhibited a tensile strength that was similar to GPS, but had better elongation at break and break energy. It was also found that plasticizers containing mixtures of urea and ethanolamine improved thermal stability, mechanical properties, and starch recrystallization [30]. As proven by FTIR and DSC (differential scanning calorimetry) analyses, the mixture of urea and ethanolamine formed more stable and stronger hydrogen bonds with starch

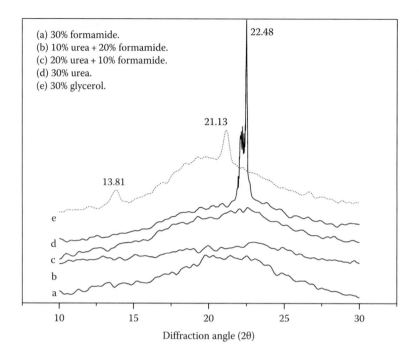

Figure 4.1 Diffractograms of five PS made with different plasticizers.

hydroxyl groups than glycerol did. Ethanolamine was also a good solvent for urea, allowing both of them to exist in their molecular forms in PS.

Yang et al. [18,31] designed a novel plasticizer, ethylenebisformamide (as shown in Figure 4.2a), containing two acyl-amine groups for preparing ethylenebisformamide-plasticized starch (EPS) based on the hydrogen bond interaction between plasticizer and starch. Similar to urea, ethylenebisformamide is a solid at room temperature. When EPS was stored for a period of time, ethylenebisformamide crystallized and separated out from the EPS. Yang et al. [32] used a mixture of ethylenebisformamide and sorbitol as the plasticizer and found that the interaction between sorbitol and ethylenebisformamide, at the appropriate ratio (e.g., 1:1), limited the separation of plasticizer out of the PS and inhibited the recrystallization of starch in the PS [33].

Zhang et al. [34] synthesized aliphatic amidediol as plasticizer for PS. Although aliphatic amidediol contains both amide and hydroxyl groups, the large number of nonpolar groups (methyl and methylene) decreased the efficiency of bond formation between plasticizer and starch. Aliphatic amidediol alone is not regarded as a plasticizer, but 2-hydroxy-N-[2-(2-hydroxy-propionylamino)-ethyl] propionamide (HPEP) (as shown in Figure 4.2b), an aliphatic amidediol, mixed with formamide or glycerol, both good solvents for HPEP, worked as a mixed plasticizer. The formamide/HPEP mixture effectively suppressed starch recrystallization [35], while the glycerol/HPEP mixture did not [34].

Hydroxyalkylformamides (HAF) can be synthesized from alcohol amine and ethyl formate without the use of solvent or catalyst. The equation for the chemical reaction is shown in Figure 4.2c. HAF commonly serves as a physiologically harmless humectant for cosmetics [36]. The HAFs N,N-bis(2-hydroxyethyl)formamide (BHF) [37] and N-(2-hydroxyethyl)formamide (HF) [38] can be used as efficient plasticizers in the preparation of PS. In order to show that HAF could be used extensively as plasticizer for corn starch, another two HAF plasticizers, N-(2-hydroxypropyl)-formamide (HPF) and N-(2-hydroxyethyl)-N-methylformamide (HMF), were studied [39]. HPF

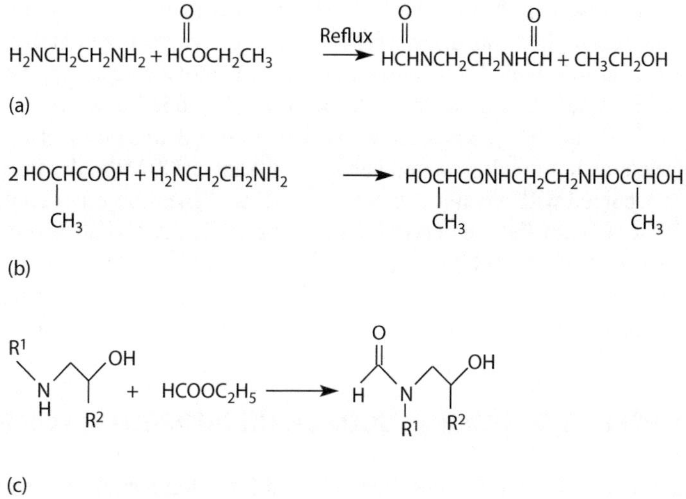

$$H_2NCH_2CH_2NH_2 + HCOCH_2CH_3 \xrightarrow{\text{Reflux}} HCHNCH_2CH_2NHCH + CH_3CH_2OH$$

(a)

$$2\ HOCHCOOH + H_2NCH_2CH_2NH_2 \longrightarrow HOCHCONHCH_2CH_2NHOCCHOH$$
$$\qquad\quad |\qquad\qquad\qquad\qquad\qquad\qquad\qquad |\qquad\qquad\qquad\qquad |$$
$$\qquad\quad CH_3 \qquad\qquad\qquad\qquad\qquad\qquad\quad CH_3 \qquad\qquad\quad CH_3$$

(b)

(c)

Figure 4.2 Chemical reaction equations for synthesis of several plasticizers. (a) The synthesis of ethylenebisformamide. (b) The synthesis of 2-hydroxy-*N*-[2-(2-hydroxy-propionylamino)-ethyl] propionamide (HPEP). (c) The synthesis of HAF from alcohol amine and ethyl formate (BHF: R^1–$HOCH_2CH_2$, R^2–H; HF: R^1–H, R^2–H; HMF: R^1–CH_3, R^2–H; HPF: R^1–H, R^2–CH_3).

(or HMF) formed strong, stable hydrogen bonds with starch. When starch was plasticized by these four HAF plasticizers, the A-type crystallinity of the native corn starch disappeared. E_H crystallinity of starch formed in the HMF-plasticized and BHF-plasticized starches, and only V_A crystallinity formed in the HPF-plasticized and HF-plasticized starches. Generally, HAF did not completely suppress the recrystallization of starch in the PS.

Ionic liquids (ILs), such as 1-allyl-3-methylimidazolium chloride ([amim]Cl), have been found to have strong hydrogen-bond-forming abilities with starch [40]. Using the casting method, PS was made from [amim]Cl/glycerol mixtures of different proportions as well as from ILs comprising [amim]Cl with various concentrations of lithium chloride (LiCl) [41]. Melt extrusion was used to plasticize starch with *N,N*-dimethylacetamide (DMAC) and different concentrations of LiCl. FTIR revealed that LiCl increased the interaction between starch and DMAC and destroyed the supramolecular structure of the starch [42]. TPS plasticized by 1-butyl-3-methylimidazolium chloride ([BMIM]Cl) was also obtained by melt processing [43]. [BMIM]Cl-plasticized TPS samples show less hygroscopicity and a much higher elongation at break in the rubbery state than the glycerol-plasticized TPS samples. The unusually low rubbery Young's modulus (0.5 MPa) in TPS suggests that [BMIM]Cl can greatly reduce hydrogen bonds between the starch chains.

4.3 EFFECT OF WATER ON PROPERTIES OF PS

As stated in the previous section, water is a good plasticizer for starch. If dry starch is employed in the preparation of PS, the dosage of plasticizer required is much higher. However, PS plasticized only by water has been proven to be a fragile material due to the volatilization of water. A small quantity of water (about 10 wt.% contained in starch) is usually necessary to prepare PS. Both the initial water content and the absorbed moisture from the storage environment will significantly affect the properties of PS. The isotherm moisture sorption of PS can be described by many models such as the BET (Brunauer, Emmet, and Teller) model, GAB (Guggenheim, Anderson, and deBoer) model, Smith model, and Flory–Huggins model [14].

When moisture is absorbed from a high-RH environment, the PS exhibits a lower tensile strength, Young's modulus, glass transition temperature, and recrystallization of starch. Water can weaken the interaction among starch molecules and make the molecular motion flexible. The moisture absorbance of PS prepared with different plasticizers is listed in Table 4.1.

As proven by a GPS study [44], when the water content of PS was below 9%, the PS exhibited a glassy state and the tensile modulus ranged between 400 and 1000 MPa. A transition from brittle to ductile behavior occurred with a water content of 9%–10%. When the water content was between 9% and 15%, the PS was tough and the maximum elongation at break was observed. When the water content was greater than 15%, the PS became weak and soft and the elongation at break decreased. Almost all PSs with different plasticizers had a similar dependence on water content; when water content was increased, the PS transformed from a fragile glassy state to a rubbery state [16,28].

4.4 EFFECT OF STARCH SOURCE ON PROPERTIES OF PS

Both native and modified starches have been used in PSs. Starches from various sources have different amylose/amylopectin ratios, molecular weights, and molecular weight distributions, which determine the properties of the PS. Many starches, such as corn starch, high-amylose corn starch (with 70% amylose), waxy corn starch (almost pure amylopectin) [45], wheat starch [46], barley starch [47], oat starch [13,48], rice starch [49], sago starch [50], amylose potato starch [51] high-amylopectin potato starch [52], cassava starch, yam starch [53], and pea starch [54], have been used to prepare PS. Table 4.2 shows the properties of GPSs from different botanical origins. T_{sec} and T_g are determined by DSC at the midpoint of the increase in heat capacity while $T\alpha$ and $T\beta$ are determined at the maximum of the tan delta peak. $T\alpha$ and $T\beta$ are ascribed to the glass transition of PS, while T_{sec} and $T\beta$ are ascribed to secondary relaxation of PS.

The effects of the amylose/amylopectin ratio on the properties of the PS have been studied in detail. The linear amylose chains effectively tangle and form strong physical cross-links, mainly through hydrogen bonding, resulting in a strong, coherent PS. Amylopectin is effective at increasing the flexibility and elongation of the materials because of its high molecular mass and its highly branched structure [29]. The effect of amylose/amylopectin ratios on glass transition temperature (T_g) of starch was studied using DSC. It was found that the higher the amylose content was, the higher the T_g was in corn starches [55]. de Graaf et al. [56] studied the effects of the amylose/amylopectin ratios on the mechanical properties of four starches. They found that the increases in amylose/amylopectin ratio resulted in decreased modulus and tensile strengths and increased elongations.

Yu and Christie [57] elucidated the effects of crystalline structures and the amylose/amylopectin ratio on the properties of PS including mechanical properties, glass transition temperature, and retrogradation. Gelatinization destroyed the crystalline structure of the starch granules and partially segregated amylose and amylopectin from each other. The short-branched chains of amylopectin remained in a regular pattern by retaining a certain "memory." These chains formed gel-balls that comprised mainly chains from the same submain chain. Alternatively, amylopectin might form superglobes after gelatinization. The molecular entanglements between the gel-balls and superglobes were much less than those between linear polymer chains due to the size and length of the chains (four to six glucose units).

The movement of these gel-balls required less energy than that for long linear chains, especially when these balls were lubricated by plasticizer (water). The gel-balls and superglobes could be lubricated more efficiently with the same amount of plasticizer as linear amylose. This explained why the amylopectin-rich starch had a lower viscosity during extrusion [58] and lower modulus and higher elongation in freshly prepared PS (prior to aging) and could also explain why high-amylopectin starches exhibited lower T_g [55].

Table 4.1 Effect of Plasticizers and Moisture on the Properties of PS (30 wt.% Plasticizer Content)

Plasticizer	Tensile Strength (MPa)	Elongation at Break (%)	Moisture Content (%)	Starch Retrogradation (2θ) Style		
				RH 0%	RH 50%	RH 100%
Glycerol	4.7 [17]	49.9 [17]	50 [17]	21.1 [16]	13.9, 21.6 [16]	17, 19.7 [16]
Urea	10.6 [17]	4.3 [17]	36.2 [17] (unsaturated)	None [16]	Urea crystal [16]	None [16]
Acetamide	5.0 [17]	2.4 [17]	55.7 [17]	13.9, 21.4 [16]	Acetamide crystal [16]	None [16]
Formamide	1.76 [17]	80.2 [17]	44.7 [17]	None [16]	None [16]	None [16]
Urea and formamide (2:1)	4.8 [28]	104.7 [28]	46.0 [28]	—	None [28]	—
Ethanolamine and urea (1:1)	9.00 [30]	34.4 [30]		—	None [30]	
Ethylenebisformamide	3.86 [18]	177 [18]	37.8 [18]	—	Ethylenebisformamide crystal [19]	None [19]
Ethylenebisformamide and sorbitol (1:1)	7.8 [32]	117.5 [32]	37.5 [32]	None [33]	None [33]	None [33]
Formamide and HPEP (2:1)	4.3 [35] (water content 12%)	80 [35] (water content 12%)	78 [35]		HPEP crystal [35]	None [35]
BHF	1.3 (RH 50%) [37] 4.9 (RH 33%) [37]	107.7 (RH 50%) [37] 21.6 (RH 33%) [37]			16.7, 19.6 (V_H) [37]	—
HF	2.5 (RH 44%) [38] 4.2 (RH 33%) [38]	75.3 (RH 44%) [38] 66.5 (RH 33%) [38]			12.8, 19.7, 22.3 (V_H) [38]	—
HPF	3.9 (RH 44%) [39]	34.7 (RH 44%) [39]			13.6, 20.9 [39]	—
HMF	2.8 (RH 44%) [39]	17.2 (RH 44%) [39]			11.8, 17.9 [39]	—

Table 4.2 Properties of GPS from Different Botanical Origins

Starch	Amylose Contents (%)	Starch/Glycerol/Water	Tβ (DMTA) (°C)	T_{sec} (DSC) (°C)	Tα (DMTA) (°C)	T_g (DSC) (°C)	Tensile Strength (MPa)	Elongation at Break (%)
Normal corn [45]	26–28						5.2	35
High-amylose corn [45]	70						11.2	23
High-amylopectin corn [51]	<2	80/20		−60		50	16	2.1
Wheat [46]	25	74/0/16	−54.8	−66.7	63.2	43.4	21.4	3.8
		70/18/12	−54.1	−72.2	31.0	8.4	52	3.3
		67/24/9	−58.1	−77.4	17.2	−7.3	2.6	110.0
		65/35	−61.6		1.4	−20.1	0.61	90.7
Barley [47]	28	70.3/20/9.7		−69		21		
		68/29/3		−56		34		
		70/14/16		−80		12		
Oat [47]	27	70.4/21.1/8.5	−53	−73	13		1	130
Rice [49]		100/20					10.9	2.8
		100/30					1.6	59.8
Amylose potato [51]	100	80/20		−70		50	27	4
Cassava [53]	19	100/20/9.4				42.42	10	33
		100/40/30				26.96	4	46
Yam [53]	29	100/20/9.4				23.26	30	13
		100/40/30				18.7	10	25
Pea [54]	35	70/30/8	−37		47		7	100

Recrystallization increased with storage time, in particular, for the amylopectin-rich starch, and resulted in an increase in modulus and a decrease in elongation. The crystallinity of the linear-amylose-rich material was relatively stable, so the mechanical properties were also stable. During aging, the amylose and amylopectin form a cocrystallinity resulting in physical cross-links between amylopectin and amylose that increased the modulus. Entanglement of polymer chains in amylopectin-rich starch was much less than that in linear-amylose-rich starch because of the highly branched microstructure and the formation of the gel-ball and superglobe after gelatinization. Orientation further reduced entanglement between the short branches. This may explain the poor mechanical properties of PS, particularly the lower elongation in the cross direction [57].

4.5 PS FROM MODIFIED STARCH

Compared to synthetic polymers, PS often exhibits poor mechanical properties, relatively huge water uptake, and brittleness. Chemical modifications, which introduce functional groups into the starch molecule using derivatization (etherification, esterification, cross-linking, and grafting) or decomposition (acid or enzymatic hydrolysis and oxidation) reactions, are necessary to significantly improve the properties. Many papers have reported the use of chemically modified starches to prepare PS.

4.5.1 PS from Starch Acetates

Acetylation of starch results in good thermoplastic processing, enhanced mechanical properties, and increased hydrophobicity. A high degree of substitution (DS) in starch acetates, the maximum of which is three since there are three hydroxyl groups in each anhydroglucose unit, is necessary for good mechanical properties. Starch acetates from high-amylose corn starch with 20 wt.% triacetin showed a tensile strength ranging from 20 to 30 MPa with different DSs (2.63–2.73) [59]. López et al. [60] studied the effect of glycerol content on the mechanical behavior of plasticized acetylated starch films.

Foam materials from starch acetate were extruded with a volatile solvent which increased the volume of materials by evaporating. Shogren [61] prepared starch acetate (DS = 2.5)–based foams with water as the volatile solvent. The foam had a higher bulk density, higher compressive strength, and lower resiliency than traditional polystyrene foam. Miladinov and Hanna [62] compared the physical and molecular properties of starch acetates (DS = 2.0 and 3.0) extruded with water or ethanol and found the foam materials extruded with water had lower spring indices (SIs), lower water absorption, and higher solid densities. Foam materials with a DS of 2.0 required less specific mechanical energy and had higher solid density and water absorption indices than starch with a DS of 3.0. Xu and Hanna [63] studied starch acetate foams with a broader DS range (from 0.5 to 2.5). A high DS increased the hydrophobicity of starch acetate, resulting in low compatibility with water but high compatibility with ethanol. Starch acetates with higher DSs extruded with water gave dense foams with thicker cell walls, lower radial expansion ratios (RERs), lower spring indices (SIs), higher unit density, and higher compressibility. When extruded with ethanol, the higher DS resulted in uniform cells with thinner and smoother cell walls, higher RER, high SI, and lower unit density and compressibility.

4.5.2 PS from Dialdehyde Starch

Oxidation is another way of chemically modifying starch. When the hydroxyl groups of starch are replaced with aldehyde groups, starch is more hydrophobic and recrystallization of the starch is inhibited. The most valuable oxidized starch is dialdehyde starch (DAS). DAS is widely used in applications such as foods, paper coatings, and biomaterials. DAS is produced by controlled periodate oxidative cleavage of the C-2 and C-3 bond of the anhydroglucose units of native starch. Compared to PS, plasticized DAS (PDAS) showed resistance to starch recrystallization during storage. DAS with higher aldehyde content contributed to better mechanical properties and decreases in glass transition temperature, hydrophilicity, water absorption, and water vapor permeability [64,65]. The biodegradability of PDAS under controlled composting conditions was evaluated. The degradation rate and final biodegradation percentage were closely related to the DAS structure. Intramolecular cross-linking of DAS can result in starch being more hydrophobic, which causes conformational changes within the starch molecule and reduces susceptibility to microorganisms. The higher the carbonyl content in DAS, the more likely it was to form intramolecular cross-linking; hence, as the oxidation degree of PDAS increased, the biodegradation percentage decreased [66].

PDAS derivatives were also used to prepare thermoplastics. Zhang et al. [67] reacted DASs containing different carbonyl contents with glycol in the presence of dense phosphoric acid to obtain glycol–DAS. Although the molecular weights of the DASs decreased during preparation, the DAS with a carbonyl content of 95% still exhibited acceptable intrinsic viscosity (as high as 0.46). This glycol–DAS could be pressed directly into film without any plasticizer and exhibited enhanced water resistance and mechanical properties. In another study [68], DAS with 35.2% carbonyl content was reacted with different alcohols (methanol, ethanol, glycol, and octanol) to prepare a series of starch derivatives. The degree of acetal formation (DAF) of methanol was 84.3%, while that of ethanol was 48.6%, and that of octanol was only 10.2% even after a reaction time of 72 h. Glycol was more reactive than ethanol, and the DAF of glycol was as high as 78.9%. Among the DAS derivatives, plasticized methanol–DAS and plasticized glycol–DAS had the lowest water absorption due to their high DAF values. The PDAS derivatives had better mechanical properties, especially elongation at break, than PS. Both PDAS and PDAS derivatives could be applied in the medical, agricultural, drug release, and packaging fields as edible films, disposable packaging, or food packaging.

4.5.3 PS and Fatty Acid

The hydrophobicity of PS is improved when starch is extruded in the presence of fatty acid. It was found that the role of glycerol as a plasticizer was enhanced by the presence of fatty acid (myristate or palmitate) in the starch matrix making the extrudates more flexible and less glassy than their controls, which contained glycerol but no fatty acids. This behavior was attributed to the interaction of the amylose–fatty acid complexes with glycerol molecules [69].

At lower extrusion temperatures (100°C) and at fairly low shear rates, the viscosities of the two starch–fatty acid systems were higher than that of the control, while at very high shear rates, the viscosity of the palmitic acid containing a sample was lower than that of the control. This could be attributed to interaction of the fatty acid with amylose and the formation of helical complexes. At higher extrusion temperatures (120°C–160°C), the viscosity of the starch–fatty acid extrudates was higher than that of the control, especially at fairly low shear rates. It must be noted that the pseudoplasticity of starch–fatty acid extrudates was much higher than that of the control, especially at higher extrusion temperatures. This could be attributed to dissociation of the formed supermolecular structure and the orderly arrangement of elongated amylose helices at high shear rates causing a significant reduction in viscosity [70].

A series of starch and amylose esters with different degrees of substitution and side-chain lengths were synthesized. The esters were prepared by acylation of the polysaccharide with the appropriate acid chlorides, such as octanoic, dodecanoic, and octadecanoic. The degrees of substitution were 0.54, 1.8, and 2.7, respectively. The tensile strength and elongation at break depended on the side-chain length and the DS. The extent of their biodegradability was assessed by weight loss measurements when they were exposed to activated sludge. The biodegradation rates decreased with increasing degree of esterification [71].

Modification of potato starch with acyl chlorides derived from palmitic and oleic acids and recovered vegetable oil (RVO) generated some interesting thermoplastic materials. The extreme brittleness of the palmitoyl starch and the rather poor mechanical properties of the oleoyl starch precluded their use as thermoplastic materials without considering blends. The improved mechanical properties of the modified starch with RVO-derived acid chlorides indicate that further investigation into its use for the preparation of thermoplastics is worthwhile [72].

The surface of corn starch films was modified with octenyl succinic anhydride as reactant. Physical properties of the films, moisture absorption, and water contact angle were measured to characterize the effect of surface esterification modification. These measurements showed that after surface esterification, the equilibrium moisture content of the starch film decreased by up to 29% at RH 95% and the surface water contact angle of the film increased by up to 83% [73].

4.5.4 Surface Modification of PS

Cross-linking technologies have also been used to modify PS. Bulk cross-linking treatments were applied to solution cast PDAS films, where DAS was the cross-linking agent [64]. Compared to bulk cross-linking, surface cross-linking modification of PS provided an approach that reduced surface hydrophilic characteristics of the materials without changing their bulk composition. In UV photo-cross-linking, the amounts of photosensitizers used in a surface modification would be significantly lower than modification of a bulk material. Furthermore, the intermolecular bridge networks in the surface layer induced by photo-cross-linking are formed more easily than those throughout the entire bulk of the material. Zhou et al. [74] found that surface photo-cross-linking, using sodium benzoate as photosensitizer, reduced hydrophilicity and improved water resistance of plasticized corn starch sheets. Surface photo-cross-linking modification increased tensile strength and Young's modulus but decreased the elongation at break of the PS sheets [75].

Polymer coating technology can enhance bulk and surface properties. Simao et al. [76] produced 1,3-butadiene plasma coatings on starch thermoplastic films that showed an enhanced resistance to humidity that was strongly dependent on coating thickness. A maximum of 80% reduction in water absorption was observed for layers 110 nm thick. On the contrary, the thickness of the deposited coatings had no influence on the initial contact angle of the coated films, although the water contact angle of the cornstarch films increased by more than 100% after plasma deposition. Edible bilayer membranes, composed of glycerol-plasticized cassava starch as a cohesive structural layer and an ethanol-cast shellac layer as a moisture barrier, were investigated for their potential use in food preservation as biopackaging films, membranes, or coatings [77]. The incorporation of PEG 200 (plasticizer) into shellac improved the flexibility and prevented defects in structure and reinforced adhesion between the shellac and PS layer. Chitosan solutions varying from 1 to 4 wt.% were coated onto GPS films [78]. The chitosan coatings contributed to the hydrophobicity and also led to improved tensile stress and modulus and a remarkable decrease in water uptake of starch-based films.

4.6 CONCLUSIONS AND PROSPECTS

PS can be processed industrially, however, due to its hydrophilic nature, PS alone cannot be pelletized easily after melt extrusion processing, and no PS resins are being produced commercially at the present time. Conversely, starch/biodegradable polyester blends have been commercialized due to the simple fact that blends can be pelletized right off the cooling trough of the extrusion line. Consequently, starch and plasticizers, rather than their pelletized resins, are immediately blended during the production of plastic bags, films, and sheets.

Quality, source, and cost of starch and plasticizer will undoubtedly affect the success of PS. The requirements for practical plasticizers are summarized as follows: (1) Plasticizers should contain hydroxyl and/or amide groups, which easily form hydrogen bonds with starch. (2) The acidity or alkalinity should be weak, so as to not cause serious starch degradation at high temperature during melt processing. (3) The processing temperature of PS is below 150°C because higher temperature causes starch degradation; therefore, the melting point of plasticizers should be below 150°C, while the boiling point and decomposition temperatures of plasticizers should be above 150°C. (4) Low cost and low toxicity of the plasticizer is essential for practical applications of PS. (5) The plasticizers themselves should be biodegradable and not pollute the natural environment after the biodegradation of starch. Procuring crude glycerol economically from the recently established biodiesel industry may improve the bottom line of both industries.

Additionally, chemical modification decreases the hydroxyl groups and the hydrogen interaction in starch granules, so it can replace the use of plasticizers. Nevertheless, chemical modification will add additional cost to the PS. Finally, environmentally friendly and facile processing and manufacturing must be sought and/or developed in order to reduce cost and improve the performance of PS. Public concerns over the environment, the depletion of fossil fuels, and climate change will continually prompt the research and development of PS. Upcoming developments of next-generation PS represents well-justified challenges and opportunities.

REFERENCES

1. Petersen K, Væggemose NP, Bertelsen G, Lawther M, Olsen MB, Nilsson NH, and Mortensen G. 1999. Potential of biobased materials for food packaging. *Trends in Food Science and Technology* 10(2): 52–68.
2. Eisenhaber F and Schulz W. 1992. Monte Carlo simulation of the hydration shell of double-helical amylose: A left-handed antiparallel double helix fits best into liquid water structure. *Biopolymers* 32(12): 1643–1664.
3. Zobel HF. 1988. Molecules to granules: A comprehensive starch review. *Starch/Stärke* 40(2): 44–50.
4. Van Soest JJG, Hulleman SHD, de Wit D, and Vliegenthart JFG. 1996. Crystallinity in starch bioplastics. *Industrial Crops and Products* 5(1): 11–22.
5. Cheetham NWH and Tao L. 1998. Variation in crystalline type with amylose content in maize starch granules: An X-ray powder diffraction study. *Carbohydrate Polymers* 36(4): 277–284.
6. Ma XF, Jian RJ, Chang PR, and Yu JG. 2008. Fabrication and characterization of citric acid-modified starch nanoparticles/plasticized-starch composites. *Biomacromolecules* 9(11): 3314–3320.
7. Curvelo AAS, de Carvalho AJF, and Agnelli JAM. 2001.Thermoplastic starch-cellulosic fibers composites: Preliminary results. *Carbohydrate Polymers* 45(2): 183–188.
8. Wang L, Shogren RL, and Carriere C. 2000. Preparation and properties of thermoplastic starch-polyester laminate sheets by coextrusion. *Polymer Engineering and Science* 40(2): 499–506.
9. Chaudhary D, Dong Y, and Kar KK. 2010. Hydrophilic plasticized biopolymers: Morphological influence on physical properties. *Materials Letters* 64(7): 872–875.
10. Zhang YC and Han JH. 2006. Plasticization of pea starch films with monosaccharides and polyols. *Journal of Food Science* 71(6): E253–E261.
11. Da Róz AL, Carvalho AJF, Gandini A, and Curvelo AAS. 2006. The effect of plasticizers on thermoplastic starch compositions obtained by melt processing. *Carbohydrate Polymers* 63(3): 417–424.
12. Veiga-Santos P, Oliveira LM, Cereda MP, and Scamparini ARP. 2007. Sucrose and inverted sugar as plasticizer. Effect on cassava starch-gelatin film mechanical properties, hydrophilicity and water activity. *Food Chemistry* 103(2): 255–262.
13. Galdeano MC, Mali S, Grossmann MVE, Yamashita F, and García MA. 2009. Effects of plasticizers on the properties of oat starch films. *Materials Science and Engineering C* 29(2): 532–538.
14. Zhang Y and Han JH. 2008. Sorption isotherm and plasticization effect of moisture and plasticizers in pea starch film. *Journal of Food Science* 73(7): E313–E324.

15. Ma XF and Yu JG. 2004. Formamide as the plasticizer for thermoplastic starch. *Journal of Applied Polymer Science* 93(4): 1769–1773.
16. Ma XF and Yu JG. 2004. The plasticizers containing amide groups for thermoplastic starch. *Carbohydrate Polymers* 57(2): 197–203.
17. Ma XF and Yu JG. 2004. The effects of plasticizers containing amide groups on the properties of thermoplastic starch. *Starch-Starke* 56(11): 545–551.
18. Yang JH, Yu JG, and Ma XF. 2006. Preparation of a novel thermoplastic starch (TPS) material using ethylenebisformamide as the plasticizer. *Starch-Starke* 58(7): 330–337.
19. Yang JH, Yu JG, Feng Y, and Ma XF. 2007. Study on the properties of ethylenebisformamide plasticized corn starch (EPTPS) with various original water contents of corn starch. *Carbohydrate Polymers* 69(2): 256–261.
20. Huang MF, Yu JG, and Ma XF. 2005. Ethanolamine as a novel plasticiser for thermoplastic starch. *Polymer Degradation and Stability* 90(3): 501–507.
21. Van Soest JJG and Kortleve PM. 1999. The influence of maltodextrins on the structure and properties of compression-molded starch plastic sheets. *Journal of Applied Polymer Science* 74(9): 2207–2219.
22. Kuutti L, Peltonen J, Myllärinen P, Teleman O, and Forssell P. 1998. AFM in studies of thermoplastic starches during ageing. *Carbohydrate Polymers* 37(1): 7–12.
23. Shi R, Liu QY, Ding T, Han YM, Zhang LQ, Chen DF, and Tian W. 2007. Ageing of soft thermoplastic starch with high glycerol content. *Journal of Applied Polymer Science* 103: 574–586.
24. Stein TM and Greene RV. 1997. Amino acids as plasticizers for starch-based plastics. *Starch-Starke* 49(6): 245–249.
25. Walenta E, Fink HP, Weigel P, Ganster J, and Schaaf E. 2001. Structure-property relationships of extruded starch, 2 Extrusion products from native starch. *Macromolecular Materials and Engineering* 286(8): 462–471.
26. Ma XF and Yu JG. 2004. Hydrogen bond of thermoplastic starch and effects on its properties. *Acta Chimica Sinica* 62(12): 1180–1184 (in Chinese).
27. Pawlak A and Mucha M. 2003. Thermogravimetric and FTIR studies of chitosan blends. *Thermochimica Acta* 396(1–2): 153–166.
28. Ma XF, Yu JG, and Feng J. 2004. Urea and formamide as a mixed plasticizer for thermoplastic starch. *Polymer International* 53(11): 1780–1785.
29. van Soest JJG and Vliegenthart JFG. 1997. Crystallinity in starch plastics: Consequences for material properties. *Trends in Biotechnology* 15(6): 208–213.
30. Ma XF, Yu JG, and Wan JJ. 2006. Urea and ethanolamine as a mixed plasticizer for thermoplastic starch. *Carbohydrate Polymers* 64(2): 267–273.
31. Yang JH, Yu JG, and Ma XF. 2006. Preparation and properties of ethylenebisformamide plasticized potato starch (EPTPS). *Carbohydrate Polymers* 63(2): 218–223.
32. Yang JH, Yu JG, and Ma XF. 2006. Study on the properties of ethylenebisformamide and sorbitol plasticized corn starch (ESPTPS). *Carbohydrate Polymers* 66(1): 110–116.
33. Yang JH, Yu JG, and Ma XF. 2006. Retrogradation of ethylenebisformamide and sorbitol plasticized corn starch (ESPTPS). *Starch-Starke* 58(11): 580–586.
34. Zhang JS, Chang PR, Wu Y, Yu JG, and Ma XF. 2008. Aliphatic amidediol and glycerol as a mixed plasticizer for the preparation of thermoplastic starch. *Starch-Starke* 60(11): 617–623.
35. Zheng PW, Chang PR, Yu JG, and Ma XF. 2009. Formamide and 2-hydroxy-*N*-[2-(2-hydroxy-propionylamino)-ethyl] propionamide (HPEP) as a mixed plasticizer for thermoplastic starch. *Carbohydrate Polymers* 78(2): 296–301.
36. Coupland K and Smith PJ. 1986. *N*-Acylalkanolamine humectants—An alternative approach to moisturisation. *Specialty Chemicals* 6: 10–17.
37. Dai HG, Chang PR, Geng FY, Yu JG, and Ma XF. 2010. Preparation and properties of starch-based film using *N,N*-bis(2-hydroxyethyl)formamide as a new plasticizer. *Carbohydrate Polymers* 79(2): 306–311.
38. Dai HG, Chang PR, Peng F, Yu JG, and Ma XF. 2009. *N*-(2-Hydroxyethyl)formamide as a new plasticizer for thermoplastic starch. *Journal of Polymer Research* 16(5): 529–535.
39. Dai HG, Chang PR, Yu JG, Geng FY, and Ma XF. 2010. *N*-(2-Hydroxypropyl)formamide and *N*-(2-hydroxyethyl)- *N*-methylformamide as two new plasticizers for thermoplastic starch. *Carbohydrate Polymers* 80(1): 139–144.

40. Wang N, Zhang XX, Liu HH, and He BQ. 2009. 1-Allyl-3-methylimidazolium chloride plasticized-corn starch as solid biopolymer electrolytes. *Carbohydrate Polymers* 76(3): 482–484.
41. Wang N, Zhang XX, Liu HH, and Han N. 2010. Ionically conducting polymers based on ionic liquid-plasticized starch containing lithium chloride. *Polymers and Polymer Composites* 18(1): 53–58.
42. Wang N, Zhang XX, Liu HH, and Wang JP. 2009. *N,N*-dimethylacetamide/lithium chloride plasticized starch as solid biopolymer electrolytes. *Carbohydrate Polymers* 77(3): 607–611.
43. Sankri A, Arhaliass A, Dez I, Gaumont AC, Grohens Y, Lourdin D, Pillin I, Rolland-Sabaté A, and Leroy E. 2010. Thermoplastic starch plasticized by an ionic liquid. *Carbohydrate Polymers* 82(2): 256–263.
44. van Soest JJG, Benes GK, de Wit D, and Vliegenthart JFG. 1996. The influence of starch molecular mass on the properties of extruded thermoplastic starch. *Polymer* 37(16): 3543–3522.
45. Mondragon M, Mancilla JE, and Rodriguez-Gonzalez FJ. 2008. Nanocomposites from plasticized high-amylopectin, normal and high-amylose maize starches. *Polymer Engineering and Science* 48(7): 1261–1267.
46. Avérous L, Moro L, Dole P, and Fringant C. 2000. Properties of thermoplastic blends: Starch-polycaprolactone. *Polymer* 41(11): 4157–4167.
47. Forssell PM, Mikkilä JM, Moates GK, and Parker R. 1997. Phase and glass transition behaviour of concentrated barley starch-glycerol-water mixtures, a model for thermoplastic starch. *Carbohydrate Polymers* 34(4): 275–282.
48. Forssell PM, Hulleman SHD, Myllärinen PJ, Moates GK, and Parker R. 1999. Ageing of rubbery thermoplastic barley and oat starches. *Carbohydrate Polymers* 39(1): 43–51.
49. Dias AB, Müller CMO, Larotonda FDS, and Laurindo JB. 2010. Biodegradable films based on rice starch and rice flour. *Journal of Cereal Science* 51(2): 213–219.
50. Ishiaku US, Pang KW, Lee WS, and Ishak ZAM. 2002. Mechanical properties and enzymic degradation of thermoplastic and granular sago starch filled poly(epsilon-caprolactone). *European Polymer Journal* 38(2): 393–401.
51. Myllärinen P, Partanen R, Seppälä J, and Forssell P. 2002. Effect of glycerol on behaviour of amylose and amylopectin films. *Carbohydrate Polymers* 50(4): 355–361.
52. Smits ALM, Kruiskamp PH, van Soest JJG, and Vliegenthart JFG. 2003. Interaction between dry starch and plasticisers glycerol or ethylene glycol, measured by differential scanning calorimetry and solid state NMR spectroscopy. *Carbohydrate Polymers* 53(4): 409–416.
53. Mali S, Grossmann MVE, García MA, Martino MN, and Zaritzky NE. 2006. Effects of controlled storage on thermal, mechanical and barrier properties of plasticized films from different starch sources. *Journal of Food Engineering* 75(4): 453–460.
54. Ma XF, Chang PR, Yu JG, and Stumborg M. 2009. Properties of biodegradable citric acid-modified granular starch/thermoplastic pea starch composites. *Carbohydrate Polymers* 75(1): 1–8.
55. Liu P, Yu L, Wang XY, Li D, Chen L, and Li XX. 2010. Glass transition temperature of starches with different amylose/amylopectin ratios. *Journal of Cereal Science* 51(3): 388–391.
56. de Graaf RA, Karman AP, and Janssen LPBM. 2003. Material properties and glass transition temperatures of different thermoplastic starches after extrusion processing. *Starch-Starke* 55(2): 80–86.
57. Yu L and Christie G. 2005. Microstructure and mechanical properties of orientated thermoplastic starches. *Journal of Materials Science* 40: 111–116.
58. Xie F, Yu L, Su B, Liu P, Wang J, Liu HS, and Chen L. 2009. Rheological properties of starch with different amylose/amylopectin ratios. *Journal of Cereal Science* 49(3): 371–377.
59. Volkert B, Lehmann A, Greco T, and Nejad MH. 2010. A comparison of different synthesis routes for starch acetates and the resulting mechanical properties. *Carbohydrate Polymers* 79(3): 571–577.
60. López OV, García MA, and Zaritzky NE. 2008. Film forming capacity of chemically modified corn starches. *Carbohydrate Polymers* 73(4): 573–581.
61. Shogren RL. 1996. Preparation, thermal properties, and extrusion of high-amylose starch acetates. *Carbohydrate Polymers* 29(1): 57–62.
62. Miladinov VD and Hanna MA. 1999. Physical and molecular properties of starch acetates extruded with water and ethanol. *Industrial & Engineering Chemistry Research* 38(10): 3892–3897.
63. Xu YX and Hanna MA. 2005. Physical, mechanical, and morphological characteristics of extruded starch acetate foams. *Journal of Polymers and the Environment* 13(3): 221–230.
64. Zhang YR, Zhang SD, Wang XL, Chen RY, and Wang YZ. 2009. Effect of carbonyl content on the properties of thermoplastic oxidized starch. *Carbohydrate Polymers* 78(1): 157–161.

65. Yu JG, Chang PR, and Ma XF. 2010. The preparation and properties of dialdehyde starch and thermo-plastic dialdehyde starch. *Carbohydrate Polymers* 79(2): 296–300.
66. Du YL, Cao Y, Lu F, Li F, Cao Y, Wang XL, and Wang YZ. 2008. Biodegradation behaviors of thermo-plastic starch (TPS) and thermoplastic dialdehyde starch (TPDAS) under controlled composting condi-tions. *Polymer Testing* 27(8): 924–930.
67. Zhang SD, Wang XL, Zhang YR, Yang KK, and Wang YZ. 2010. Preparation of a new dialdehyde starch derivative and investigation of its thermoplastic properties. *Journal of Polymer Research* 17: 439–446.
68. Zhang SD, Zhang YR, Zhu J, Wang XL, Yang KK, and Wang YZ. 2007. Modified corn starches with improved comprehensive properties for preparing thermoplastics. *Starch-Starke* 59(6): 258–268.
69. Raphaelides SN, Dimitreli G, Exarhopoulos S, Kokonidis G, and Tzani E. 2010. Effect of processing his-tory on the physicochemical and structural characteristics of starch-fatty acid extrudates plasticized with glycerol. *Carbohydrate Polymers* 83:727–736. DOI:10.1016/j.carbpol.2010.08.041 (available online).
70. Raphaelides SN, Arsenoudi K, Exarhopoulos S, and Xu ZM. 2010. Effect of processing history on the functional and structural characteristics of starch-fatty acid extrudates. *Food Research International* 43(1): 329–341.
71. Aburto J, Alric I, Thiebaud S, Borredon E, Bikiaris D, Prinos J, and Panayiotou C. 1999. Synthesis, char-acterization, and biodegradability of fatty-acid esters of amylose and starch. *Journal of Applied Polymer Science* 74(6): 1440–1451.
72. Fang JM, Fowler PA, Tomkinson J, and Hill CAS. 2002. An investigation of the use of recovered veg-etable oil for the preparation of starch thermoplastics. *Carbohydrate Polymers* 50(4): 429–434.
73. Zhou J, Ren LL, Tong J, and Ma YH. 2009. Effect of surface esterification with octenyl succinic anhy-dride on hydrophilicity of corn starch films. *Journal of Applied Polymer Science* 114: 940–947.
74. Zhou J, Zhang J, Ma YH, and Tong J. 2008. Surface photo-crosslinking of corn starch sheets. *Carbohydrate Polymers* 74(3): 405–410.
75. Zhou J, Ma YH, Zhang J, and Tong J. 2009. Influence of surface photocrosslinking on properties of ther-moplastic starch sheets. *Journal of Applied Polymer Science* 112(1): 99–106.
76. Simao RA, Thiré RMSM, Coutinho PR, de Araújo PJG, Achete CA, and Andrade CT. 2006. Application of glow discharge butadiene coatings on plasticized cornstarch substrates. *Thin Solid Films* 515(4): 1714–1720.
77. Phan TD, Debeaufort F, Luu D, and Voilley A. 2005. Functional properties of edible agar-based and starch-based films for food quality preservation. *Journal of Agricultural and Food Chemistry* 53(4): 973–981.
78. Bangyekan C, Aht-Ong D, and Srikulkit K. 2006. Preparation and properties evaluation of chitosan-coated cassava starch films. *Carbohydrate Polymers* 63(1): 61–71.

CHAPTER 5

Rheological and Thermal Properties of Starch and Starch-Based Biopolymers

M.A. Rao, Jirarat Tattiyakul, and Hung-Ju Liao

CONTENTS

5.1 INTRODUCTION

In addition to being a major source of energy in many higher plants, starch is very diverse in its functionality. Its native and modified forms have played an important part in food and nonfood industries for decades. The versatility of starch has been demonstrated in a number of applications including viscosity and texture modification, fat replacement, adhesives, food preservation, chemical synthesis, gum replacement, binders, sizing for textiles and fiberglass, bulking agents, paper coating, dextrin and sweetener production, and many more (Balagopalan et al. 1988; Brown et al. 1995; Singh et al. 2003, 2007).

The major sources of starch include cereal grains (maize, rice, wheat, barley, oat, and sorghum), root crops (sweet potatoes, cassava, arrowroots, and yam), tubers (potatoes), stems (sago palm), and legume seeds (peas and beans). Starches from corn, potato, rice, and wheat are used widely in foods around the world. However, in specific regions of the world, starches from other sources, such as tapioca (cassava) and cowpea, are used extensively. Much has been written about starch structure, including good reviews, such as that of Buléon et al. (1998). The two principal components of starch are amylose and amylopectin, both of which are polymers of α-D-glucose units. In amylose, they are linked α-$(1 \rightarrow 4)$-, with the ring oxygen atoms all on the same side, while in amylopectin about 1 residue in every 20 is also linked α-$(1 \rightarrow 6)$- forming branch points.

5.1.1 Objectives

The main objective of this chapter is to review the principal physicochemical, especially the thermal and rheological, characteristics of starches and the products derived from them. Specifically, the thermal characteristics deal with the gelatinization and glass transition properties of starch polymers. With respect to rheological characteristics, we consider two classes of starch polymer products: (1) the intrinsic viscosity–molecular weight (MW) relationships when the starch product can be dissolved in a solvent, especially water, and (2) the important role of the starch granule which is discussed in granular starch dispersions (SDs), irrespective of whether it is a native or modified starch.

5.2 NATIVE AND MODIFIED STARCHES

It is noted that starch exists in the form of granules whose average size varies widely, 20–100 μm, depending on the source of starch. Singh et al. (2003) summarized the sizes of the granules from different sources. The average granule size of potato starch granules ranges from 1 to 20 mm for small and 20 to 110 mm for large, while for cornstarch granules, the ranges are from 1 to 7 mm for small and from 15 to 20 mm for large. Rice starch granules are relatively smaller, ranging from 3 to 5 mm in size. With respect to the shape, potato starch granules are oval and irregular, corn starch granules are angular shaped, and rice starch granules are pentagonal and angular shaped. Wheat starch granules are made up of large granules, disk-like or lenticular in shape, with diameters ranging from 10 to 35 mm, and small starch granules that are roughly spherical or polygonal in shape, 1–10 mm in diameter (Singh et al. 2003).

In addition to the main sources of starch indicated earlier, there are mutant genotypes of some cereals that contain starches with a higher content of either amylose or amylopectin; the latter is called waxy. Native starch granules are insoluble in water, and when heated, they expand up to an extent after which they rupture. Further, starch can be modified physically or chemically to produce a wide range of modified starches (Singh et al. 2007).

The main objective of a modification technique is to induce changes in the characteristics of the starch granule in order to create a starch polymer with desirable functional characteristics. These modifications, described in detail by Singh et al. (2007), are summarized in Table 5.1. Among the products of modified starches, cross-linked waxy starches resist rupturing when heated to high temperatures of the order of 121°C, and maltodextrins and cyclodextrins (CDs) are soluble in water. With respect to native and modified starches that are granular in nature, the granules play an important role in the rheological properties of their dispersions. However, when a starch polymer can be dissolved in a solvent, much information on their molecular characteristics can be obtained from their solution viscosities.

5.3 PHASE TRANSITIONS IN STARCH

The thermodynamic definition of a phase transition is based on changes occurring in Gibbs free energy, G, and chemical potential, μ, at the transition temperature (Rao 2003). A first-order transition is defined as one in which the first derivatives of G and μ with respect to temperature exhibit

Table 5.1   Types of Starch Modification and Their Goals

Modification Type	Manufacturing Techniques and Goals
Heat/moisture treatment (physical)	Heat–moisture treatment: heat starch at a temperature above its gelatinization point with insufficient moisture to prevent complete gelatinization
	Annealing: heat a slurry of granular starch at a temperature below its gelatinization point for long periods of time
Pregelatinization (physical)	Pregel/instant/cold-water swelling starches prepared using drum drying/spray cooking/extrusion/solvent-based processing; goal: cold-water dispersibility
Partial acid hydrolysis (conversion)	Treatment with hydrochloric acid or orthophosphoric acid or sulfuric acid; goal: reduced MW polymers
Partial enzymatic hydrolysis (conversion)	Treatment in an aqueous solution at a temperature below the gelatinization point with one or more food-grade amylolytic enzymes; goal: reduced MW polymers
Alkali treatment (conversion)	Treatment with sodium hydroxide or potassium hydroxide
Oxidation/bleaching (conversion)	Treatment with peracetic acid and/or hydrogen peroxide, or sodium hypochlorite or sodium chlorite, or sculpture dioxide, or potassium permanganate or ammonium persulfate; goal: low-viscosity, high-clarity, and low-temperature stability
Pyroconversion (conversion) sugar (dextrinization)	Pyrodextrins: produced by dry roasting acidified starch; goal: low-viscosity, high-reducing content polymers
Etherification (derivatization)	Hydroxypropyl starch: esterification with propylene oxide; goal: starch pastes with improved clarity and greater viscosity
Esterification (derivatization)	Starch acetate: esterification with acetic anhydride or vinyl acetate
	Acetylated distarch adipate: esterification with acetic anhydride and adipic anhydride
	Starch sodium octenylsuccinate: esterification by octenylsuccinic anhydride; goal: lower gelatinization temperature
Cross-linking (derivatization)	Monostarch phosphate: esterification with orthophosphoric acid, or sodium or potassium orthophosphate, or sodium tripolyphosphate
	Distarch phosphate: esterification with sodium trimetaphosphate or phosphorus oxychloride
	Phosphated distarch phosphate: combination of treatments for monostarch phosphate and distarch phosphate; goal: increased stability of granules
Dual modification (derivatization)	Acetylated distarch phosphate: esterification by sodium trimetaphosphate or phosphorus oxychloride combined with esterification by acetic anhydride or vinyl acetate
	Hydroxypropyl distarch phosphate: esterification by sodium trimetaphosphate or phosphorus oxychloride combined with etherification by propylene oxide; goal: stability against acid, thermal, and mechanical degradation

Source: Adapted from Singh, J. et al., *Food Hydrocolloids*, 21(1), 1, 2007.

discontinuities at the transition temperature. Concomitantly, a step change occurs in enthalpy, entropy, and volume at the transition temperature.

Important first-order transitions in foods include crystallization, melting, protein denaturation, and starch gelatinization. In pure materials, first-order transitions occur at well-defined and material-specific temperatures. However, invariably in a food, many compounds are present, so that a transition may occur over a range of temperatures instead of a fixed temperature.

A second-order transition is defined as one in which the second derivatives of G and μ with respect to temperature exhibit discontinuities at the transition temperature. Although glass transition of amorphous foods has the properties of a second-order transition, there are no well-defined second-order transitions in foods (Rao 2003).

Phase transition temperatures. In order to study and understand a phase transition, it is important to know magnitudes of the temperatures over which it takes place. The differential scanning calorimeter (DSC) is used extensively to determine first-order and glass transition temperatures,

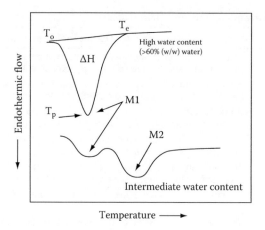

Figure 5.1 DSC output for gelatinization of starch in excess water and an idealized output curve of starch with intermediate water content are observed. Often, the temperature at the end of gelatinization is called starch melting temperature.

more so than other techniques. A DSC measures the rate of heat flow into or out of a cell containing a sample in comparison to a reference cell in which no thermal events occur. It maintains a programmed sample cell temperature by adjusting heat flow rates. It should be noted that data obtained with a DSC depend to some extent on the heating/cooling rate that should be specified when discussing the data; common heating rates are 5°C/min and 10°C/min. The heat flow versus temperature diagrams are known as thermograms.

The DSC output for gelatinization of starch in excess water and an idealized output curve of starch with intermediate water content are shown in Figure 5.1 (Hoseney 1998). As the amount of water is reduced, the DSC peak for starch gelatinization widens. For example, for wheat starch, the range of temperatures for gelatinization is about 7°C in excess water, and at low water content (water:starch, 0.35:1), it is greater than 30°C, and the endotherm shows two peaks (Hoseney 1998), M1 and M2, respectively. In addition, the starch does not gelatinize completely even at 100°C. For the excess moisture curve, drawing a tangent at the start and end points of transition, one can determine well-defined gelatinization initiation, T_o; peak, T_p; and end, T_e, temperatures, respectively. Further, the area under the curve is a measure of the energy required for the transition, ΔH. For gelatinization at intermediate water condition, two endothermic peaks are observed.

Data related to the gelatinization of starches from different botanical sources from Singh et al. (2003) are given in Table 5.2.

5.4 RHEOLOGICAL TECHNIQUES

5.4.1 Intrinsic Viscosity

In dilute solutions, polymer chains are separate, and the intrinsic viscosity, denoted as $[\eta]$, of a polymer in solution depends only on the dimensions of the polymer chain (Rao 2007).

$$[\eta] = \lim_{c \to 0} \left(\frac{\eta_{sp}}{c} \right) \tag{5.1}$$

where

$\eta_{sp} = [(\eta - \eta_s)/\eta_s]$

η and η_s are the viscosities of the solution and the solvent, respectively

Table 5.2   Gelatinization Parameters of Some Starches Using DSC

Source	Starch:Water Ratio	T_o (°C)	T_p (°C)	T_e (°C)	ΔH (J/g)
Potato	1:2:3	59.72–66.2	62.9–69.6	67.28–75.4	12.55–17.9
Potato	1:3:3	57.0–68.3	60.6–72.4	66.5–78.0	13.0–15.8
Normal corn	1:1.5	62.3	67.7	84.3	14.0
Normal corn	1:3	64.1	69.4	74.9	12.3
Waxy corn	1:9	66.6	73.6	—	14.2
Waxy corn	1:3	64.2	69.2	74.6	15.4
High-amylose corn	1:9	66.8	73.7	—	13.7
Rice	1:1.5	62.0	67.4	97.5	11.0
Rice	1:9	57.7	65.1	—	11.5
Waxy rice	Not known	66.1–74.9	70.4–78.8	—	7.7–12.1
Wheat	1:1.5	51.2	56.0	76.0	9.0
Wheat	1:2:3	46.0–52.4	52.2–57.6	57.8–66.1	14.8–17.9

Source: Adapted from Singh, J. et al., *Food Chem.*, 81(2), 219, 2003.
Note: T_o, T_p, and, T_e are gelatinization initiation, peak, and end temperatures, respectively; ΔH is enthalpy of gelatinization.

When a dilute solution exhibits shear-thinning behavior, its zero shear viscosity, η_0, at very low shear rates, sometimes referred to as "vanishing shear rates," may be used in place of the Newtonian viscosity, η. The viscosities of low-viscosity Newtonian fluids can be determined using glass capillary viscometers. Further, for determination of intrinsic viscosity, viscosity data can be obtained at several polymer concentrations using a dilution, also known as Ubbelohde, viscometer.

There are several ways of determining the intrinsic viscosity $[\eta]$ from experimental dilute solution viscosity data. The two equations commonly employed for determining $[\eta]$ of food polymers are those of Huggins (Equation 5.2) and Kraemer (Equation 5.3):

$$\frac{\eta_{sp}}{c} = [\eta] + k_1 [\eta]^2 c \tag{5.2}$$

$$\ln \frac{(\eta_r)}{c} = [\eta] + k_2 [\eta]^2 c \tag{5.3}$$

For polyelectrolytes (charged polymers), a plot of η_{sp}/c versus c may be a curve. An alternate expression of Fuoss and Strauss (1948) can be used:

$$\frac{\eta_{sp}}{c} = \frac{[\eta]}{\left(1 + Bc^{1/2}\right)} \tag{5.4}$$

When (c/η_{sp}) is plotted against $c^{1/2}$, a straight line is obtained with an intercept of $1/[\eta]$ and a slope of $B/[\eta]$.

The empirical Mark–Houwink equation is used to relate the intrinsic viscosity and the MW of a polymer:

$$[\eta] = K (M)^a \tag{5.5}$$

where K and a are constants at a specific temperature for a given polymer-solvent system. The Mark–Houwink exponent a is found to vary from 0.8 for polymers with coil-like behavior in a good solvent to 0.5 in a theta solvent, a solvent in which polymer coils act like ideal chains. Values higher than unity are characteristic of rodlike polymers.

5.4.2 Intrinsic Viscosity and Mark–Houwink Constants of Starch Polymers

For waxy maize starches in 90% dimethyl sulfoxide (DMSO) and H_2O, Millard et al. (1997) determined the Mark–Houwink constants to be K = 0.59 and a = 0.29 ± 0.04.

The low exponent value (0.29) is indicative of a compact hydrodynamic shape in solution. Intrinsic viscosity measurements on wheat amylopectin that had been cooked by extrusion and drum dried (Colonna et al. 1984) displayed a similar Mark–Houwink relationship to Millard et al. (1997): K = 0.414 and a = 0.306.

Values of intrinsic viscosity and MW of amylose and amylopectin from sago starch in 1 M KOH (Ahmad et al. 1999), and high-amylopectin starch fractions in 90% DMSO/10% water (Chamberlain and Rao 2000) are given in Tables 5.3 and 5.4, respectively. Intrinsic viscosity for amylose ranged from 310 to 460 mL g^{-1} in 1 M KOH, while for amylopectin under similar conditions, the intrinsic viscosity was 210–250 mL g^{-1}. The intrinsic viscosities for amylose obtained from sago starch were high compared to that from cornstarch but quite similar to potato, 250–570 mL g^{-1}, and cassava, 367 mL g^{-1}. The intrinsic viscosity values for amylopectin from sago starch were much lower than amylose but agree well with the values for amylopectin from other sources (Ahmad et al. 1999). The starch polymers that were acid hydrolyzed and then neutralized

Table 5.3 Intrinsic Viscosity and MW of Sago Starch Amylose and Amylopectin Dissolved in 1 M KOH Solution at 25°C

Sample	Intrinsic Viscosity, mL g^{-1}	MW × 10^{-6}, Da (Based on Mark–Houwink Equation)	MW × 10^{-6}, Da (Based on Light Scattering)
Amylose 1	350	1.39	1.41
Amylose 2	310	1.23	1.24
Amylose 5	460	1.91	2.07
Amylopectin 1	210	nd[a]	6.92
Amylopectin 2	220	nd	8.64
Amylopectin 5	210	nd	9.23

Source: From Ahmad, F.B. et al., Carbohyd. Polym., 38, 361, 1999.
[a] nd stands for not determined.

Table 5.4 Intrinsic Viscosity (mL g^{-1}) of High-Amylopectin Starch in 90% DMSO and 10% Water, and Estimated MW

Sample	Intrinsic Viscosity	MW[a]
Amioca	140	155
25 min conversion	100	48.6
45 min conversion	85	27.8
90 min conversion	83	25.6

Source: From Chamberlain, E.K. and Rao, M.A., Food Hydrocolloids, 14(2), 163, 2000.
[a] MW based on the Mark–Houwink constants of Millard et al. (1997).

behaved like polyelectrolytes (Chamberlain and Rao 2000). This might have been due to the positive and negative charges imparted to the starch by the acid and base used for the hydrolysis and subsequent neutralization. Polyelectrolytes show strong interactions in solutions due to intramolecular and intermolecular forces. The Fuoss–Strauss relationship was used to estimate the intrinsic viscosities.

5.4.3 Steady-Shear Rheological Measurements

Measurement of rheological properties of foods has been well discussed by Rao (2007) and Steffe (1996) and will be discussed only in brief here. Steady-shear rheological measurements on fluid foods can be conducted using several well-defined geometries: concentric cylinder, cone–plate, parallel plate, and capillary/tube. In addition, for highly viscous semisolid materials, like dough, lubricated biaxial deformation is used.

5.4.4 Viscoelastic Parameters of Starch Dispersions and Gels

Small-amplitude oscillatory shear (SAOS), also called dynamic rheological experiment, can be used to determine viscoelastic properties of foods without altering a material's structure (Rao 2007). In a SAOS experiment, a sinusoidal oscillating stress or strain with a frequency ω is applied to the material, and the phase difference between the oscillating stress and strain, as well as the amplitude ratio, is measured. For shear deformation within the linear viscoelastic range, the generated stress (σ_0) is expressed in terms of the storage modulus G' (Pa) and a loss modulus G'' (Pa); G' is a measure of the magnitude of the energy that is stored in the material, and G'' is a measure of the energy which is lost as viscous dissipation per cycle of deformation, respectively. For a viscoelastic material, the resultant stress is also sinusoidal but shows a phase lag of δ radians when compared with the strain. The phase angle δ covers the range of $0-\pi/2$ as the viscous component increases. The dependence of the storage modulus, G', and the loss modulus, G'', on the oscillatory frequency (ω) in the linear viscoelastic region is one set of useful information. If $G'' > G'$, the material is behaving predominantly as viscous liquid; if $G' > G''$, the material is behaving predominantly as a solid (Tattiyakul and Rao 2009).

5.5 STRUCTURAL AND RHEOLOGICAL CHARACTERISTICS OF STARCH DISPERSIONS

In excess water content, starch granules start to swell and imbibe water when energy needed to break some of the intermolecular hydrogen bonds in amorphous regions is applied. After a critical temperature is reached, the water molecules disrupt hydrogen bonding and penetrate the granules, and the starch granules swell. The swelling of starch granules results in increase in the granule volume fraction and, in turn, causes an increase in SD viscosity. Several SDs were prepared by heating them at near isothermal conditions for various lengths of time. For example, in corn (Okechukwu and Rao 1995), cowpea (Okechukwu and Rao 1996), cross-linked waxy maize (Chamberlain 1996), and tapioca (Rao and Tattiyakul 1999) SDs, increase in the average starch granule size resulted in increase in the notional volume fraction, cQ, where c is the concentration of starch (g dry starch/g dispersion) and Q is the swelling power of the starch granules (g swollen starch/g dry starch) (Christianson and Bagley 1983), and the power law (Equation 5.6) consistency index of the SDs increased.

$$\sigma = K\dot{\gamma}^n \tag{5.6}$$

where

σ is the shear stress

$\dot{\gamma}$ is the shear rate

n is the flow behavior index (dimensionless)

K is the consistency index (Pa s^n)

Christianson and Bagley (1984) and Okechukwu and Rao (1996) found that the yield stress of gelatinized SDs determined using the Casson model (Equation 5.7) increased with cQ:

$$(\sigma)^{0.5} = (\sigma_{0c})^{0.5} + (\eta_\infty \dot{\gamma})^{0.5} \tag{5.7}$$

Heating of SDs also results in loss of radial orientation of the micelles and birefringence. Further heating of the starch suspension results in more loosening of the micelle's network, enhanced water absorption, and enlarged granule size. Raising temperature to above the end temperature of gelatinization or applying high-shear agitation results in granule disruption and decrease in dispersion viscosity. Thus, information on changes in granule size during the gelatinization process would lead to an understanding of the rheological changes of SDs during gelatinization.

5.5.1 Thermorheological Models of Starch

Yang and Rao (1998) have proposed a thermorheological (TR) model for a starch-containing fluid food that was based on dynamic rheological data obtained as a function of temperature at several oscillatory frequencies. The model was reasonably close to the situations of thermal processing and useful for studying the rheological properties of fluid foods under the process conditions. They obtained a set of experimental curves at different oscillatory frequencies using temperature sweep tests, which were superposed by translation along the viscosity coordinate with a shift factor $(\omega/\omega_r)^\beta$ resulting in a master curve. Because η^* versus temperature curves had similar profiles, it was possible to reduce them to a single curve in terms of the reduced complex viscosity (η_R^*):

$$\eta_R^* = \eta^* \left(\frac{\omega}{\omega_r} \right)^\beta \tag{5.8}$$

TR data on a 4% waxy rice SD (Liao et al. 1999, 2000) and a 5% cross-linked waxy maize SD (Tattiyakul and Rao 2000; Tattiyakul et al. 2002) are shown in Figures 5.2 and 5.3, respectively; they were used in computer simulation of heat transfer during sterilization (Liao et al. 2000; Tattiyakul et al. 2002). It should be noted that at high temperatures, while the viscosity of native SD reached low values at high values of temperature that of cross-linked waxy maize, SD was still high. Using the same technique as the superposition of SD, the η_R^* versus temperature master curve of 45.5% waxy rice starch dough was derived with the shift factor suitable for the experimental data at all the dynamic frequencies employed (Liao et al. 1999).

5.6 RHEOLOGY OF DOUGH

It has been recognized for about 80 years that wheat flour doughs were complex and its rheological properties are time dependent. Excellent contributions by many cereal scientists have been made over the years. In simple terms, the components of dough may be described as consisting of a viscoelastic matrix of gluten, which is a branched polymer, filled with hydrated starch particles (Ng and McKinley 2008). The starch increases the G' of the sample by forming physical and

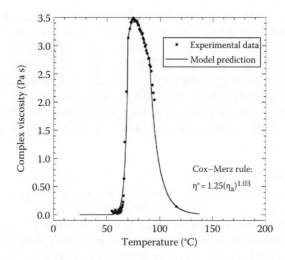

Figure 5.2 TR data on a 4% waxy rice starch dispersion.

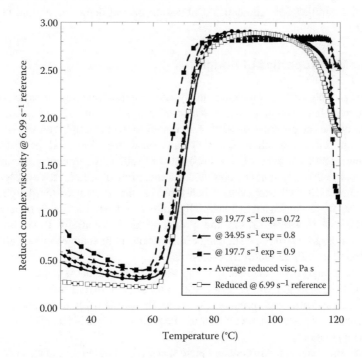

Figure 5.3 TR data on a 5% cross-linked waxy maize starch dispersion. The open squares indicate values predicted by the model and used in computer simulation.

chemical bonds with the polymer. The interactions between the gluten and starch constituents lead to severe rheological nonlinearities and complex responses (Hibberd 1970; Uthayakumaran and Lukow 2003). Even the precise microstructure of the gluten forming the dough matrix itself is still poorly understood due to its very high MW and the high degree of chain branching; the latter two characteristics contribute to poor solubility and the lack of solution properties.

It should be recognized that even though flour–water doughs can be produced with little effort, obtaining repeatable measurements and understanding the material rheology are difficult due to the strong time-dependent and nonlinear softening effects observed in doughs (Bagley et al. 1998; Ng and McKinley 2008). Dough properties depend on many factors, such as time, work input, and rate of work, as well as the extreme sensitivity of dough properties to water level; and biological activity (e.g., activity due to the presence of specific enzymes) complicates the task of obtaining reliable and meaningful data. Comprehensive studies on these systems, which can lead to insight into the role of microstructure and which, in turn, are affected by various added components, are hampered by the difficulty in collecting reproducible data; variations of up to ±50% in modulus have been observed (Ng and McKinley 2008) just by varying the mixing and testing schedule.

Nevertheless, the importance of fundamental investigations into rheological behavior of dough has been emphasized by several authors. However, it is not the intention here to review all the efforts. Instead, the objective here is to review recent efforts that have shed light into the rheological characteristics of doughs.

Bagley et al. (1998) pointed out that in capillary flow of dough, the end corrections are very large, being more than half the length of the viscometer barrel. Thus, the reliability and significance of data obtained when more than half of the dough in the barrel has been extruded were questioned. Because of the complex nature of dough and the large number of factors that influence its rheological behavior, the experimental approaches such as steady-shear measurements over low-shear rates in cone–plate and parallel-plate geometries, lubricated biaxial deformation, and oscillatory shear have been taken.

5.6.1 Rheological Properties of Dough

Berland and Launay (1995) conducted steady and dynamic rheological experiments on wheat dough using a cone–plate geometry over the shear rate range 10^{-4} and 6.10^{-2} s^{-1}. This low-shear rate range is typical of the ranges used in studies on dough to avoid sample roll out of the gap before a steady-state condition was reached. The shear rate–shear stress data and the dynamic shear data followed the power law with respect to the shear rate or oscillatory frequency. An interesting observation is that because data could not be obtained at a shear rate of 1.0 s^{-1}, the consistency coefficient, K′, at a shear rate of 10^{-3} s^{-1} was used. When the moisture content of the dough was increased from 45.2% to 49.4%, K′ decreased exponentially. Likewise, when the dough moisture content was increased from 43.8% to 49.4%, G′, G″, and η* decreased exponentially. It was concluded that in normal dough, water has a strong plasticizing effect, resulting in changes in the values of the rheological parameters but without modifying the structure.

Kouassi-Koffi et al. (2010) conducted biaxial shear experiments on a dough: 41.3 w/w% water, 0.2 g of salt, and 0.162 g of glucose prepared with a mixograph (5 min mixing at 85 rpm). Samples of thicknesses 1.2 ± 0.15 mm were prepared by manually rolling using a Chopin alveograph. Disks were cut with a circular punch (diameter 18 mm). Paraffin oil ($\eta = 0.15$ Pa s) was used as a lubricant in a parallel-plate geometry.

Experiments were conducted in constant plate speed (CPS) mode (i.e., at a continuously increasing extension rate) and constant extension rate (CER) modes using one of the very few commercial rheometers that allow operation in the CER mode.

The consistency index, K, and flow behavior index, n, were calculated for each biaxial strain. The consistency index, K, increased linearly with ε_b under both modes:

$$K = K_0 + B\varepsilon_b \qquad (5.9)$$

with $K_0 = 2.5 \times 10^3$ Pa s^n and $B = 23 \times 10^3$ Pa s^n for the CPS mode or $K_0 = 1.3 \times 10^3$ Pa s^n and $B = 9.3 \times 10^3$ Pa s^n for the CER mode.

The stress versus biaxial strain data could then be described by the equation:

$$\sigma_b = \left(K_0 + B\varepsilon b\right) \times \varepsilon_b^n \tag{5.10}$$

In both modes, the parameter n was approximately constant up to $\varepsilon_b \sim 0.2$. It was systematically lower in the CER mode than under CPS, with a mean value in the latter case of $n = 0.28 \pm 0.03$ being obtained for $\varepsilon_b < 0.2$

Leroy et al. (2010) conducted small-amplitude oscillatory and low-intensity ultrasonic shear wave measurements on dough made from Canadian Western Hard White Spring wheat flour (13.8% protein, 14% moisture) with 2.40 g of salt and 60 mL of distilled water. They reported the power relationships for the moduli, Pa:

$$G' = 10,900 \times \omega^{0.234 \pm 0.004} \tag{5.11}$$

$$G'' = 4300 \times \omega^{0.271 \pm 0.005} \tag{5.12}$$

5.6.2 Rheology of Reconstituted Dough

Another approach to dough rheology involves measurements on dough samples with altered chemical composition. This approach involves separation and fractionation of flour components. Each component or fraction is then evaluated by adding to a base flour or by interchanging between flours to establish the role of each fraction. An essential requirement of this approach is that neither the isolation nor the reconstitution procedures affect the functional properties of the components (Uthayakumaran and Lukow 2003).

The starch in dough, which acts as a filler, increases the G' of the sample by forming physical and chemical bonds with the polymer. A dough made from low protein flour contains a weaker gluten matrix and a higher quantity of starch resulting in a high value of G', but the dough expands less. However, the trend of decrease in storage modulus G' and loss modulus G'' as protein concentration is increased would be reversed if the doughs were to be mixed at a constant water level. At a constant protein content, significant variation in the dough quality properties, despite similar high-MW glutenin subunits composition and Glu-1 score, was observed within the varieties. Some of the differences observed in dough properties between cultivars may be due to differences in low-MW glutenin subunits and gliadin composition (Table 5.5).

Table 5.5   Values of G' and G'' at 1 Hz of Doughs from Three Wheat Cultivars and Reconstituted Samples

Sample	Type	Protein Content (%)	G' (Pa)	G'' (Pa)
Dough A	Control	11.2	7,893	4,330
Dough A	Reconstituted	11.2	10,963	5,534
Dough B	Control	12.8	7,449	3,710
Dough B	Reconstituted	12.8	10,316	5,119
Dough C	Control	15.7	7,263	4,286
Dough C	Reconstituted	15.7	7,508	4,339

Source: Adapted from Uthayakumaran, S. and Lukow, O.M., *J. Sci. Food Agric.*, 83, 889, 2003.

5.7 MALTODEXTRINS

By definition, maltodextrins are hydrolyzed starch products made up and built up by units of α-D-glucose bound together, mainly, by glycosidical (1.4) linkages with a general formula $[(C_6H_{10}O_5)H_2O].[1]$. They consist of a mixture of saccharides, mainly D-glucose, maltose, and a series of oligosaccharides and polysaccharides, such as maltotriose and mixtures of malto-tetraose with a wide distribution of molecular masses. They are produced from native starch through partial hydrolysis, purification, and spray drying. The physical and enzymatic treatments result in the loss of the granular starch structure and make them soluble in water. They are used as texture modifier, gelling agent, fat replacer, volume enhancers, cryoprotectors, and to extend the product shelf life, mainly as encapsulation matrix. Maltodextrins are less expensive than other edible hydrocolloids, and their aqueous solutions have more subtle flavor and mouth feel (Takeiti et al. 2010). Because of their low molecular mass, their aqueous solutions are Newtonian in nature.

Maltodextrins are usually classified by their values of dextrose equivalent (DE), ranging up to 20. DE expresses the number of reducing-end aldehyde groups relative to pure glucose at the same concentration, so that high DE indicates high hydrolytic conversion and lower average molecular mass. DE is a rough estimate of the similarity of maltodextrin to glucose. In addition, the extent of starch degradation in a maltodextrin is indicated by the parameter DE that evaluates the content of reducing-end groups. The chemical structure of maltodextrins, by virtue of the hydrolysis process, is between the complex polysaccharide chains of starch itself and the simpler molecules of sugars. However, the DE of maltodextrins has been shown to be inadequate to pre-dict product performances in various applications (Chronakis 1998). Further, it has been shown that the MW distribution is a more accurate tool to predict maltodextrin fundamental properties (Avaltroni et al. 2004).

Intrinsic viscosity of maltodextrins. Using maltodextrins of known values of DE, Avaltroni et al. (2004) determined values of [η] between 3.5 and 11.0 dL g⁻¹. MW distributions of the malto-dextrin samples dissolved overnight at room temperature. Using the MW and intrinsic viscosity data, the magnitudes of the parameters K and a in the Mark–Houwink equation were determined to be $K = 2.43 \times 10^{-3}$ and $a = 0.337$. They suggested that the structure of maltodextrins in solution is between coil structures in good solvent for samples with the lowest MWs and ovoid-like shape rather than rodlike shape for samples with the highest MWs.

Glass transition temperature of maltodextrins. Using a phase transition analyzer, the glass transition temperatures of the maltodextrin powders were determined. Using the obtained data on glass transition temperatures and the MWs, together with the linear relationship that links the T_G value to the inverse of the MW of a monodisperse polymer,

$$T_G = T_{G\infty} - \frac{A}{M}$$

(5.13)

where
  A is a constant
  M is the MW of the polymer
  $T_{G\infty}$ is the T_G for infinite MW

The values of the parameters $T_{G\infty}$ and the constant A were found to be 520 K and 53,003 K g mol⁻¹, respectively.

5.8 CYCLODEXTRINS

CDs are cyclic oligosaccharides produced using *Bacillus macerans* amylase on starch. They contain 6–12 units of 1,4 linked D-glucose (Paduano et al. 1990). The most common CDs consist of six, seven, or eight glucose monomers and are denoted as α-, β-, γ-CD, respectively (Figure 5.4); their MWs (g mol^{-1}) are 972, 1135, and 1297, respectively (Del Valle Martin 2004). CDs have rigid, truncated-cone molecular structures with a hollow interior composed of two rings of C–H groups with a ring of glycosidic oxygen in between. The internal diameters of these cavities are about 0.57, 0.78, and 0.95 nm, respectively, with a depth of about 0.78 nm. The presence of the cavity enables the CD to trap various low-MW substances, resulting in the formation of inclusion or "host–guest" complexes.

The structure and solubility of CDs and their use in foods have been well reviewed by Astray et al. (2009). They noted that at room temperature, β-CD is at least nine times less soluble (1.85 g/100 mL) than the other CDs: 14.5 g/100 mL and 23.2 g/100 mL for the α- and γ-CDs, respectively. The solubility of CDs in water depends strongly on the temperature. Denoting C as the concentration of a CD and T is the temperature in K, the following equations describe the solubility of CDs (Astray et al. 2009; Simal-Gandara 2011):

$$C_{\alpha\text{-CD}} = (112.71 \pm 0.45)\exp - \left\{(3530 \pm 31)\left[(1/T) - (1/298.1)\right]\right\}$$

$$C_{\beta\text{-CD}} = (14.386 \pm 1.166)\exp - \left\{(4911.79 \pm 177.68)\left[(1/T) - (1/298.1)\right]\right\}$$

$$C_{\gamma\text{-CD}} = (219.4 \pm 9.8)\exp - \left\{(3187 \pm 320)\left[(1/T) - (1/298.1)\right]\right\}$$

Del Valle Martin (2004) listed several useful changes in the chemical and physical properties of the guest molecules: stabilization of light- or oxygen-sensitive substances, modification of the chemical reactivity of guest molecules, fixation of very volatile substances, improvement of solubility of substances, modification of liquid substances to powders, protection against degradation of substances by microorganisms, masking of ill smells and tastes, and masking of pigments or the color of substances.

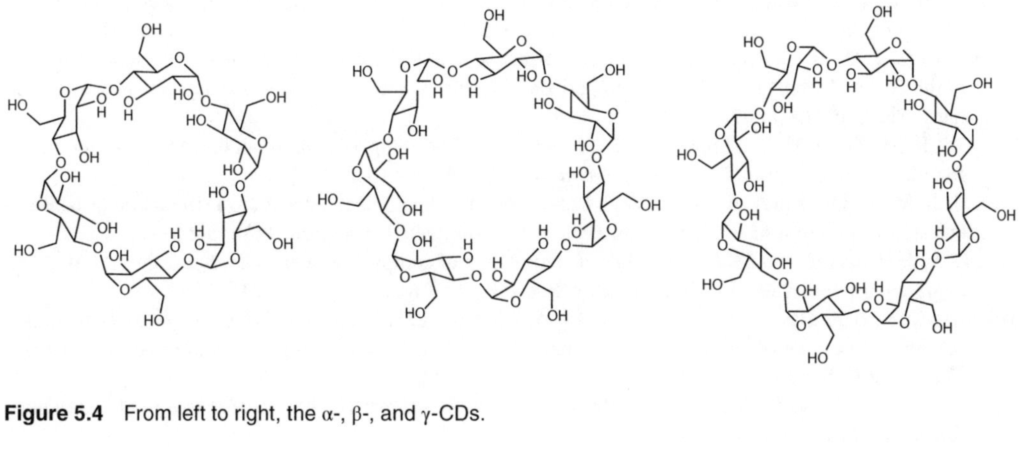

Figure 5.4 From left to right, the α-, β-, and γ-CDs.

REFERENCES

Ahmad, F. B., P. A. Williams, J.-L. Doublier, S. Durand, and A. Buleon. 1999. Physico-chemical characterisation of sago starch. *Carbohydrate Polymers* 38:361–370.

Astray, G., C. Gonzalez-Barreiro, J. C. Mejuto, R. Rial-Otero, and J. Simal-Gándara. 2009. A review on the use of cyclodextrins in foods. *Food Hydrocolloids* 23(7):1631–1640.

Avaltroni, F., P. E. Bouquerand, and V. Normand. 2004. Maltodextrin molecular weight distribution influence on the glass transition temperature and viscosity in aqueous solutions. *Carbohydrate Polymers* 58(3):323–334.

Bagley, E. B., F. R. Dintzis, and S. Chakrabarti. 1998. Experimental and conceptual problems in the rheological characterization of wheat flour doughs. *Rheologica Acta* 37:556–565.

Balagopalan, C., G. Padmaja, S. K. Nanda, and S. N. Moorthy. 1988. *Cassava in Food, Feed, and Industry.* Boca Raton, FL: CRC Press.

Berland, S. and B. Launay. 1995. Rheological properties of wheat flour doughs in steady and dynamic shear: Effect of water content and some additives. *Cereal Chemistry* 72(1):48–52.

Brown, I. L., K. J. McNaught, and E. Moloney. 1995. New direction in starch technology and nutrition. *Food Australia* 47:272–275.

Buléon, A., P. Colonna, V. Planchot, and S. Ball. 1998. Starch granules: Structure and biosynthesis. *International Journal of Biological Macromolecules* 23(2):85–112.

Chamberlain, E. K. 1996. Characterization of heated and thermally processed cross-linked waxy maize starch utilizing particle size analysis, microscopy and rheology. MS thesis, Food Science and Technology, Cornell University, Ithaca, NY.

Chamberlain, E. K. and M. A. Rao. 2000. Effect of concentration on rheological properties of acid-hydrolyzed amylopectin solutions. *Food Hydrocolloids* 14(2):163–171.

Christianson, D. D. and E. B. Bagley. 1983. Apparent viscosities of dispersions of swollen cornstarch granules. *Cereal Chemistry* 60(2):116–121.

Christianson, D. D. and E. B. Bagley. 1984. Yield stresses in dispersions of swollen, deformable cornstarch granules. *Cereal Chemistry* 61(6):500–503.

Chronakis, I. S. 1998. On the molecular characteristics, compositional properties, and structural functional mechanisms of maltodextrins: A review. *Critical Reviews in Food Science* 38(7):599–637.

Colonna, P., J. L. Doublier, J. P. Melcion, F. de Monredon, and C. Mercier. 1984. Extrusion cooking and drum drying of wheat starch. I. Physical and macromolecular modifications. *Cereal Chemistry* 61(6):538–543.

Del Valle Martin, E. M. 2004. Cyclodextrins and their uses: A review. *Process Biochemistry* 39(9):1033–1046.

Fuoss, R. M. and U. P. Strauss. 1948. Polyelectrolytes. II. poly-4-vinylpyridonium chloride and poly-4-vinyl-N-n-butylpyridonium bromide. *Journal of Polymer Science* 3(2):246–263.

Hibberd, G. E. 1970. Dynamic visco-elastic behaviour of wheat flour doughs. III. The influence of the starch granules. *Rheol Acta* 9:501–505.

Hoseney, R. C. 1998. Gelatinization phenomena of starch. In *Phase/State Transitions in Foods: Chemical, Structural, and Rheological Changes*, M. A. Rao and R. W. Hartel (eds.). New York: Marcel Dekker, Inc.

Kouassi-Koffi, J. D., B. Launay, S. Davidou, L. P. Kouamé, and C. Michon. 2010. Lubricated squeezing flow of thin slabs of wheat flour dough: Comparison of results at constant plate speed and constant extension rates. *Rheologica Acta* 49:275–283.

Leroy, V., K. M. Pitura, M. G. Scanlon, and J. H. Page. 2010. The complex shear modulus of dough over a wide frequency range. *Journal of Non-Newtonian Fluid Mechanics* 165:475–478.

Liao, H. J., M. A. Rao, and A. K. Datta. 2000. Role of thermo-rheological behaviour in simulation of continuous sterilization of a starch dispersion. *Food and Bioproducts Processing* 78(1):48–56.

Liao, H. J., J. Tattiyakul, and M. A. Rao. 1999. Superposition of complex viscosity curves during gelatinization of starch dispersion and dough. *Journal of Food Process Engineering* 22(3):215–234.

Millard, M. M., F. R. Dintzis, J. L. Willett, and J. A. Klavons. 1997. Light-scattering molecular weights and intrinsic viscosities of processed waxy maize starches in 90% dimethyl sulfoxide and H_2O. *Cereal Chemistry* 74(5):687–691.

Ng, T. S. K. and G. H. McKinley. 2008. Power law gels at finite strains: The nonlinear rheology of gluten gels. *Journal of Rheology* 52:417–449.

Okechukwu, P. E. and M. A. Rao. 1995. Influence of granule size on viscosity of cornstarch suspension. *Journal of Texture Studies* 26(5):501–516.

Okechukwu, P. E. and M. A. Rao. 1996. Role of granule size and size distribution in the viscosity of cowpea starch dispersions heated in excess water. *Journal of Texture Studies* 27(2):159–173.

Paduano, L., R. Sartorio, V. Vitagliano, and L. Costantino. 1990. Diffusion properties of cyclodextrins in aqueous solution at 25°C. *Journal of Solution Chemistry* 19(1):31–39.

Rao, M. A. 2003. Phase transitions, food texture and structure. In *Texture in Food: Semi-Solid Foods*, B. M. McKenna, (ed.). Cambridge, U.K.: Woodhead Publishing Ltd.

Rao, M. A. 2007. *Rheology of Fluid and Semisolid Foods: Principles and Applications*. 2nd edn. New York: Springer.

Rao, M. A. and J. Tattiyakul. 1999. Granule size and rheological behavior of heated tapioca starch dispersions. *Carbohydrate Polymers* 38:123–132.

Simal-Gandara, J. 2011. Personal communication.

Singh, J., L. Kaur, and O. J. McCarthy. 2007. Factors influencing the physico-chemical, morphological, thermal and rheological properties of some chemically modified starches for food applications—A review. *Food Hydrocolloids* 21(1):1–22.

Singh, N., J. Singh, L. Kaur, N. S. Sodhi, and B. S. Gill. 2003. Morphological, thermal and rheological properties of starches from different botanical sources. *Food Chemistry* 81(2):219–231.

Steffe, J. F. 1996. *Rheological Methods in Food Process Engineering*. East Lansing, MI: Freeman Press.

Takeiti, C. Y., T. G. Kieckbusch, and F. P. Collares-Queiroz. 2010. Morphological and physicochemical characterization of commercial maltodextrins with different degrees of dextrose-equivalent. *International Journal of Food Properties* 13(2):411–425.

Tattiyakul, J. and M. A. Rao. 2000. Rheological behavior of cross-linked waxy maize starch dispersions during and after heating. *Carbohydrate Polymers* 43(3):215–222.

Tattiyakul, J. and M. A. Rao. 2009. Effect of high pressure and ultrasonic processing of foods on rheological properties. In *Novel Food Processing—Effects on Rheological and Functional Properties*, J. Ahmed, H. S. Ramaswamy, S. Kasapis, and J. Boye, (eds.). Boca Raton, FL/New York: CRC Press, Inc.

Tattiyakul, J., M. A. Rao, and A. K. Datta. 2002. Heat transfer to three canned fluids of different thermo-rheological behaviour under intermittent agitation. *Food and Bioproducts Processing* 80(1):20–27.

Uthayakumaran, S. and O. M. Lukow. 2003. Functional and multiple end-use characterisation of Canadian wheat using a reconstituted dough system. *Journal of the Science of Food and Agriculture* 83:889–898.

Yang, W. H. and M. A. Rao. 1998. Complex viscosity-temperature master curve of cornstarch dispersion during gelatinization. *Journal of Food Process Engineering* 21(3):191–207.

CHAPTER 6

Modification of Biodegradable Polymers through Reactive Extrusion-I

I. Moura and A.V. Machado

CONTENTS

6.1 INTRODUCTION

Plastic materials produced from petrochemicals are used in a wide range of applications, such as packaging, automotive, health-care application, industry, and communication or electronic industries. Most of these polymers are extremely durable, requiring more than 100 years for their degradation.[1] Therefore, they may accumulate in the environment and become a significant source of environmental pollution.[1-3] After use, plastics can be disposed in different environments, such as composting facilities or soil burial, wastewater treatment facilities, and landfill. However, it has some adverse risks, like pollution of waterways due to high biochemical O_2 demand concentration, migration of plastic by-products to groundwater and surface water bodies, and soil and crop contamination.[4]

A possible solution to solve this problem could be to replace commodity synthetic polymers by biodegradable ones. However, it is necessary to take into account that any marketable plastic product must meet the performance requirements of its intended function, and most of the biodegradable polymers do not meet these functional requirements, that is, they do not have the performance specifications required for a given application.[4] Therefore, the development of biodegradable polymers with good performance, which after use would be susceptible to microbial and environmental degradation, using adequate solid waste management disposal practices, without any adverse environmental impact became a challenge.[5]

REX has been used as an attractive method to prepare new materials based on biodegradable polymers. It allows preparing new materials, in the melt, by blending, polymerization, grafting, branching, and functionalization.[6-8] Polymerization or chemical modification reactions in the melt were identified as an efficient and economic way for low-cost production, which enhances the commercial viability and cost-competitiveness of these materials.[9,10]

Thus, in this chapter, it will be discussed how REX has been used in continuous production and modification of biodegradable polymers.

6.2 BIOBASED AND BIODEGRADABLE POLYMERS

The words biobased and biodegradable both incorporate the prefix "bio," but they cannot be used indistinctly. Both biobased and biodegradable polymers can form the basis of an environmentally preferable and sustainable alternative to conventional polymers, based exclusively on petroleum feedstocks.[11]

The American Society for Testing and Material (ASTM) defines a biobased material as "an organic material, in which carbon is derived from a renewable resource, via biological processes. These materials include all plant and animal mass derived from carbon dioxide (CO_2) recently fixed via photosynthesis, per definition of a renewable resource."[11,12]

Therefore, a biobased material should be organic and contain carbon from biological sources, which is synthesized by many types of living mater (bacteria, animals, and plants), being portion of the ecosystem.[13] However, the use of a biobased material must take into account what happens to the product after the use and its impact in the environment since the most important factor of sustainability and environmental responsibility lies in the disposal of the products after use.[13] Thus, the U.S. Department of Agriculture (USDA), in the ASTM D6866, defined a percentage of carbon that is required to carry the term biobased.[14] This standard was developed to attest the biological content of bioplastics, that is, to determine exactly the amount of the material that comes from renewable resources.

For example, high-density polyethylene (HDPE) can be totally biobased, that is, containing only renewable carbon, but it is still nonbiodegradable. Thus, a polymer that contains only renewable raw materials could be or not biodegradable; it depends also on the molecular structure and on the chemical or biological methods used for polymerization. Accordingly, for single-use and short-life disposable material applications, biobased materials should be engineered to be biodegradable.[11] A product that is entitled biobased does not mean that it is based entirely on renewable resources. Rather, many of these products combine both petroleum- and natural-based materials in order to provide satisfactory properties and simultaneously reduce the overall amount of synthetic polymers contained in the products.

Biodegradable polymeric materials can be disposed in safe and ecological ways through waste management's composting, soil application, and biological wastewater treatment. According to ASTM 6400-99,[15] the common definition of biodegradable *is a degradable plastic in which the degradation results from the action of naturally occurring microorganisms such as bacteria, fungi, and algae*. Essentially, a polymer is called biodegradable when under the right conditions, the microbes in the environment can chemically break down the polymer chain and use it as a food source. The process of biodegradation essentially converts carbon into energy, taking place in many environments including soils, compost sites, waste management facilities, water treatment facilities, and marine environments. However, not all materials are biodegradable under the same conditions. While some are susceptible to microbes found in a wastewater treatment plant, others need microbes found in the soils.[16]

Biodegradable polymers can be divided into natural and nonrenewable synthetic polymers. The former are produced in nature during the growth cycles of all organisms,[17] and their synthesis generally involves enzyme-catalyzed, chain growth polymerization reactions of activated monomers, which are typically formed within cells by complex processes, and the latter are petroleum based.

The biodegradation process occurs in two different steps: First, the long polymer chains are shortened or cut at the carbon–carbon bonds.[16] This process can be started by different factors including heat, microbial enzymes, moisture, or other environmental conditions. This first step is not a synonym of biodegradation and is usually called degradation.[16] In the second step, called biodegradation, the short carbon chains are used as a food source and are converted into water (H_2O), biomass, carbon dioxide (CO_2), and methane (CH_4) (depending upon process taking place under aerobic or anaerobic conditions). Figure 6.1 illustrates the biodegradation process under aerobic conditions.

Moreover, there is also a difference between biodegradable and compostable polymers. Even if both breakdown of the polymeric chain into smaller fragments, due to the action of microorganisms, and transformation of the latter into CO_2, H_2O, minerals, and biomass and/or CH_4 occur, a

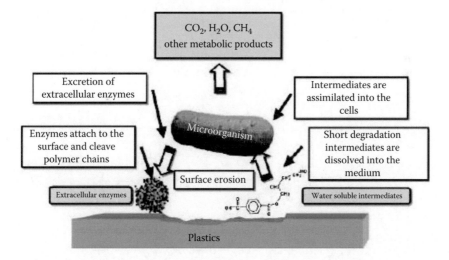

Figure 6.1 General mechanism of plastic biodegradation under aerobic conditions. (From Mueller, R.J., Biodegradability of polymers: Regulations and methods for testing, in *Biopolymers*, Steinbüchel, A., Ed., vol. 10., Wiley-VCH, Weinheim, Germany, 2003.)

compostable polymer should disintegrate and biodegrade quickly and must not leave visible, distinguishable, or toxic residues. To be called compostable, a product should meet D6400 standard,[15] which is the regulatory framework for the United States and sets a less stringent threshold of 60% biodegradation within 180 days, again within commercial composting conditions.

Unfortunately, most products are designed with limited concern in relation to its ultimate disposability. They are designed to be biobased, that is, they fragment into smaller fragments and may even degrade to residues invisible to the naked eye; thus, no products are completely biodegradable within a short period of time. These plastic residues will migrate into the water and in the ecosystem, causing damage to the environment.[19]

Biodeterioration and biodegradation of polymer substrate can rarely reach 100% because a small portion of the polymer will be incorporated into microbial biomass and other natural process.[20,21]

6.3 SYNTHETIC PLASTICS

The demand of synthetic polymeric materials has been fairly increasing during the last decades, and presently, they are one of the most attractive categories of materials.[22] This success is mainly related to their properties, namely, low cost, aesthetic qualities, and resistance to physical aging and biological attack.[23] It is estimated that global synthetic plastic production is approximately 140 million tons per year.[24,25] The most widely used plastics are polyethylene (PE), polypropylene (PP), polystyrene (PS), polyvinyl chloride (PVC), polyurethane (PU), poly(ethylene terephthalate) (PET), poly(butylene terephthalate) (PBT), and nylons (Figure 6.2 and Table 6.1).

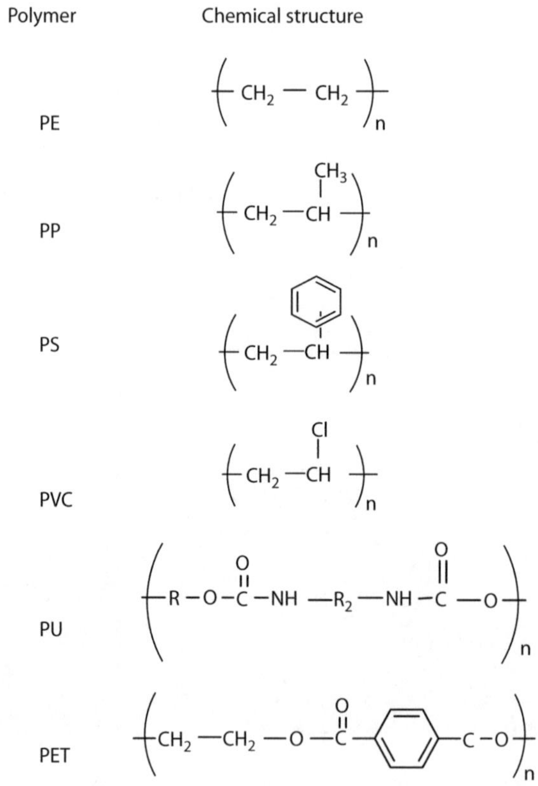

Figure 6.2 Structures of conventional plastics.

Table 6.1 Global Plastic Market

Plastic Type	Market Share by Volume Produced (%)	Use
PE	29	Plastic bags, milk and water bottles, food packaging, motor oil bottles
PP	12	Bottle caps, drinking straws, medicine bottles, car seats, car batteries, bumpers, disposable syringes, carpet backings
PS	9	Disposable cups, packaging materials, laboratory ware, certain electronic uses
PVC	17	Automobile seat covers, shower curtains, raincoats, bottles, visors, shoe soles, garden hoses, and electricity pipes
PU	5	Tires, gaskets, bumpers, in refrigerator insulation, sponges, furniture, cushioning, and life jackets
Others	28	
Total	100	

Polyolefins are the synthetic polymers with the highest commercial success, accounting for more than 47% of Western Europe's total consumption, 24.1 million ton per year. They present a combination of physical properties (flexibility, strength, lightness, stability, impermeability, and easiness of sterilization) that are ideally suited to a wide variety of applications such as food and drinks packaging.[26]

Synthetic polymers have an undesirable influence on the environment and a well-known resistance to degradation,[27] which became a problem with waste disposal. Once such material became part of the natural ecosystem, the negative effect is its long-lasting contribution to environmental contamination.[28] The growing environmental awareness and the new environmental regulations are forcing the industries to seek for more ecologically friendly materials for their products, namely, in applications where they are used for a short period of time before becoming waste.[22]

Under natural conditions, the degradation of synthetic plastics is a very slow process that involves environmental factors, followed by the action of wild microorganisms.[29–31] The degradation depends on physical and chemical properties, with hydrolysis or oxidation being the main mechanism.[32] Hydrolysis occurs by penetrating H_2O into the polymer backbone, attacking the chemical bonds in the amorphous phase, and converting them into shorter H_2O-soluble fragments, promoting a reduction in molecular weight. Then, metabolization of the fragments and bulk erosion also occur, leading also to the loss in the physical properties, making it more accessible for further microbial assimilation.[33–35] Some synthetic polymers, generally vinyl polymers, are not susceptible to hydrolysis. Therefore, the prevailing degradation mechanism occurs by oxidation due to the presence of an oxidizable functional group.[36]

Due to the lower sensitivity to biodegradation, there is a tendency to replace such polymers by polymers that could undergo easily the biodegradable process. The use of these materials, namely, in applications with short life cycle, such as packaging, would be an ecological alternative for reducing the solid plastic waste.[37]

Some examples of synthetic polymers that are biodegradable are polylactic acid (PLA), poly(ε-caprolactone) (PCL), polyamides (PA), and poly(vinyl alcohol) (PVA) and also some oligomeric structures, like ethylene, styrene, isoprene, butadiene, acrylonitrile, and acrylate.[38]

6.3.1 Polyethylene

Synthetic materials like polyolefins are difficult to be biodegraded by microorganisms and have a long lifetime.[36,39,40] In its natural form, PE is not biodegradable due to the higher hydrophobic character and also high molecular weight, but a comprehensive study of polyolefins biodegradation has shown that some microorganisms could use polyolefins with low molecular weight.[41] Thus, to convert conventional PE into biodegradable PE, it is necessary to modify their characteristics, such as molecular weight and crystallinity degree, which contribute to the high resistance to degradation.[31]

Bonhomme et al.[42] and Wang et al.[43] performed biodegradation studies of PE. The results indicated that chemical degradation occurred by two different pathways: hydro- and oxobiodegradation.[42] Other researchers also observed that the oxidation products of polyolefins are biodegradable.[44–51] The explanation is that these products have low molecular weight values and incorporate O_2-containing groups, such as acid, alcohol, and ketone. This is the basis of the term oxobiodegradable polyolefins. Oxobiodegradation involves two stages. First, oxidative degradation occurs followed by the biodegradation of the oxidized products. When a molecule undergoes oxidative degradation, the size is reduced, and at a given size, the microbial degradation starts. It has been demonstrated that the biodegradation of polar molecular fragments from PE occurs quite quickly.[52] Another alternative to accelerate the attack of microorganisms to polyolefins is by blending biodegradable polymers, like starch, PCL, and PLA, to guarantee at least a partial biodegradation. This effect will be discussed later in this chapter.

6.3.2 Ethylene Vinyl Acetate

Another synthetic polymer widely used in the packaging industry is ethylene vinyl acetate, which is a copolymer of ethylene and vinyl acetate. The weight percent of vinyl acetate usually varies between 10% and 40%, and the remainder is ethylene. It behaves like an elastomeric material in softness and flexibility and can be processed like other thermoplastics. The material has good transparency and gloss, barrier properties, low-temperature toughness, stress-crack resistance, hot-melt adhesive waterproof properties, and resistance to UV radiation. EVA copolymers have a broad range of industrial applications, such as packaging, adhesives, wire, cable, and health care. Also, due to the mechanical properties, these copolymers are used perhaps in a broadest spectrum of applications of any synthetic polymeric material.[53,54] Therefore, it would be interesting to have products made from this polymer with biodegradable potential.

6.4 BIODEGRADABLE POLYMERS

Limited resources of petroleum-based polymers and increased environmental awareness have attracted a higher interest toward biodegradable and biobased polymers for industrial applications.[55] The use of these materials is an alternative to conventional nonbiodegradable plastics, which could contribute to the solution of the environmental problem.[56]

The consumption of biodegradable polymers has increased in the last decades. The target market is mainly packaging materials, hygiene products, agricultural tools, and consumer goods. Nevertheless, there is still a competition between commodity plastics and biodegradable ones due to the low cost of the former.[57]

Biodegradable polymers can be derived from renewable or petroleum resources (Figure 6.3). Thus, industry, beyond nonbiodegradable petroleum-based plastics and the renewable source-based biodegradable polymers, is also thinking in terms of aliphatic/aromatic ratio by using chemical processes to achieve petroleum-based biodegradable plastics.[58]

According to Narayan,[12] biodegradable polymers can be divided as follows:

1. Biopolymers or natural biodegradable polymers formed in nature during the growth cycles of all organisms. The synthesis implies enzyme-catalyzed, chain growth polymerization reactions of activated monomers, which are formed within cells by some metabolic processes (e.g., starch and cellulose).

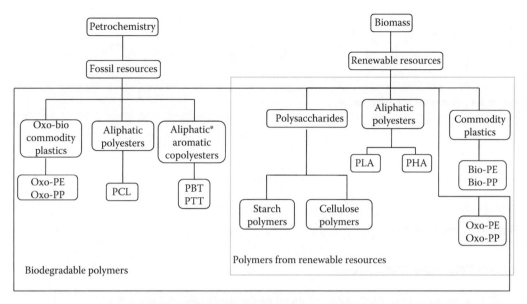

Figure 6.3 Resources of biodegradable polymers. *Partially from chemistry. (From Camino, G., *Conference of Bioplastics and Related Materials*, Gargnano, Italy, 2009.)

2. Polymers with hydrolyzable backbones—these polymers are susceptible to biodegradation, which includes aliphatic polyesters, PA, and PU.
3. Polymers with carbon backbones—the biodegradation of this kind of materials involves first an oxidation process. An example is PVA, which is not susceptible to hydrolysis. Biodegradable vinyl polymers contain functional groups that are easily oxidizable, and a catalyst is added to promote their oxidation or photooxidation, or both.[60]

6.4.1 Aliphatic Polyesters

Aliphatic polyesters or aliphatic–aromatic copolyesters are the most known petroleum source-derived biodegradable polymers.[61] In recent years, there is a growing interest on the synthesis and development of fully biodegradable polymers, such as PCL, polyhydroxybutyrate (PHB) and its copolymer with hydroxyvaleric acid, PLA, and aliphatic polyesters from different lactic acid derivatives.[62–64]

Aliphatic polyesters made from dimethyl esters and diols are expected to be of the most economically competitive biodegradable polymers.[65,66] Moreover, it was found that polyesters derived from diacids of medium-sized monomers (C_6–C_{12}) are more readily degraded by microorganisms than those derived from longer monomers.[67] A synthetic polymer can only be biodegradable by enzyme catalysts if the polymer chains are able to fit into the enzymes active site. This is the reason why flexible aliphatic polyesters are degradable and the rigid aromatic polyesters are not.[60,68,69] Another major feature of these polymers is their compatibility with the natural environment and their ability to undergo hydrolytic and biological degradation.[70] Their biodegradability depends mainly on their chemical structure and especially of the hydrolyzable ester bonds in the main chain, which are susceptible to microbial attack. Other factors such as molecular weight, crystallinity degree, stereoregularity, and morphology also affect the rate of biodegradation.[65,71,72]

The most important synthetic aliphatic polyesters are PLA and PCL, which are usually prepared by ring-opening polymerization (ROP) of the respective cyclic monomers (L,L-lactide [LA] and ε-caprolactone [CL]). This method provides sufficient polymerization control, resulting in polymers with required molecular weight and with the desired end groups.

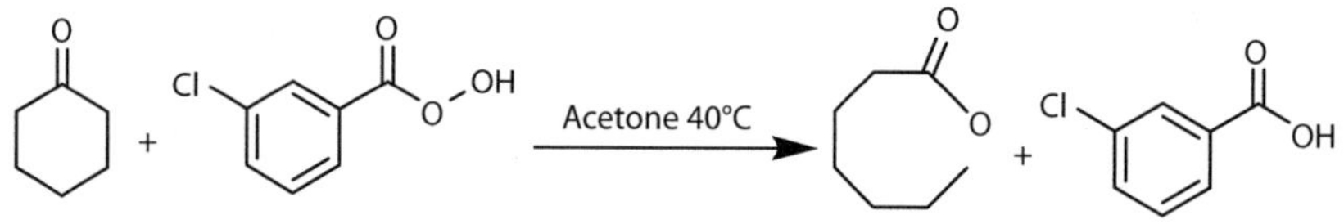

Figure 6.4 Synthesis of CL monomer.

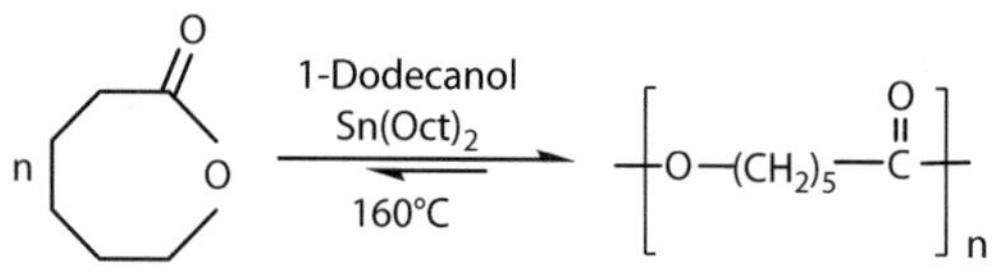

Figure 6.5 Synthesis of PCL.

6.4.1.1 Poly(ε-Caprolactone)

PCL is prepared from ROP of CL, as illustrated in the Figures 6.4 and 6.5.

PCL is appreciated by its biodegradable properties; it can be biodegraded aerobically by a large number of microorganisms in various microbiological environments.[73] Moreover, due to its flexibility, it has been found to be miscible with many other polymers.[60,73] However, the high cost and low performance of PCL for some applications have prevented its widespread industrial use.[74]

6.4.1.2 Polylactic Acid

PLA can be derived from renewable and petroleum-based resources.[75] The production of PLA presents advantages over other synthetic materials: (1) PLA can be obtained from renewable agricultural sources (e.g., corn), (2) its production consumes CO_2, providing significant energy savings, and (3) PLA is recyclable and compostable.[76-78]

Early economic studies have shown that PLA is an economically feasible material that can be used as a packaging material.[79] PLA properties are determined both by the polymer architecture (stereochemical makeup of the backbone) and the molecular weight, the latter being controlled by the addition of hydroxylic compounds. The control of the polymer stereochemical architecture allows precise control over the crystallization rate and the crystallinity degree, mechanical properties, and processing temperature.[80] PLA is a polyester with one of the highest melting temperatures, around 160°C–180°C. PLA can exist as two stereoisomers, designated as D and L, or as a racemic mixture, designated as DL. While the D and L forms are optically active, the DL form is inactive. Poly(L-lactic acid) (PLLA) and poly(D-lactic acid) (PDLA) are semicrystalline while poly(DL-lactic acid) (PDLLA) is amorphous.[81]

Bacterial fermentation is used to produce LA from corn or cane sugar. However, LA cannot be polymerized as a useful product because during polymerization reaction, molecules of H_2O are generated, and its presence degrades the forming polymer chain. Thus, PLA of high molecular weight is produced from ROP of LA using a catalyst (Figure 6.6), by solvent-free continuous process and distillation method.[38] This mechanism does not generate additional H_2O, and thus, a wide range of molecular weights are accessible.

PLA is currently used in industrial packaging and biomedical applications.[82] Nevertheless, it has been demonstrated that it is not suitable for hard tissue regeneration due to its weak mechanical properties.[83-87]

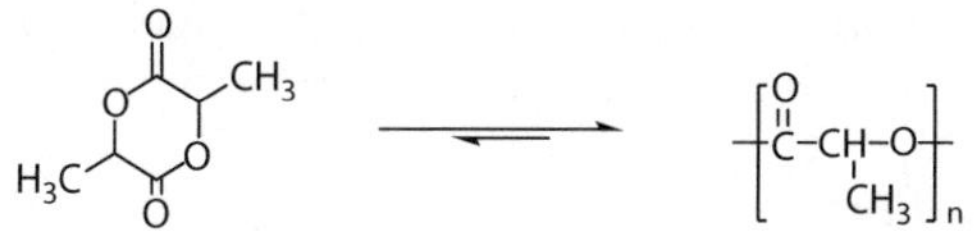

Figure 6.6 Synthesis of PLA.

6.4.1.3 Starch

Starch is a carbohydrate consisting of a large number of glucose units joined together by glycosidic bonds containing, generally, 20%–25% amylase and 75%–80% amylopectin. It occurs widely in plants, like rice, corn, cassava, and potatoes. In all of these plants, starch is produced in form of granules, varying in size, and in composition according to the plant used. Starch granules are hydrophilic, and the H_2O content of starch varies with relative humidity changes. While the branched amylopectin component contains crystalline areas, the linear amylase is mostly amorphous. Starch granules can be gelatinized in H_2O at lower temperatures in alkaline solution and can be used as a thickening, stiffening, and gluing agent, giving wheat paste.[88]

There are several degradable plastics made from starch.[12,60,89] For instance, a fully biodegradable starch-based polymer is prepared from corn or potato starch, along with smaller amounts of food-grade additives. The resin is suitable for manufacturing injection-molded pieces, films, and starch-based loose-fill packaging material.[90] These pieces degrade in an active biological environment.

Starch has a wide variety of applications including adhesives and industrial emulsions, construction, glass fiber, medical gloves, personal care, packaging, and agricultural.

The interest in this biopolymer has been recently renewed due to its abundance, low cost, availability, biodegradability, and possibility of blending with conventional polymers, and it can be processed using conventional polymer processing equipment, such as extrusion and injection molding.[91]

6.5 MODES OF BIODEGRADATION

6.5.1 Enzyme Mechanisms

Two categories of enzymes are involved in biological degradation of polymers: extracellular and intracellular depolymerases.[92,93] Such kind of enzyme has different action mechanism; some enzymes change the substrate through a free radical mechanism, while others follow alternative chemical routes (typical examples are biological oxidation and hydrolysis). During degradation, exoenzymes from microorganisms break down complex polymers, yielding smaller molecules of short chains, like oligomers, dimers, and monomers. These are small enough to pass the semipermeable outer bacterial membranes to be utilized as carbon and energy source. However, the biodegradative pathways associated with polymers are determined by environmental conditions. When oxygen (O_2) is available, aerobic microorganisms are mostly responsible for destruction of complex materials, with the formation of biomass, CO_2, and H_2O. Contrarily, under anaerobic conditions, microorganisms are responsible for polymer deterioration, being the primary products microbial biomass, CO_2, H_2O, and CH_4.[94]

Polymer degradation involves changes in physical properties due to the chain scission along the polymer backbone.[95,96] Since the degradation mode depends on the initiation process, it can be classified as thermal, mechanical, photochemical, biological, or chemical.[96] Additionally, the environmental conditions such as moisture, temperature, and type of microorganisms influence polymer degradation. Moreover, it also depends on the structural properties of the polymer, as chain orientation, stereochemical configuration, crystallinity degree, molecular weight, molecular weight distribution, and degree of cross-linking are among the important ones.[97–99]

6.5.1.1 Biological Oxidation

Many enzymes can react directly with O_2, which has a special role in the metabolism of aerobic organisms. The enzyme can be hydroxylases (Equation 6.1), which is responsible for the hydroxylation, that is, a chemical process that introduces hydroxyl groups (–OH) into the organic compound, or can be oxygenases (Equation 6.2), in this case transfers O_2 from molecular O_2 to the substrate. The substrate has another type of biological oxidation when the O_2 molecule is not incorporated into the substrate but rather functions as a hydrogen acceptor. Enzymes of this type are called oxidases, and one type produces H_2O (Equation 6.3) and the other peroxides (Equation 6.4).

$$AH_2 + O_2 \longrightarrow AHOH + H_2O \qquad (6.1)$$
$$BH_2 \nearrow \qquad \searrow B$$

$$AH_2 + O_2 \longrightarrow A(OH)_2 \qquad (6.2)$$

$$AH_2 + 1/2\,O_2 \longrightarrow A + H_2O_2 \qquad (6.3)$$

$$AH_2 + O_2 \longrightarrow A + H_2O_2 \qquad (6.4)$$

Biodegradable polymers disposed in bioactive environments degrade not only by the enzymatic action of microorganisms such as bacteria, fungi, and algae but also by nonenzymatic processes, such as chemical hydrolysis that breaks down the polymer chains.

6.5.1.2 Biological Hydrolysis

Polymers with hydrolyzable backbones have been found to be susceptible to biodegradation. Among others are polyglycolide (PGA), PCL, poly(lactic-co-glycolic) (PLGA), polyether–polyurethane, and poly(amide-enamine)s.

Hydrolysis occurs by scission of chemical bond in the main chain by reaction with H_2O.[81] The hydrolysis of esters can occur through both acid- and base-catalyzed mechanisms. While in the base-catalyzed mechanism (Figure 6.7a), the reactant goes from a neutral species to a negatively charged intermediate; in the acid-catalyzed (Figure 6.7b), a positively charged reactant goes to a

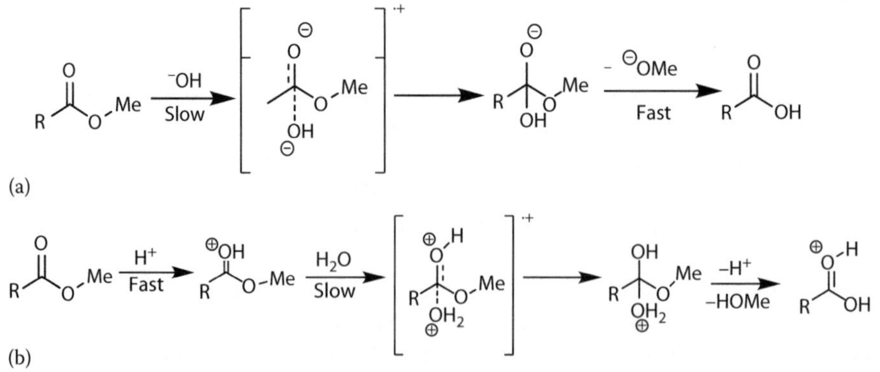

(a)

(b)

Figure 6.7 Base-catalyzed (a) and acid-catalyzed (b) ester hydrolysis mechanisms.

$$R_1 - COOR_2 + H_2O \longrightarrow R_1 - COOH + R_2OH$$

Figure 6.8 General equation of ester hydrolysis.

positively charged intermediate. Additionally, the mechanism associated with hydrolysis of ester linkage in neutral or acidic media is different from the one in alkaline media.[100] Both in neutral and acidic media, the hydrolysis is initiated by protonation and is followed by the addition of H_2O and the cleavage of the ester linkage.[100] In alkaline media, hydroxyl ions (OH^-) are attached to the carbonyl carbons and followed by the breaking of the ester linkages.

Several different hydrolysis reactions can occur in biological organisms, being the general equation represented in Figure 6.8. The degradation kinetics of different raw materials changes substantially, which might be attributed to the hydrophilic or hydrophobic nature of the different polymers.

6.6 BIODEGRADATION PARAMETERS

Biodegradation is a very complex process, which is affected by different factors, including type of microorganism, polymer features, and nature of pretreatment. Furthermore, polymer molecular weight, mobility, crystallinity degree, type of functional groups, tacticity, and additives play an important role.[93,101]

6.6.1 Effect of Polymer Structure

Biodegradability is mainly determined by the molecular structure and the length of the polymer chains.[102–104] Natural polymers, as starch, are generally degraded in biological systems by hydrolysis followed by oxidation.[105] Most of the synthetic biodegradable polymers contain hydrolyzable backbones. For instance, ester linkages are susceptible to biodegradation by microorganisms and hydrolytic enzymes. Since many proteolytic enzymes specifically catalyze the hydrolysis of peptide linkages adjacent to substituents in proteins, polymers containing substituents, such as benzyl, hydroxyl, carboxyl, methyl, and phenyl groups, have been prepared, aiming that the introduction of these substituents might increase biodegradability.[106] Huang et al.[107] investigated the effect of stereochemistry on the biodegradation using monomeric and polymeric ester–ureas that were synthesized from D-, L-, and D,L- phenylalanines. They found out that after enzyme-catalyzed degradation, the pure L-isomer degraded faster. Shuichi Matsumura et al.[108] studied the effects of stereoregularity on biodegradation of PVA by *Alcaligenes faecalis* and observed that the biodegradability of PVA was influenced by its stereoregularity, being the isocratic moiety more biodegraded. Another parameter that influences the degree of biodegradation is the hydrophilic and hydrophobic character of polymers, since most enzyme-catalyzed processes occur in aqueous medium. Machado et al.[61] performed biodegradability studies in aqueous medium, and they found that a polymer containing both hydrophobic and hydrophilic segments (HDPE/PCL or HDPE/PLA) seems to have a higher biodegradability than HDPE, which contains only hydrophobic segments on its structure. Additionally, HDPE/PCL system showed higher biodegradability than HDPE/PLA. This result can be explained once more, based on the hydrophilic character of both polymers. It is well known, that PLA is more hydrophobic than PCL, which explains the smaller result obtained for the biodegradability results.

6.6.2 Effect of Polymer Morphology and Crystallinity Degree

Synthetic polymers can have short repeating units, and due to this regularity, the crystallization is enhanced, making the hydrolyzable groups less accessible to enzymes.[109–111] On the contrary, if the repeating units are long, the polymer has less tendency to crystallize and consequently is more susceptible to biodegradation. It is well known that during degradation, semicrystalline polymers suffer some changes, namely, concerning the crystallinity degree. Firstly, the crystallinity of the polymers increases rapidly then levels off when the rate of crystallinity approaches 100%. This occurrence is due to the disappearance of the amorphous phase because biodegradation occurs preferably in the amorphous regions of the polymer that have a higher mobility of the polymeric chains and, therefore, are more accessible to the microorganisms.[22] Moura et al.[112] investigated the miscibility and biodegradability of blends of EVA and PCL or PLA. They observed that the blend with coarser morphology (EVA/PLA) showed a smaller degree of biodegradability than the one of EVA/PCL, which exhibited finer morphology. Jenkins and Harrison[113] found that an increase in the crystallinity degree of PCL reduced the rate of biodegradation.

Chin-San Wu[114] dedicated his study to the physical properties of maleated-PCL/starch blend and its relationship with biodegradability. The results indicated that even though PCL-g-MAH/starch shows higher compatibility, a slightly lower biodegradation rate was observed in a soil environment compared to the uncompatibilized one. J. K. Pandey et al.[115] found out that the biodegradability of polyester increases with compatibilization within PCL-starch compositions. Also, Chang-Sik Ha and Cho[116] observed that the rate of enzymatic degradation of poly(3-hydroxybutyrate) (P(3HB)) films decreases with an increase in crystallinity, and it was also influenced by the size of P(3HB) spherulites. It was suggested that the PHB depolymerizes, firstly hydrolyzes the amorphous P(3HB) chains on the surface of the film and subsequently erodes P(3HB) chains in the crystalline state.[117] Other factors, such as shape, size, and number of the crystallites, have also a significant effect on the chain mobility of the amorphous regions and thus affect the rate of the degradation.

6.6.3 Effect of Molecular Weight

A lot of studies have been performed on the effect of molecular weight on the biodegradation rate.[56,108,118] An increase of molecular weight results in a decline of degradability by microorganisms. Contrarily, monomers, dimers, and oligomers of a polymer's repeating units are much easier to degrade and mineralize.[38]

Some polymers remain relatively immune to microbial attack as long as their molecular weight remains high. While some plastics, such as polyolefins and PS, do not support microbial growth, low-molecular-weight hydrocarbons can be degraded by microorganisms. Some natural molecules, such as starch and cellulose, suffer conversions to low-molecular-weight components by enzyme reactions, which occur outside the cells.[60] Nevertheless, this process cannot be applied to some polymers, when their molecules are too big to enter into the cells.

Photodegradation and chemical degradation may decrease sufficiently the molecular weight to enable microbial attack. For instance, low-density polyethylene (LDPE) with an average molecular weight of $M_w = 150,000 \, g/mol$ contains about 11,000 carbons.[60] Decreasing molecules of this size to biologically acceptable dimensions requires extensive destruction of PE chains. This destruction could be partly accomplished by blending PE with biodegradable or natural polymers.

6.7 REACTIVE EXTRUSION

6.7.1 Reactive Process

This method differs from conventional ones, where synthesis was made separately and extruders were used only for processing (melting, pumping, and shaping).[119] REX is receiving much attention as an industrial technique because it has several advantages, such as continuous process, versatility, low cost, good heat transfer, short residence time, wide range of temperatures, high viscosity, and it is a solvent-free process.[10,120] Moreover, it is an attractive route for melt blending, filler dispersion, and various reactions (e.g., (co)polymerization, grafting, branching, and functionalization),[6–8,121] combining polymer processing and chemical reaction.

Nevertheless, there are also some drawbacks in using an extruder as a chemical reactor, including limited residence time, efficient heat transfer, medium polarity, and high viscosity leading to possible strong viscous dissipation, which can promote side reactions, like thermal degradation.[121]

REX is a very complex process since it has to deal with several parameters, such as processing, chemical reaction, and heat transfer. Due to the mixing capability and higher heat and mass transfer, twin-screw extruders are generally used for REX.[122] Twin-screw extruders can operate in counterrotating and corotating way, being the latter preferred in the REX process. The main interest on using the corotating system is, namely, because of the high speed and throughputs, better temperature control, adjustable residence time distribution, and continuous stable flow through the die.[10,123]

According to Xanthos,[124] many types of reactions can be performed in an extruder, including bulk polymerization, grafting reactions, interchain copolymer formation, coupling/cross-linking reactions, controlled degradation, and functionalization.

6.7.2 Reactive Blending of Immiscible Polymer Blends

Blending of polymers has become an attractive method to prepare new polymeric materials with enhanced properties and relative low cost.[125–130] However, there are two different types of polymer blends, miscible and immiscible blends. The former are characterized by the presence of only one phase and the existence of only one glass transition temperature (T_g). Contrarily, immiscible blends are phase separated, exhibiting the T_g and melting temperatures (T_m) of each blend component. It is well known that the blend features strongly depend on the properties of the individual components, but morphology is a key factor for producing polymer blends with enhanced properties.[129]

Immiscible blends can have important industrial application if they are compatibilized.[131] The main challenge of compatibilization is to generate good adhesion between the phases and fine morphology.[130,131–135] Essentially, three methods have been used to compatibilize immiscible polymer blends, which are the following[121,136–141]:

1. *Ex situ compatibilization*—it consists of the addition of a presynthesized block or a graft copolymer, which has blocks or grafts identical to the ones existing in the polymers of the blend.
2. *In situ compatibilization*—block of grated copolymers and synthesized at the interface during blending, in this cases, both polymers should have reactive groups.
3. *Cross-linking or "dynamic vulcanization"*—one of the phases cross-links, which stabilizes the morphology and avoids coalescence.

Ex situ compatibilization allows controlling the molecular architecture of the copolymer added.[121] This copolymer, called compatibilizer, should locate at the interface, reduce the interfacial tension, improve dispersion, and stabilize the morphology.[121] A major drawback of this method is that each polymer blend requires a specific copolymer, whose preparation requires specific chemical routes and reaction conditions.[121,142] Besides, due to thermodynamic and dynamic reasons, there are always some copolymer chains which cannot get to the interface where they are most needed. Dispersion of the copolymer in matrix is not simple, and its diffusion to the interface is generally a slow process.

In situ compatibilization of immiscible polymers produces desired copolymers, through interfacial reactions between reactive polymers during blending. This method is more attractive and cost effective because it allows producing the copolymer at the interface without separate preparation step.[142] When one of the polymers does not contain reactive groups, it needs to be to be functionalized previously. Generally, polymers grafted with maleic anhydride are extensively used as compatibilizers.[143-145]

Other parameters like thermodynamic and rheological properties and composition and processing conditions (screw configuration, time, screw rotation speed, temperature, throughput, etc.) have a strong influence on the morphology development during blending.[146]

6.7.3 Preparation of Blends of Nonbiodegradable and Biodegradable Polymers

As stated before, polyolefins constitute the majority of thermoplastics currently used as packaging materials. Since the use of plastics continuously increases, the problem of postconsumer recycling has become an important issue for economic and environmental reasons.[147] Nevertheless, recycling would be neither practical nor economical for certain applications such as bags, agricultural mulch films, and food packaging, since these materials contain many organic residues and have a low lifetime. For these applications, it would be better to use plastics that could degrade into safe by-products under normal composting conditions.[147] Thus, blending biodegradable polymers, such as starch, PCL, and PLA, with nonbiodegradable polymers, such as PE, has received considerable attention.[27,61,148,149] The reasoning behind this approach is that if the biodegradable component is presented in sufficient amount and if it is removed by microorganisms in the waste disposal environment, the plastic containing the remaining inert components should disintegrate and disappear.[60]

Starch can be used like an additive in two different ways in biodegradable plastics: it can be compounded into plastics in the form of biodegradable filler,[90] which is added to various resin systems to make films that were impermeable to H_2O but permeable to water vapor,[150] and it can be plasticized with H_2O (5%–20%) and compatibilized with other polymers to become part of the polymeric matrix. Since thermoplastic starch (TPS) is a very hydrophilic product, research has been performed to modify the starch structure by acetylation to reduce the hydrophilic character of the chains.[151-153] Avérous et al.[152] described changes in mechanical properties of TPS and its relationship with crystallinity, plasticizer content, and H_2O during aging. They found that the moisture sensitivity and the critical aging have led to the necessity to associate TPS with another biopolymer. Association between polymers can be as a form of blends or multilayer products. Nevertheless, most of the times, compatibilization is required in order to promote adhesion between the polymers and to achieve the product specification. Many biodegradable TPS blends have been developed, such as starch/PCL, starch/cellulose acetate, and starch/PLA.[154,155] Also, this kind of materials can to be mixed with synthetic polymers (such as PE and PP) in order to create plastic products more degradable than conventional synthetic plastics.

Blends of PE and starch can be melt-processed to obtain products with PE-like properties. Starch, either in its virgin form or chemically modified, has been used to increase its compatibility with the polymer matrix in order to produce this type of blends. It was found that the effective accessibility of the starch, which is required for extensive enzymatic hydrolysis and removal, is achieved only if the starch content exceeds 30%.[60] However, increasing the amount of starch leads to a decrease in mechanical properties, and the resulting material has poor properties when compared to conventional polyolefins. These worsening properties arise from the different polar characteristics of starch and most of the synthetic polymers, which leads to poor interfacial adhesion. Nakamura et al.[27] investigated the incorporation of different starches (native, adipate, acetylated, and cassava starch) in an LDPE matrix to study the possibility to obtain partially biodegradable materials. The results indicated that the increase of the starch into the LDPE matrix was responsible for the reduction on mechanical properties of the products when compared with conventional LDPE (Table 6.2).

Table 6.2   Tensile Test for LDPE/Starch Compounds

Sample	Tensile Strength (MPa)	Elongation (%)	Young Modulus (MPa)
Pure LDPE	12.9 ± 0.2	131.9 ± 4.8	139.3 ± 6.8
LDPE + 5 wt.% native starch	12.7 ± 0.2	58.0 ± 1.5	113.1 ± 7.0
LDPE + 10 wt.% native starch	11.9 ± 0.1	50.1 ± 1.5	122.6 ± 5.1
LDPE + 20 wt.% native starch	11.3 ± 0.1	30.9 ± 1.6	151.7 ± 11.2
LDPE + 5 wt.% RD125	12.5 ± 0.2	55.6 ± 3.7	118.9 ± 6.1
LDPE + 10 wt.% RD125	12.1 ± 0.1	50.3 ± 1.5	131.9 ± 5.6
LDPE + 20 wt.% RD125	11.1 ± 0.1	35.0 ± 1.8	151.2 ± 9.7
LDPE + 5 wt.% adipate starch	12.9 ± 0.1	52.6 ± 1.4	118.5 ± 6.5
LDPE + 10 wt.% adipate starch	12.2 ± 0.1	43.5 ± 1.0	129.3 ± 9.7
LDPE + 20 wt.% adipate starch	11.3 ± 0.1	33.7 ± 1.1	150.8 ± 3.6
LDPE + 5 wt.% cassava starch	12.7 ± 0.1	55.6 ± 1.9	113.3 ± 5.3
LDPE + 10 wt.% cassava starch	12.0 ± 0.2	49.2 ± 1.4	119.0 ± 7.8
LDPE + 20 wt.% cassava starch	11.4 ± 0.2	36.9 ± 1.3	149.9 ± 6.2

Therefore, a lot of research work has been done in order to improve the compatibility/adhesion between starch and PE, including the modification of starch,[156–158] modification of PE,[60] and/or the introduction of a compatibilizer.[159–163] The compatibilizers include, among others, ethylene–acrylic acid (EAA) copolymer, PE grafted with maleic anhydride (PE-g-MA), and EVA. EAA is one of the most effective compatibilizer used, but it must be used in high amounts to achieve satisfactory mechanical properties. Unfortunately, EAA lowers the biodegradation rate of starch, while at the same time, it accelerates the thermo-oxidative degradation of LDPE/starch blends when used in low amounts together with a pro-oxidant.[159] The results showed that using PE-g-MA as compatibilizer, a much better dispersion of starch within the PE matrix together with a significant reduction in the phase size was achieved.[60,159,162] Moreover, concerning the biodegradability results, it was observed that the compatibilized blends showed a slightly lower biodegradation than the uncompatibilized ones.[159,160]

Blending PE with other biopolymers, such as PLA and PCL, has also been studied.[148,149] Matzinos et al.[149] observed that the effect of PCL on the mechanical properties of LDPE/TPS/PCL materials depends not only on its content but also on the final morphology. Machado et al.[61] investigated the mechanical and rheological properties and the potential for biodeterioration of HDPE blended with biodegradable polymers, such as PLA, PCL, and Mater-Bi (TPS with PLA or PCL), in a corotating twin-screw extruder. They observed that adhesion between PLA and HDPE, even with the addition of PE-g-MA, was not good enough to improve the mechanical properties, which were similar to the value observed to pure PLA.

Better adhesion was observed for PCL and HDPE blend. The differences observed in morphology when either PCL or PLA were used were attributed to their chemical structures. The ratio between ester links/aliphatic chains is higher for PLA than for PCL, affecting the polarity.[57] Thus, the polarity of PLA is higher than that of PCL, and consequently, the compatibility of PCL with HDPE is better. For this reason, also mechanical properties obtained for the HDPE/PCL system were enhanced compared to HDPE/PLA system. The morphology of all HDPE/Mater-Bi blends (Figure 6.9a through c) similar to the individual components of the blends can be detected, which can be associated with the interfacial tension among the components. The addition of Mater-Bi (including PLA/TPS or PCL/TPS) has a minor effect on mechanical properties.

The biodegradation potential, based on bacterial counts in the biofilm surface of the blends (ASTM G22-76), showed that HDPE/PCL has a lower resistance to bacterial attack than HDPE/PLA. Moreover, the addition of 30% starch to HDPE/PLA blend enhanced its biodeterioration

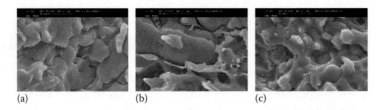

(a)   (b)   (c)

Figure 6.9   SEM micrographs of blends of HDPE with (a) SPLA 50 (50 wt.% TPS + 50 wt.% PLA), (b) SPLA 70 (30 wt.% TPS + 70 wt.% PLA), and (c) with SPCL (30 wt.% TPS + 70 wt.% PCL).

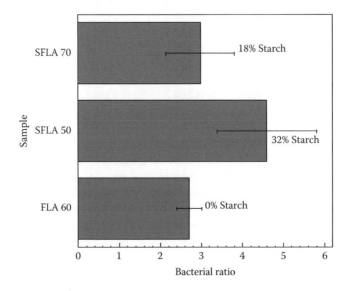

Figure 6.10   Bacterial ratio as a function of % starch (0%, 18%, and 32%).

potential, but the same was not observed in the case of the HDPE/PCL blend containing only 18% starch.

Concerning the effect of the amount of TPS (0%, 18%, and 30%) on the biodeterioration potential of PLA blends, the study indicated that the ratio between bacterial counts obtained was not significantly different in the cases of the blends containing 0% and 18% of starch but increased significantly in the case of 30% (Figure 6.10). The results suggested that the amount of starch might have been to low or simply not available at the polymer surface for bacterial growth in the blend containing 18% starch. At 30%, starch decreased the resistance of the blend to bacterial attack and promoted microbial growth. This result may be attributed to crystallinity and hydrophobicity of starch as biodegradation occurs preferably in the amorphous regions because of the higher mobility of the chains and their accessibility to the microorganisms. Also, starch, being less crystalline compared to PLA, is more prone to microbial attack. Additionally, its hydrophilic nature characterized by a higher number of –OH groups in structure compared to PLA promotes swelling in the culture medium, enhancing biodeterioration.

M. Mihai et al.[164] and N. Ljungberg et al.[165] studied the miscibility of polyolefin/PLA blends and found out that due to the differences in their chemical structures, a weak interfacial adhesion and poor dispersion were achieved.

A similar system was investigated by A. Kramschuster et al.[166] and M. Shibata et al.[167] using a different approach. They used PE-b-PLLA as a compatibilizer, and it was possible to improve the dispersion and achieve smaller PE particles in the PLLA phase (Figure 6.11b).

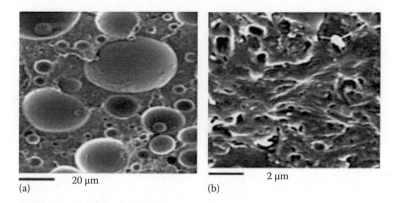

(a)   20 μm   (b)   2 μm

Figure 6.11   SEM micrographs of (a) 80:20 PLLA/LDPE and (b) 80:20:10 PLLA/LDPE/PE-b-PLLA blends.

6.7.4 Polymer Modification

Instead of blending polymers A and B, the monomer of A can be polymerized in the presence of polymer B. The functional groups of polymer B, located along the chain or at the end, can be initiating sites, from which an A chain could grow (Figure 6.12). This way, grafted or block copolymers can be formed.

Generally, there are three main ways of synthesizing block or graft copolymers:

1. Living copolymerization
2. Chemical modification by postpolymerization
3. Coupling between two appropriately functionalized polymer chains

Figure 6.13 illustrates schematically approaches 1 and 2, which can be associated with the "grafting from" method and the approach 3 with "grafting onto" method.

Structures of copolymers obtained through methods 1 and 2 are specifically relevant to REX since they could not be obtained by classical copolymerization method. As referred before, the chemical reaction occurs at the interface, and thus, a large quantity of copolymer is difficult to obtain. This interfacial reaction leads to compatibilization of the blends during mixing.[168]

6.7.4.1 Living Polymerization

Living polymerization is also called controlled polymerization. This method was developed by Michael Szwarc[169] in 1956 in the anionic polymerization of styrene with an alkali metal/naphthalene system in tetrahydrofuran (THF). This method is used for synthesizing block copolymers

ROP of various cyclic biodegradable monomers, such as lactams, lactones, 1,4-dioxane-2-one, lactides, and also carbonates, in a twin-screw extruder has been widely studied due to the reaction kinetics being compatible with the process conditions (high monomer conversion in a very short range of time at high temperature and good control of the structure through the judicious choice of

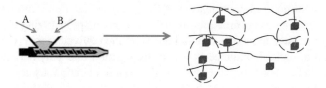

Figure 6.12   Schematic representation of in situ polymerization.

(a)   (b)

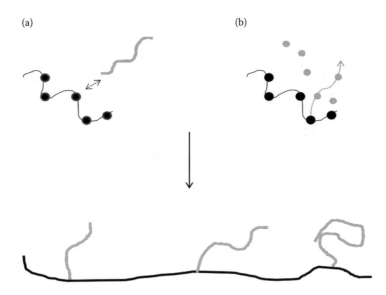

Figure 6.13 Schematic representation of (a) "grafting onto" and (b) "grafting from" methods. (From Machado, A.V. et al., Reactive polymer processing and design of stable micro- and nanostructures, in *Advances in Polymer Processing from Macro to Nano Scales*, Thomas, S. and Weimin, Y., Eds., Woodhead Publishing Limited, Cambridge, U.K., pp. 579–615, 2009.)

the polymerization catalyst). The major interest in these copolymers is also based on their potential to participate in the development of biodegradable polymeric materials.

Raquez et al.[170] published a review on specific homo- and copolymerization carried out by REX, where it was showed, for the copolymerization of CL with 1,4-dioxan-2-one, that in approximately 2 min, 100% of conversion was obtained at 130°C using $Al(O_{sec}Bu)_3$ as catalyst. It was observed that an increase in the copolymerization yields an increase in the molar fraction of CL. The ROP of CL by $Al(O_{sec}Bu)_3$ active species is well controlled and proceeds via so-called coordination–insertion mechanism, which yields polyester chains endcapped by an active aluminum alkoxide bond.[171] As a result of the trifunctionality of $Al(O_{sec}Bu)_3$, it allows the initiation and propagation of three growing polyester chains per one aluminum atom. A three-arm star-shaped PCL with an average molecular weight of each arm around 200,000 g/mol has been successfully produced in the extruder within a mean residence time of less than 5 min (monomer conversions in excess of 95%).[172] Similarly, a new process has been developed for the continuous production of PLA using REX, using tin octoate ($Sn(Oct_2)$) added with one equivalent of triphenylphosphine (as cocatalyst), which not only enhances the kinetics of LA by ROP but also eliminates any side degradation reactions, such as transesterification reactions.[173] Actually, it has been shown that the addition of one equimolar amount of a Lewis base like triphenylphosphine on 2-ethylhexanoic tin(II) salt ($Sn(Oct)_2$) significantly enhances the LA polymerization rate in bulk. This kinetic effect has been accounted for the coordination of the Lewis base onto the tin atom, making easier the insertion of the monomer into the metal alkoxide bond of the initiator/propagation active species.[174] This tin alkoxide bond is formed in situ by reaction of alcohol and the tin(II) dicarboxylate. As reported in polymerization of CL catalyzed by aluminum trialkoxides, the LA ROP proceeds via the same "coordination–insertion" mechanism involving the selective O_2–acyl cleavage of the cyclic ester monomer. The addition of one equivalent of $P(C_6H_5)_3$ onto $Sn(Oct)_2$ allows reaching an acceptable balance between propagation and polymerization rates, so that the polymerization is fast enough to be performed through a continuous one-stage process in an extruder.[175] Using this process is also possible to produce PLA with controlled molecular weight by the addition of alcohol.

Raquez et al.[9] investigated the PLA production, based on molecular parameters, using batch bulk polymerization and a single-stage continuous REX. Even though the conversion was similar (98.5% and 99%, respectively), the time necessary to reach this conversion was very different (40 vs. 7 min). Moreover, the molecular weight obtained was different, which was related with the diffusion and the reactivity of the monomer. Byong Jun Kim and James L. White[176] described how feed rate and feed order of comonomers influenced the formation of lactam–lactone copolymers, their structure, and molecular weight.

The polymerization of CL and LLA, using calcium ammoniate catalyst treated with ethylene oxide (EO) and propylene oxide (PO), was studied by Longhai Piao et al.[177] Both exhibited high activity, and they found that the living ROP behaved a quasi-living characteristic.

6.7.4.2 Chemical Modification by Postpolymerization

Another way to prepare block or graft copolymers through the "grafting from" method consists of polymerizing a monomer in an extruder in the presence of functionalized prepolymer or polymer (end or pendant functional groups initiating the monomer polymerization). Postpolymerization modification to incorporate monomer units focuses on two types of reactions. One is the removal of the protecting groups, where monomers with the desired functionality are incompatible with one or more components of the selected polymerization process. The functional monomers are polymerized with a protected functional group, which is deprotected to provide the desired functionality after the reaction is complete. The other approach is to copolymerize monomers with one functional group then convert that functional group into the desired functional group after the first polymerization is complete.[178]

The postpolymerization modification of monomer units method has some advantages, namely, it allows incorporation of functionality that is incompatible with the polymerization process, allows also the characterization of the initial copolymer prior to further functionalization, and facilitates "grafting from" reactions.[178]

The preparation of prepolymers or macromonomers with functional end groups, so-called telechelic polymers, is another approach to structurally unconventional architecture.[179] The functional end groups are introduced either by functional initiation or endcapping of living polymers, or by a combination of the two. Therefore, monomers that were not able to copolymerize can be incorporated in a copolymer. Telechelic prepolymers can be linked together using chain extenders such as diisocyanates.[180] In this process, it is essential that the structure and end groups of the prepolymers can be quantitatively and qualitatively controlled.[181] REX has been used as a simple way of producing segmented copolymers.[182] Lee Bet and White[183] investigated the in situ polymerization of caprolactam using isocyanate-terminated telechelic poly(tetramethyl ether glycol) (PTMEG). The analysis of polyetheramide triblock copolymer indicated that the conversion of caprolactam was around 95%.

A method to produce biodegradable aliphatic polyesters by REX was developed by Jacobsen et al.[175] They dedicated their studies to the effect of triphenylphosphine on the efficiency of $Sn(Oct)_2$ as a catalyst for the ROP of LA to produce PLA. A corotating closely intermeshing twin-screw extruders has often been used for polymerization reactions, but in any case, the reaction time was sufficiently smaller than the residence time in the extruder. In this case, a sophisticated screw design has been used to ensure further enhancement of the polymerization reaction by using mixing elements. Under these conditions, it was possible to realize a single-stage process to polymerize LA and to produce a PLA that can be used right away from the process for any known polymer processing technology.

Stevels et al.[182] reported the polymerization of L-lactide initiated by both a hydroxyl-terminated PCL and a polyethylene glycol (PEG). More recently, a new process has been developed for the production of PLA using REX, based on a new catalytic system that not only enhances the ROP kinetics of L-lactide but also suppresses side and degradation reactions. This process can be used to produce PLA continuously in larger quantities and at lower costs than before.[175]

The ROP of lactones in the extruder under anhydrous conditions has also been reported and can be catalyzed by Lewis acids (CL),[172,176,184,185] LA,[175,186] or base (CL with sodium hydride).[187] Lewis acid–catalyzed (aluminum tri-*sec*-butoxide) grafting of CL on starch has been carried out under anhydrous conditions in the extruder to form high-molecular-weight grafts.[185,188,189] A similar process has been used to graft CL on poly(ethylene-*co*-vinyl alcohol) by REX under anhydrous conditions.[190,191]

Becquart et al.[192] studied the functionalized poly(vinyl alcohol-*co*-vinyl acetate)-g-CL copolymers resulting from the in situ polymerization of the lactone ring. They found that the –OH groups were essential to initiate the polymerization.

Jae et al.[193] dedicated their work to the synthesis of triblock copolymers composed of PPG and PCL. The degree of CL conversion and the molecular weight of PCL increased linearly with the polymerization time or with the feed ratio of CL. The study of the ROP of the CL initiated by titanium phenoxide (Ti(OPh)$_4$) evidenced that, on average, one phenoxide ligand initiates the ROP.[194] Second, an increase of the polymer molecular weight was observed after complete monomer conversion, with a decrease of phenoxyl ester end groups concentration. Actually, this phenomenon is due to transesterification reactions favored with end groups in case of polymerization with Ti(OPh)$_4$. In fact, the C–O bonds of phenoxyl ester terminal is more prone to nucleophilic substitution than the C–O bond in repetitive unit due to the influence of phenyl group on electronic delocalization. This leads to more selective transfer reactions and consequently to more efficiency for grafting reactions (Figure 6.14).

Recently, Moura et al.[112] prepared grafted copolymers of EVA/PLA and EVA/PCL using in situ polymerization of LA and CL in the presence of molten EVA. The process takes the advantage of the living character of PLA and PCL chains growing from LA and CL monomers by ROP to increase, through the specific exchange reaction between the living PLA or PCL end chain and the acetate groups of EVA, the probability of grafting and consequently the concentration of the formed copolymer. When polymerization of the cyclic monomer initiated by the Ti(OPh)$_4$ takes place in the presence of molten EVA, two reactions occur, leading to the grafting process shown in Figure 6.15 for CL (the mechanism being the same for LA).

First, a transfer reaction between the acetate group of the EVA and the living Ti-O-polyester end bond results in EVA chain functionalized by Ti(OPh)$_3$ and polyester functionalized by an acetate group (Figure 6.15a). Then, this new titanate species would react either on an ester function of

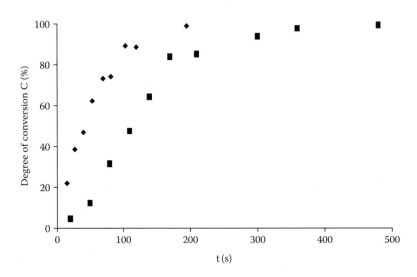

Figure 6.14 Time conversion curves for the bulk polymerization of CL between the rheometer plates at T = 100°C (M$_0$/I$_0$ = 300) initiated with titanium *n*-propoxide (♦) or titanium phenoxide (■). (From Cayuela, J. et al., *Macromolecules*, 39, 1338, 2006.)

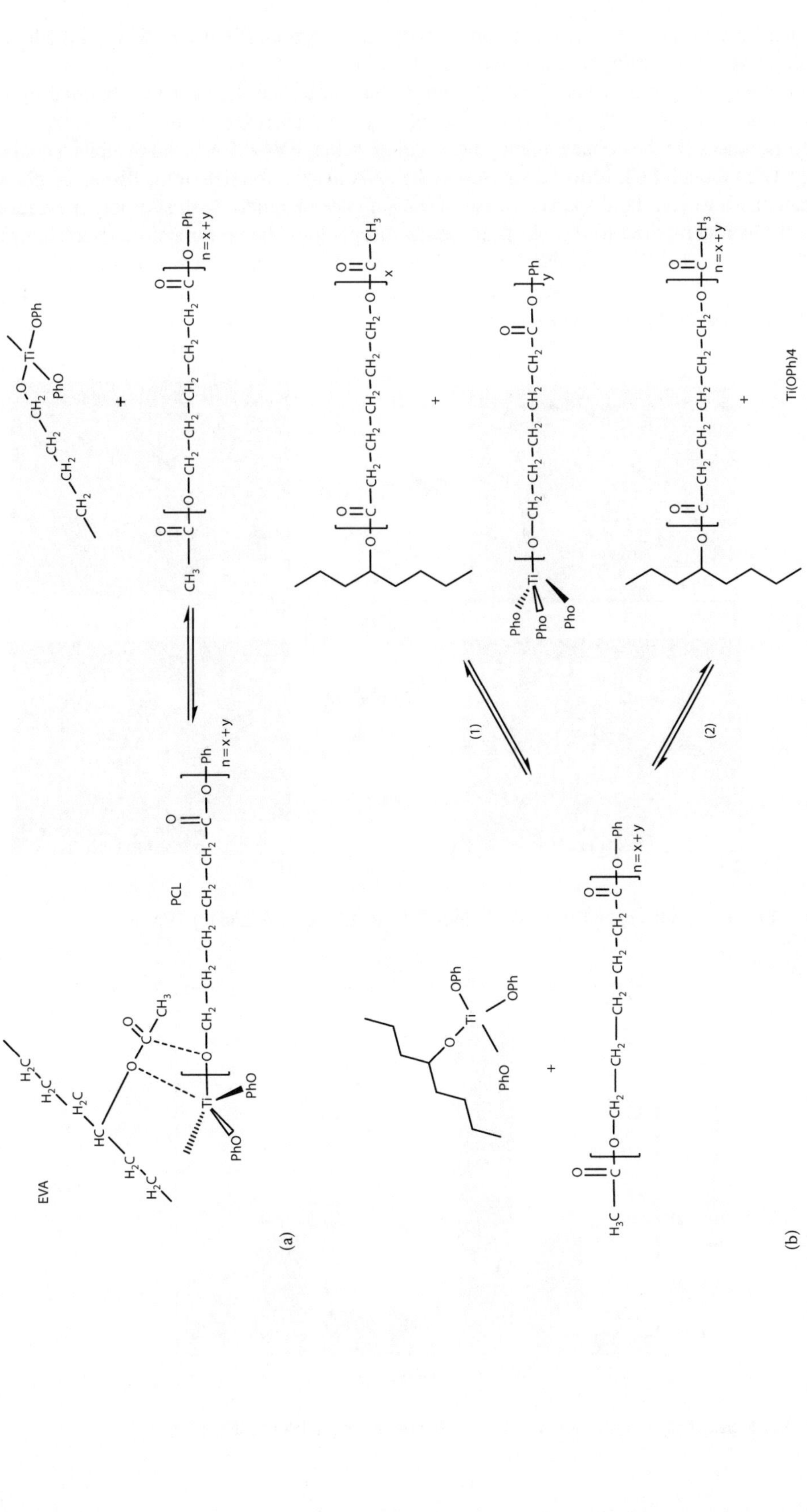

Figure 6.15 Reaction mechanism of EVA-g-PCL copolymer formation. (a) Transfer reaction between the acetate group of the EVA and the Ti-O-Polyester bond; (b)-(1) Transfer reaction between the new species and either an ester function of polyester chain; (b)-(2) Transfer reaction between the new titanate species on the phenoxyl ester end-group.

polyester chain (Figure 6.15b [1]) or on the phenoxyl ester end group (Figure 6.15b [2]), the latter reaction being favored according to the previous explanations.

The morphology of the polymer blends (Figure 6.16a and b) and the samples obtained by in situ polymerization (Figure 6.16c and d) consists of dispersed particles in the EVA matrix, but significant differences can be noticed among them. Figure 6.16a, EVA/PLA blend exhibits a coarse morphology. Even though PCL is also dispersed in the EVA matrix, the size of the dispersed phase is much smaller. However, the dispersed phase of the samples obtained by in situ polymerization is very small when compared to the physical blends morphology, being almost undetectable for EVA-g-PCL sample. This decrease can be explained by the copolymer formed during reaction, which acts as compatibilizer, decreasing the interfacial tension between blends components and, consequently, the size of the dispersed phase.

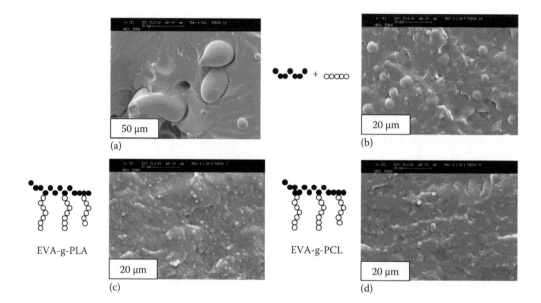

Figure 6.16 SEM micrographs of (a) EVA/PLA, (b) EVA/PCL, (c) EVA-g-PLA, and (d) EVA-g-PCL.

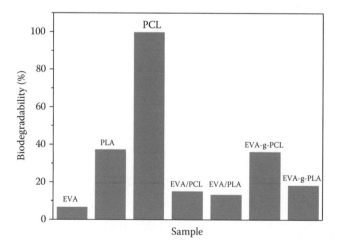

Figure 6.17 Biodegradability of polymers and all samples according to ISO 14851(1999).

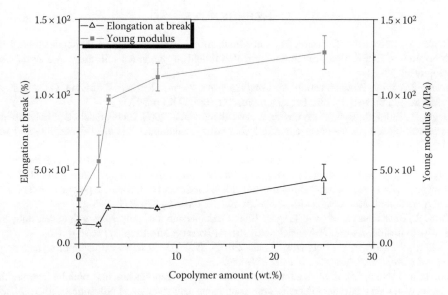

Figure 6.18   Young modulus and elongation at break as a function of copolymer amount.

The samples prepared by in situ polymerization, mainly for EVA-g-PCL sample, exhibit the better mechanical performance.[112] Moreover, differences in biodegradability behavior were observed, being EVA-g-PCL the more biodegradable sample (Figure 6.17).

6.7.4.3 Coupling between Two Appropriately Functionalized Polymer Chains

Moura et al.[195] also dedicated their studies to the synthesis of EVA-g-PLA copolymers by transesterification reactions between EVA and PLA catalyzed by titanium propoxide ($Ti(OPr)_4$). They found out that the amount of grafted copolymer (EVA-g-PLA) had a significant effect on biodegradation, mechanical properties, and other physical properties as well (Figure 6.18).

Elongation at break and Young modulus as a function of copolymer amount indicated that an increase of copolymer amount increases these properties. This enhancement can be attributed to the formation of EVA-g-PLA copolymer and its compatibility effect, which are related to the chemical structure of the copolymer formed, its amount, and its effect as compatibilizer.

6.8 CONCLUSIONS

This chapter focused of the preparation of biodegradable polymers using different approaches by reactive extrusion. It is possible to adapt the chemistry, by the right selection of the catalytic system, and the processing parameters to use REX for (1) production of compatibilized blends of biodegradable polyesters and synthetic polymers, (2) synthesis of aliphatic polyesters by catalyzed ROP, and (3) chemical modification of polymers. There is no doubt that the earlier examples are only some illustrations of the huge potential of REX, a solvent-free melt process, in the field of the biodegradables polymers.

ACKNOWLEDGMENTS

The authors would like to thank the Portuguese Foundation of Science and Technology (FCT) for financial support (PTDC/AMB/73854/2006).

REFERENCES

1. Ohtake, Y., Kobayashi, T., Asabe, H., and Murakami, N. 1998. Studies on biodegradation of LDPE-observation of LDPE films scattered in agricultural fields or in garden soil. *Polymer Degradation and Stability* 60: 79–84.
2. Huang, S.J. 1989. Biodegradation. In: *Comprehensive Polymer Science*, Eastmond, G.C., Ledwith, A., Russo, S., and Sigwalt, P., Eds. Pergamon Press, Oxford, U.K., p. 597.
3. Iovino, R., Zullo, R., Rao, M.A., Cassar, L., and Gianfreda, L. 2008. Biodegradation of poly(lactic acid)/starch/coir biocomposites under controlled composting conditions. *Polymer Degradation and Stability* 93: 147–157.
4. Lee, S. and Lee, J.W. 2005. Characterization and processing of biodegradable polymer blends of poly(lactic acid) with poly(butylene succinate adipate). *Australian Rheology Journal* 17: 71–77.
5. Mohanty, A.K., Misra, M., and Hinrichsen, G. 2000. Biofibres, biodegradable polymers and biocomposites: An overview. *Macromolecular Materials and Engineering* 276–277: 1–24.
6. Mani, R., Bhattacharya, M., and Tang, J. 1999. Fictionalization of polyesters with maleic anhydride by reactive extrusion. *Journal Polymer Science Part A: Polymer Chemistry* 37: 1693–1702.
7. Michaeli, W., Höcker, H., Berghaus, U., and Frings, W. 1993. Reactive extrusion of styrene polymers. *Journal of Applied Polymer Science* 48: 871–886.
8. Kye, H. and White, J.L. 1994. Continuous polymerization of caprolactam in a modular intermeshing co-rotating twin screw extruder integrated with continuous melt spinning of polyamide 6 fiber: Influence of screw design and process conditions. *Journal of Applied Polymer Science* 52: 1249–1262.
9. Raquez, J.M., Degée, Ph., Nabar, Y., Narayan, R., and Dubois, Ph. 2006. Biodegradable materials by reactive extrusion: From catalyzed polymerization to functionalization and blend compatibilization. *Comptes Rendus Chimie* 9: 1370–1379.
10. Raquez, J.M., Narayan, R., and Dubois, Ph. 2008. Recent advances in reactive extrusion processing of biodegradable polymer-based compositions. *Macromolecular Engineering Materials* 293: 447–470.
11. Narayan, R. 2006. Biobased and biodegradable polymer materials: Rationale, drivers, and technology exemplars. *ACS Symposium Series* 939: 282–306.
12. Narayan, R. 1994. Impact of governmental policies, regulations, and standards activities on an emerging plastics industry. In: *Biodegradable Plastics and Polymers*, Doi, Y. and Fukuda, K., Eds. Elsevier, New York, pp. 261–271.
13. Graziani, M. and Fornasiero, P. 2007. *Renewable Resources and Renewable Energy: A Global Challenge*. CRC Press, Boca Raton, FL.
14. ASTM D6866-10 Standard test methods for determining the biobased content of solid, liquid, and gaseous samples using radiocarbon analysis.
15. ASTM D6400-04 Standard specification for compostable plastics.
16. Bloembergen, S., David, J., Geyer, D., Gustafson, A., Snook, J., and Narayan, R. 1994. Biodegradation and composting studies of polymeric materials. In: *Biodegradable Plastics and Polymers*, Doi, Y. and Fukuda, K., Eds. Elsevier, New York, pp. 601–606.
17. Lenz, R.W. 1993. Biodegradable polymers. *Advances in Polymer Science* 107: 1–40.
18. Mueller, R.J. 2003. Biodegradability of polymers: Regulations and methods for testing. In: *Biopolymers*, Steinbüchel, A., Ed., vol. 10. Wiley-VCH, Weinheim, Germany.
19. Reynnells, R. 2005. *Food and Biobased Cafeteria Ware Composting for Federal Facilities in WDC. A Perspectives Roundtable Discussion*. USDA/CSREES/PAS Pat Millner, USDA/ARS Carmela Bailey, USDA/CSREES/PAS.
20. Atlas, R.M. and Bartha, R. 1997. *Microbial Ecology: Fundamentals and Applications*, 4th edn. Benjamin/Cummings Publishing Company, Menlo Park, CA.
21. Narayan, R. 1993. Biodegradation of polymeric materials (anthropogenic macromolecules) during composting. In: *Science and Engineering of Composting: Design, Environmental, Microbiological and Utilization Aspects*, Hoitink, H.A.J. and Keener H.M., Eds. Renaissance Publishers, Worthington, OH, pp. 339–362.
22. Moura, I., Machado, A.V., Duarte, F.M., and Nogueira, R. 2011. Biodegradability assessment of aliphatic polyesters-based blends using standard methods. *Journal of Applied Polymer Science* 19: 3338–3346.
23. Vert, M. 2005. Aliphatic polyesters: Great degradable polymers that cannot do everything. *Biomacromolecules* 6: 538–546.

24. Shimao, M. 2001. Biodegradation of plastics. *Current Opinion in Biotechnology* 12: 242–247.
25. Pavia, D.L., Lampman, G.M., and Kriz, G.S. 1998. *Introduction to Organic Laboratory Techniques*, 3rd edn. Saunders, Fort Worth, TX.
26. Simon, C.M., Kaminsky, W., and Schlesselmann, B. 1996. Pyrolysis of polyolefins with steam to yield olefins. *Journal of Analytical and Applied Pyrolysis* 38: 75–87.
27. Nakamura, E.M., Cordi, L., Almeida, G.S.G., Duran, N., and Mei, L.H.I. 2005. Study and development of LDPE/starch partially biodegradable compounds. *Journal of Materials Processing Technology* 162–163: 236–241.
28. Balzano, L. 2008. Flow induced crystallization of polymers. PhD thesis, Eindhoven University of Technology, Eindhoven, the Netherlands.
29. Albertsson, A.C. and Bánhidi, Z.G. 1980. Microbial and oxidative effects in degradation of polyethylene. *Journal of Applied Polymer Science* 25: 1655–1671.
30. Cruz-Pinto, J.J.C., Carvalho, M.E.S., and Ferreira, J.F.A. 1994. The kinetics and mechanism of polyethylene photo-oxidation. *Angewandte Makromoleculare Chemie* 216: 113–133.
31. Albertsson, A.C., Barenstedt, C., and Karlsson, S. 1994. Degradation of enhanced environmentally degradable polyethylene in biological aqueous media: Mechanisms during the first stages. *Journal of Applied Polymer Science* 51: 1097–1105.
32. El-Rehim, H.A.A., El-Sayed, A., Hegazy, A.M.A., and Rabie, A.M. 2004. Synergistic effect of combining UV-sunlight–soil burial treatment on the biodegradation rate of LDPE/starch blends. *Journal of Photochemistry and Photobiology A: Chemistry* 163: 547–556.
33. Albertsson, A.C. and Karlsson, S. 1990. The influence of biotic and abiotic environments on the degradation of polyethylene. *Progress in Polymer Science* 15: 177–192.
34. Albertsson, A.C., Andersson, S.O., and Karlsson, S. 1987. The mechanism of biodegradation of polyethylene. *Polymer Degradation and Stability* 18: 73–87.
35. Huang, J., Shetty, A.S., and Wang, M. 1990. Biodegradable plastics: A review. *Advanced in Polymers Technology* 10: 23–30.
36. Chandra, R. and Rustgi, R. 1997. Biodegradation of maleated linear low-density polyethylene and starch blends. *Polymer Degradation and Stability* 56: 185–202.
37. Sasek, V., Vitásek, J., Chromcová, D., Prokopová, I., Brozek, J., and Náhlík, J. 2006. Biodegradation of synthetic polymer by composting and fungal treatment. *Folia Microbiology* 51: 425–430.
38. Shah, A.A., Hasan, F., Hameed, A., and Ahmed, S. 2008. Biological degradation of plastics: A comprehensive review. *Biotechnology Advances* 26: 246–265.
39. Sastry, P.K., Satyanarayana, D., and Rao, D.V.M. 1998. Accelerated and environmental weathering studies on polyethylene-starch blend films. *Journal of Applied Polymer Science* 70: 2251–2257.
40. Seidenstücker, T. and Fritz, H.G. 1998. Innovative biodegradable materials based upon starch and thermoplastic poly(ester-urethane) (TPU). *Polymer Degradation and Stability* 59: 279–285.
41. Yamada-Onodera, K., Mukumoto, H., Katsuyaya, Y., Saiganji, A., and Tani, Y. 2001. Degradation of polyethylene by a fungus *Penicillium simplicissimum* YK. *Polymer Degradation and Stability* 72: 323–327.
42. Bonhomme, S., Cuer, A., Delort, A.M., Lemaire, J., Sancelme, M., and Scott, G. 2003. Environmental biodegradation of polyethylene. *Polymer Degradation and Stability* 81: 441–452.
43. Wang, Y.Z., Yang, K.K., Wang, X.L., Zhou, Q., Zheng, C.Y., and Chen, Z.F. 2004. Agricultural application and environmental degradation of photo-biodegradable polyethylene mulching films. *Journal of Polymer Environment* 12: 7–10.
44. Arnaud, R., Dabin, P., Lemaire, J., Al-Malaika, S., Chohan, S., and Coker, M. 1994. Photooxidation and biodegradation of commercial photodegradable polyethylenes. *Polymer Degradation and Stability* 46: 211–224.
45. Weiland, M., Daro, A., and David, C. 1995. Biodegradation of thermally oxidized polyethylene. *Polymer Degradation and Stability* 48: 275–289.
46. Scott, G. 1997. Abiotic control of polymer biodegradation. *Trends in Polymer Science* 5: 361–368.
47. Albertsson, A.C., Barenstedt, C., Karlsson, S., and Lindberg, T. 1995. Degradation product pattern and morphology changes as means to differentiate abiotically and biotically aged degradable polyethylene. *Polymer* 36: 3075–3083.
48. Karlsson, S. and Albertsson, A.C. 1998. Biodegradable polymers and environmental interaction. *Polymer Engineering Science* 38: 1251–1253.

49. Chiellini, E., Corti, A., and Swift, G. 2003. Biodegradation of thermally-oxidized fragmented low density polyethylenes. *Polymer Degradation and Stability* 81: 341–351.
50. Jakubowicz, I. 2003. Evaluation of degradability of biodegradable polyethylene (PE). *Polymer Degradation and Stability* 80: 39–43.
51. Wiles, D.M. 2005. Oxo-biodegradable polyolefins. In: *Biodegradable Polymers for Industrial Applications*, Smith, R., Ed. Woodhead Publishing, Ltd., Cambridge, U.K.
52. Scott, G. and Wiles, D.M. 2001. Programmed-life plastics from polyolefins: A new look at sustainability. *Biomacromolecules* 2: 615–622.
53. Tambe, S.P., Singh, S.K., Patri, M., and Kumar, D. 2008. Ethylene vinyl acetate and ethylene vinyl alcohol copolymer for thermal spray coating application. *Progress Organic Coatings* 62: 382–386.
54. Henderson, A.M. 1993. Ethylene—Vinyl acetate (EVA) copolymers: A general review. *American Society for Testing and Materials* 9: 30–38.
55. Nabar, Y., Raquez, J.M., Dubois, P., and Narayan, R. 2005. Production of starch foams by twin-screw extrusion: Effect of maleated poly(butylene adipate-co-terephthalate) as a compatibilizer. *Biomacromolecules* 6: 807–817.
56. Tserki, V., Matzinos, P., Pavlidou, E., Vachliotis, D., and Panayiotou, C. 2006. Biodegradable aliphatic polyesters. Part I. Properties and biodegradation of poly(butylenesuccinate-co-butylene adipate). *Polymer Degradation and Stability* 91: 367–376.
57. Gross, A. and Kalra, B. 2002. Biodegradable polymers for the environment. *Science* 297: 803–806.
58. Cranston, E., Kawada, J., Raymond, S., Morin, F.G., and Marchessault, R.H. 2003. Cocrystallization model for synthetic biodegradable poly(butylene adipate-co-butylene terephthalate). *Biomacromolecules* 4: 995–999.
59. Camino, G. 2009. *Conference of Bioplastics and Related Materials*. Gargnano, Italy.
60. Chandra, R. and Rustgi, R. 1998. Biodegradable polymers. *Progress in Polymer Science* 23: 1273–1335.
61. Machado, A.V., Moura, I., Duarte, F.M., Botelho, G., Nogueira, R., and Brito, A.G. 2007. Evaluation of properties and biodeterioration's potential of polyethylene and aliphatic polyesters blends. *International Polymer Processing* XXII: 512–518.
62. Scherer, T.M., Fuller, R.C., Lenz, R.W., and Goodwin, S. 1999. Hydrolase activity of an extracellular depolymerase from *Aspergillus fumigatus* with bacterial and synthetic polyesters. *Polymer Degradation and Stability* 64: 267–275.
63. Edlund, U. and Albertsson, A.C. 2003. Polyesters based on diacid monomers. *Advanced Drug Delivery Reviews* 55: 585–609.
64. Penco, M., Sartore, L., Bignotti, F., D'Antone, S., and Landro, L. 2000. Thermal properties of a new class of block copolymer based on segments of poly(D,L—lactic-glycolic acid) and poly(epsilon-caprolactone). *European Polymer Journal* 36: 901–908.
65. Kim, M.N., Kim, K.H., Jin, H.J., Park, J.K., and Yoon, J.S. 2001. Biodegradability of ethyl and n-octyl branched poly(ethylene adipate) and poly(butylene succinate). *European Polymer Journal* 37: 1843–1847.
66. Jin, H.J., Kim, D.S., Lee, B.Y., and Atul, J.S. 2000. Chain extension and biodegradation of poly(butylenes succinate) with maleic acid units. *Journal of Polymer Science Part B: Polymer Physics* 38: 2240–2246.
67. Bitritto, M.M., Bell, J.P., Brenchle, G.M., Huang, S.J., and Knox, J.R. 1979. Synthesis and biodegradation of polymers derived from hydroxyacids. *Journal of Applied Polymer Science Applied Polymer Symposium* 35: 405–414.
68. Tokiwa, Y. and Suzuki, T. 1981. Hydrolysis of copolyesters containing aromatic and aliphatic ester blocks by lipase. *Journal of Applied Polymer Science* 26: 441–448.
69. Tokiwa, Y., Ando, T., and Suzuki, T. 1976. Degradation of polycaprolactone by a fungus. *Journal of Fermentation Technology* 54: 603–608.
70. Duda, A. and Penczek, S. 2002. Mechanisms of aliphatic polyester formation. In: *Biopolymers*, Steinbüchel, A. and Doi, Y., Eds. Wiley-VCH, Weinheim, Germany, pp. 371–383.
71. Jin, H.J., Lee, B.Y., Kim, M.N., and Yoon, J.S. 2000. Thermal and mechanical properties of mandelic acid-copolymerized poly(butylenes-succinate) and poly(ethylene adipate). *Journal of Polymer Science Part B: Polymer Physics* 38: 1504–1511.
72. Kang, H.J. and Park, S.S. 1999. Characterization and biodegradability of poly(butylenes adipate-co-succinate)/poly(butylenes terephthalate) copolyester. *Journal of Applied Polymer Science* 72: 593–608.

73. Salom, C., Nava, D., Prolongo, M.G., and Masegosa, R.M. 2006. Poly(β-caprolactone) + unsaturated isophthalic polyester blends: Thermal properties and morphology. *European Polymer Journal* 42: 1798–1810.

74. Calil, M.R., Gaboardi, F., Guedes, C.G.F., and Rosa, D.S. 2006. Comparison of the biodegradation of poly(ε-caprolactone), cellulose acetate and their blends by the Sturm test and selected cultured fungi. *Polymer Testing* 25: 597–604.

75. Ray, S.S. and Bousmina, M. 2005. Biodegradable polymers and their layered silicate nanocomposites: In greening the 21st century materials world. *Progress in Materials Science* 50: 962–1079.

76. Drumright, R.E., Gruber, P.R., and Henton, D.E. 2000. Polylactic acid technology. *Advanced Materials* 12: 1841–1846.

77. Whiteman, N. 2000. *Polymers, Laminations and Coatings Conference*. Chicago, IL, 2000, pp. 631–635.

78. Massadier-Nageotte, V., Pestre, C., Cruard-Pradet, T., and Bayard, R. 2006. Aerobic and anaerobic biodegradability of polymer films and physico-chemical characterization. *Polymer Degradation and Stability* 91: 620–627.

79. Datta, R., Tsai, S.P., Bonsignore, P., Moon, S.H., and Frank, J.R. 2006. Technological and economic potential of poly(lactic acid) and lactic acid derivatives. *FEMS Microbiology Reviews* 16: 221–231.

80. Auras, R., Harte, B., and Selke, S. 2004. An overview of polylactides as packaging materials. *Macromolecular Bioscience* 4: 835–864.

81. Vasanthan, N. and Ly, O. 2009. Effect of microstructure on hydrolytic degradation studies of poly(L-lactic acid) by FTIR spectroscopy and differential scanning calorimetry. *Polymer Degradation and Stability* 94: 1364–1372.

82. Nair, L.S. and Laurencin, C.T. 2006. Polymers as biomaterials for tissue engineering and controlled drug delivery. *Advanced Biochemical Engineering/Biotechnology* 102: 47–90.

83. Nam, Y.S. and Park, T.G. 1999. Biodegradable polymeric microcellular foams by modified thermally induced phase separation method. *Biomaterials* 20: 1783–1790.

84. Okuzaki, H., Kubota, I., and Kunugi, T. 1999. Mechanical properties and structure of the zone-drawn poly (L-lactic acid) fibers. *Journal of Polymer Science Part B: Polymer Physics* 37: 991–996.

85. Kulkarni, R.K., Moore, E.G., Hegyeli, A.F., and Leonard, F. 1971. Biodegradable poly(lactic acid) polymers. *Journal Biomedical Materials Resistance* 5: 169–181.

86. Tsuji, H. and Nakahara, K. 2002. Poly(L-lactide). IX. Hydrolysis in acid media. *Journal of Applied Polymer Science* 86: 186–194.

87. Ajioka, M., Suizu, H., Higuchi, C., and Kashima, T. 1998. Aliphatic polyesters and their copolymers synthesized through direct condensation polymerization. *Polymer Degradation and Stability* 59: 137–143.

88. Brown, W.H. and Poon, T. 2005. *Introduction to Organic Chemistry*. Wiley, Hoboken, NJ, ISBN 0-471-44451-0.

89. Lawton, J.W. 1996. Effect of starch type on the properties of starch containing films. *Carbohydrate Polymers* 29: 203–208.

90. Briassoulis, D. 2004. An overview on the mechanical behavior of biodegradable agricultural films. *Journal of Polymers and the Environment* 12: 65–81.

91. Yew, G.H., Yusof, A.M.M., Ishak, Z.A.M., and Ishiaku, U.S. 2005. Natural weathering effects on mechanical properties of Polylactic Acid/Rice Starch Composites. *Proceedings of the Eighth Polymers for Advanced Technologies International Symposium Budapest*. Hungary, Europe, September 13–16.

92. Doi, Y. 1990. *Microbial Polyesters*. VCH Publishers, New York.

93. Gu, J.D., Ford, T.E., Mitton, D.B., and Mitchell, R. 2000. Microbial degradation and deterioration of polymeric materials. In: *The Uhlig Corrosion Handbook*, Review, Ed., 2nd edn. Wiley, New York, pp. 439–460.

94. Barlaz, M.A., Ham, R.K., and Schaefer, D.M. 1989. Mass-balance analysis of anaerobically decomposed refuse. *Journal of Environmental and Engineering* 115: 1088–1102.

95. Ohtaki, A. and Nakasaki, N. 1998. Biodegradation of poly-ε-caprolactone under controlled composting conditions. *Polymer Degradation and Stability* 61: 499–505.

96. Albertsson, A.C. 2002. *Degradable Aliphatic Polyesters; Advances in Polymer Science*, vol. 157. Springer, Berlin, Germany.

97. Weir, N.A., Buchanan, F.J., Orr, J.F., Farrar, D.F., and Boyol, A. 2004. Processing, annealing and sterilisation of poly-L-lactide. *Biomaterials* 25: 3939–3949.

98. Ikada, Y. and Tsuji, H. 2000. Biodegradable polyesters for medical and ecological applications. *Macromolecular Rapid Communications* 21: 117–132.

99. Scott, G. 1999. Antioxidant control of polymer biodegradation. *Macromolecular Symposia* 144: 113–125.

100. Tokiwa, Y. and Calabia, B.P. 2004. Degradation of microbial polyesters. *Biotechnology Letters* 26: 1181–1189.

101. Artham, T. and Doble, M. 2008. Biodegradation of aliphatic and aromatic polycarbonates. *Macromolecular Bioscience* 8: 14–24.

102. Tokiwa, Y. and Suzuki, T. 2003. Hydrolysis of copolyesters containing aromatic and aliphatic ester blocks by lipase. *Journal of Applied Polymer Science* 26: 441–448.

103. Müller, R.J., Witt, U., Rantze, E., and Deckwer, W.D. 1998. Architecture of biodegradable copolyesters containing aromatic constituents. *Polymer Degradation and Stability* 59: 203–208.

104. Chen, Y., Tan, L., Chen, L., Yang, Y., and Wang, X. 2008. Study on biodegradable aromatic/aliphatic copolyesters. *Brazilian Journal of Chemical Engineering* 25: 321–335.

105. Sahoo, S.K. and Prusty, A.K. 2010. Two important biodegradable polymers and their role in nanoparticle preparation by complex coacervation method—A review. *International Journal of Pharmaceutical and Applied Sciences* 1(2): 1–8.

106. Huang, S.J., Bitritto, M., Leong, K.W., Paulisko, J., Roby, M., and Knox, J.R. 1978. The effects of some structural variations on the biodegradability of step-growth polymers. *Advances in Chemistry Series* 169: 205–214.

107. Huang, S.J., Bansleben, D.A., and Knox, J.R. 1979. Biodegradable polymers: Chymotrypsin degradation of a low-molecular weight poly(ester-urea) containing phenylalanine. *Journal of Applied Polymer Science* 23: 429–437.

108. Matsumura, S., Shimura, Y., Terayama, K., and Kiyohara, T. 1994. Effects of stereoregularity on biodegradation of poly(vinyl alcohol) (PVA) by *Alcaligenes faecalis*. *Biotechnology Letters* 16: 1205–1210.

109. Tokiwa, Y. and Suzuki, T. 1977. Degradation of polycaprolactone by a fungus. *Agricultural and Biological Chemistry* 41: 265–274.

110. Potts, J.E. 1984. Plastics, environmentally degradable. In: *Kirk-Othmer Encyclopedia of Chemical Technology*, Grayson, M., Ed. Wiley-Interscience, New York, p. 626.

111. Toncheva, V., Bulcke, A.V.D., Schacht, E., Mergaert, J., and Swings, J. 1996. Synthesis and environmental degradation of polyesters based on poly (ε-caprolactone). *Journal Environmental Polymer Degradation* 4: 71–83.

112. Moura, I., Machado, A.V., Nogueira, R., and Bounor-Legare, V. 2011. Biobased grafted polyesters prepared by in situ ring-opening polymerization. *Reactive and Functional Polymers Journal* (in press).

113. Jenkins, M.J. and Harrison, K.L. 2008. The effect of crystalline morphology on the degradation of poly(ε-caprolactone) in a solution of phosphate buffer and lipase. *Polymer for Advanced Technologies* 19: 1901–1906.

114. Wu, C.S. 2003. Physical properties and biodegradability of maleated-polycaprolactone starch composite. *Polymer Degradation and Stability* 80: 127–134.

115. Pandey, J.K., Rutot, D., Degée, Ph., and Dubois, Ph. 2003. Biodegradation of poly(ε-caprolactone)/starch blends and composites in composting and culture environments: The effect of compatibilization on the inherent biodegradability of the host polymer. *Carbohydrate Research* 338: 1759–1769.

116. Ha, C.S. and Cho, W.J. 2002. Miscibility, properties and biodegradability of microbial polyester containing blends. *Progress in Polymer Science* 27: 759–809.

117. Kumagai, Y., Kanesawa, Y., and Doi, Y. 1992. Enzymatic degradation of microbial poly(3-hydroxybutyrate) films. *Makromolecular Chemistry* 193: 53–57.

118. Arutchelvi, J., Sudhakar, M., Arkatkar, A., Doble, M., Bhaduri, S., and Uppara, P.V. 2008. Biodegradation of polyethylene and polypropylene. *Indian Journal of Biotechnology* 7: 9–12.

119. Taşdemir, M., Topsakaloglu, M., and Caneba, G.T. 2010. Dynamically vulcanized EPDM (ethylene-propylene-diene/PP) (Polypropylkene) polymer blend performance index (IP). *Journal of Polymer Materials* 27: 347–359.

120. Bikales, M. and Menges, O. 1991. *Encyclopedia of Polymer Science and Engineering*, 2nd edn. Wiley Inter-Science Publication, New York, pp. 169–177.

121. Machado, A.V., Covas, J.A., Bounor-Legare, V., and Cassagnau, P. 2009. Reactive polymer processing and design of stable micro- and nanostructures. In: *Advances in Polymer Processing from Macro to Nano Scales*, Thomas, S. and Weimin, Y., Eds. Woodhead Publishing Limited, Cambridge, U.K., pp. 579–615.

122. Cartier, H. and Hu, G.H. 2001. A novel reactive extrusion process for compatibilizing immiscible polymer blends. *Polymer* 42: 8807–8816.
123. Riaz, M. 1997. *Extruders in Food Applications*. Technomic Publishing Co., Lancaster, PA.
124. Xanthos, M. 1992. *Reactive Extrusion, Principles and Practice*. Hanser Publishers, New York.
125. Utracki, L.A. 1989. *Polymer Alloys and Blends*. Hanser Publishers, New York.
126. Paul, D.R., Barlow, J.W., and Keskula, H. 1988. *Polymer Blends, Encyclopedia Polymer Science Engineering*, 2nd edn., vol. 12. Wiley Interscience, New York, p. 399.
127. Paul, D.R. and Newman, S. 1979. *Polymer Blends*. Academic Press, New York.
128. Utracki, L.A. 1994. *Encyclopaedic Dictionary of Commercial Polymer Blends*. ChemTec Publishing, Toronto, Canada.
129. Datta, S. and Lohse, D. 1996. *Polymeric Compatibilizers*. Hanser Publishers, New York.
130. Koning, C., Duin, M., Pagnoulle, C., and Jérôme, R. 1998. Strategies for compatibilization of polymer blends. *Progress in Polymer Science* 23: 707–757.
131. Zhang, Q., Yang, H., and Fu, Q. 2004. Kinetics-controlled compatibilization of immiscible polypropylene/polystyrene blends using nano-SiO_2 particles. *Polymer* 45: 1913–1922.
132. Gaylord, N. 1989. Compatibilizing agents: Structure and function in polyblends. *Journal of Macromolecular Science* 26: 1211.
133. Xanthos, M. and Dagli, S. 1991. Compatibilization of polymer blends by reactive processing. *Polymer Engineering and Science* 31: 929–935.
134. Triacca, V.J., Ziaee, S., Barlow, J.W., Keskkula, H., and Paul, D.R. 1991. Reactive compatibilization of blends of nylon 6 and ABS materials. *Polymer* 32: 1401–1413.
135. Dali, S.S., Xanthos, M., and Biesenberger, J.A. 1994. Kinetic studies and process analysis of the reactive compatibilization of nylon 6/polypropylene blends. *Polymer Engineering and Science*. 1720.
136. Faircloth, B., Rohrs, H., Tiberio, R., Ruolff, R., and Krchnavek, R.R. 2000. Bilayer, nanoimprint lithography. *Journal of Vacuum Science & Technology B* 18: 1866–1873.
137. Xia, Y. and Whitesides, G.M. 1998. Soft lithography. *Angewandte Chemie International Edition* 37: 550–575.
138. Michel, B., Bernard, A., Bietsch, A., Dalamarche, E., Geissler, M., Juncker, D., Kind, H., Renault, J.P., Rothuizen, H., Schmid, H., Schmidt-Winkel, P., Stutz, R., and Wolf, H. 2001. Printing meets lithography: Soft approaches to high-resolution patterning. *Journal of Research and Development* 45: 697–719.
139. Delamarche, E., Schmid, H., Michel, B., and Biebuyck, H. 1997. Stability of molded polydimethylsiloxane microstructures. *Advanced Materials* 9: 741–746.
140. Bietsch, A. and Michel, B. 2000. Conformal contact and pattern stability of stamps used for soft lithography. *Journal of Applied Physics* 88: 4310–4318.
141. Suh, K.Y., Kim, Y.S., and Lee, H.H. 2001. Capillary force lithography. *Advanced Materials* 13: 1386–1389.
142. Cartier, H. and Hu, G.H. 2000. Compatibilization of polypropylene and poly(butylenes terephthalate) blends by reactive extrusion: Effects of the molecular structure of a reactive compatibilizer. *Journal of Materials Science* 35: 1985–1996.
143. Ide, F. and Hasegawa, A. 1974. Studies on polymer blend of nylon 6 and polypropylene or nylon 6 and polystyrene using the reaction of polymer. *Journal of Applied Polymer Science* 18: 963–974.
144. Xiaomin, Z., Gang, L., Dongmei, W., Zhihui, Y., Jinghua, Y., and Jingshu, L. 1998. Morphological studies of polyamide 1010/polypropylene blends. *Polymer* 39: 15–21.
145. Cartier, H. and Hu, G.H. 1999. Morphology development of in situ compatibilized semicrystalline polymer blends in a co-rotating twin-screw extruder. *Polymer Engineering and Science* 39: 996–1013.
146. Meijer, H.E.M. and Elemans, P.H.M. 1988. The modeling of continuous mixers. Part I: The corotating twin-screw extruder. *Polymer Engineering and Science* 28: 275–290.
147. Pawlak, A., Morawiec, J., Pazzagli, F., Pracella, M., and Galeski, A. 2000. Recycling of postconsumer poly(ethylene terephthalate) and high-density polyethylene by compatibilized blending. *Journal of Applied Polymer Science* 86: 1473–1485.
148. Contat-Rodrigo, L. and Greus, A.R. 2002. Biodegradation studies of LDPE filled with biodegradable additives: Morphological changes. I. *Journal Applied Polymer Science* 83: 1683–1691.
149. Matzinos, P., Tserki, V., Gianikouris, C., Pavlidou, E., and Panayiotou, C. 2002. Processing and characterization of LDPE/starch/PCL blends. *European Polymer Journal* 38: 1713–1720.
150. Shulman, J. and Howarth, J.T. 1964. US Patent 137: 664.

151. Fringant, C., Desbrières, J., and Rinaudo, M. 1996. Physical properties of acetylated starch-based materials: Relation with their molecular characteristics. *Polymer* 37: 2663–2673.
152. Avérous, L., Moro, L., and Fringant, C. 1999. Properties of plasticized starch acetates. *ICBT Conference.* Coimbra, Portugal, September 28–30.
153. van Soest, J. 1996. *Starch Plastics Structure–Property Relationships.* P&L Press, Wageningen, Germany/ Utrecht University, Utrecht, the Netherlands.
154. Rodriguez-Gonzalez, F.J., Ramsay, B.A., and Favis, B.D. 2003. High performance LDPE/thermoplastic starch blends: A sustainable alternative to pure polyethylene. *Polymer* 44: 1517–1526.
155. Shin, B.Y., Lee, S.I., Shin, Y.S., Balakrishnan, S., and Narayan, R. 2004. Rheological, mechanical and biodegradation studies of blends of thermoplastic starch and polycaprolactone. *Polymer Engineering & Science* 44: 1429–1438.
156. Thakore, I.M., Desai, S., Sarawade, B.D., and Devi, S. 2001. Studies on biodegradability, morphology and thermomechanical properties of LDPE—Modified starch blends. *European Polymer Journal* 37: 151–160.
157. Abdul-Khalil, H.P.S., Chow, W.C., Rozman, H.D., Ismail, H., Ahmad, M.N., and Kumar, R.N. 2001. The effect of anhydride modification of sago starch on the tensile and water absorption properties of sago-filled linear low-density polyethylene (LLDPE). *Polymer-Plastics Technology and Engineering* 40: 249–263.
158. Kiatkamjornwong, S., Thakeow, P., and Sonsuk, M. 2001. Chemical modification of cassava starch for degradable polyethylene sheets. *Polymer Degradation and Stability* 73: 363–375.
159. Bikiaris, D., Prinos, J., Koutsopoulos, K., Vouroutzis, N., Pavlidou, E., Frangi, N., and Panayiotou, C. 1998. LDPE/plasticized starch blends containing PE-g-MA copolymer as compatibilizer. *Polymer Degradation and Stability* 59: 287–291.
160. Yoo, S.I., Lee, T.Y., Yoon, J.S., Lee, I.M., Kim, M.N., and Lee, H.S. 2002. Interfacial adhesion reaction of polyethylene and starch blends using maleated polyethylene reactive compatibilizer. *Journal of Applied Polymer Science* 83: 767–776.
161. Matzinos, P., Bikiaris, D., Kokkou, S., and Panayiotou, C. 2001. Processing and characterization of LDPE/starch products. *Journal of Applied Polymer Science* 79: 2548–2557.
162. Nikazar, M., Milani, Z., Safari, B., and Bonakdarpour, B. 2005. Improving the biodegradability and mechanical strength of corn starch-LDPE Blends through formulation modification. *Iranian Polymer Journal* 14: 1050–1057.
163. Swanson, C.L., Shogren, R.L., Fanta, G.F., and Imam, S.H. 1993. Starch-plastic materials—Preparation, physical properties, and biodegradability. *Journal of Environmental Polymer Degradation* 1: 155–166.
164. Mihai, M., Huneault, M.A., Favis, B.D., and Li, H. 2007. Foaming of PLA/thermoplastic starch blends. *Macromolecular Bioscience* 7: 907–920.
165. Ljungberg, N., Anderson, T., and Wesslen, B. 2003. Film extrusion and film weldability of poly(lactic acid) plasticized with triacetine and tributyl citrate. *Journal of Applied Polymer Science* 88: 3239–3247.
166. Kramschuster, A., Pilla, S., Gong, S., Chandra, A., and Turng, L.S. 2007. Injection molded solid and microcellular polylactide compounded with recycled paper shopping bag fibers. *International Polymer Processing* 12: 436–445.
167. Shibata, M., Ozawa, K., Teramoto, K., Yosomya, R., and Takeishi, H. 2003. Biocomposites made from short abaca fiber and biodegradable polyesters. *Macromolecular Materials Engineering* 288: 35–43.
168. Baker, W., Scott, C., and Hug, H. 2001. *Reactive Polymer Blending.* Hanser Publishers, Connecticut, OH.
169. Szwarc, M. 1956. Living polymers. *Nature* 178: 1168–1169.
170. Raquez, J.M., Degée, P., Dubois, P., Balakrishnan, S., and Narayan, R. 2005. Melt-stable poly(1,4-dioxan-2-one) (co)polymers by ring-opening polymerization via continuous reactive extrusion. *Polymer Engineering and Science* 45: 622–629.
171. Mecerreyes, D., Jérôme, R., and Dubois P. 1999. Novel macromolecular architectures based on aliphatic polyesters: Relevance of the "coordination-insertion" ring-opening polymerization. *Advanced in Polymer Science* 147: 2–59.
172. Balakrishnan, S., Krishnan, M., Dubois, P., and Narayan, R. 2006. Three-arm poly (ε-caprolactone) by extrusion polymerization. *Polymer Engineering and Science* 46: 235–240.
173. Jacobsen, S., Fritz, H.G., Degée, P., Dubois, P., and Jérôme, R. 2000. New developments on the ring opening polymerisation of polylactide. *Industrial Crops and Products* 11: 265–275.

174. Degée, P., Dubois, P., Jérôme, R., Jacobsen, S., and Fritz, H.G. 1999. New catalysis for fast bulk ring-opening polymerization of lactide monomers. *Macromolecular Symposia* 144: 289–302.

175. Jacobsen, S., Fritz, H.G., Degée, Ph., Dubois, Ph., and Jérôme, R. 2000. Single-step reactive extrusion of PLLA in a corotating twin-screw extruder promoted by 2-ethylhexanoic acid tin(II) salt and triphenylphosphine. *Polymer* 41: 3395–3403.

176. Kim, B.J. and White, J.L. 2003. Continuous polymerization of lactam-lactone block copolymers in a twin screw extruder. *Journal of Applied Polymer Science* 8: 1429–1437.

177. Piao, L., Deng, M., Chen, X., Jiang, L., and Jing, X. 2003. Ring-opening polymerization of ε-caprolactone and L-lactic acid using organic amino calcium catalyst. *Polymer* 44: 2331–2336.

178. Matyjaszewski, K., Shipp, D.A., Wang, J.L., Grimaud, T., and Patten, T.A. 1998. Utilizing halide exchange to improve control of atom transfer radical polymerization. *Macromolecules* 31: 6836–6840.

179. Hiltunen, K., Härkönen, M., Seppälä, J., and Väänänen, T. 1996. Synthesis and characterization of lactic acid based telechelic prepolymers. *Macromolecules* 29: 8677–8682.

180. Kylmä, J. and Seppälä, J. 1997. Synthesis and characterization of biodegradable thermoplastic poly(ester urethane) elastomer. *Macromolecules* 30: 2876–2882.

181. Hiltunen, K., Seppälä, J., and Härkönen, M. 1997. Lactic acid based poly(ester-urethane)s: The effect of different polymerization conditions on the polymer structure and properties. *Journal of Applied Polymer Science* 64: 865–873.

182. Stevels, W.M., Bernard, A., Van de Witte, P., Dijkstra, P.J., and Feijen, J. 1996. Block copolymers of poly(L-lactide) and poly(ε-caprolactone) or poly(ethylene glycol) prepared by reactive extrusion. *Journal of Applied Polymer Science* 62: 1295–1301.

183. Lee, B.H. and White, J.L. 2002. Formation of a polyetheramide triblock copolymer by reactive extrusion; process and properties. *Polymer Engineering & Science* 42: 1710–1723.

184. Machado, A.V., Bounor-Legare, V., Goncalves, N.D., Melis, F., Cassagnau, P., and Michel, A. 2008. Continuous polymerization of ε-caprolactone initiated by titanium phenoxide in a twin-screw extruder. *Journal of Applied Polymer Science* 110: 3480–3487.

185. Narayan, R., Krishnan, M., Snook, J.B., Gupta, A., and Dubois, P. 1997. Bulk-reactive-extrusion polymerization process producing aliphatic polyester compositions. WO9741165A1, Michigan State University, Ypsilanti, MI.

186. Jacobsen, S., Fritz, H.G., Degee, P., Dubois, P., and Jerome, R. 2000. Continuous reactive extrusion polymerization of L-lactide—An engineering view. *Macromolecular Symposium* 153: 261–273.

187. Kim, I. and White, J.L. 2005. Reactive copolymerization of various monomers based on lactams and lactones in a twin-screw extruder. *Journal Applied Polymer Science* 96: 1875–1887.

188. Dubois, P., Krishnan, M., and Narayan, R. 1999. Aliphatic polyester-grafted starch-like polysaccharides by ring-opening polymerization. *Polymer* 40: 3091–3100.

189. Dubois, P. and Narayan, R. 2003. Biodegradable compositions by reactive processing of aliphatic polyester/polysaccharide blends. *Macromolecular Symposia* 198: 233–243.

190. Becquart, F., Chalamet, Y., Chen, J., Zhao, Y., and Taha, M. 2009. Poly[ethylene-co-(vinyl alcohol)]-graft-poly(ε-caprolactone) synthesis by reactive extrusion, 1–structural and kinetics study. *Macromolecular Materials Engineering* 294: 643–650.

191. Zhao, Y., Becquart, F., Chalamet, Y., Chen, J.D., and Taha, M. 2009. Poly[ethylene-co-(vinyl alcohol)]-graft-poly(e-caprolactone) by reactive extrusion, 2–parameter analysis. *Macromolecular Materials Engineering* 294: 651–657.

192. Becquart, F., Taha, M., Zerroukhi, A., Kaczun, J., and Llauro, M.F. 2007. Microstructure and properties of poly(vinyl alcohol-co-vinyl acetate)-g-ε-caprolactone. *European Polymer Journal* 43: 1549–1556.

193. Oh, J.M., Lee, S.H., Son, J.S., Khang, G., Kim, C.H., Chun, H.J., Min, B.H., Kim, J.H., and Kim, M.S. 2009. Ring-opening polymerization of 3-caprolactone by poly(propyleneglycol) in the presence of a monomer activator. *Polymer* 50: 6019–6023.

194. Cayuela, J., Bounor-Legaré, V., Cassagnou, P., and Michel, A. 2006. Ring-opening polymerization of ε-caprolactone initiated with titanium *n*-propoxide or titanium phenoxide. *Macromolecules* 39: 1338–1346.

195. Moura, I., Machado, A.V., Nogueira, R., and Bounor-Legare, V. Influence of EVA-g-PLA grafted copolymers on physical properties and biodegradability. (Submitted to *Materials Chemistry and Physics*).

CHAPTER 7

Modification of Biodegradable Polymers through Reactive Extrusion-II

Sathya B. Kalambur

CONTENTS

7.1 CURRENT STATUS OF BIODEGRADABLE POLYMERS

About 100 billion tons of biomass is produced through photosynthesis annually, and the majority of this material comprises starch, cellulose, proteins, and other polysaccharides. In comparison, about 25 billion pounds of petroleum-based plastics were produced in the United States in 2008 [1,2]. The application of petroleum-based plastics exploded after World War II due to cheaper fossil fuel prices and better functionality including mechanical strength and barrier properties. However, after the fossil fuel crisis in 1973, development and application of biodegradable polymers gained fresh impetus.

Biodegradable polymers like rubber, proteins, starch, and cellulose have had a unique history of applications before synthetic plastics replaced them. Gluten proteins from wheat and zein proteins from corn have been used in the manufacture of wood composites [3]. Gliadin protein fibers have been used in medical applications to design materials that adhere and aid muscle growth [4]. In the 1950s, zein proteins were in high demand for producing various items including buttons, clothing, and furniture stuffing and sold under the brand name Vicara. In the 1940s, soy protein–based plastics were made famous by the Ford Motor Company who filed a patent for making car panels from soy meal. Similarly modified cellulosics including nitrocellulose were used in late nineteenth century to make billiard balls, photographic film, and subsequently windshields. Cellophane is another example of modified cellulose that is still being used in certain food packaging applications. Recently, one of the well-known commercial applications of starch is the Mater-Bi products produced by Italian company Novamont. By combining starches with different polymers including cellulosics, ethylene–vinyl alcohol, and polycaprolactone (PCL), plastics with different properties are produced and used in suitable applications like food trays, compostable garden waste bags, and food packaging [5]. In addition to earlier examples, significant work has been done to explore use of biodegradable polymers in different applications that have not been commercialized yet. Biodegradable polymers like polylactic acid (PLA) and polyhydroxy alkanoates (PHAs) are showing greater promise in expanding the application of biodegradable polymers. Properties of these polymers can be tailored to match those of synthetic plastics like polypropylene (PP). Also it is important to note that while selecting biodegradable polymers for various applications, it is necessary to consider the energy costs associated with manufacturing, modification, and disposal of these polymers.

Biodegradable polymers like starch and cellulose possess certain disadvantages that prevent them from replacing fossil fuel–based plastics in many applications including packaging, construction, and consumer products. They have to be modified in order to replace synthetic plastics in certain applications. However, these modifications can impair biodegradability of these polymers. For example, native cellulose has to be modified into its acetate before it can be used in applications including apparels and high absorbency products [1]. But modification of cellulose consumes energy and chemicals (from fossil fuels) and also reduces biodegradability of modified cellulose.

Polymers mentioned in the previous paragraph are derived from plants or other renewable sources; however, there are other biodegradable polymers that can be synthesized from petrochemicals. These include aliphatic polyesters (e.g., polybutylenes succinate—PBS), polyesteramides (copolymer of ε-caprolactam, adipic acid, and butanediol), and aliphatic–aromatic polyesters (e.g., copolymer of adipic acid and terephthalic acid). Because these polymers are derived from petroleum sources, interest in these polymers has decreased compared to previous years. However, significant quantities are still produced by large companies like BASF, Showa Denko, Dupont, and Eastman Chemical. These polymers have been used in commercial applications including flower pots, cutlery, agricultural mulch films, compostable packaging, and hydrophobic coatings on starch-based food trays [1].

This chapter will highlight the important role of reactive extrusion technology in expanding applications of biodegradable polymers. For example, relatively inexpensive polymers like starch or cellulose can be blended with expensive polymers like PHAs and PLA without significant property degradation using this technology.

7.1.1 Standards for Biodegradable Polymers

Many countries including the United States and European Union (EU) have standards in place to certify biodegradable plastics. In the United States, the American Society of Testing and Materials (ASTM) has relevant standards and testing methods for biodegradable materials. Similarly in EU, the same function is performed by the European Committee for Standardization (CEN). Specifications for biodegradability and compostability of plastics are covered in the ASTM 6400-99 and EN13432 standards. ASTM also provides other standards that include biodegradability testing methods for plastics in different environments [6].

At the end of their application, plastics are disposed in the environment or in some cases incinerated to regenerate energy. When disposed in the environment, they can be captured by active recycling or composting programs in the case of recyclable/compostable materials or go to landfill areas. Irrespective of where they end up, biodegradable plastics must be able to degrade under fully aerobic and anaerobic or limited aerobic conditions (e.g., landfills). Thus, biodegradable and compostable plastics must satisfy the following definitions:

Degradable plastic: A degradable plastic in which the degradation results from the action of naturally occurring microorganisms such as bacteria, fungi, and algae. The rate of biodegradation could also be an important standard in certain countries. In Germany, for example, a requirement for biodegradable plastics is that under laboratory conditions, at least 60% of the carbon must be converted into carbon dioxide (CO_2) within 6 months.

Compostable plastic: A plastic that undergoes degradation by biological processes during composting to yield carbon dioxide, water, inorganic compounds, and biomass at a rate consistent with other known, compostable materials and leaves no visually distinguishable or toxic residue. The absence of toxic residues is a key requirement that separates these plastics from other synthetic plastics that may be degradable but still leave behind low-molecular-weight oligomers.

7.1.2 Challenges to Widespread Use of Biodegradable Polymers

As discussed earlier, there are a wide variety of biodegradable polymer types derived from different sources and with different properties. The nature of disadvantages or drawbacks depends on each polymer type. Polymers derived from natural sources including starch and cellulose have limitations that are different from other polymers including PHA and PLA. Biodegradable polymers depending on type may have the following limitations:

1. Brittleness in the absence of suitable plasticizers (e.g., starch, cellulose, hydrophilic proteins).
2. Hydrophilic nature of polymers from natural sources and poor water resistance (same examples as mentioned earlier).
3. Deterioration of mechanical properties upon exposure to environmental conditions like humidity and microbial attack.
4. Soft and weak mechanical properties of starch in the presence of plasticizers. If sufficient amount of plasticizers are present, then retrogradation of starch amylose and amylopectin branches can weaken mechanical properties over time.
5. Cost per lb of biodegradable polymers compared to polymers from fossil fuels. Commodity synthetic plastics like PP or polyethylene (PE) cost approximately $0.40–$0.70/lb depending on grade, whereas costs of biodegradable polymers like PLA and PHA are between $1 and $10/lb, depending on manufacturing processes and polymer properties. However, other biodegradable polymers like starch and cellulose are relatively cheaper with starch costs ranging from 15 to 70 cents/lb. Thus, it is an attractive proposition to blend cheaper biodegradable polymers with more expensive ones.

Blending of two or more polymers invariably leads to deterioration in properties because of immiscibility of the polymers. This limitation can be overcome by compatibilizing the different polymers through covalent bonds that can be synthesized through organic reactions between functional groups on these polymers. These functional groups can be activated; and organic addition, substitution, or free radical chain reactions can be carried out to improve interfacial adhesion between polymers. Reactive extrusion then becomes a commercially attractive technology as it provides economies of scale, good degree of mixing between various reactants/polymers, and an ability to produce various shapes of the final extruded product. Reactive extrusion may not be only limited to blends of polymers but also can be employed to modify single polymers to improve properties. Subsequent sections of this chapter describe some applications of reactive extrusion technology to carry out reactions of the aforementioned nature.

7.2 BIODEGRADABLE POLYMERS

Before we dive into the technology of reactive extrusion, it is important to understand the molecular structures of various biodegradable polymers and their monomers. The following sections will attempt to provide insights on the source and presence of functional groups on various biodegradable polymers. Because of large number of biodegradable polymer types, only a few but important polymers will be covered in this section.

7.2.1 Starch

7.2.1.1 Source

Starch is one of the major components of cereal grains and isolated from cereal kernels or vegetable tubers or roots through the wet milling process or variations thereof. Corn and wheat are major sources of commercial starches in the United States. Other sources include rice, potato, peas, and

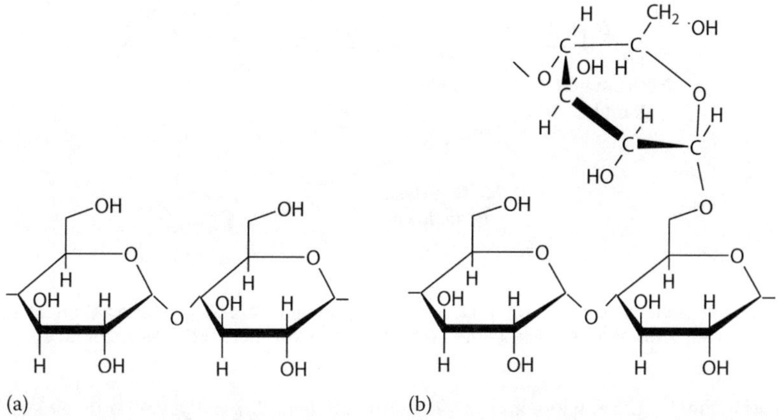

Figure 7.1   (a) Two basic linear 1,4-α-glucopyranosyl repeating units of starch (e.g., amylose) and (b) two basic linear 1,4-α-glucopyranosyl repeating units with a 1,6-branching (e.g., amylopectin).

tapioca. Special varieties of cereal and vegetable starches include waxy maize (100% amylopectin), high-amylose cornstarch (50% or 70% amylose), and waxy potato starch (100% amylopectin). Cost of typical cereal starches varies between $0.15 and $0.70/lb depending on the source and starch types. Due to rising prices of fossil fuels, on a dry weight basis, the cost of starches can compare favorably with bulk fossil fuel–based resins like PP or low-density PE (LDPE).

7.2.1.2 Molecular Structure of Starch

Starch is a mixture of two polysaccharides—amylose (linear 1,4-α-glucopyranosyl units) and amylopectin (linear 1,4-α-glucopyranosyl units and branched 1,6-α-glucopyranosyl units) [7]. The amylose fraction has a degree of polymerization (DP) of $1 \times 10^2 - 4 \times 10^5$, and amylopectin has a DP of $1 \times 10^4 - 4 \times 10^7$ with branches after every 19–25 linear units. Amylopectin has one of largest molecular weights of all naturally occurring polymers [7]. Significant amounts of amylopectin (approx. 72%) are present in native wheat or cornstarch, the rest being made up of amylose. The basic repeating unit of starch is the 1,4-α-glucopyranosyl unit and is shown in Figure 7.1. Figure 7.1a shows the linear chain present in amylose, while Figure 7.1b shows branching points characteristic of amylopectin.

7.2.1.3 Physicochemical Changes in Starch during Processing

Native starches are granular (2–55 μm) in nature with both amylose and amylopectin chains arranged in a radial direction. The granules contain both crystalline and noncrystalline regions in distinct regions, and only amylopectin chains contribute to crystallinity. Native corn and wheat starches which form a major percentage of all starch production consist of ca. 28% amylose and 72% amylopectin [7].

Native starch granules show unique behavior during processing. Depending on process conditions and formulation, starch granules can behave in different ways:

1. *Melting*: The melting point of native starch granules is very high at approximately 170°C–190°C [8]. Starch decomposition will occur before melting, and since mechanical properties are proportional to molecular weights, starch granule melting is not desirable.
2. *Gelatinization*: In the presence of heat and excess amounts of low-molecular-weight plasticizers like water and glycerol, starch granules swell. Upon application of shear, these swollen granules break

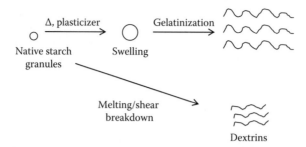

Figure 7.2 Effect of heat and shear on native starch granules inducing gelatinization at high moisture/
plasticizer content and melting/shear breakdown at low moisture/plasticizer levels.

up to release most of the amylose and amylopectin polymers. Typically, very little molecular weight
changes happen during the gelatinization process [8,9].

3. *Shear breakdown*: Application of shear energy during processing operations (e.g., extrusion, injection molding) results in granule size reduction and release of polymers from granule to the surrounding dispersed polymer matrix. Like melting, this process is also accompanied by starch polymer
molecular weight reduction. Figure 7.2 illustrates this concept of starch gelatinization and melting/
shear breakdown.

Wang et al. [8] have reported that when water was present at >60 wt%, starch granules underwent
gelatinization at 52°C–85°C (depending on starch type) [7] with little molecular weight degradation. At lower levels of water, gelatinization took place at higher temperatures. In the absence of any
water or plasticizers, there was only melting and shear-induced granular rupture. Thus, synthesis of
starch-based bioplastics requires processing operations that bring about gelatinization with minimal
melting or molecular weight degradation.

7.2.1.4 Functional Groups in Starch

As seen in Figure 7.1, one can notice immediately the presence of hydroxyl functional groups on
both amylose and amylopectin molecules in starch. In fact, there are three hydroxyl groups on each
monomer anhydroglucose unit in the polymer with one primary and two secondary hydroxyls. Thus,
there are several possible reactions that starch can be expected to undergo. Figures 7.3 through 7.6
illustrate the different types of reactions: esterification (Figure 7.3), oxidation (Figure 7.4), etherification (Figure 7.5), and cross-linking (Figure 7.6). The degree of substitution (DS) is an important
characteristic of modified starches. It is a measure of the average number of hydroxyl groups on
each glucopyranosyl unit of starch that is derivatized by substituent groups. DS is expressed as
moles of substituent per glucopyranosyl unit. Since the monomeric unit in starch has three hydroxyl
groups available for substitution, the maximum possible DS is 3. Starch derivatization can be carried out without degrading the starch granules so that the resulting derivatized or modified starch
is still in granular form. Modification can also be done on degraded (maltodextrins or dextrins) or
gelatinized starches.

Starch esters (Figure 7.3) can be prepared by reacting starch with acetic anhydride, chloroacetic acid and disodium hydrogen phosphate. Oxidized starches (Figure 7.4) are prepared by
reaction with hypochlorites or periodates. Starch ethers (Figure 7.5) are prepared by reactions
with ethylene or propylene oxide. Cross-linked starches (Figure 7.6) are produced by reaction
with epichlorohydrin. The readers are referred to excellent reviews of batch manufacturing processes and properties of various derivatized starches by different authors—oxidation [10–12],
cross-linking and esterification [13], and etherification [14–16]. It should be noted that while

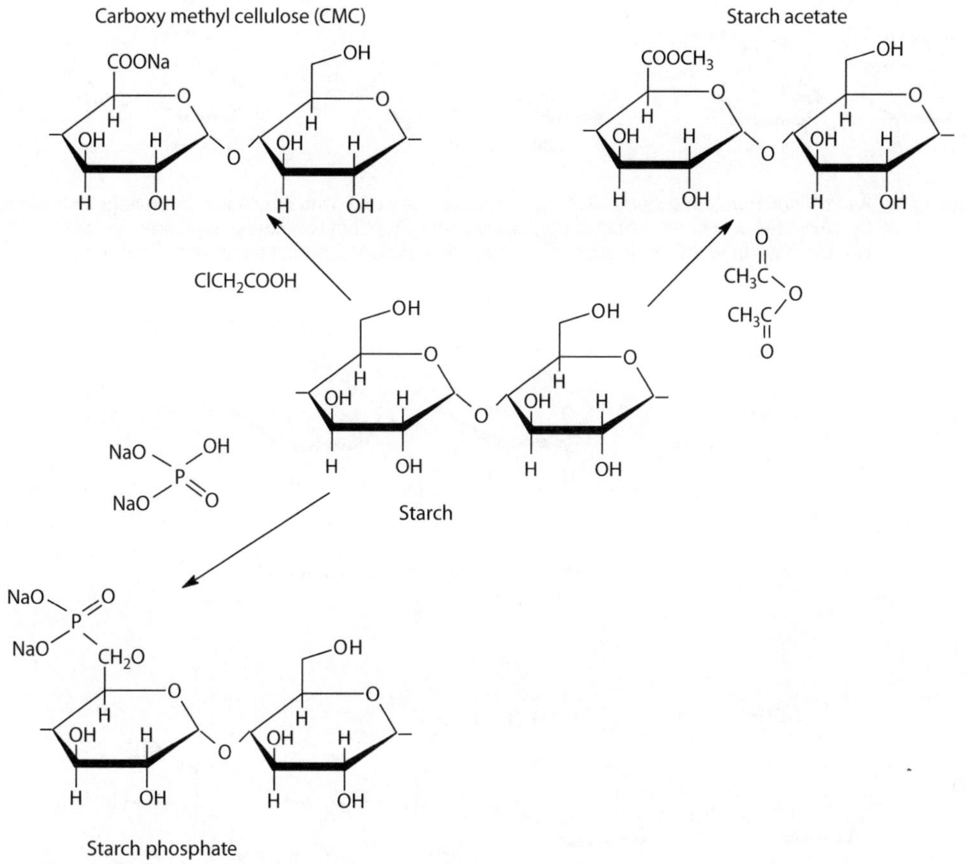

Figure 7.3   A simplified illustration of reaction of starch with chloroacetic acid, acetic anhydride, and disodium hydrogen phosphate to form carboxy methyl starch, starch acetate, and starch phosphate, respectively.

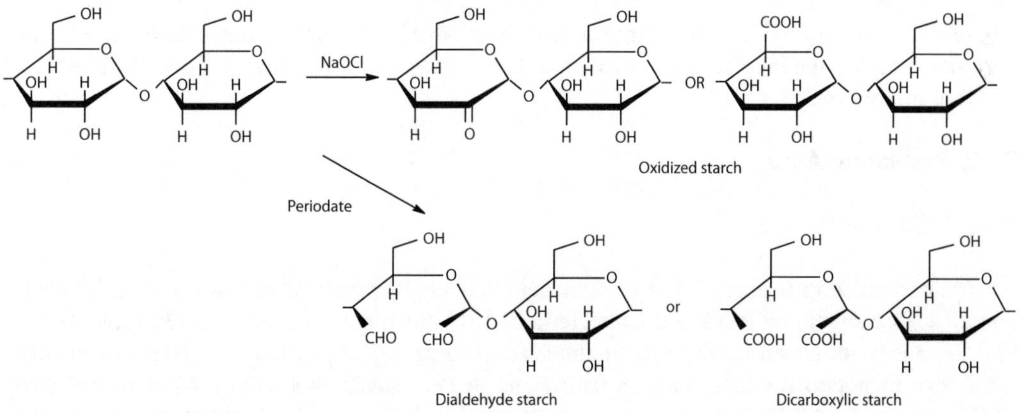

Figure 7.4   A simplified reaction scheme illustrating reaction of starch with hypochlorites to form oxidized starches with keto or carboxyl groups and periodate to form dialdehyde or dicarboxylic starches.

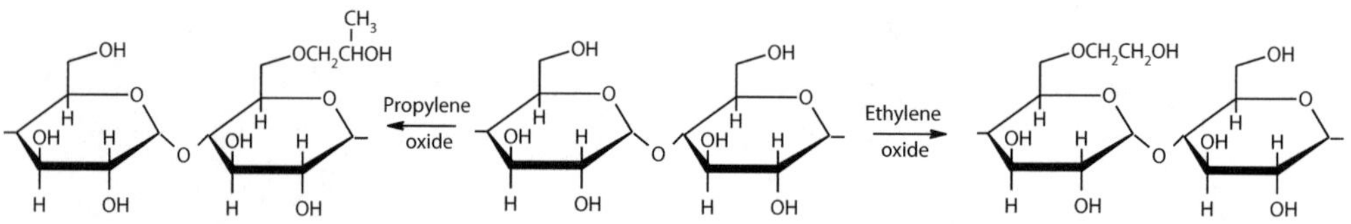

Figure 7.5 A simplified reaction scheme illustrating reaction of starch with ethylene or propylene oxide to form hydroxyethyl or hydroxypropyl starches, respectively. Propyl or ethyl groups on the modified starch can increase in length by reacting with additional moles of propylene or ethylene oxides.

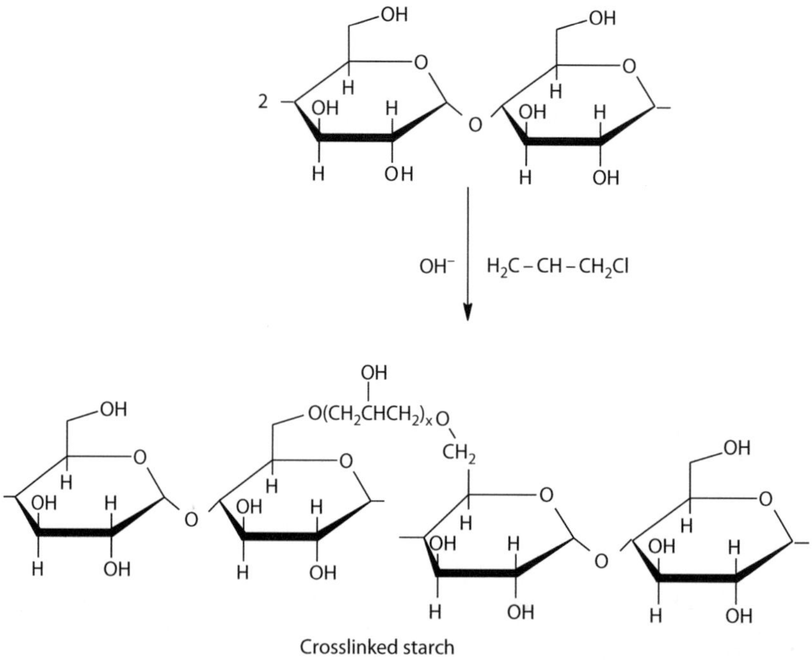

Crosslinked starch

Figure 7.6 A simplified reaction scheme illustrating the reaction of 2 mol of starch with epichlorohydrin to form cross-linked starches. X indicates the number of epichlorohydrin moles involved in the reaction.

Figures 7.3 through 7.6 illustrate a simplified mechanism of various reactions, the actual reactions are quite complex and involve side reactions and presence of additives (e.g., pH control agents and swelling inhibitors).

7.2.2 Polylactic Acid

7.2.2.1 Source

The basic monomer unit of PLA is lactic acid that can be produced in two ways: (a) fermentation of hexose sugars by lactobacillus bacteria and (b) chemical synthesis from lactonitrile [17]. PLA is mostly produced by the first method, and sources of sugars include glucose or maltose from corn or potato starches, sucrose from cane or beet sugar, and lactose from cheese whey [18]. Lactic acid produced by bacteria can be present in both D- and L-enantiomeric forms. D or L forms or racemic mixtures of lactic acid are subsequently polymerized to form low- or high-molecular-weight PLA. High-molecular-weight PLA can be produced from lactic acid by three different routes: (a) direct condensation polymerization, (b) azeotropic dehydrative condensation,

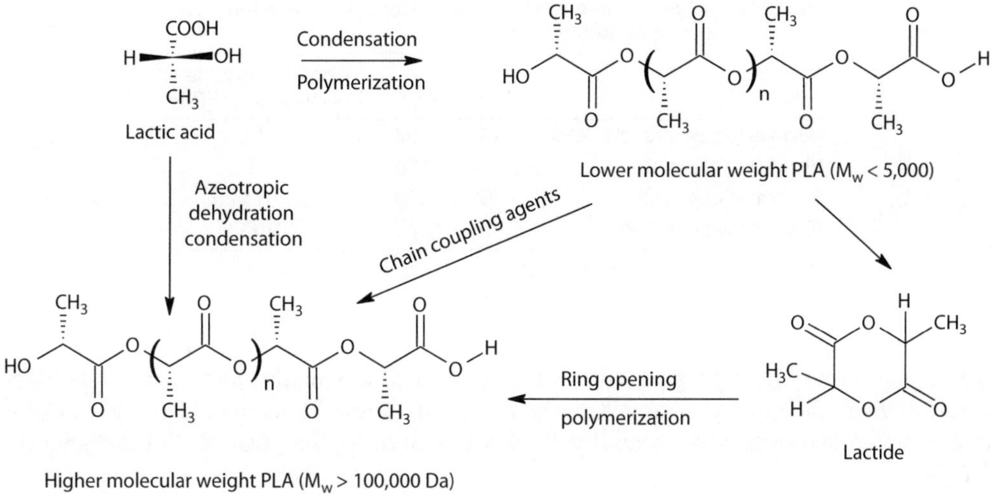

Figure 7.7 Production of lower- and higher-molecular-weight PLAs from different polymerization routes. (From Auras, R. et al., *Macromol. Biosci.*, 4, 835, 2004.)

and (c) lactide formation and polymerization of lactide to high-molecular-weight polymers. The third route is now the most commonly used commercial method to produce high-molecular-weight PLA. In some publications, polymers from lactide are referred to as polylactides. Here, polylactide polymer is used interchangeably with polymers produced from direct polymerization of lactic acid. A simplified reaction scheme illustrating the different PLA polymer production routes is shown in Figure 7.7 [18].

7.2.2.2 Molecular Structure

The molecular structure of PLA is illustrated in Figure 7.7. Most of the lactic acid produced by bacteria is the L-lactic acid although some bacteria produce the D-type. Thus, the lactide dimer can be present in L,L- or D,D-enantiomeric forms. In addition, they can also be present as meso-lactides and racemic mixtures of L,L- and D,D-enantiomers. Carbon chirality present in the original lactic acid monomer or lactide dimer is preserved during polymerization. Typically, meso-lactides formed by racemization during lactic acid oligomerization and cyclization are removed by distillation before polymerization is carried out. However, some amounts of meso-lactides remain as part of ring-opening polymerization process. The presence of D-enantiomers in poly(L-lactide) (PLLA) chain imparts a twist in otherwise regular molecular structure.

7.2.2.3 PLA Properties

Properties of PLA are determined by molecular weights and amounts of D-lactic acid units in the PLLA structure. PLAs containing >93% L-lactide units are semicrystalline, while those with 50%–93% L-lactide units are completely amorphous [18]. PLLA has a melting point of 173°C–183°C and a glass transition temperature of 55°C–65°C, while poly(D-lactide) (PDLA) has a glass transition of approximately 59°C [19–21]. Due to crystalline nature of PLLA, these polymers are generally preferred over PDLA polymers because of their unique mechanical properties. A comparison of mechanical properties between PLLA and other commodity plastics is given in Table 7.1 [18,21]. The values depend on molecular weight and % crystallinity, but it

Table 7.1 A Comparison of Mechanical Properties between PLA and Commodity Plastics

	PLA	PET	General Purpose Polystyrene (GPPS)
Ultimate tensile strength, MPa	68	57	45
Elongation at break, %	4	300	3
Flexural strength, MPa	98	88	76
Flexural modulus, MPa	3700	2700	3000

Source: From Auras, R. et al., *Macromol. Biosci.*, 4, 835, 2004; Gupta, A.P. and Kumar, V., *Eur. Polym. J.*, 43, 4053, 2007.

gives a good idea of how the properties compare. PLA films can also have better light barrier properties than LDPE and comparable mechanical properties to PE terephthalate (PET). Carbon dioxide, water, and oxygen permeability of PLA is also lower than that of PS but higher than PET [18].

7.2.3 Cellulose

7.2.3.1 Source

Cellulose is widely distributed in the plant kingdom where it forms a major component of cell walls in plants. The proportion of cellulose in cell walls varies with plant type and may range from 10% to 80%. Though it is possible to produce cellulose from cell wall extraction, it has been economically viable to produce cellulose from wood pulps and cotton linters. Cotton linters are the short fibers from cotton ball that are too short to be suitable for use in weaving. The wood pulping process has been covered in detail by other authors [22,23], but usually, this process leads to degradation in molecular weights and consumption of significant amounts of energy.

7.2.3.2 Molecular Structure

Figure 7.8 shows the structure of cellulose in a chair configuration. As seen in the figure, cellulose is made of glucose units where the C1 of one unit is bonded to C4 of adjacent unit in a β-conformation. Because of this conformation, native cellulose polymers have high degree of hydrogen bonding between hydroxyl groups of adjacent molecules. Cellulose is also a semi-crystalline polymer where the % crystallinity depends on the source and/or manufacturing process. Because of high degree of crystallinity and hydrogen bonding in cellulose, the native polymer undergoes thermal decomposition before it melts. Thus, it is important to change the molecular structure of cellulose in order to convert native cellulose into a thermoplastic polymer. Thermoplastic cellulose polymers are typically produced by introducing ester groups including acetate, butyrate, or propionate [24,25]. Cellulose acetates, for example, are prepared from cellulose by its reaction with acetic acid and acetic anhydride in the presence of sulfuric acid. The

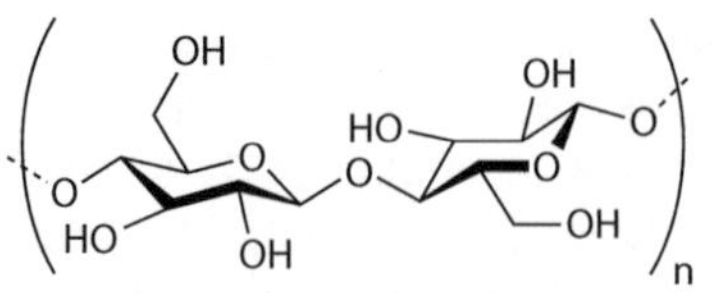

Figure 7.8 Structure of cellulose repeating unit in chair conformation. The two adjacent repeating units are bonded by a glycosidic linkage between C1 of one unit and C4 of adjacent unit.

structure of cellulose acetate resembles that of starch acetate in Figure 7.3 except that the glyco-sidic linkages are in β-conformation.

The properties of these thermoplastic cellulose polymers (e.g., cellulose acetate, cellulose pro-pionate, cellulose acetate butyrate) will depend on the molecular weight of cellulose backbone, DS, and type of plasticizers used. Even though thermoplastic cellulose esters can be processed using typical processes including extrusion and molding, they require plasticizers to impart significant elongational properties and toughness to the polymer. Examples of plasticizers that are typically used with cellulose esters include dimethyl phthalate, phosphate esters, triacetin, and polymeric plasticizers [23,24].

7.2.4 Proteins

7.2.4.1 Source

Proteins are found in both plant and animal species. In cereals and legumes, storage proteins in endosperm provide source of nitrogen during seed germination. Some of the plant-based proteins include gluten, soy, sunflower, and corn proteins. In animals, proteins are important components of skin, muscles, tissues, blood, and enzymes. Proteins from dairy processing are also important raw materials for industrial applications including biodegradable plastics. Some examples of animal proteins include gelatin, keratin, casein, and whey.

7.2.4.2 Molecular Structure

The repeating unit in proteins is the amino acid. The basic structure of amino acids comprises amino and carboxylic groups on each side of the molecule. Thus, being a bifunctional molecule, amino acids can join together through peptide bonds to form protein polymers. The structure of a simple amino acid and a peptide linkage in a dipeptide is given in Figure 7.9.

Most of proteins found in nature consist of one or more of 20 different types of amino acids, and each is characterized by the presence of amino and carboxyl groups [26]. The differences in structure between these amino acids are then driven by the type of groups attached to the α-carbon atom. An example of this is illustrated in Figure 7.10 that shows examples of hydrophobic, acidic, basic, and aromatic amino acids.

The molecular conformation of protein polymers is quite complex as the number of amino acid units in a protein polymer can range from 50 to several thousands. Depending on type of amino acids present in the polymer chain, the conformation of native protein chains can be completely extended or present as random coil, helixes, β-sheets, or highly folded globular proteins.

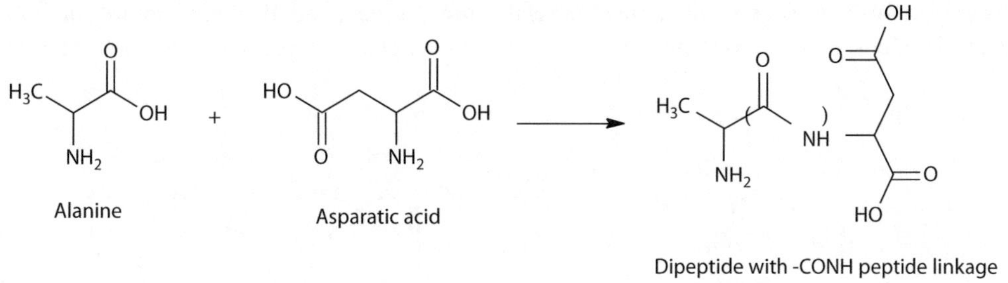

Alanine   Asparatic acid

Dipeptide with -CONH peptide linkage

Figure 7.9 Structure of amino acids alanine and aspartic acid and their dipeptide through a peptide (amide) linkage.

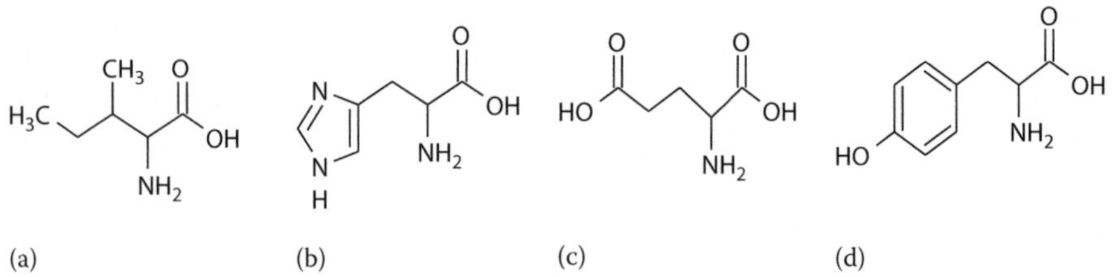

(a)　　　　　　(b)　　　　　　(c)　　　　　　(d)

Figure 7.10　Structure of amino acids with different properties: (a) hydrophobic: isoleucine, (b) basic: histidine, (c) acidic: glutamic acid, and (d) aromatic: tyrosine.

Because of 20 different types of amino acids that can be combined in different ways to form a protein chain, different interactions can exist within and between protein chains. These interactions include covalent, ionic, hydrogen bonding, and van der Waals or hydrophobic interactions. The flow behavior of proteins is governed by changes in protein conformation and inter-/intraprotein interactions happening due to application of thermal energy and shear (e.g., extrusion). Presence of certain functional groups on amino acids can cause cross-linking of protein making these polymers behave as thermoset polymers that are difficult to process in an extrusion or molding-type process [27–30]. Additives are then required to break down covalent and other interactions to enable thermoplastic flow behavior. Thus, it is important to control the extent of interactions between protein molecules in order to make them "extrusion friendly." Interactions including hydrogen bonding and hydrophobic interactions can increase or decrease during the high temperature and shear conditions encountered during extrusion. These same conditions, however, are favorable toward creating new covalent bonds between certain functional groups on protein molecules. One example of a functional group involved in such reactions is the sulfhydryl (SH) group on cysteine amino acids. The SH groups can form disulfide bonds with other SH groups on the same molecule or adjacent molecules. These disulfide bonds then impede the thermoplastic behavior of these polymers, and subsequently reducing agents (e.g., sodium sulfites) have to be used to achieve desirable flow behavior [30–33]. An example of typical disulfide reaction is illustrated in Figure 7.11.

Plasticizers are also required to process protein polymers in typical plastic processing operations like extrusion. Most proteins are made of amino acids with different degrees of hydrophobicity/hydrophilicity on the same chain. Thus, the protein molecule will have different domains on the chain that are compatible with different types of plasticizers, and using a single plasticizer for a protein may not be an effective processing aid. Various researchers have used amphiphilic compounds or surfactants like sodium dodecyl sulfate (SDS) as plasticizers [30,34,35] to overcome this challenge.

Proteins have an abundance of functional groups available to undergo different reactions during extrusion. However, there are also significant challenges to maintaining thermoplastic behavior, preventing molecular degradation or hydrolysis due to high shear and high temperature conditions. Thus, it is important to understand the molecular structure and predominant interactions present in

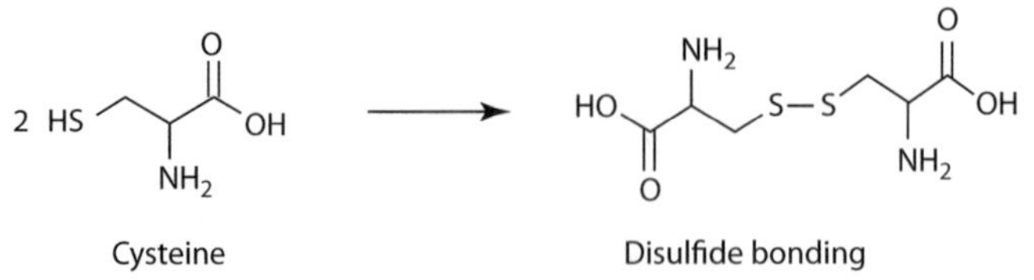

Cysteine　　　　　　　　　　　Disulfide bonding

Figure 7.11　Formation of disulfide bond between cysteine amino acids.

the native protein of interest. This will help in selecting the right plasticizers, understanding possible reaction chemistries, and solving flow-related challenges during extrusion.

7.2.4.3 Properties

7.2.4.3.1 Corn Zein

Corn zein is a prolamin fraction of corn proteins. It has a high content of hydrophobic amino acids including leucine, alanine, and proline making it insoluble in water. The molecular weight of corn zein ranges between 18 and 45 kDa [36].

7.2.4.3.2 Wheat Gluten

Wheat gluten consists of the prolamin fraction (gliadin) and glutenin. Glutenin polymers are of higher molecular weight than gliadins and are present in an extended conformation, while gliadin proteins are globular with a higher degree of secondary and tertiary structure. Glutenin fractions are characterized by a number of disulfide bonds formed between the cysteine amino acid residues [36–38].

7.2.4.3.3 Soy Proteins

There are four main fractions in soy proteins that differ in molecular weight and number of intramolecular disulfide bonds with glycinin and conglycinin being the two major proteins. These two proteins belong to globulin protein family. An interesting property of soy proteins is their ability to form fibrous structures in the presence of alkali and heat. Alkaline denaturation of soy proteins unfolds the globular structure and facilitates disulfide bonds between chains. Soy proteins have high degree of linear symmetry, and absence of bulky side groups on constituent amino acids enables them to undergo texturization by denaturation and polymer orientation [37,39].

7.2.4.3.4 Milk Proteins

Caseins and whey proteins are two major milk protein fractions. Caseins constitute 80% of the milk proteins, and three main forms exist: α-casein, β-casein, and κ-casein. The structure of individual casein molecule is similar to a random coil [40]. Whey proteins, the other class of milk protein, consist of β-lactoglobulin, α-lactalbumin, bovine serum albumin, immunoglobulins, and proteose-peptone proteins [37].

7.2.5 Polyhydroxy Alkanoates from Microbial Sources

PHA is another class of biodegradable polymers that are produced by microbial fermentation. PHAs are polyesters of hydroxy alkanoic acid units produced by natural or genetically modified bacteria using different carbohydrates or carbon-based compounds as substrates.

7.2.5.1 Source

PHAs can be produced by wild-type or recombinant bacteria using a variety of carbon sources. Some of these carbon sources include glucose, organic acids (lactic and acetic acids), vegetable oils, methanol, alkanoic acids, etc. Sustainable carbon-based compounds have been studied as a possible substrate for PHA production by bacteria. These materials include whey, rice bran, molasses, starch,

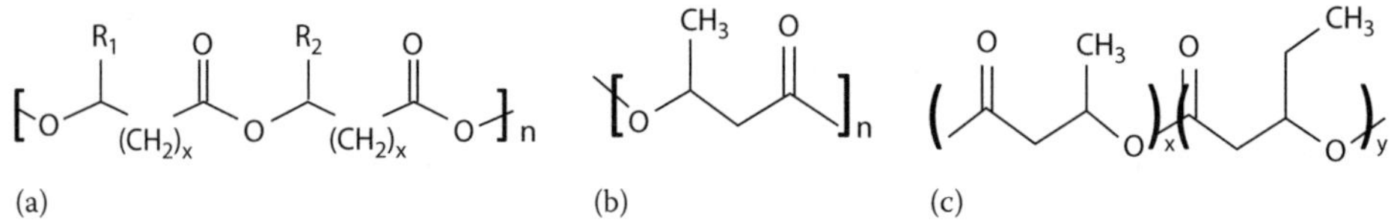

Figure 7.12 Molecular structures of PHAs: (a) general structure of PHAs with n > 100, R1/R2 alkyl group with C1–C13, and x = 1–4; (b) molecular structure of PHB; and (c) molecular structure of PHB-co-valerate, x-hydroxy butyrate units, y-hydroxy valerate units.

and carbon dioxide and hydrogen [41]. The cost of PHAs can range from $7 to $15/lb depending on molecular weight and PHA type [1–2].

7.2.5.2 Molecular Structure

Polyhydroxy butyrate (PHB) and polyhydroxy valerate (PHV) are important polymer types in the PHA family. The general structure of PHAs is given in Figure 7.12. The valerate unit is typically found as a copolymer with butyrate units. Structures of PHB and PHV polymers are also shown in Figure 7.12. PH-co-PV is produced by selecting the appropriate bacterial species and carbon source [42,43]. Typical molecular weights of PHAs range from 2×10^5 to 3×10^6 Da depending on bacterial source and carbon substrate [41]. The hydroxy alkanoic acid monomeric unit produced by bacteria is in the R isomeric configuration. This translates into the polymer as well, and this is important for biodegradability of the polymer in the environment.

7.2.5.3 Properties

Molecular weight, length of side chain (R1 and R2 in Figure 7.12), position of hydroxyl group, and number of carbons in the monomeric unit are key factors affecting polymer properties. Short chain length monomers consist of 3–5 carbon atoms, while medium chain length monomers consist of 6–14 carbon atoms [41]. Certain PHA types including P(3HB) and P(3HB-co-3HV) are crystalline polymers with relatively high melting points and glass transition temperatures. Thus, these polymers are quite stiff and brittle. On the other hand, P(4HB) is a thermoplastic polyester that is pliable.

There is significant interest in PHAs because they have properties similar to that of PP. PHAs with desired properties can be tailored by selecting appropriate bacterial species, carbon substrate, and fermentation conditions. A comparison of properties of various types of PHAs with PP is given in Table 7.2 [41,44].

Table 7.2   A Comparison of Physical and Mechanical Properties of PHAs with PP

Polymer	Melting Point, °C	Glass Transition, °C	Ultimate Tensile Strength, MPa	% Elongation to Break, %	Young's Modulus, GPa
Poly (4HB)	53	−48	104	1000	149
Poly (3HB)	180	4	40	5	3.5
Poly (3HB-co-20% 3HV)	145	−1	20	50	1.2
Poly (3HB-co-16% 4HB)	150	−7	26	444	—
PP	176	−10	34.5	400	1.7

Source: From Akaraonye, E. et al., *J. Chem. Techn. Biotechnol.*, 85, 732, 2010; Lee, S.Y., *Biotechnol. Bioeng.*, 49, 1, 1995.

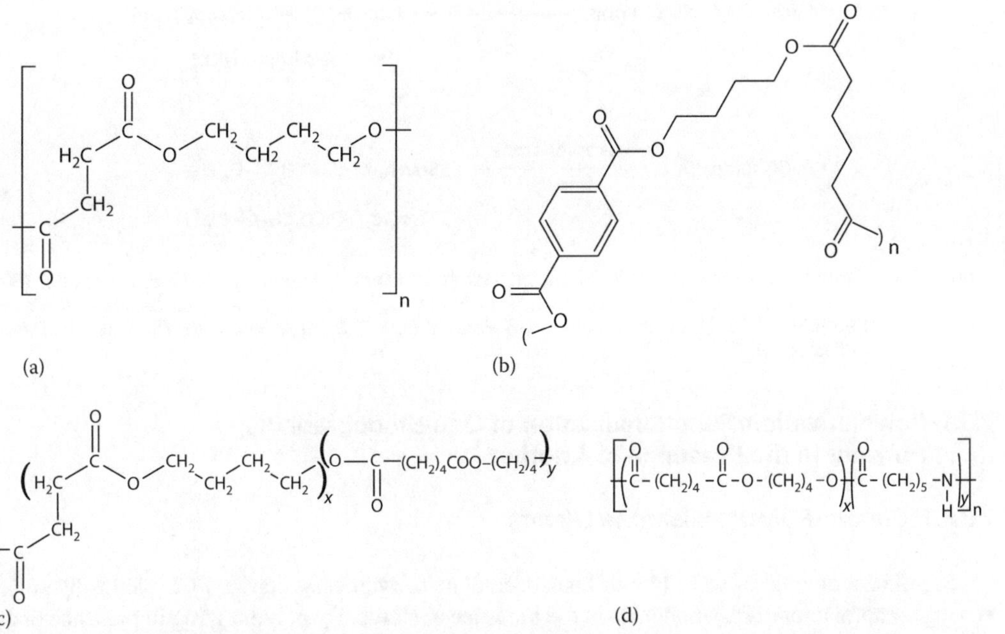

Figure 7.13 Molecular structures of (a) aliphatic polyester (PBS), (b) aliphatic–aromatic polyester (PAT), (c) aliphatic polyester (PBSA), and (d) polyesteramide.

7.2.6 Other Biodegradable Polymers

Other biodegradable polymers that need to be mentioned here belong to category of aliphatic polyesters, polyesteramides, and aliphatic-aromatic polyesters. These are manufactured from fossil fuels; however, they are manufactured in significant quantities. PBS (aliphatic polyester), PBS adipate (PBSA—aliphatic polyester), copolymers of ε-caprolactam, adipic acid, and butanediol (polyesteramide), and PAT (aliphatic–aromatic polyester) are some examples of biodegradable and/ or compostable polymers from fossil fuels [1,2]. The molecular structures of these polymers are illustrated in Figure 7.13.

7.3 REACTIVE EXTRUSION AS A PROCESSING TOOL

Reactive extrusion can be a commercially viable technology to compatibilize two polymers that are otherwise thermodynamically incompatible. These blends can be functionally superior to materials that are made by just physically mixing the two polymers together. By selecting suitable polymer blends and plasticizers and adjusting blend proportions, material properties can be tailored to satisfy relevant applications. For example, this technology platform can be used to blend one cheap (e.g., starch) and one expensive (e.g., PHA) biodegradable polymer without significant degradation of PHA properties. The otherwise incompatible polymers can be reactively blended using several approaches. All approaches take advantage of the availability of suitable functional groups on the polymers. The following sections will illustrate the applications of reactive extrusion process in three main areas: (a) polymerization/functionalization of one biodegradable polymer in the presence of another, (b) chemical modification of a polymer, and (c) homopolymerization.

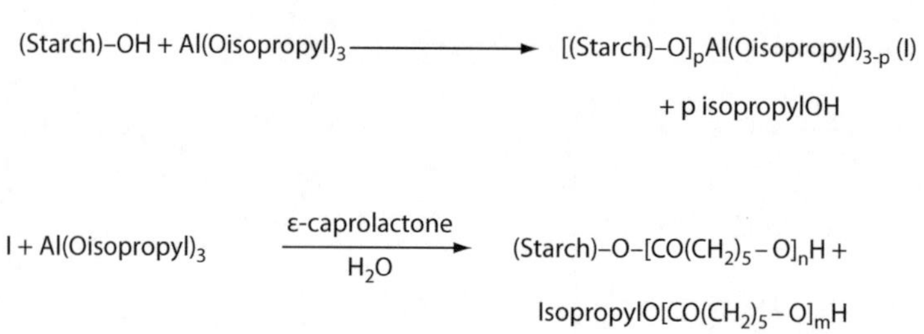

$$(Starch)–OH + Al(Oisopropyl)_3 \longrightarrow [(Starch)–O]_pAl(Oisopropyl)_{3-p} (I)$$

$$+ p\ isopropylOH$$

$$I + Al(Oisopropyl)_3 \xrightarrow[\text{H}_2\text{O}]{\text{ε-caprolactone}} (Starch)–O–[CO(CH_2)_5 – O]_nH +$$

$$IsopropylO[CO(CH_2)_5 – O]_mH$$

Figure 7.14 Grafting and polymerization of CL on starch and their initiation by aluminum alkoxide. (From Narayan, R. et al., *Polymer*, 40, 3091, 1999; Narayan, R. et al., *Polymer Compositions*, US Patent# 5969089, 1999; Narayan, R. et al., US Patent# 5906783, 1999; Narayan, R. et al., US Patent# 5801224, 1999.)

7.3.1 Polymerization/Functionalization of One Biodegradable Polymer in the Presence of Another

7.3.1.1 Starch–Polycaprolactone Blends

Significant amount of work [45–48] has been done to synthesize starch–PCL blends by polymerizing ε-caprolactone (CL) monomer in the presence of starch. By polymerizing in presence of aluminum or titanium alkoxide catalysts, a low conversion of 30% was achieved in batch preparation methods in the presence of starch and aluminum alkoxides. However, the use of titanium alkoxide catalysts led to 98.5% conversion. High conversions were achieved by reactive extrusion process with PCL molecular weights (M_w) ranging from 25,000 to 400,000 at residence times of 1–3 min. Relevant patents [46–47] mentioned a temperature of 180°C during the extrusion process and discussed effects of screw configuration, residence time, and acid value of the monomer on molecular weight. The patents indicated that the starch–PCL blends were synthesized in three extrusion steps: (a) homopolymerization of ε-CL to PCL in the presence of aluminum alkoxides, (b) extrusion of starch–plasticizer pellets, and (c) extrusion blending of materials from (a) and (b). Figure 7.14 illustrates the reactions occurring during this process [45–48]. The mechanical properties of films produced by this method are given in Table 7.3. These films are currently marketed under the name of "Envar."

Another example of reactive extrusion process in this category is one in which high amounts of starch (approx. 40 wt%) were blended with a biodegradable polyester (PCL) in the presence of nanoclay resulting in tough nanocomposite blends with elongational properties approaching that of 100% PCL [49,50]. Starch, PCL, plasticizer, modified montmorillonite (MMT) organoclay, and Fenton's reagent (H_2O_2 and ferrous sulfate) were extruded in a conical corotating twin-screw extruder at 120°C and injection molded at 150°C. The authors hypothesized starch oxidation and cross-linking reactions happening during the extrusion process, as illustrated in Figure 7.15. Native

Table 7.3 Mechanical Properties of 100% PCL and Starch-g-PCL Produced from Reactive Extrusion

Material	Starch Weight %	Ultimate Tensile Strength, MPa	Ultimate Elongation, %
100% PCL	0	25	360
Starch-g-PCL	30	16	257

Source: From Narayan, R. et al., *Polymer*, 40, 3091, 1999; Narayan, R. et al., *Polymer Compositions*, US Patent# 5969089, 1999; Narayan, R. et al., US Patent# 5906783, 1999; Narayan, R. et al., US Patent# 5801224, 1999.

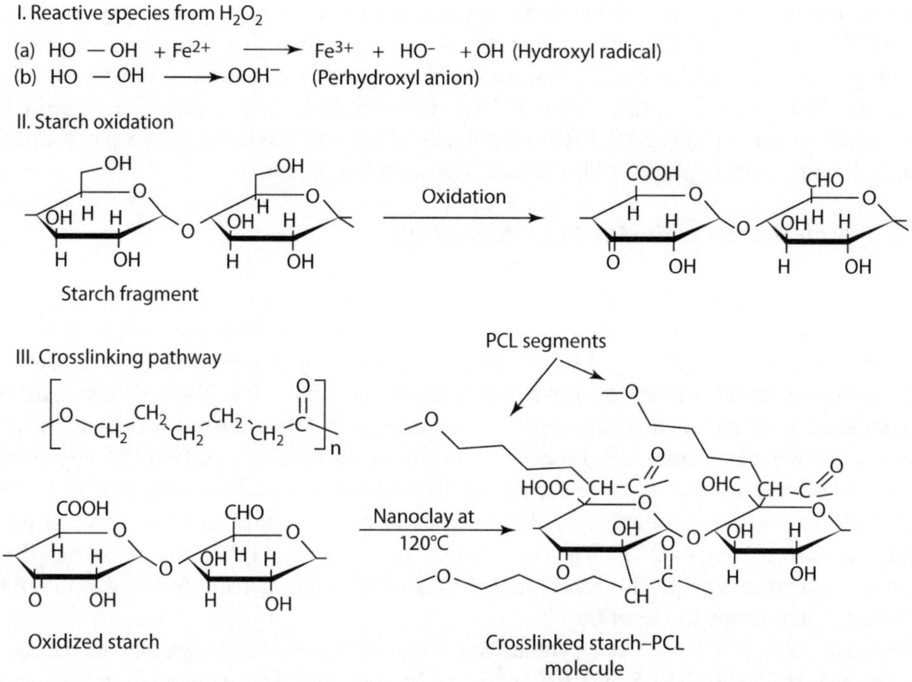

I. Reactive species from H_2O_2

(a) HO — OH + Fe^{2+} ⟶ Fe^{3+} + HO^- + OH (Hydroxyl radical)

(b) HO — OH ⟶ OOH^- (Perhydroxyl anion)

II. Starch oxidation

Starch fragment

III. Crosslinking pathway

PCL segments

Oxidized starch

Crosslinked starch–PCL molecule

Figure 7.15 Simplified reaction scheme showing oxidation and cross-linking between starch and PCL using Fenton's reagent. (Kalambur, S. and Rizui, S.S.H., *J. Appl. Polym. Sci.*, 96, 1072, 2004.)

starch was partially oxidized by the peroxide, and ester groups in PCL can cross-link with the carbonyl and/or carboxyl groups in oxidized starch through a peroxide-initiated free radical process. This cross-linked starch–PCL fraction acted as a compatibilizer between unmodified starch and PCL. A single extrusion step with low residence time (approx. 74 s) was found to accomplish the earlier mentioned set of reactions. The properties of these materials are illustrated in Table 7.4 [50]. Elongation of these reactively extruded starch–PCL nanocomposite blends with 3 and 6 wt% organoclay approached that of 100% PCL. However, strength and modulus remained the same as that of starch–PCL composites prepared from simple physical mixing. X-ray diffraction results showed mainly intercalated flocculated behavior of clay at 1, 3, 6, and 9 wt% organoclay. Scanning electron microscopy (SEM) pictures showed that there was improved starch–PCL interfacial adhesion in reactively extruded blends compared to starch–PCL composites. Dynamic mechanical analysis showed changes in primary α-transition temperatures for both the starch and PCL fractions,

Table 7.4   Tensile Properties of Starch–PCL Blends from Reactive and Nonreactive Extrusion

Material	Max. Strength, MPa	Ultimate Elongation, %	Tensile Modulus (Secant) at 1% Strain, MPa
100% PCL	30.8	1242	303.3
STPCL (nonreactive extrusion)	12.6	265	211.0
STPCLPER-3	13.9	1197	263.4
STPCLPER-6	14.0	1074	253.4
STPCLPER-9	9.5	913	210.5

Source: From Kalambur, S.B., Starch polycaprolactone nanocomposites from reactive extrusion: Synthesis, characterization, properties and scale-up considerations, PhD thesis, Cornell University, New York, 2005.

reflecting cross-linking changes in the nanocomposite blends at different organoclay content. Also, starch–PAT blends prepared by the aforementioned reactive extrusion process exhibited elongational properties approaching that of 100% PAT. This reactive extrusion chemistry can thus be extended to other starch–PCL-like polymer blends with polyhydroxy polymers like polyvinyl alcohol and starch on one side and PBS, PHB–valerate, and PLA on the other to create cheap, novel, and compatible biodegradable polymer blends with increased toughness.

7.3.1.2 Starch Blends with Maleated Polyesters

Addition of maleic anhydride (MA) has been done on many nonbiodegradable synthetic plastic polymers, many of which were blends with polyamides. These include poly(ethylene–propylene) rubber (EPR) elastomer [51–54], PE [55–58], PP [59–61], poly(phenylene ether) (PPE) [62], poly(acrylonitrile–butadiene–styrene) (ABS) [63], and polystyrene (PS) [64–66]. The main reason for wide use of MA-functionalized polymers is the ease with which the anhydride can be grafted onto many polymers at normal melt-processing temperatures without significant homopolymerization and undesirable scission-type side reactions. The addition reaction can be done in solution or melt states. The reaction is initiated by the presence of peroxide initiators like benzoyl peroxide (BPO) or dicumyl peroxide (DCP). The maleated polymers can react with starch through the presence of free anhydride groups; this reaction of anhydride with starch hydroxyl to form an ester does not produce water during the reaction.

Maleated polyesters have been synthesized from biodegradable polyesters including high-molecular-weight PCL [67], PBS [68], PAT [68], and PLA [68,69]. These materials are subsequently used as compatibilizers in corresponding starch–polyester blends [70,71]. A simplified reaction scheme for polyester maleation [67] is given in Figure 7.16 for starch–PCL blends with maleated PCL as compatibilizer. Maleated polyesters were prepared from reactive extrusion in the presence of organic peroxide initiators like BPO, DCP, and dimethyl dibutylperoxy hexane [68]. Grafted MA% ranged from 0.4% to 1.6% in PCL [67] and 0.2% to 1.2% for PBS and PAT polyesters [68]. Little change was reported in the molecular weight of polyesters before and after the MA addition process. Extrusion parameters like temperature, initiator and MA concentration, and residence time were optimized to avoid cross-linking and scission side reactions during extrusion. Mani and Bhattacharya [70] evaluated the properties of starch–polyester blends containing small amounts of maleated polyesters (5%). For starch–PCL blends with 5% compatibilizer and 50 wt% starch, a threefold increase in strength over starch–PCL composites was observed (Table 7.5) although there was no change in elongation or modulus. Similarly, addition of compatibilizer to starch–PBS blends resulted in a twofold increase in tensile strength approaching that of 100% PBS with no effect on modulus or elongation (Table 7.5). No significant effect of compatibilizer on tensile strength and elongation of starch–PAT blends was observed. DMA studies showed a decrease in T_g of starch from 74°C to 65°C in compatibilized starch–PBS blends, and it was related to better miscibility between starch and PBS in the presence of maleated PBS.

Narayan et al. [69] studied the maleation of PLA by reactive extrusion and evaluated the effects of initiator concentration and temperature on % maleation. Figure 7.17 illustrates a simple scheme of maleation of PLA. Up to 0.6% maleation was achieved with 2 wt% MA concentration in the presence of peroxide initiators at 180°C–200°C. Interfacial adhesion by SEM was also evaluated in blends containing 70 wt% maleated PLA and 30 wt% starch [71]. Though mechanical properties were not evaluated, starch particles did not show surface dewetting in these blends, thus indicating good adhesion between starch and maleated PLA. Molecular weights of PLA remained stable, and it was hypothesized to be due to competition between chain scission and chain cross-linking.

In addition to MA, other anhydrides have also been used to prepare anhydride-modified polyesters. Avella et al. [72] reported grafting of pyromellitic anhydride on PCL and its use as a

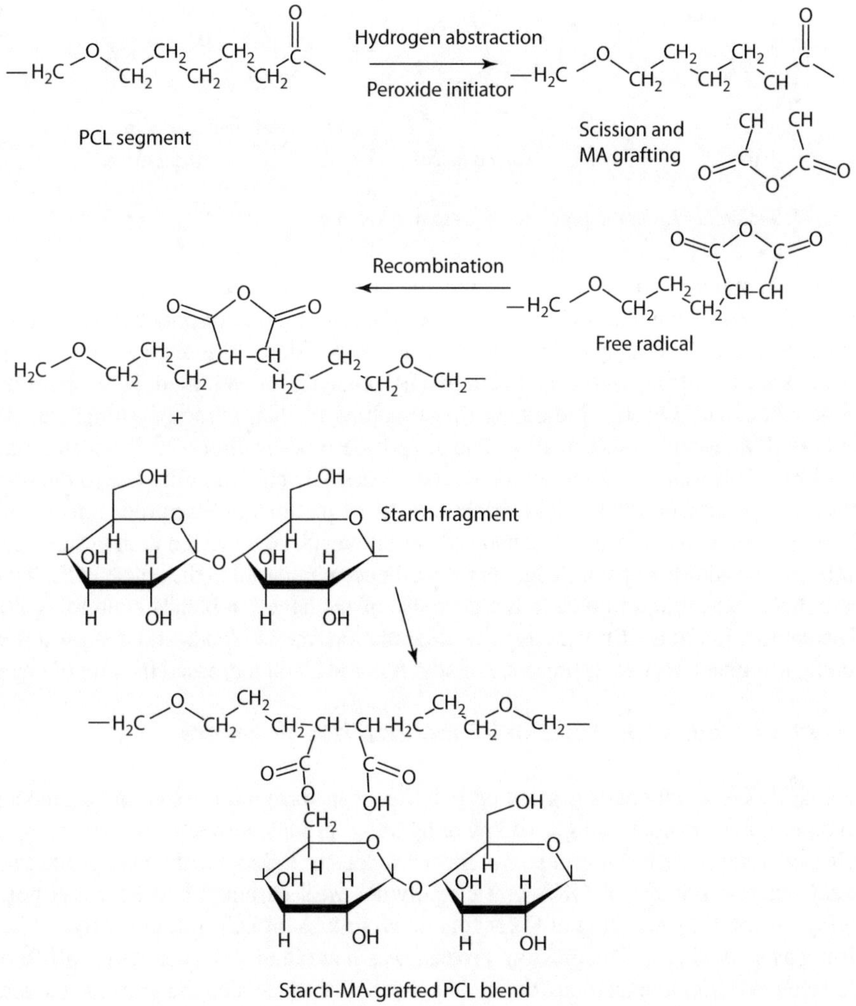

PCL segment

Hydrogen abstraction

Peroxide initiator

Scission and MA grafting

Recombination

Free radical

Starch fragment

+

Starch-MA-grafted PCL blend

Figure 7.16 Simplified reaction scheme illustrating addition of MA to PCL and subsequent reaction of maleated PCL with starch.

Table 7.5 A Comparison of Mechanical Properties between Starch-Maleated Polyester Blends versus Starch-Unmodified Polyester Blends

Material	Tensile Strength at Break, MPa		% Elongation at Break	
	With MA	Without MA	With MA	Without MA
100% PBS	37	37	—	300
Starch–PBS	33	15	10–20	10–20
100% PCL	25	25	—	650
Starch–PCL	21	8	10–20	10–20

Source: From Mani, R. and Bhattacharya, M., *Eur. Polym. J.*, 37, 515, 2001.

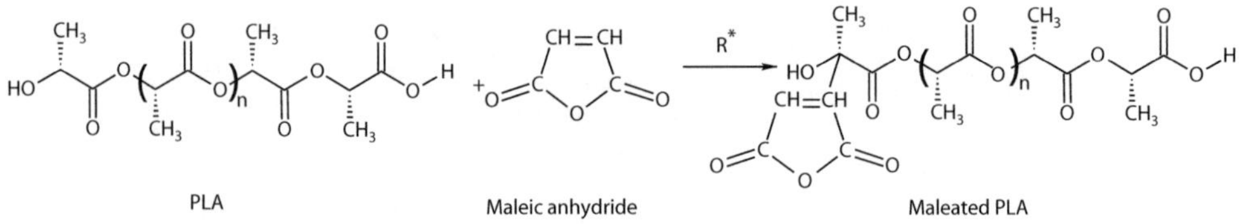

Figure 7.17 Maleation of PLA in the presence of peroxide initiator.

compatibilizer in starch–PCL blends. However, they used a low-molecular-weight (Mw ~ 20,000) PCL for anhydride addition and high-molecular-weight (Mw ~ 80,000) polymer as the unmodified polyester in the blend. The anhydride addition process was done in molten state in the presence of small amounts of tetrahydrofuran. Infrared (IR) analysis of modified PCL showed appearance of a new band at $3200\,cm^{-1}$ indicating the stretching of PCL carbonyl groups due to chemical bonding with anhydride functionality. The anhydride modification of PCL shifted its T_g from −66°C to −55°C. This was correlated with reduced segmental chain mobility due to the presence of pyromellitic groups attached to the PCL backbone. SEM pictures of fractured samples of starch–PCL (70:30 wt ratio) with compatibilizer showed better interfacial adhesion than in samples without compatibilizer. The addition of starch decreased resilience compared to the 100% PCL, but addition of compatibilizers was found to reduce the decrease of resilience in blends containing 30%–50% starch. The authors found no differences in biodegradation rate of blends and composites with and without compatibilizers. However, presence of starch in the blend increased the rate of degradation.

7.3.1.3 Starch Blends with Oxazoline-Functionalized Polymers

Oxazoline (OXA) compounds grafted on polymers can form amidoester and amidoether linkages with other polymers containing carboxyl or hydroxyl groups, respectively. For example, modified starches containing carboxyl groups can form amidoester linkage with OXA-grafted polymers. The general reaction scheme of OXA-grafted polymer with carboxyl and hydroxyl polymers is shown in Figure 7.18 [73]. Grafting of OXA functional groups on PCL through a free radical chain mechanism was studied [74]. The grafting process was done in an extruder, and graft % of 0.9%–2.2% was achieved. Ricinoloxazoline, which is a 2-OXA, was used as the grafting monomer. The structure of ricinoloxazoline is given in Figure 7.19. The double bonds on carbon atoms indicated by a and b were found to react with the polyester chain similar to MA grafting on PCL shown in Figure 7.16. When used as an initiator, DCP was found to produce minimum homopolymerization, high grafting efficiency, and minimum decrease in PCL molecular weight. Though no blending studies with starch were carried out, the authors predicted reactions with starch and modified starch containing carboxyl groups and wheat gluten because of the presence of carboxyl and hydroxyl groups.

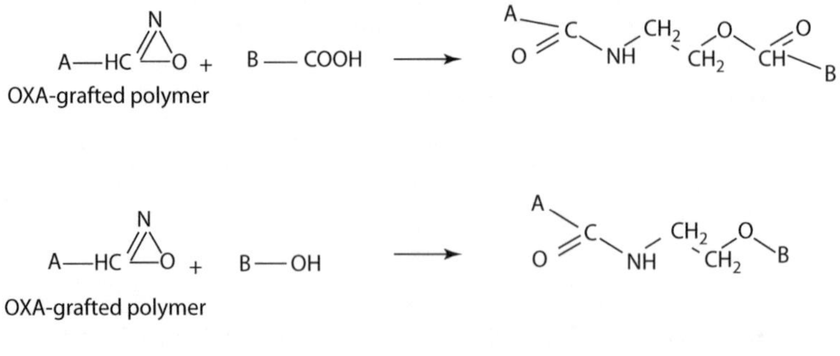

Figure 7.18 Reaction of OXA-grafted polymers with other polymers containing carboxyl or hydroxyl groups.

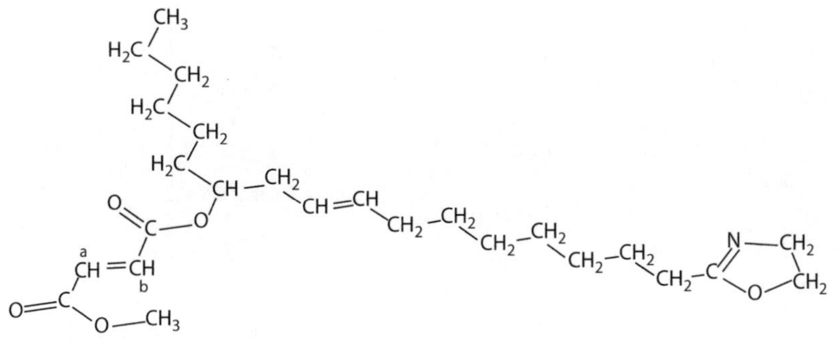

Figure 7.19 Structure of ricinoloxazoline used for OXA grafting on PCL. Carbon atoms marked a and b function as chain extenders during grafting reaction.

7.3.2 Chemical Modification of a Polymer

A prominent example of chemical modification of biodegradable polymers is addition of various functional groups to starch through reactive extrusion. This section will attempt to discuss few important reaction types that have been used to modify starch polymer through reactive extrusion.

7.3.2.1 Graft Copolymerization of Starch with Polyacrylonitrile

A well-known application of these polymers is their use as water-absorbing materials. The market name for these materials is the "Super Slurper." These materials are produced by a batch process from starch and acrylonitrile in the presence of ceric ions. The resulting starch-g-polyacrylonitrile material was then saponified to convert the nitrile groups into carbamoyl and alkali metal carboxylate groups. Removal of water from this saponified material yields a solid that can absorb several hundred times its weight of water [75,76]. A simplified reaction scheme of this grafting process is illustrated in Figure 7.20.

Starch-g-polyacrylonitrile polymers were synthesized from reactive extrusion, and 72% of acrylonitrile was polymerized compared to 74% from the batch process [77]. Similar results were found from another study [78] that confirmed feasibility of reactive extrusion process to replace conventional batch process. The residence time in extruders was approximately 7 min compared to 2 h in a batch reactor.

7.3.2.2 Cationic Starches

Cationic starches are important starch derivatives that are used in the paper industry for various functional reasons that include providing drainage and strength improvements. Della Valle et al. [79] and Carr [80] studied the preparation of cationic starches from a reactive extrusion process. The

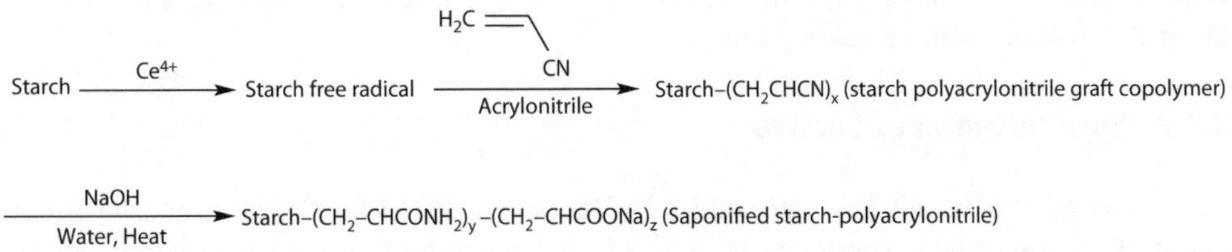

Figure 7.20 A simplified reaction scheme illustrating grafting of acrylonitrile on starch and subsequent saponification.

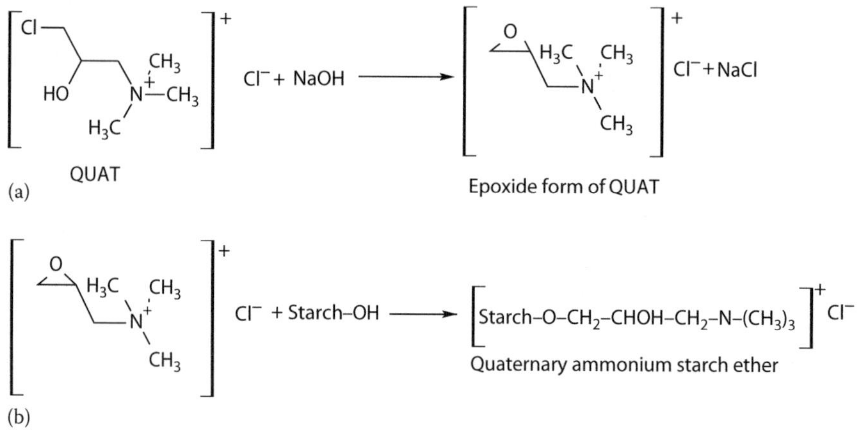

(a)

QUAT $Cl^- + NaOH \longrightarrow$ Epoxide form of QUAT $Cl^- + NaCl$

(b)

$Cl^- +$ Starch–OH $\longrightarrow$ [Starch–O–CH$_2$–CHOH–CH$_2$–N–(CH$_3$)$_3$]$^+$ Cl$^-$

Quaternary ammonium starch ether

Figure 7.21 Reactions occurring during preparation of cationic starches: (a) conversion of QUAT into its epoxide form in the presence of alkali and (b) subsequent reaction of epoxide with starch to produce quaternary ammonium starch ether.

compounds involved in the reaction were cornstarch, 3-chloro-2-hydroxypropyltrimethyl ammonium chloride (QUAT), and caustic soda. At certain extrusion conditions, a DS of 0.05 and high reaction efficiencies were achieved. Reactions occurring during extrusion are illustrated in Figure 7.21 [13].

7.3.2.3 Oxidized Starch

Figure 7.4 showed the creation of carbonyl and carboxyl groups on starch molecule through oxidation. Wing and Willett [81] used hydrogen peroxide along with cuprous and cupric acid catalysts (Fenton's reagent) to introduce carboxyl and carbonyl groups. Presence of residual granule structures after reactive extrusion depended on starch type and reaction conditions. Extrusion conditions studied were also found to decrease solution viscosities and molecular weights that were comparable to maltodextrins of dextrose equivalent (DE) 5–10.

7.3.3 Homopolymerization

7.3.3.1 Polymerization of CL

Coordination–insertion polymerization of CL was studied in a twin-screw extruder [82]. Aluminum trialkoxide was used as the catalyst. The resulting polymer chain is a 3-arm chain with each arm originating from the aluminum atom. Molecular weights on the order of 200,000 were achieved. Residence time of less than 5 min was found to convert at least 95% of the monomer. This anionic polymerization scheme is illustrated in Figure 7.22. The acyl carbon–oxygen bond in CL is cleaved and replaced with the alkoxy group.

7.3.3.2 Polymerization of Lactide

Polymerization of lactide has been studied by Dubois et al. [83,84] using tin octanoate as initiator. The reaction scheme is illustrated in Figure 7.23. Tin octanoate is converted into its alkoxide form that subsequently reacts with lactide in a manner similar to one described earlier with aluminum alkoxides. Polylactides produced from this process were compared to that made from a batch

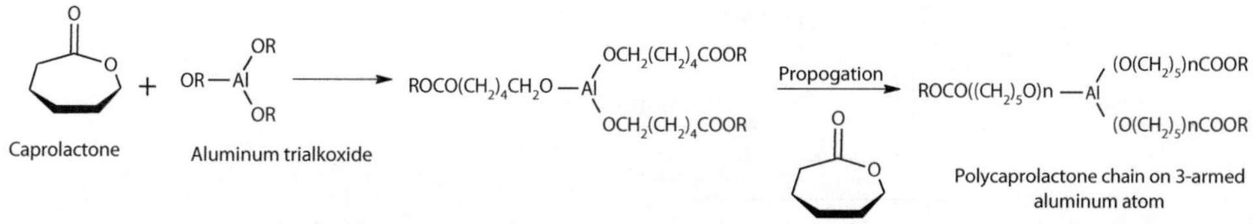

Figure 7.22 Polymerization of CL initiated by aluminum trialkoxide. Acyl bond in CL is cleaved with insertion of alkoxide group. Propagation with CL monomer forms PCL chain on three arms of aluminum atom.

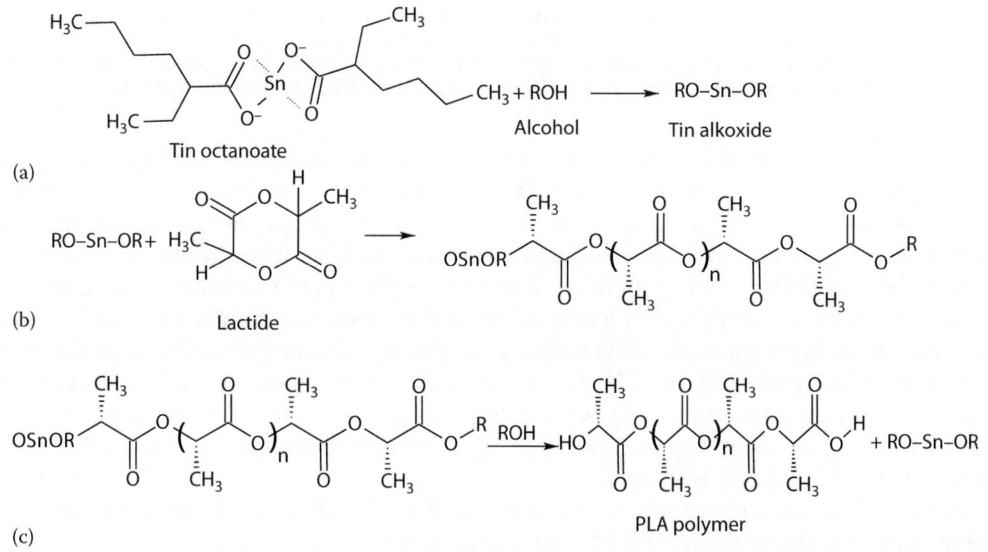

Figure 7.23 A simplified scheme of reactions occurring during reactive extrusion polymerization of polylactide: (a) activation of tin octanoate into its alkoxide form in the presence of alcohol, (b) reaction of tin alkoxide with lactide and insertion of growing polymer chain in between tin atom and alkoxide group, and (c) formation of polylactide and liberation of tin alkoxide. (From Degée, P. et al., *Macromol. Symp.*, 144, 289, 1999; Degée, P. et al., *Polymer*, 41, 3395, 2000.)

process. Greater than 90% conversions were achieved in both processes with M_n for batch and extrusion processes being 246,000 and 91,100 respectively.

The field of reactive extrusion of biodegradable polymers though relatively new offers intriguing and complex challenges that come from the study of a large number of reaction types, different catalysts/initiators for synthesizing same polymer, and different monomers and copolymers. The earlier section has attempted to illustrate some well-known applications of reactive extrusion. Readers can refer to other excellent published material on bulk polymerization or reactive extrusion of polymers of their interest [73,85–88].

7.3.4 Considerations for Using Extruders as a Continuous Reactor

In reactive extrusion, extruders function as a continuous chemical reactor in which the desired reactions are carried out. As dwell times in commercial extruders are in the order of minutes, it is necessary to achieve high conversions within this short period of time for the process to be economically feasible. Thus, selection of appropriate extrusion conditions and reaction initiators

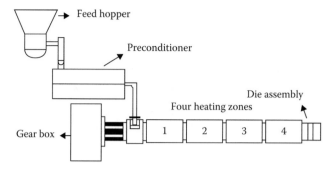

Figure 7.24 Setup of a typical twin-screw extruder consisting of feed hopper to deliver volumetric or gravimetric feed (dry) material, preconditioner with heated jackets for conditioning certain polymers with plasticizer, and extruder modular head that consists of multiple heating zones/barrels. Die assembly typically consists of a die plate, spacer, and a die head containing the die.

is important. In addition to functioning as a reactor, extruders also enable melting, mixing, and pumping of polymers along the length of the extruder. In the case of binary or multiple polymer blend systems where at least one polymer is semicrystalline, it is important to determine when the desired reaction between polymers needs to occur—during melting, immediately after melting, or in the melted homogenous system. Because of short dwell times in these extruders and their scale, it may be preferable to carry out preliminary studies in small lab-scale extruders where dwell times and other extrusion variables can be controlled. Examples of a commercial and bench-top extruder are illustrated in Figures 7.24 and 7.25. Key components of a commercial scale extruder (Wenger twin-screw corotating extruder) are shown in Figure 7.24 [89]. An example of one type of a bench-top extruder is illustrated in Figure 7.25. Figure 7.25a through c illustrates the design of a conical microextruder and its screw elements [89].

In this section, we shall consider some factors that affect the efficiency of reactions during reactive extrusion and properties of reactively extruded material.

7.3.4.1 Polymer Crystallinity

During reactive extrusion blending of two polymers where at least one polymer is crystalline or semicrystalline, it is important to understand the impact of crystallinity. One example of a polymer in which crystallinity plays an important role during reactive extrusion is starch. As discussed in Section 7.2.1, starch is a semicrystalline polymer that gelatinizes in the presence of excess plasticizer (e.g., water) and heat. Gelatinization is the disruption of crystalline order within the starch granule preceded by granule swelling and increase in the volume of amorphous regions in the granule. In the presence of limited amounts of plasticizers which is the case in many food and industrial applications, granule swelling is limited, and there is partial disruption of crystalline order within the granule. Shear forces encountered in the extruder and heat are then key factors in breaking down the granular structure by mechanical means.

A successful starch-based reactive extrusion (e.g., chemical modification of starch) that results in material with desired properties will depend on the physicochemical state of starch granule when the modification reagent is most reactive. Reaction rates will be faster in an amorphous and gelatinized starch matrix than in a semicrystalline one. Also there could be a difference in where new functional groups are inserted. A simplistic scenario that can explain this difference will be the insertion of more functional groups on few polymer chains versus few groups introduced on many chains with the same degree of reactant conversion. Also the viscosity of the starch matrix will depend on the state of starch granule inside the extruder. Viscosity is an important factor in

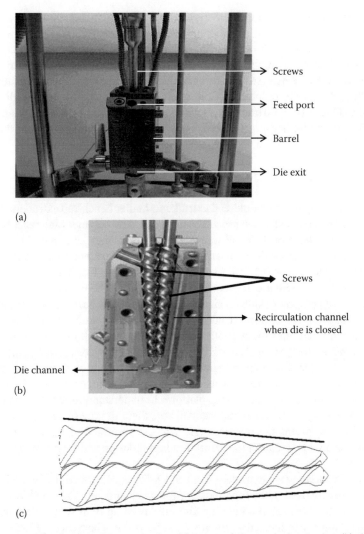

Figure 7.25 An example of a bench-top extruder: (a) setup of the conical extruder, (b) the inside of the extruder, and (c) design of screw elements in the extruder with decreasing pitch and diameter from feed port to die exit.

determining the rate of diffusion of monomeric or oligomeric reactants, thus affecting reaction conversion. Thus, it is important to have the correct extruder screw configurations, temperature profile, and proportion of reactants so that granule disruption can occur before the reagent is allowed to get in contact with molten starch. The extruder should also have a delivery setup that allows desirable reagents to be dosed after starch is converted to its molten state.

During starch modification by reactive extrusion (e.g., cationic modification of starch), it is possible to enable the reactive low-molecular-weight chemical reagent (e.g., QUAT) to react with starch before the granule is completely disrupted or while the granule is partially swollen. Because of lesser volume of amorphous regions in a partially swollen granule, the reaction is limited to polymer chains accessible from the surface of starch granule compared to that in molten state when all polymer chains are completely dispersed. Depending on starch source, size of starch granules, and nature of polymer chains near granule surface (amylose or amylopectin, molecular weights, etc.), reactive extrusion process may result in materials with different properties. Thus, it is important to understand the crystalline nature of raw materials and study their impact on conversion and final material properties.

In the case of extruders with modular designs, it may be possible to extract material (with appropriate sample handling/storage protocols) from the extruder and evaluate the dynamics of polymer conversion from semicrystalline or crystalline phase to amorphous domains and relation to conversion % and material properties. Commonly used techniques including differential scanning calorimetry (DSC), x-ray diffraction, and IR spectroscopy can be used to quantify amount of crystallinity, % conversion, and grafting efficiency of the low-molecular-weight-modifying agent during different stages of extrusion.

7.3.4.2 Extruder Screw Configuration

Because of size and scale of commercial extruders, smaller bench-top extruders may be preferable in order to study reactive extrusion processes involving new polymers and reaction mechanisms. It is important to remember that different screw configurations may be required in extruders of different scale to synthesize materials with same properties. One example of a smaller scale extruder is the bench-top microextruder that can extrude several grams of product. An illustration of this type of extruder is shown in Figure 7.25 earlier [89]. As seen in the figure, this extruder is set up in such a way that molten polymer can be contained in the extruder until the die opening is manually opened. As seen in the picture, this extruder uses a conical corotating twin-screw configuration. The screw diameter, pitch, and flight depth decrease from top to bottom (feed to die exit). Raw materials are premixed and fed into the extruder that is heated to desired temperature. Because of the design of this extruder, any volatile material present in the raw material (e.g., water in starch/cellulose) or generated during reactive extrusion can escape out of the raw material feed port. Thus, this extruder may be an excellent bench-top platform to understand molecular changes happening during the extrusion process. It is also important to note that this type of an extruder may work well for processes where no premature reactions occur in the premixed material. Other smaller scale extruders are also available that are more similar in barrel design and screw configurations to commercial extruders.

An example of a larger scale extruder (Wenger TX-52) is illustrated in Figure 7.24 [89]. These extruders are commonly used in the food industry for snack and pet food extrusion. Wenger TX-52 is a continuous parallel horizontal corotating and intermeshing twin-screw extruder. The number 52 stands for the diameter of the screw. Figure 7.24 shows the schematic of TX-52 extruder setup. Formulations or polymers in dry form (pellets/powder) are filled in a single-screw feed system known as the feed hopper or live bin. From the feed hopper, the powder was conveyed to a twin-screw preconditioner. The speed of the single-screw feeder can be varied to obtain variable feed rates. From the preconditioner, material flowed into the extruder barrel. The extruder barrel contained two screws that are rotated by a motor configured with a gearbox. Figure 7.24 also shows the classification of entire extruder barrel length into four different heating zones or heads. The working principle in TX-52 is similar to twin-screw extruders used for plastic applications. It is also a versatile system in which the degree of mixing could be varied by using suitable segmented screw configurations. Also the modular design allows the number of heads to be decreased from 4 to 2 thus reducing the L/D ratio. The extruder is equipped with a jacketed heating/cooling system in which high-pressure steam or cold fluid can be circulated to obtain high/low temperatures. It is also possible to install venting ports on any of the heating zone sections that can remove any undesirable volatiles formed during reactive extrusion.

Figure 7.26 illustrates typical forward screw elements, reverse screw elements, and kneading disks used in extruders. As the name suggests, forward screw elements help in conveying polymer material (melt/dry forms) forward toward the die exit. The pitch and flight depth will affect the pressure drop inside the extruder as well as the degree of transverse mixing. Kneading disks

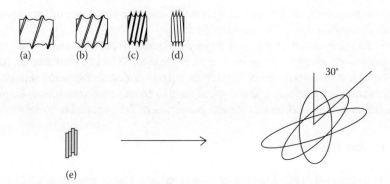

Figure 7.26 Examples of different screw configurations (a–c): forward (depending on correct direction of rotation) screw elements with different pitch and flight depth, (d) reverse screw element with material flow in the reverse direction to facilitate mixing and pressure development, and (e) mixing paddles with each paddle offset by an angle.

are a special kind of screw element which offers a higher degree of dispersive mixing than other screw elements. Chemical reactions occurring in an extruder require collisions between relevant functional groups. The rate of these collisions is then governed by the type and intensity of mixing occurring in the extruder. Thus, it is evident that screw configurations can play a critical role during reactive extrusion. The readers are referred to other literature to get a fundamental understanding of polymer flow inside an extruder [90–93].

7.3.4.3 Specific Mechanical Energy

Specific mechanical energy (SME) is a measure of mechanical energy applied to the polymer melt by the rotation of screws. SME is calculated as follows:

$$SME(kJ/kg) = \frac{2\pi\omega T}{1000\,\dot{m}}$$

where
 ω is screw speed in revolutions per second
 T is torque in N-m
 $\dot{m}$ is mass flow rate in kg/s

In addition to SME, screw speed will also affect the amount of heat generated by viscous dissipation. In a reactive extrusion process, melt and product temperatures are important factors that affect reaction pathways and material properties. Thus, it is important to quantify the amount of heat generated by viscous dissipation. Viscous dissipation is given by [94]

$$q_v = K\dot{\gamma}^{1+n}$$

where
 q_v is dissipation in power per unit volume
 K (Pa-s^n) and n are consistency coefficient and power law index from the power law model
 $\dot{\gamma}$ is shear rate (s^{-1})

Thus, for a given polymer melt, higher screw speeds resulted in higher shear rate and subsequently higher viscous dissipation and higher melt temperatures.

SMEs are also a measure of polymer depolymerization especially among biopolymers like starch or cellulose. For example, it has been reported that an SME of >100 kJ/kg (at limited levels of plasticizers) begins to degrade starch molecular weights during an extrusion process [88]. Thus, requirements of mixing by adjusting screw speed need to be balanced with impact on polymer degradation and subsequent impact on mechanical properties of the extruded material.

7.3.4.4 Extrusion Temperature

In addition to external applied heat to extrusion barrels, heat is also generated by viscous dissipation due to applied shear, as discussed earlier. In order to create a homogenous melt especially in case of polymer blends or semicrystalline polymers, it is necessary to provide enough heat to melt the crystals as well as to lower the viscosity to enable proper mixing.

The well-known Arrhenius relationship illustrates the impact of temperature on the rate of chemical reactions. As illustrated in the relation as follows, temperature has a direct relation to rate of reaction,

$$K = PZ \exp\left(-\frac{Ea}{RT}\right)$$

where
   K is reaction rate
   P is the probability factor that accounts for orientation of reacting molecules
   Z is collision frequency factor
   R is gas constant
   T is temperature in Kelvin
   Ea is the activation energy

As seen in the preceding equation, besides temperature, orientation of reactant molecules, activation energy, and collision frequency will affect reaction rate. Alternatively, extrusion variables like screw speed, screw type, and configuration will impact non-temperature-related factors that affect reaction rate. Thus, rate of reaction and final material properties will depend on the interplay between extrusion variables and polymer properties including molecular weights, physical state of polymer, i.e., amorphous, crystalline, degree of branching, etc., concentration of functional groups on polymer chain, and viscosity.

7.3.4.5 Presence of Other Additives

Plasticizers play an important role in the extrusion of biodegradable polymers that are extremely brittle and stiff. Examples of such polymers include starch and cellulose. Functional groups present on plasticizers can then be expected to participate in reactions that are otherwise targeted towards the macromolecule.

Nanoclays belong to another class of additives or functional compounds that are being used in the extrusion of biopolymers. Nanoclays impart many beneficial properties to polymer matrices including improved water barrier, stiffness in plasticized polymer matrices, and resistance to thermal degradation. In certain studies, nanoclays have been found to play an interesting role during the reactive extrusion process [49–50,73,89]. Figure 7.27 illustrates the structure of one type of nanoclay, i.e., montmorillonite nanoclays. These materials are characterized by high surface

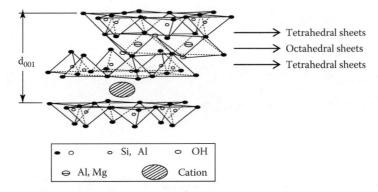

Figure 7.27 Structure of MMT clays with intercalated sodium cation that can be exchanged with quaternary ammonium compounds to design organoclays.

area (approx. $700 \, m^2/g$) and abundance of hydroxyl groups that can form hydrogen bonds with surrounding molecules (e.g., water). They are also characterized by presence of intercalated separation between two particles that is filled with counteracting cations like sodium. These nanoclays can be modified so that the intercalated sodium ions are replaced with hydrophobic quaternary ammonium alkyl chains. Thus, when polymers are extruded in the presence of suitable nanoclays, polymer molecules can intercalate between two nanoclay particles thus impacting polymer flow behavior, mechanical, and other properties. Because of its high surface area, effects on polymer rheology, and presence of charged cations, these nanoclays can affect kinetics and reaction mechanisms during a reactive extrusion process [95].

7.3.4.6 Undesirable Side Reactions

Many reactive extrusion processes involve free radical reactions that may be difficult to control due to several factors present in a typical extrusion process. Undesirable reactions during the extrusion process can lead to unexpected and inferior material properties. These undesirable reactions can happen because of impurities, wrong selection of extrusion conditions (e.g., temperature), and nature of polymer itself (e.g., PLA). Examples of reactions that can be undesirable include chain transfer, back biting, hydrolysis, and cross-linking [86–87]. Figure 7.28 illustrates some of these reactions.

7.3.5 Characterizing Products from Reactive Extrusion Process

Many challenges remain in understanding chemical and morphological changes happening during the reactive extrusion process. Many reactive extrusion chemistries employ free radical reaction mechanisms to compatibilize two polymer blends or to initiate monomer polymerization in an extruder. In order to fundamentally understand reaction mechanisms and material properties, it is necessary to employ a wide gamut of analytical and material characterization tools. Postreactive extrusion, researchers have employed techniques including IR spectroscopy, nuclear magnetic resonance (NMR), and chemical analyses to study changes in functional groups; calorimetric and dynamic testing tools to quantify crystallinity and glass transition changes; microscopy techniques like SEM to characterize changes in polymer blend interfaces; and measurements to characterize material properties including mechanical and barrier properties. Figure 7.29 illustrates an example of using SEM to understand the impact of reactive extrusion on stability of starch domains in a starch–PCL polymer matrix [49].

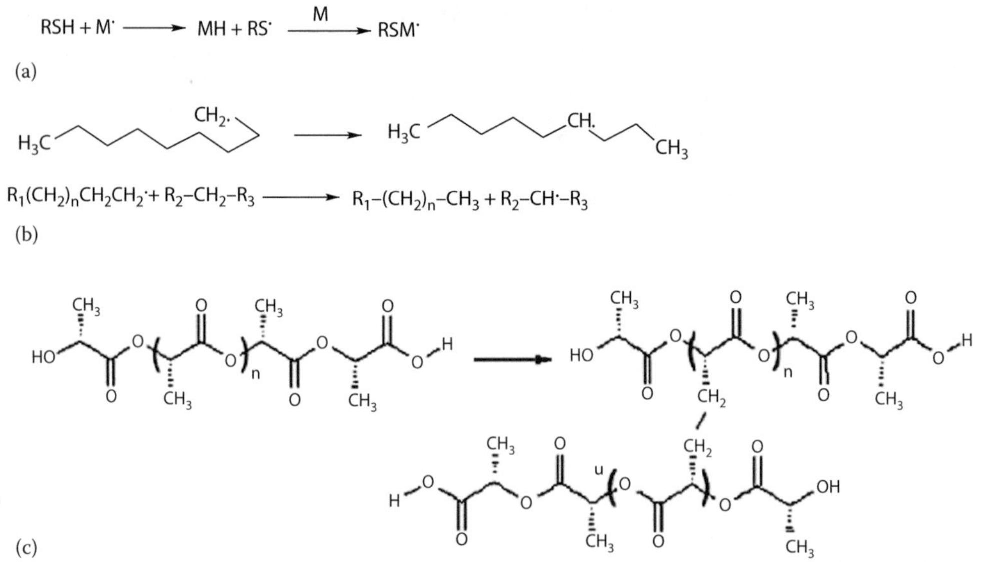

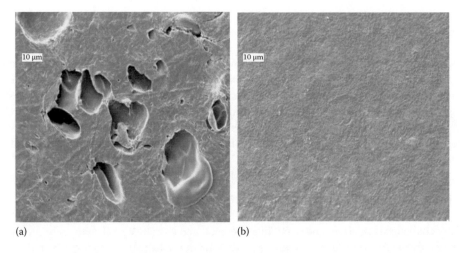

Figure 7.28 Examples of reactions that can be undesirable during reactive extrusion: (a) decrease in poly-mer molecular weights by presence of chain transfer agents (e.g., mercaptans), M is the mono-mer radical or propagating free radical polymer chain (b) short and long branching brought about by back-biting reactions at certain temperature or pressure conditions in extruder, and (c) cross-linking of PLA as a side reaction during free radical reactions involving PLA (e.g., maleation of PLA).

Figure 7.29 SEM pictures of starch–PCL blends ultrasonicated in warm water: (a) blends mixed and extruded without any compatibilization showing separation of soluble starch from the matrix and (b) reac-tively extruded blends that indicate better adhesion of starch to PCL matrix due to compatibili-zation between two polymers. (Kalambur, S. and Rizui, S.S.H., *J. Appl. Polym. Sci.*, 96, 1072, 2004.)

7.4 SUMMARY

This chapter attempts to describe some of the interesting modifications of biodegradable poly-mers that have been carried out in a continuous extrusion process. Understanding polymer reaction mechanisms happening in an extruder is an exciting area of study that is interdisciplinary in nature and requires knowledge pooled from the fields of engineering, organic and analytical chemistry, and materials science. Given the number of biodegradable polymer types and a variety of functional groups on these polymers, reactive extrusion technology can offer an economically feasible and environment-friendly platform to carry out homopolymerization, polymer functionalization, and/or polymer blending processes. This technology will play an important role in expanding the application of biodegradable polymers.

REFERENCES

1. Mecking, S., *Angewandte Chemie International Edition*, 43, 1078, 2004.
2. Mooney, B.P., *Biochemical Journal*, 418, 219, 2009.
3. Wu, Q., Sakabe, H., and Isobe, S., *Industrial and Engineering Chemistry Research*, 42, 6765, 2003.
4. Reddy, N. and Yang, Y., *Journal of Materials Science: Materials in Medicine*, 19, 2055, 2008.
5. Bastioli, C., *Polymer Degradation and Stability*, 59, 263, 1998.
6. Narayan, R. and Pettigrew, C.P., *ASTM Standardization News*, 36, December,1999.
7. Bemiller, J.N. and Whistler, R.L., *Carbohydrates in Food Chemistry*, ed. Fennema, O.R., Marcel Dekker, New York, 1996.
8. Wang, S.S. and Qu, D., *Starch*, 46, 225, 1994.
9. Jaluria, P. and Wang, S.S., *Advances in Extrusion Technology*, eds. Chang, Y.K. and Wang, S.S., Technomic Publishing Inc., Lancaster, PA, 1998.
10. Whistler, R.L., Linke, E.G., and Kazeniac, S., *Journal of American Chemical Society*, 78, 4704, 1956.
11. Hullinger, C.H. and Whistler, R.L., *Cereal Chemistry*, 28, 153, 1951.
12. Whistler, R.L. and Schweiger, R., *Journal of American Chemical Society*, 79, 6460, 1957.
13. Rutenberg, M.W. and Solarek, D., *Starch Derivatives: Production and Uses in Starch: Chemistry and Technology*, eds. Whistler, R.L., Bemiller, J.N., and Paschall, E.F., Academic Press Inc., New York, 1984.
14. Radley, J.A., *Starch Production Technology*, Applied Science Publishers, London, U.K., 1976.
15. Hjermstad, E.T., in *Industrial Gums*, eds. Whistler, R.L. and Bemiller, J.N., Academic Press, New York, 1973.
16. Hjermstad, E.T., in *Starch: Chemistry and Technology*, vol. 2, eds. Whistler, R.L. and Paschall, E.F., Academic Press, New York, 1967.
17. Kricheldorf, H.R., Kreiser-Saunders, I., Jurgens, C., and Wolter, D., *Macromolecular Symposia*, 103, 85, 1996.
18. Auras, R., Harte, B., and Selke, S., *Macromolecular Bioscience*, 4, 835, 2004.
19. Cohn, D. and Salomon, A.H., *Biomaterials*, 26, 2297, 2005.
20. Kim, S.Y., Shin, I.G., and Lee, Y.M., *Journal of Controlled Release*, 56, 197, 1998.
21. Gupta, A.P. and Kumar, V., *European Polymer Journal*, 43, 4053, 2007.
22. Casey, J.P., in *Pulp and Paper: Chemistry and Chemical Technology*, 3rd edn., John Wiley & Sons, New York, 1980.
23. Hon, D.N.S., Cellulose plastics, in *Handbook of Thermoplastics*, ed. Olabisi, O., Marcel Dekker Inc., New York, 1997.
24. Malm, C.J. and Haitt, G.D., *Cellulose and Cellulose Derivatives*, eds. Ott, E. and Grafflin, M.W., John Wiley & Sons, New York, 1954.
25. *Ullman's Encyclopedia of Industrial Chemistry*, vol. 5, p. 419, VCH, New York, 1996.
26. Damodaran, S., Amino acids, peptides and proteins, in *Food Chemistry*, ed. Fennema, O.R., Marcel Dekker Inc., New York, 1996.
27. Redl, A., Morel, M.H., Bonicel, J., Vergnes, B., and Guilbert, S., *Cereal Chemistry*, 76, 361, 1999.
28. Toufeili, I., Lambert, I.A., and Kokini, J.L., *Cereal Chemistry*, 79, 138, 2002.
29. Rouilly, A., Mériaux, A., Geneau, C., Silvestre, F., and Rigal, L., *Polymer Engineering and Science*, 46, 1635, 2006.
30. Verbeek, C.J.R. and Van Den Berg, L.E., *Macromolecular Materials and Engineering*, 295, 10, 2010.
31. Barone, J.R., Schmidt, W.F., and Gregoire, N.T., *Journal of Applied Polymer Science*, 100, 1432, 2006.
32. Orliac, O., Silvestre, F., Rouilly, A., and Rigal, L., *Industrial and Engineering Chemistry Research*, 42, 1674, 2003.

33. Ralston, B. and Osswald, T., *Journal of Polymers and Environment*, 16, 169, 2008.
34. Mo, X. and Sun, X., *Journal of Polymers and Environment*, 8, 161, 2000.
35. Sessa, D.J., Selling, G.W., Willett, J.L., and Palmquist, D.E., *Industrial Crops and Products*, 23, 15, 2006.
36. Kokini, J.L., Cocero, A.M., Madeka, H., and DeGraaf, E., *Trends in Food Science and Technology*, 5, 281, 1994.
37. Krochta, J.M., Edible protein films and coatings, in *Food Proteins and Their Applications*, eds. Damodaran, S. and Paraf, A., Marcel Dekker Inc., New York, 1997.
38. Hernandez-Izquierdo, H.M. and Krochta, J.M., *Journal of Food Science*, 3, R30, 2008.
39. Khorshid, N., Hossain, M., and Farid, M.M., *Journal of Food Engineering*, 79, 1214, 2007.
40. Khwaldia, K., Perez, C., Banon, S., Desobry, S., and Hardy, J., *Critical Reviews in Food Science*, 44, 239, 2004.
41. Akaraonye, E., Keshavarz, T., and Roy, I., *Journal of Chemical Technology and Biotechnology*, 85, 732, 2010.
42. Futui, T. and Doi, Y., *Applied Microbiology and Biotechnology*, 49, 333, 1998.
43. Futui, T., Kichise, T., Yoshida, Y., and Doi, Y., *Biotechnology Letters*, 19, 1093, 1997.
44. Lee, S.Y., *Biotechnology and Bioengineering*, 49, 1, 1995.
45. Narayan, R., Krishnan, M., and Dubois, P., *Polymer*, 40, 3091, 1999.
46. Narayan, R., Krishnan, M., Snook, J.B., Gupta, A., and Dubois, P., *Polymer Compositions*, US Patent# 5969089, 1999.
47. Narayan, R., Krishnan, M., Snook, J.B., Gupta, A., and Dubois, P., Bulk reaction extrusion polymerization process producing aliphatic ester polymer compositions, US Patent# 5906783, 1999.
48. Narayan, R., Krishnan, M., Snook, J.B., Gupta, A., and Dubois, P., Bulk reactive extrusion polymerization process producing aliphatic ester polymer compositions. US Patent# 5801224, 1999.
49. Kalambur, S. and Rizvi, S.S.H., *Journal of Applied Polymer Science*, 96, 1072, 2004.
50. Kalambur, S.B., Starch polycaprolactone nanocomposites from reactive extrusion: Synthesis, characterization, properties and scale-up considerations, PhD thesis, Cornell University, New York, 2005.
51. Borggreve, R.J.M., Gaymans, R.J., and Schuijer, J., *Polymer*, 30, 71, 1989.
52. Greco, R., Malinconico, M., Martuscelli, E., Ragosta, G., and Scarinzi, G., *Polymer*, 28, 1185, 1987.
53. Cimmino, S., Coppola, F., D'Orazio, L., Greco, R., Maglio, G., Malinconico, M., Mancarella, C., Martuscelli, E., and Ragosta, G., *Polymer*, 27, 1874, 1986.
54. Cimmino, S., D'Orazio, L., Greco, R., Maglio, G., Malinconico, M., Mancarella, C., Martuscelli, E., Palumbo, R., and Ragosta, G., *Polymer Engineering and Science*, 24, 48, 1984.
55. Chuang, H.K. and Han, C.D., *Journal of Applied Polymer Science*, 30, 2431, 1985.
56. Hobbs, S.Y., Bopp, R.C., and Watkins, V.H., *Polymer Engineering and Science*, 23, 80, 1983.
57. Kim, B.K., Park, S.Y., and Park, S.J., *European Polymer Journal*, 27, 349, 1991.
58. Liu, N.C., Baker, W.E., and Russell, K.E., *Journal of Applied Polymer Science*, 41, 2285, 1990.
59. Ide, F. and Hasegawa, A., *Journal of Applied Polymer Science*, 18, 963, 1974.
60. Park, S.J., Kim, B.K., and Jeong, H.M., *European Polymer Journal*, 26, 131, 1990.
61. Dagli, S.S., Xanthos, M., and Biesenberger, J.A., *Proceedings of the Annual Technical Conference of SPE*, Dallas, TX, p. 969, 1991.
62. Campbell, J.R., Hobbs, S.Y., Shea, T.J., and Watkins, V.H., *Polymer Engineering and Science*, 30, 1056, 1990.
63. Carrot, C., Guillet, J., and May, J.F., *Plastics, Rubber and Composites Processing and Applications*, 16, 61, 1991.
64. Kim, B.K. and Park, S.J., *Journal of Applied Polymer Science*, 43, 357, 1991.
65. Angola, J., Fujita, Y., Sakai, T., and Inoue, T., *Journal of Polymer Science-Part B: Polymer Physics*, 26, 807, 1988.
66. Triacca, V.J., Ziaee, S., Barlow, J.W., Keskkula, H., and Paul, D.R., *Polymer*, 32, 1401, 1991.
67. Bhattacharya, M., John, J., Tang, J., and Yang, Z., *Journal of Polymer Science-Part A: Polymer Chemistry*, 35, 1139, 1997.
68. Bhattacharya, M., Mani, R., and Tang, J., *Journal of Polymer Science-Part A: Polymer Chemistry*, 37, 1693, 1999.
69. Narayan, R., Carlson, D., Nie, L., and Dubois, P., *Journal of Applied Polymer Science*, 72, 477, 1999.
70. Mani, R. and Bhattacharya, M., *European Polymer Journal*, 37, 515, 2001.

71. Dubois, P. and Narayan, R., *Macromolecular Symposia*, 198, 233, 2003.
72. Avella, M., Errico, M.E., Rimedio, R., and Sadocco, P., *Journal of Applied Polymer Science*, 83, 1432, 2002.
73. Kalambur, S.B. and Rizvi, S.S.H., *Journal of Plastic Film and Sheeting*, 22, 39, 2006.
74. Bhattacharya, M., John, J., and Tang, J., *Journal of Applied Polymer Science*, 67, 1947, 1998.
75. Otey, F.H. and Doane, W.H., *Chemicals from Starch in Starch: Chemistry and Technology*, eds. Whistler, R.L., BeMiller, J.N., and Paschall, E.F., Academic Press Inc., New York, 1984.
76. Weaver, M.O., Bagley, E.B., Fanta, G.F., and Doane, W.M., US Patent# 3935099, 1976.
77. Carr, M.E., Kim, S., Yoon, K.J., and Stanley, K.D., *Cereal Chemistry*, 69, 70, 1992.
78. Yoon, K.J., Carr, M.E., and Bagley, E.B., *Journal of Applied Polymer Science*, 45, 1093, 1994.
79. Della Valle, G., Colonna, P., and Tayeb, J., *Starch*, 43, 300, 1991.
80. Carr, M.E., *Journal of Applied Polymer Science*, 54, 1855, 1994.
81. Wing, R.E. and Willett, J.L., *Industrial Crops and Products*, 7, 45, 1997.
82. Mecerreyes, D., Jérôme, R., and Dubois, P., *Advances in Polymer Science*, 147, 1, 1999.
83. Degée, P., Dubois, P., Jérôme, R., Jacobsen, S., and Fritz, H.G., *Macromolecular Symposia*, 144, 289, 1999.
84. Degée, P., Dubois, P., Jérôme, R., Jacobsen, S., and Fritz, H.G., *Polymer*, 41, 3395, 2000.
85. Racquez, J.M., Narayan, R., and Dubois, P., *Macromolecular Materials and Engineering*, 0293, 447, 2008.
86. Racques, J.M., Degee, P., Nabar, Y., Narayan, R., and Dubois, P., *Comptes Rendus Chimie*, 9, 1370, 2006.
87. Rodriguez, F., Polymer formation, in *Principles of Polymer Systems*, Taylor & Francis, Washington, DC, 1996.
88. Xie, F., Yu, L., Liu, H., and Chen, L., *Starch*, 58, 131, 2006.
89. Kalambur, S. and Rizvi, S.S.H., *Polymer International*, 53, 1413, 2004.
90. Riaz, M., *Extruders in Food Applications*, Technomic Publishing Co., Lancaster, PA, 1997.
91. Tadmor, Z. and Gogos, C.G., Pressurization and pumping, in *Principles of Polymer Processing*, John Wiley & Sons Inc, New York, 1979.
92. Riande, E., Diaz-Calleja, R., Prolongo, M.G., Masegosa, R.M., and Salom, C., Flow behavior of polymer melts and solutions, in *Polymer Viscoelasticity-Stress and Strain in Practice*, Marcel Dekker, New York, 2000.
93. Rauwendaal, C., *Polymer Extrusion*, Hanser Publishers, Munich, Germany, 1986.
94. Rauwendaal, C. and Ponzielli, G., *Temperature Development in Screw Extruders*, Antec, 2003 (see a copy of publication in http://www.rauwendaal.com).
95. Kalambur, S.B. and Rizvi, S.S.H., *Polymer Engineering and Science*, 46, 650, 2006.

BIOGRAPHY

Sathya B. Kalambur received his BS technology from Harcourt Butler Technological Institute in Kanpur, India, and MS in food science from the University of Maryland before obtaining his PhD in food science with a minor in chemical engineering from Cornell University. In addition to his research efforts at Maryland and Cornell, he worked for Galaxy Surfactants and is now working as a product development scientist at Frito-Lay, Inc., United States.

CHAPTER 8

Starch-Derived Cyclodextrins and Their Future in the Food Biopolymer Industry

C. González-Barreiro, R. Rial-Otero, J. Simal-Gándara, G. Astray,
A. Cid, J.C. Mejuto, J.A. Manso, and J. Morales

CONTENTS

8.1 INTRODUCTION TO NATURE AND PROPERTIES OF CYCLODEXTRINS

Cyclodextrins (CDs) are natural cyclic oligosaccharides formed by the enzymatic degradation of starch or starch derivatives, composed of glucose units, using cyclodextrin glycosyltransferase (CGTase) (Bender and Komiyama, 1978; Saenger, 1980). The preparation process of CDs consists of four principal phases: (1) culturing of the microorganism that produces the CGTase; (2) separation, concentration, and purification of the enzyme from the fermentation medium; (3) enzymatical conversion of prehydrolyzed starch in mixture of cyclic and acyclic dextrins; and (4) separation of CDs from the mixture, their purification, and crystallization. CGTase enzymes degrade the starch and produce intramolecular reactions without the water participation. The enzymatic product is usually a mixture of CDs.

The most common forms are α-, β-, and γ-CDs with 6, 7, or 8 D-glucopyranosyl residues, respectively, linked by α-1,4 glycosidic bonds (Figure 8.1). These three major CDs are crystalline,

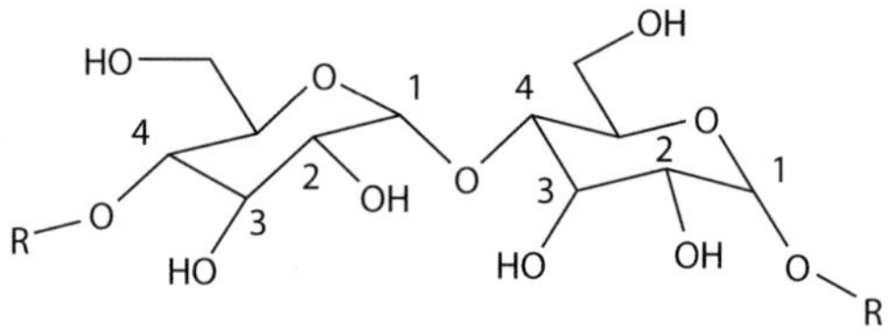

Figure 8.1 Glycosidic oxygen bridge α (1,4) between two molecules of glucopyranose.

homogeneous, and nonhygroscopic substances. CDs with fewer than six glucopyranose residues do not exist, probably for steric reasons. Larger CDs have already been identified: δ- (delta), ε- (epsilon), ζ- (zeta), and η- (eta) CDs containing 9, 10, 11, and 12 glucopyranose units, respectively, and, more recently, the θ- (theta) composed of 13 glucopyranose units (Cabral Marques, 2010). However, their yields are extremely small, and their complexing properties are not as good as natural CDs (Larsen, 2002).

The CD molecules are often described as a torus but somewhat more realistically pictured as a *"shallow truncated cone"* with hydrophilic exteriors and comparably more hydrophobic cavities. The depths of CD cavities are all the same (approximately 7.5 Å), being determined by the width of a glucose unit, but the size of their cavities differs in diameter (4.5 Å for α-CD, 7.0 Å for β-CD, and 8.5 Å for γ-CD), given rise to a gradation in binding affinity (Connors, 1997), as it can be seen in Figure 8.2. Regardless of the finer details of their structure, the most important feature of CDs is their cavity because this enables them to form inclusion complexes (ICs) with small molecular guests of an appropriate size, shape, and polarity (Connors, 1997).

CDs have no well-defined melting point and start to decompose at temperatures above 270°C, and at 300°C, a sharp endothermic process is detected by differential scanning calorimetry, which indicates that melting is accompanied by decomposition (Cabral Marques, 2010).

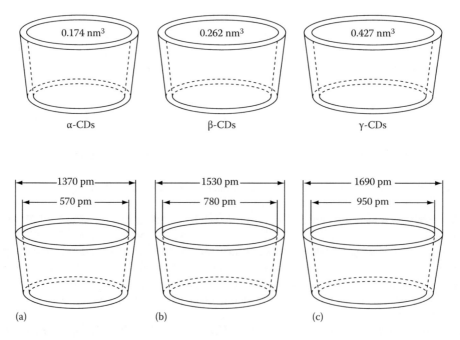

Figure 8.2 Geometric dimensions of (a) α-CDs, (b) β-CDs and (c) γ-CDs.

The water solubility of CDs is unusual. β-CD is at least nine times less soluble (1.85 g/100 mL at room temperature) than the other CDs (14.5 g/100 mL and 23.2 g/100 mL for α- and γ-CDs, respectively). The low solubilities of CDs are a consequence of the relatively unfavorable enthalpies of solution (more positive), partially offset by the more favorable entropies of solution (more negative). The thermodynamic properties of α- and γ-CDs are similar. The decreased solubility of β-CD in water appears to be due to the marked structure of water arising from water-β-CD interactions, causing a compensation of the favorable enthalpy by the unfavorable entropy of solution (Linert et al., 1992). The solubility of CDs depends strongly on the temperature (Astray et al., 2009), and in general, their solubility in water is increased at higher temperatures (Cabral Marques, 2010).

8.2 REGULATIONS AFFECTING CYCLODEXTRINS

In order to protect the health of consumers, toxicological data of CDs are examined, and the industrial use of these compounds is subject to governmental regulatory requirements. The general trend is toward a wider acceptance of CDs as food additives. In fact, CDs are the only supramolecular compounds tested and approved for the use in foodstuff. Other supramolecular ligands, for example, crown ethers, calixarenes, or cucurbiturils, are not used in the food industry due to their unknown toxicological data. However, regulations for CDs differ between countries. The so-called α-, β-, and γ-CDs have been deemed acceptable by the Joint Expert Committee on Food Additives (JECFA) of the United Nations Food and Agriculture Organization/World Health Organization (UN FAO/WHO, 1995, 2000, 2002) and are also approved for use in various countries. Since November 2000, β-CD is authorized in Germany as food additive (E 459). In Japan, CDs have been approved as "modified starch" for food applications for more than two decades, serving to mask odors (Hara et al., 2002) in fresh food and to stabilize fish oils. In Korea, α- and γ-CDs are approved for dietary supplement. In addition, the US Food and Drug Administration (FDA) decided to include the so-called α-, β-, and γ-CDs in the GRAS list (Generally Recognized as Safe) for use as additives in food products (FDA, 2000, 2001, 2004). Thus, α-CD is GRAS for use in selected foods for fiber supplementation; as a carrier or stabilizer for flavors (flavor adjuvant), colors, vitamins, and fatty acids; and to improve mouth-feel in beverages (Table 8.1). β-CD is GRAS for use as a flavor carrier or protectant (Table 8.2) and γ-CD for use as a stabilizer, emulsifier, carrier, and formulation aid (Table 8.3).

In the European Union, the use of α-CD as a novel food ingredient was authorized (Commission Decision, 2008/413/EC), and at this moment, the use of γ-CD is being studied. However, some countries such as Hungary have approved γ-CD for use in certain applications because of its low toxicity (Kohata et al., 2010).

8.3 FUNCTIONAL AND TECHNOLOGICAL INTERESTS OF CYCLODEXTRINS

Due to the steric arrangement of the glucose units, the inner side of the torus-like CD molecules is less polar than the outer side. This enables CDs to form ICs with various organic compounds (Munro et al., 2004). ICs are entities composed of two or more molecules. One of the molecules, the *host*, includes, totally or partly, the *guest* molecules by physical forces. Therefore, CDs are considered typical *host* molecules. The driving forces of ICs formation between the CDs cavity and the *guest* molecule include geometric compatibility, Van der Waals forces, and hydrophobic interactions (Harada et al., 1994; He et al., 2005). To stabilize the ICs, hydrogen bonding may occur between the CD and the *guest* molecule and between the hydroxyl groups on the rims of neighboring CDs (Loethen et al., 2007; Yuen and Tam, 2010).

Table 8.1 Food Categories and Use Levels for α-CD

Food Category	Maximum Use Level Percent (w/w)
Breads, rolls, cakes, baking mixes, refrigerated dough	5
Brownies and bars	7
Crackers (sweet and nonsweet)	10
Diet soft drinks, beverage mixes, fruit juices, instant coffees and teas, coffee whiteners (dry), formula diets, meal replacements, and nutritional supplements	1
Vegetable juices, soy milk, and nonsoy (imitation) milk	2
Ready-to-eat breakfast cereals	2–9[a]
Instant rice, pasta, and noodles (prepared)	2
Condiments	3
Reduced fat spreads	20
Dressings and mayonnaise	5
Yogurt, milk beverage mixes, and frozen dairy desserts	2.5
Pudding mixes (dry)	1
Snack foods	1
Canned and dry soups (prepared)	2
Hard candy	15
Chewing gum	10

[a] The notifier states that use level in ready-to-eat cereals will vary based on weight of serving size (i.e., if less than 20 g/cup, the level is 2%; 20–43 g/cup, the level is 9%; greater than or equal to 43 g/cup, the level is 5%).

Table 8.2 Food Categories and Use Levels for β-CD

Food Category	Maximum Use Level Percent (w/w)
Baked goods prepared from dry mixes	2
Breakfast cereal	
Chewing gum	
Compressed candies	
Gelatins and puddings	1
Flavored coffee and tea	
Processed cheese products	
Dry mix for beverages	
Flavored savory snacks and crackers	0.5
Dry mixes for soups	0.2

The formation of *host–guest* complexes is composed of several steps and occurs through desolvation of the species (Figure 8.3). Nevertheless, the stability of the complex is related to the amount of water, which may be released by the CD upon the encapsulation of the *guest* molecule (García-Río et al., 2007, 2010; Inoue et al., 1993). Therefore, the formation of complexes between CDs and *guest* molecules is reversible, and excess water would in most cases result in a dissociation of the complex (Hedges et al., 1995).

The formation of an IC with a *guest* molecule is the basis for many applications of CDs in food, cosmetics, and pharmaceutical preparations (Astray et al., 2010; Munro et al., 2004). The suitability of the different CDs for these applications varies in relation to the size of the *guest* molecule which the CD ring should accommodate. It is important that the geometrical dimensions of the *guest* molecules are rather close to those of substituted benzene ring or its condensed

Table 8.3   Intended Use of γ-CD

Food	Maximum Use Level Percent (w/w)
Carrier for flavors, sweeteners, and colors	1
Dry mixes for beverages, soups, dressings, gravies, sauces, puddings, gelatins, and fillings	
Instant coffee, instant tea, and coffee whiteners	
Compressed candies and chewing gum	
Breakfast cereals (ready-to-eat)	
Savory snacks and crackers	
Spices and seasonings	
Carrier for vitamins in dry food mixes and dietary supplement products	90
Carrier for polyunsaturated fatty acids (including meal replacements and dietary supplement products)	80
Flavor modifier—soya milk	2
Stabilizer	
Bread spreads (fat reduced)	20
Fat-based fillings	5
Fruit-based fillings	3
Frozen dairy desserts	3
Processed cheese	3
Dairy desserts	3
Baked goods (excluding bread, but including dough and baking mixes)	2
Bread	1

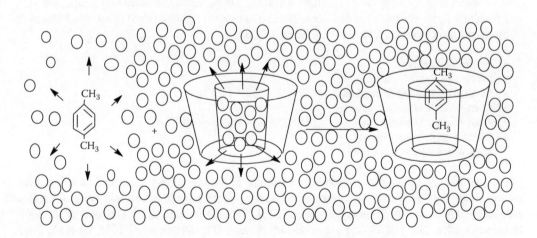

Figure 8.3   Schematic illustration of inclusion complexation of p-xylene by a CD.

homologues (Astray et al., 2009). The charge and polarity of the *guest* play also an important role in the CD-substrate *host–guest* interaction; however, it is less decisive than that of the geometric fitting. CDs can form ICs with molecules less hydrophilic or less polar than water, and there is a positive correlation between stability of CD complexes and the hydrophilic character of molecules or certain parts of the *guest* molecules. In the case of the charge, the complexation of neutral molecules is easier than the ionized counterpart.

In addition, CD structures and specially their cavity diameters also influence the formation of ICs. Compared with α- and β-CDs, γ-CD exhibits more favorable properties in terms of the size of its internal cavity, water solubility, and bioavailability, so it has even wider applications in many fields, especially in the food and pharmaceutical industries (Munro et al., 2004; Szejtli, 2004). However, to date, the most marketed CD is β-CD and to lesser extent α-CD, while the markets share of γ-CD is considerably small because of its low yield and high price (Szejtli, 2004).

Several classes of compounds that can form ICs with natural CDs have been subjected to systematic studies. These cover almost every class of compounds such as aliphatic alcohols (Cabaleiro-Lago et al., 2006; García-Río et al., 2005), amines and acids (Gadosy et al., 2000), purine nucleosides (Tian-Xiang and Anderson, 1990), amino acids (Tee et al., 1997), ketones (Iglesias, 2003), surfactants (Dorrego et al., 2000; Fernández et al., 2000; García-Río et al., 2004), and other compounds (García-Río et al., 2006; Iglesias et al., 2007; Rekharsky and Inoue, 1998). The formation of these ICs greatly modifies the physical and chemical properties of the *guest* molecules. Due to their highly polar exteriors and less polar interiors, CDs are best known for their ability to increase the solubility of low-polarity *guest* molecules in water as the encapsulated compound is shielded from the aqueous solvent. CDs are used to encapsulate hydrophobic drugs in drug-delivery systems, to enhance the scrubbing efficiency of low-polarity volatile organic compounds, and to mask toxic substances by converting them into nontoxic ICs (Yuen and Tam, 2010).

8.4 APPLICATIONS OF CYCLODEXTRINS IN DIFFERENT FOOD SECTORS

One of the main goals of the food industry is to develop new additives that improve the color and flavor of foods. In this respect, there has been a renewal of interest in natural additives that improve the sensory properties of foods, and among the most promising of such agents are CDs.

CDs as multifunctional food ingredients can be utilized mainly as carriers for molecular encapsulation improving the stability of sensitive ingredients (e.g., flavors, aromatizing agents, and others), both in the physical and chemical sense leading to extended product shelf life (Cabral Marques, 2010). In the food industry, CDs can be used mainly to

- Increase solubility
- Increase the rate of solubilization
- Increase the stability of emulsions
- Change the rheological properties
- Protect against oxidation, and light- and heat-induced decomposition
- Stabilize against evaporation
- Mask or reduce undesired taste

The use of CDs as food additives has been reviewed in the literature extensively. Detailed information about their specific usage can be found in the literature given in several books and review articles (Astray et al., 2009; Buschman et al., 2001; Cabral Marques, 2010; Cravotto et al., 2006; Hashimoto, 1996; Hedges, 1998; Hedges and McBride, 1999; Hedges et al., 1995; Li et al., 2007; Martin Del Valle, 2004; Qi and Hedges, 1995; Samant and Pai, 1991; Singh et al., 2002; Szejtli, 1982, 1988, 2004; Szejtli and Fenyvesi, 2005; Szejtli and Szente, 2005; Szente and Szejtli, 2004).

To provide a different perspective from those published so far, we will focus on very recent works (mainly the last 3 years) in the field of food industry, differentiating by sectors.

8.4.1 Plant Food

CDs are primarily used in the plant food sector in order to improve the chemical stability of these products (Table 8.4).

Table 8.4   Recent Applications of CDs in the Plant Food Sector

Application	CD Type	*Guest* Molecules	References
Protection against oxidation and light- and heat-induced decomposition of flavors	β-CD	Monoterpene hydrocarbons: carvacrol, thymol, and eugenol	Locci et al. (2004)
	β-CD	Vanillin	Karathanos et al. (2007)
	β-CD	Monoterpenes: thymol and geraniol	Mourtzinos et al. (2008)
	β-CD	Terpineol	Mazzobre et al. (2011)
Improvement of the chemical stability	γ-CD	Ferulic acid	Anselmi et al. (2006)
	β-CD, γ-CD	Tocopherol and quercetin	Koontz et al. (2009)
	β-CD, heptakis (2,6-di-O-methyl) β-cyclodextrin (DM-β-CD) and 2-hydroxypropyl-β-cyclodextrin (HP-β-CD)	Flavone aglycones: chrysin, apigenin, and luteolin	Kim et al. (2008)
	β-CD, HP-β-CD	Resveratrol	López-Nicolás and García-Carmona (2008, 2010); López-Nicolás et al. (2006)
	β-CD	Oxyresveratrol	Rodríguez-Bonilla et al. (2010)
	α-CD, β-CD, γ-CD, HP-β-CD, methyl-β-CD, and ethyl-β-CD	Pterostilbene	López-Nicolás et al. (2009c)
	α-CD, β-CD, γ-CD, HP-β-CD, methyl-β-CD, and ethyl-β-CD	Pinosylvin	López-Nicolás et al. (2009b)
	β-CD	Amylose and conjugated linoleic acid	Yang et al. (2010)
	β-CD	Astaxanthin	Kim et al. (2010)
	β-CD	Curcumin	Paramera et al. (2011)

Generally, most flavor components are highly volatile and chemically unstable in the presence of air, light, moisture, and heat; complexation with CDs provides a promising alternative to the conventional encapsulation technologies for flavor protection. β-CD as a molecular encapsulant allows the flavor quality and quantity to be preserved to a greater extent and longer period compared to other encapsulants and provides longevity to the food item (Muñoz-Botella et al., 1995). Natural and synthetic coffee flavors or other food flavors (anethol, anis oil, citral, citronellal, linalool, menthol, sage oil, cinnamon oil, jasmine oil, bergamot oil, orange oil, lemon oil, lime oil, onion oil, garlic oil, mustard oil, marjoram oil, other monoterpenes such as thymol and geraniol, etc.) are stabilized with CDs to avoid the loss of flavors during storage of the product or as a result of exposure to light or oxygen; upon contact with water, the complex-bound flavor substances were released immediately. These ICs have a good reputation for their high stability exhibited when they are heated during industrial food processing, and they are stable for a longer period than liquid essence or the components themselves (Cabral Marques, 2010). In this sense,

- Carvacrol, thymol, and eugenol (components of essential oils of vegetable origin) are oxidized, decomposed, or evaporated when exposed to the air, light, or heat. If prepared as β-CD ICs, they are stabilized as CDs and greatly reduce volatility, oxidation, and heat decomposition (Locci et al., 2004).

- CD complexation can provide oxidative protection to vanillin (Karathanos et al., 2007), thymol, and geraniol (Mourtzinos et al., 2008).
- A common flavor, terpineol, has been recently studied (Mazzobre et al., 2011) to form ICs with β-CD. The limited water solubility of terpineol could be overcome by the formation of β-CD ICs, and the complexes were stable at different storage conditions (relative humidities 11%–97% and 25°C).

CDs are able to increase the bioavailability of different compounds with proven health properties. Among the *guest* molecules which have been complexed by CDs, several works have studied the inclusion of different antioxidant molecules to facilitate their use as ingredients of functional foods or nutraceuticals. In recent years, ferulic acid (Anselmi et al., 2006), tocopherol and quercetin (Koontz et al., 2009), flavone aglycones (Kim et al., 2008), and different stilbenoids with high antioxidant activity such as resveratrol (López-Nicolás and García-Carmona, 2008, 2010; López-Nicolás et al., 2006), oxyresveratrol (Rodríguez-Bonilla et al., 2010), pterostilbene (López-Nicolás et al., 2009c), and pinosylvin (López-Nicolás et al., 2009b) have been complexed by natural and modified CDs.

Encapsulation in CDs can enhance the stability of other food components such as amylase and conjugated linoleic acid (Yang et al., 2010) or food coloring additives such as astaxanthin (Kim et al., 2010), curcumin (Paramera et al., 2011), and phycoerythrin and phycocyanin (Kohata et al., 2010) under various storage conditions of pH, temperature, ultraviolet irradiation, and the presence of oxygen.

8.4.2 Animal Food

The main application of CDs in animal food is its use as a cholesterol sequestrant due to the growing interest in the manufacture of cholesterol-reduced dairy products (Table 8.5). Several studies have indicated that the cholesterol removal in milk, cream, cheese, butter, lard, and egg yolks was effectively conducted by β-CD (Astray et al., 2009). More recently, Alonso et al. (2009) determined the optimum conditions (β-CD concentration, mixing time, and holding time) for cholesterol removal from pasteurized nonhomogenized milk at 4°C on a commercial scale by adding β-CD in a specially designed bulk mixer tank. Seon et al. (2009) investigated the influence of salt content on cholesterol-reduced Cheddar cheese obtained by a treatment with cross-linked β-CD and if the ripening process was accelerated by the cross-linked β-CD treatment.

8.4.3 Beverages

The improvement of chemical stability and the modification of tastes and odors are the major applications of CDs in the beverage sector (Table 8.5).

8.4.3.1 Improvement of Chemical Stability

CDs improve the shelf life of beverages by protecting against light-, air-, and heat-induced decomposition. In the production of juices, the mechanical damage suffered by vegetable and fruit tissues often leads to rapid enzyme-catalyzed browning reactions. Thus, polyphenol oxidase converts the colorless polyphenols to color compounds. In order to avoid this effect, fruits and vegetable juices can be treated with CDs, which removes polyphenol oxidase from juices by complexation (Del Valle, 2004). In recent years, the color of fruit juices, such as apple, banana, pear, or grape, has been evaluated in the presence of natural and modified CDs, and the effect of these compounds as browning inhibitors has been determined using the CIEL*a*b* color space system (López-Nicolás and García-Carmona, 2007; López-Nicolás et al., 2007a–c; Núñez-Delicado et al., 2005). These studies showed that the fruit juice enzymatic browning could be slowed down (or activated) when

Table 8.5   Recent Applications of CDs in the Animal Food Sector, Beverages, and Food Packaging

Application	Sample Matrix	CD Type	*Guest* Molecules	References
Animal food				
Cholesterol sequestrant	Milk	β-CD	Cholesterol	Alonso et al. (2009)
	Cheddar cheese	Crosslinked β-CD	Cholesterol	Seon et al. (2009)
Beverages				
Browning inhibitor	Dominga grape juice	Maltosyl-β-CD	4-*tert*-butylcatechol	Núñez-Delicado et al. (2005)
	Pear juice	α-CD and β-CD, and maltosyl-β-CD	Diphenols	López-Nicolás and García-Carmona (2007)
	Apple juice	α-CD, β-CD, γ-CD, and maltosyl-β-CD	Diphenols	López-Nicolás et al. (2007a)
	Peach juice	α-CD, β-CD, and maltosyl-β-CD	Diphenols	López-Nicolás et al. (2007c)
	Pear juice	α-CD	Diphenols	López-Nicolás et al. (2009a)
Browning activator	Banana juice	α-CD, β-CD, and maltosyl-β-CD	The natural browning inhibitors present in banana	López-Nicolás et al. (2007b)
Modification of tastes: bitterness	Ginseng	γ-CD	Ginsenosides	Lee et al. (2008)
	Ginseng	β-CD, γ-CD	Ginsenosides	Tamamoto et al. (2010)
Food packaging				
Retain or scavenge substances		β-CD	Hexanal	Almenar et al. 2007
		β-CD	α-tocopherol	Siró et al. (2006)
		β-CD	Aldehydes and cholesterol	López-de-Dicastillo et al. (2011)

increasing concentrations of different CDs were added. More recently, the same authors (López-Nicolás et al., 2009a) have reported the effect of the addition of CDs on the flavor profile of a pear juice on both odor and aroma, using sensory and instrumental protocols. The addition of α-CD at 90 mM resulted in pear juices with the best color but with low aromatic intensity and low sensory quality. On the other hand, the addition of α-CD at 15 mM led to a pear juice also with an acceptable color but at the same time with a high intensity of fruity and pear-like odors/aromas, making it the best appreciated juice by the panel.

α-CD-cinnamic acid (CA) ICs may provide an alternative to traditional heat processes coupled with additional processing steps. Truong et al. (2010) determined the effectiveness of solubility-enhancing α-CD-CA ICs against *Escherichia coli* 0157:H7 and *Salmonella entérica serovars* suspended in apple cider or orange juice at two different incubation temperatures (4°C and 26°C).

8.4.3.2 Modification of Tastes and Odors

It was proved that CDs may alter the sensory profile of a food. The flavor release depends on the CD type (Reineccius et al., 2002) and the temperature (Reineccius et al., 2003), and it may also be solvent dependent (Reineccius et al., 2005). Their beneficial effects essentially derive from the ability to form stable ICs with sensitive lipophilic nutrients and constituents of flavor and taste, making it easy to prepare powdered flavor materials (Liu et al., 2000; Tobitsuka et al., 2005, 2006), and even to release such flavors during cooking (Shiga et al., 2003).

The complexation of bitter molecules using CDs is a potential bitterness-minimizing treatment. CDs form ICs with the bitter compounds (e.g., naringin in citrus fruit juice, or chlorogenic acid and polyphenols in coffee), resulting in reduced bitterness (Szejtli and Szente, 2005; Szente and Szejtli, 2004). The inclusions are formed through hydrogen bonding and Van der Waals forces between the bitter compound and the CDs (Szejtli, 1988). The bitter taste is reduced because of the inability of the complexed molecules to bind to the taste receptors on the tongue (Szejtli and Szente, 2005).

The bitter taste of grapefruit or mandarin juices decreased substantially when 0.3% β-CD was added prior to a heat treatment of canned juices. CDs have efficiently reduced the bitter taste of some plant extracts such as guava tea extract, gingko extract, and gymnema extract in the order γ-CD≫β-CD≫α-CD (Cabral Marques, 2010). It was reported in a patent by Lee et al. (2008) that the bitterness of 100 g of ginseng extract in solution could be removed with the addition of approximately 1 g of γ-CD. In a subsequent study, Tamamoto et al. (2010), based on the results of a series of pilot studies, found that 0.09 g of γ-CD significantly reduce the bitter taste and aftertaste of ginseng in 0.052 g ginseng/100 mL solution.

On the other hand, the complexation of CDs with sweetening agents such as aspartame stabilizes and improves the taste. It also eliminates the aftertaste of other sweeteners such as stevioside, glycyrrhizin, and rubusoside (Astray et al., 2009). CD itself is a promising new sweetener. Enhancement of flavor by CDs has been also claimed for alcoholic beverages such as whisky and beer (Singh et al., 2002).

8.4.4 Food Packaging

The advantages of CD application in plastic packaging can be inclusion of by-products of polyethylene generated by heat seal, decreased release of impurities and undesired volatile by-products formed during manufacture of the packaging material into the food or beverages, improvement of the barrier function of the packaging material entrapping both the penetrating volatiles, atmospheric pollutants migrating inward as well as the aroma substances escaping outward, odor absorption when "empty" CD is used and controlled release of the active component (antimicrobial, antioxidant, etc.) when CD complex is applied (Fenyvesi et al., 2007). With regard to this last application, several papers have reported on the complexation of different antimicrobial compounds by CDs (Ayala-Zavala et al., 2008). When a CD–antimicrobial compound molecular complex is exposed to water molecules, its interaction is weakened, and the antimicrobial agent is passively released to the environment. These mechanisms could be used to generate an antimicrobial-active packaging and therefore protect fresh-cut product against bacterial and fungal growth. In this sense, complexation by β-CD was found to be an effective tool for controlling the release of α-tocopherol from antioxidative-active packaging (Siró et al., 2006). Almenar et al. (2007) used β-CD-hexanal to reduce or avoid postharvest berry diseases because of their capacity to provide an antifungal volatile during storage, distribution, and consumer purchasing. Novel ethylene–vinyl alcohol copolymer (EVOH) films containing β-CDs were used to reduce the presence of aldehydes (substances which develop as a result of oxidative processes) in packaged fried peanuts and cholesterol in UHT milk (López-de-Dicastillo et al., 2011).

Several recent applications have been described in which CDs or CD derivatives have been immobilized in different polymeric supports. CDs have been immobilized in polyvinyl alcohol–based membranes for enantiomer separation and sorption of xylene (Touil et al., 2008) and in an ethylene–vinyl alcohol copolymer with a 44% molar percentage of ethylene (EVOH44) by using regular extrusion with glycerol as an adjuvant (López-de-Dicastillo et al., 2010). In addition, the surface of cellulose membranes were modified by covalent bonding of β-CD for chiral separation of tryptophan (Xiao and Chung, 2007).

8.5 FUTURE TRENDS FOR BIOPOLYMERS AND ESPECIALLY CYCLODEXTRINS

Today, the renewable nature of biopolymers is helping biopolymers experience a renaissance. In the past 20 years, interest in sustainable products has driven the development of new biopolymers from renewable feedstocks. Biopolymers must compete with existing fossil-fuel-derived polymers not only in their functional properties but also in price. Biopolymers become competitive with fossil-fuel-derived polymers when the price of oil is high and the price of feedstocks, such as cornstarch, is low. These drivers act as a stimulus for R&D activity in microbiology, genetic engineering, plant sciences, fermentation, and purification technologies. Many biopolymers are today in commercial use. Optimization of the performance of biopolymers expands the range of applications in which biopolymers can replace fossil-fuel-derived polymers. Biopolymers that biodegrade are a focus for bioplastic companies that develop materials for single-use, biodegradable packaging. Manufacturers of biopolymers form partnerships with downstream products' manufacturers to commercialize their biopolymers. Other opportunity areas include industrial, medical, food, consumer products, and pharmaceutical applications. In these applications, biopolymers act as stabilizers, thickeners, gallants, binders, dispersants, lubricants, adhesives, and drug-delivery agents.

Many types of encapsulation are available which coat a fine particle of an active core with an outer shell. Encapsulation also can occur on a molecular level. This can be accomplished using a category of carbohydrates called CDs; encapsulates made with these molecules may possibly hold the key for many future encapsulated formulation solutions. The ability of CDs to form ICs with many *guest* molecules by taking up a whole molecule or some part of it into the cavity of CDs is a unique class of encapsulation technique. This type of molecular encapsulation will affect many of the physicochemical properties of the *guest* molecules. The ability of CDs to form complexes with a wide variety of organic compounds helps to alter the apparent solubility of the molecule; to increase the stability of compound in the presence of light, heat, and oxidizing conditions; and to decrease volatility of compound. These properties have resulted in the growing importance of the applications of CDs in food, pharmaceutical, agriculture, and chromatographic techniques. The versatility of CDs and modified CDs is demonstrated in their range of applications from cosmetics and food to drugs. Recent biotechnological advancements have resulted in dramatic improvements in the efficient manufacture of CDs, lowering the cost of these materials and making highly purified CDs and CD derivatives available. Accelerated and long-term storage stability test results showed that the stability of CD-entrapped food ingredients surpassed that of the traditionally formulated ones. Technological advantages of the use of CDs in foods and food processing technologies are also manifested in improved sensory, nutritional, and performance properties. An example of the growing interest in their potential applications is the steadily increasing number of publications in the last 40 years (Figure 8.4).

8.6 OVERALL CONCLUSIONS

The ability to hold, orient, conceal, and separate *guest* molecules, together with CDs' chirality and low toxicity, places CDs in a unique class of building blocks for constructing novel molecular architecture. CDs are not merely another group of excipients, extenders, or bulking agents, but they are multipurpose technological tools that can be finely honed by chemical modification. CDs act as molecular chelating agents of growing importance in food and agriculture, but also in pharmaceuticals and chromatographic techniques. The versatility of CDs and modified CDs is demonstrated in

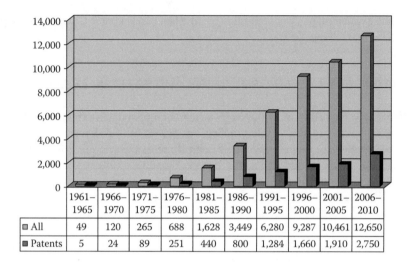

	1961–1965	1966–1970	1971–1975	1976–1980	1981–1985	1986–1990	1991–1995	1996–2000	2001–2005	2006–2010
All	49	120	265	688	1,628	3,449	6,280	9,287	10,461	12,650
Patents	5	24	89	251	440	800	1,284	1,660	1,910	2,750

Figure 8.4 Publications and patents' frequency during 40 years of CD-related works in a 5 year sum.

their range of applications from cosmetics and food to drugs. CDs have also widened the presently limited scope of various compounds by acting as powerful toxicity modifiers. In addition, CDs broaden the horizons of catalysis by increasing enantioselectivity.

Owing to a unique architecture, CDs are now becoming an important part of the scientists' options in drug development, as enzyme mimics; in chiral separations; and as complexing agents in the food, cosmetics, and pharmaceutical industries. The growing health consciousness of consumers and expanding market for functional foods and nutraceutical products are opening up to CDs a promising future in food industry.

ACKNOWLEDGMENTS

G. Astray thanks the Government of Galicia for a training research grant (P.P. 0000 300S 14008). R. Rial-Otero and C. González-Barreiro acknowledge Xunta de Galicia for their contract sponsorship through the Isidro Parga Pondal program.

REFERENCES

Almenar, E., Auras, R., Rubino, M., and Harte, B. 2007. A new technique to prevent the main post harvest diseases in berries during storage: Inclusion complexes β-cyclodextrin-hexanal. *Int. J. Food Microbiol.* 118: 164–172.

Alonso, L., Cuesta, P., Fontecha, J., Juárez, M., and Guilliland, S. E. 2009. Use of β-cyclodextrin to decrease the level of cholesterol in milk fat. *J. Dairy Sci.* 92: 863–869.

Anselmi, C., Centini, M., Ricci, M., Buonocore, A., Granata, P., Tsuno, T., and Facino, R. M. 2006. Analytical characterization of a ferulic acid/γ-cyclodextrin inclusion complex. *J. Pharm. Biomed. Anal.* 40: 875–881.

Astray, A., González-Barreiro, C., Mejuto, J. C., Rial-Otero, R., and Simal-Gándara, J. 2009. A review on the use of cyclodextrins in foods. *Food Hydrocolloids* 23: 1631–1640.

Astray, G., Mejuto, J. C., Morales, J., Rial-Otero, R., and Simal-Gándara, J. 2010 Factors controlling flavors binding constants to cyclodextrins applications in foods. *Food Res. Int.* 43: 1212–1218.

Ayala-Zavala, J. F., Del-Toro-Sánchez, I., Álvarez-Parrilla, E., and González-Aguilar, G. A. 2008. High relative humidity in-package of fresh-cut fruits and vegetables: Advantage or disadvantage considering microbiological problems and antimicrobial delivering systems? *J. Food Sci.* 73: 41–47.

Bender, M. L. and Komiyama, M. 1978. *Cyclodextrin Chemistry*. Berlin, Germany: Springer.

Buschman, H. J., Knittel, D., and Jonas, C. 2001. Applications of cyclodextrins in food production. A short summary. *Lebensmittelchemie* 55: 54–56.

Cabaleiro-Lago, C., García-Río, L., Hervés, P., Mejuto, J. C., and Pérez-Juste, J. 2006. Characterization of alkane diol-CD complexes. Acid denitrosation of N-methyl-N-nitroso-p-toluenesulphonamide as a chemical probe. *J. Inclusion Phenom.* 54: 209–216.

Cabral Marques, H. M. 2010. A review on cyclodextrin encapsulation of essential oils and volatiles. *Flavour Fragr. J.* 25: 313–326.

Commission Decision 2008/413/EC authorizing the placing on the market of alpha-cyclodextrin as a novel food ingredient OJ L146 of June 5, 2008, p. 12.

Connors, K. A. 1997. The stability of cyclodextrin complexes in solution. *Chem. Rev.* 5: 1325–1357.

Cravotto, G., Binello, A., Baranelli, E., Carraro, P., and Trotta, F. 2006. Cyclodextrins as food additives and in food processing. *Curr. Nutr. Food Sci.* 2: 343–350.

Del Valle, E. M. M. 2004. Cyclodextrins and their uses. *Process Biochem.* 39: 1033–1046.

Dorrego, A. B., García-Río, L., Hervés, P., Leis, J. R., Mejuto, J. C., and Pérez-Juste, J. 2000. Micellization versus cyclodextrin-surfactant complexation. *Angew. Chem.* 39: 2945–2948.

FDA. 2000. Agency Response Letter GRAS Notice No. GRN 000046. CFSAN/Office of Premarket Approval. CFSAN/Office of Food Additive Safety. Available on-line at http://www.fda.gov/Food/FoodIngredientsPackaging/GenerallyRecognizedasSafeGRAS/GRASListings/ucm153769.htm (accessed February 20, 2011).

FDA. 2001. Agency Response Letter GRAS Notice No. GRN 000074. CFSAN/Office of Food Additive Safety. Available on-line at http://www.fda.gov/Food/FoodIngredientsPackaging/GenerallyRecognizedasSafeGRAS/GRASListings/ucm154182.htm (accessed February 20, 2011).

FDA. 2004. Agency Response Letter GRAS Notice No. GRN 000155. CFSAN/Office of Premarket Approval. CFSAN/Office of Food Additive Safety. Available on-line at http://www.fda.gov/Food/FoodIngredientsPackaging/GenerallyRecognizedasSafeGRAS/GRASListings/ucm154385.htm (accessed February 20, 2011).

Fenyvesi, E., Balogh, K., Siró, I., Orgoványi, J., Sényi, J. M., Otta, K., and Szente, L. 2007. Permeability and release properties of cyclodextrin-containing poly(vinyl chloride) and polyethylene films. *J. Incl. Phenom. Macrocycl. Chem.* 57: 371–374.

Fernández, I., García-Río, L., Hervés, P., Mejuto, J. C., Pérez-Juste, J., and Rodríguez-Dafonte, P. J. 2000. ß-Cyclodextrin-micelle mixed systems as a reaction-medium. Denitrosation of N-methyl-N-nitroso-p-toluenesulfonamide. *J. Phys. Org. Chem.* 13: 664–669.

Gadosy, T. A., Boyd, M. J., and Tee, O. S. 2000. Catalysis of ester aminolysis by cyclodextrins. The reaction of alkylamines with p-nitrophenyl alkanoates. *J. Org. Chem.* 65: 6879–6889.

García-Río, L., Hall, R. W., Mejuto, J. C., and Rodríguez-Dafonte, P. 2007. The solvolysis of benzoyl halides as a chemical probe determining the polarity of the cavity of dimethyl-β-cyclodextrin. *Tetrahedron* 63: 2208–2214.

García-Río, L., Hervés, P., Leis, J. R., Mejuto, J. C., Pérez-Juste, J., and Rodríguez-Dafonte, P. 2006. Evidence for complexes of different stoichiometries between organic solvents and cyclodextrins. *Org. Biomol. Chem.* 4: 1038–1048.

García-Río, L., Leis, J. R., Mejuto, J. C., Navarro-Vázquez, A., Pérez-Juste, J., and Rodríguez-Dafonte, P. 2004. Basic hydrolysis of crystal violet in β-cyclodextrin/surfactant mixed systems. *Langmuir* 20: 606–613.

García-Río, L., Mejuto, J. C., Nieto, M., Pérez-Juste, J., Pérez-Lorenzo, M., and Rodríguez-Dafonte, P. 2005. Denitrosation of N-nitrososulfonamide as chemical probe for determination of binding constants to cyclodextrins. *Supramol. Chem.* 17: 649–653.

García-Río, L., Mejuto, J. C., Rodríguez-Dafonte, P., and Hall, R. W. 2010. The role of water release from the cyclodextrin cavity in the complexation of benzoyl chlorides by dimethyl-β-cyclodextrin. *Tetrahedron* 66: 2529–2537.

Hara, K., Mikuni, K., Hara, K., and Hashimoto, H. 2002. Effects of cyclodextrins on deodoration of "Aging Odor." *J. Incl. Phenom. Macrocycl. Chem.* 44: 241–245.

Harada, A., Li, J., and Kamachi, M. 1994. Double-stranded inclusion complexes of cyclodextrin threaded on poly(ethylene glycol). *Nature* 370: 126–128.

Hashimoto, H. 1996. Comprehensive supramolecular chemistry. In *Cyclodextrins*, Vol. 3, eds. J. Szejtli and T. Osa, pp. 483–502. Oxford, U.K.: Elsevier.

He, L. H., Huang, J., Chen, Y. M., and Liu, L. P. 2005. Inclusion complexation between comblike PEO grafted polymers and α-cyclodextrin. *Macromolecules* 38: 3351–3355.

Hedges, R. A. 1998. Industrial applications of cyclodextrins. *Chem. Rev.* 98: 2035–2044.

Hedges, A. R. and McBride, C. 1999. Utilization of β-cyclodextrin in food. *Cereal Foods World* 44: 700–704.

Hedges, A. R., Shieh, W. J., and Sikorski, C. T. 1995. Use of cyclodextrins for encapsulation in the use and treatment of food products. In *Encapsulation and Controlled Release of Food Ingredients*, eds. S. J. Risch and G. A. Reineccius, pp. 60–71. Washington, DC: ACS symposium series 590, American Chemical Society.

Iglesias, E. 2003. Nitrosation of 2-Acetylcyclohexanone. 2. Reaction in water in the absence and presence of cyclodextrins. *J. Org. Chem.* 68: 2689–2697.

Iglesias, E., Brandariz, I., and Penedo, F. 2007. Cyclodextrin effects on physical–chemical properties of novocaine. *J. Incl. Phenom. Macrocycl. Chem.* 57: 573–576.

Inoue, Y., Hakushi, T., Liu, Y., Tong, L., Shen, B., and Jin, D. 1993. Thermodynamics of molecular recognition by cyclodextrins. 1. Calorimetric titration of inclusion complexation of naphthalenesulfonates with α-, β-, and γ-cyclodextrins: Enthalpy-entropy compensation. *J. Am. Chem. Soc.* 115: 475–481.

Karathanos, V. T., Mourtzinos, I., Yannakopoulou, K., and Andrikopoulos, N. K. 2007. Study of the solubility, antioxidant activity and structure of inclusion complex of vanillin with β-cyclodextrin. *Food Chem.* 101: 652–658.

Kim, S., Cho, E., Yoo, J., Cho, E., Ju Choi, S., Son, S. M., Lee, J. M., In, M. J., Kim, D. C., Kim, J. H., and Chae, H. J. 2010. β-CD-mediated encapsulation enhanced stability and solubility of astaxanthin. *J. Appl. Biol. Chem.* 53: 559–565.

Kim, H., Kim, H. W., and Jung, S. 2008. Aqueous solubility enhancement of some flavones by complexation with cyclodextrins. *Bull. Korean Chem. Soc.* 29: 590–594.

Kohata, S., Matsunaga, N., Hamabe, Y., Yumihara, K., and Sumi, T. 2010. Photo-stability of mixture of violet pigments phycoerythrin and phycocyanin extracted without separation from discolored nori seaweed. *Food Sci. Technol. Res.* 16: 617–620.

Koontz, J. L., Marcy, J. E., O'Keefe, S. F., and Duncan, S. E. 2009. Cyclodextrin inclusion complex formation and solid-state characterization of the natural antioxidants α-tocopherol and quercetin. *J. Agric. Food Chem.* 57: 1162–1171.

Larsen, K. L. 2002. Large cyclodextrins. *J. Incl. Phen. Macrocyclic. Chem.* 43: 1–13.

Lee, S. K., Yu, H. J., Cho, N. S., Park, J. H., Kim, T. H., Abdi, H., Kim, K. H., and Lee, S. K., inventors; Bioland, Ltd., assignee. 2008. A method for preparing the inclusion complex of ginseng extract with gamma-cyclodextrin, and the composition comprising the same. U.S. Patent WO/2008/127063.

Li, Z., Wang, M., Wang, F., Gu, Z, Du, G., Wu, J., and Chen, J. 2007. γ-Cyclodextrin: A review on enzymatic production and applications. *Appl. Microbiol. Biotechnol.* 77: 245–255.

Linert, W., Margl, P., and Renz, F. 1992. Solute-solvent interactions between cyclodextrin and water: A molecular mechanical study. *Chem. Phys.* 161: 327–338.

Liu, X. D., Furuta, T., Yoshii, H., Linko, P., and Coumans, W. J. 2000. Cyclodextrin encapsulation to prevent the loss of *l*-menthol and its retention during drying. *Biosci. Biotechnol. Biochem.* 64: 1608–1613.

Locci, E., Lai, S., Piras, A., Marongiu, B., and Lai, A. 2004. ^{13}C-CPMAS and ^{1}H-NMR study of the inclusion complexes of β-cyclodextrin with carvacrol, thymol, and eugenol prepared in supercritical carbon dioxide. *Chem. Biodivers.* 1: 1354–1366.

Loethen, S., Kim, J. M., and Thompson, D. H. 2007. Biomedical applications of cyclodextrin based polyrotaxanes. *Polym. Rev.* 47: 383–418.

López-de-Dicastillo, C., Catalá, R., Gavara, R., and Hernández-Muñoz, P. 2011. Food applications of active packaging EVOH films containing cyclodextrins for the preferential scavenging of undesirable compounds. *J. Food Eng.* (in press, DOI: 10.1016/j.jfoodeng.2010.12.033).

López-de-Dicastillo, C., Gallur, M., Catalá, R., Gavara, R., and Hernández-Muñoz, P. 2010. Immobilization of β–cyclodextrin in ethylene-vinyl alcohol copolymer for active food packaging applications. *J. Membr. Sci.* 353: 184–191.

López-Nicolás, J. M., Andreu-Sevilla, A. J., Carbonell-Barrachina, A. A., and García-Carmona, F. 2009a. Effects of addition of α-cyclodextrin on the sensory quality, volatile compounds, and color parameters of fresh pear juice. *J. Agric. Food Chem.* 57: 9668–9675.

López-Nicolás, J. M. and García-Carmona, F. 2007. Use of cyclodextrins as secondary antioxidants to improve the color of fresh pear juice. *J. Agric. Food Chem.* 55: 6330–6338.

López-Nicolás, J. M. and García-Carmona, F. 2008. Rapid, simple and sensitive determination of the apparent formation constants of trans-resveratrol complexes with natural cyclodextrins in aqueous medium using HPLC. *Food Chem.* 109: 868–875.

López-Nicolás, J. M. and García-Carmona, F. 2010. Effect of hydroxypropyl-β-cyclodextrin on the aggregation of (E)-resveratrol in different protonation states of the guest molecule. *Food Chem.* 118: 648–655.

López-Nicolás, J. M., Núñez-Delicado, E., Pérez-López, A. J., Carbonell, A., and Cuadra-Crespo, P. 2006. Determination of stoichiometric coefficients and apparent formation constants for β-cyclodextrin complexes of trans-resveratrol using reversed-phase liquid chromatography. *J. Chromatogr. A* 1135: 158–165.

López-Nicolás, J. M., Núñez-Delicado, E., Sánchez-Ferrer, A., and García-Carmona, F. 2007a. Kinetic model of apple juice enzymatic browning in the presence of cyclodextrins: The use of maltosyl-β-cyclodextrinas secondary antioxidant. *Food Chem.* 101: 1164–1171.

López-Nicolás, J. M., Pérez-López, A. J., Carbonell-Barrachina, A., and García-Carmona, F. 2007b. Kinetic study of the activation of banana juice enzymatic browning by the addition of maltosyl-β-cyclodextrin. *J. Agric. Food Chem.* 55: 9655–9662.

López-Nicolás, J. M., Pérez-López, A. J., Carbonell-Barrachina, A., and García-Carmona, F. 2007c. Use of natural and modified cyclodextrins as inhibiting agents of peach juice enzymatic browning. *J. Agric. Food Chem.* 55: 5312–5319.

López-Nicolás, J. M., Rodríguez-Bonilla, P., and García-Carmona, F. 2009b. Complexation of pinosylvin, an analogue of resveratrol with high antifungal and antimicrobial activity, by different types of cyclodextrins. *J. Agric. Food Chem.* 57: 10175–10180.

López-Nicolás, J. M., Rodríguez-Bonilla, P., Méndez-Cazorla, L., and García-Carmona, F. 2009c. Physicochemical study of the complexation of pterostilbene by natural and modified cyclodextrins. *J. Agric. Food Chem.* 57: 5294–5300.

Martin Del Valle, E. M. 2004. Cyclodextrins and their uses: A review. *Process Biochem.* 39: 1033–1046.

Mazzobre, M. F., dos Santos, C. I., and Buera, M. d. P. 2011. Solubility and stability of β-cyclodextrin-terpineol inclusion complex as affected by water. *Food Biophys.* (in press, DOI: 10.1007/s11483-011-9208-1).

Mourtzinos, I., Kalogeropoulos, N., Papadakis, S. E., Konstantinou, K., and Karathanos, V. T. 2008. Encapsulation of nutraceutical monoterpenes in β-cyclodextrin and modified starch. *J. Food Sci.* 73: S89–S94.

Muñoz-Botella, S., Del Castillo, B., and Martín, M. A. 1995. Cyclodextrin properties and applications of inclusion complex formation. *Ars. Pharm.* 36: 187–198.

Munro, I. C., Newberne, P. M., Young, V. R., and Bär, A. 2004. Safety assessment of γ-cyclodextrin. *Regul. Toxicol. Pharmacol.* 39: S3–S13.

Núñez-Delicado, E., Serrano-Megías, M., Pérez-López, A. J., and López-Nicolás, J. M. 2005. Polyphenol oxidase from Dominga table grape. *J. Agric. Food Chem.* 53: 6087–6093.

Paramera, E. I., Konteles, S. J., and Karathanos, V. T. 2011. Stability and release properties of curcumin encapsulated in *Saccharomyces cerevisiae*, β-cyclodextrin and modified starch. *Food Chem.* 125: 913–922.

Qi, Z. H. and Hedges, A. R. 1995. Use of cyclodextrins for flavours. In *Flavour Technology: Physical Chemistry, Modification and Process*, eds. C. T. Ho, C. T. Tan, and C. H. Tong, pp. 231–243. Washington, DC: ACS Symposium Series 610, American Chemical Society.

Reineccius, T. A., Reineccius, G. A., and Peppard, T. L. 2002. Encapsulation of flavors using cyclodextrins: Comparison of flavor retention in alpha, beta, and gamma types. *J. Food Sci.* 67: 3271–3279.

Reineccius, T. A., Reineccius, G. A., and Peppard, T. L. 2003. Flavor release from cyclodextrin complexes: Comparison of alpha, beta, and gamma types. *J. Food Sci.* 68: 1234–1239.

Reineccius, T. A., Reineccius, G. A., and Peppard, T. L. 2005. The effect of solvent interactions on α-, β-, and γ-cyclodextrin/flavor molecular inclusion complexes. *J. Agric. Food Chem.* 53: 388–392.

Rekharsky, M. V. and Inoue, Y. 1998. Complexation thermodynamics of cyclodextrins. *Chem. Rev.* 98: 1875–1917.

Rodríguez-Bonilla, P., López-Nicolás, J. M., and García-Carmona, F. 2010. Use of reversed phase high pressure liquid chromatography for the physicochemical and thermodynamic characterization of oxyresveratrol/β-cyclodextrin complexes. *J. Chromatogr. B* 878: 1569–1575

Saenger, W. 1980. Cyclodextrin inclusion compounds in research and industry. *Angew. Chem.* 19: 344–362.

Samant, S. K. and Pai, J. S. 1991. Cyclodextrins: New versatile food additive. *Indian Food Packer* 45: 55–65.

Seon, K. H., Ahn, J., and Kwak, H. S. 2009. The accelerated ripening of cholesterol-reduced Cheddar cheese by crosslinked β-cyclodextrin. *J. Dairy Sci.* 92: 49–57.

Shiga, H., Yoshii, H., Taguchi, R., Nishiyama, T., Furuta, T., and Linko, P. 2003. Release characteristics of flavor from spray-dried powder in boiling water and during rice cooking. *Biosci. Biotechnol. Biochem.* 67: 426–428.

Singh, M., Sharma, R., and Banerjee, U. C. 2002. Biotechnological applications of cyclodextrins. *Biotechnol. Adv.* 20: 341–359.

Siró, I., Fenyvesi, É., Szente, L., De Meulenaer, B., Devlieghere, F., Orgoványi, J., Sényi, J., and Barta, J. 2006. Release of alpha-tocopherol from antioxidative low-density polyethylene film into fatty food simulant: Influence of complexation in beta-cyclodextrin. *Food Addit. Contam.* 2: 845–853.

Szejtli, J. 1982. Cyclodextrins in food, cosmetics and toiletries. *Starch* 34: 379–385.

Szejtli, J. 1988. *Cyclodextrin Technology*. Boston, MA: Kluwer Academic Publishers.

Szejtli, J. 2004. Past, present, and future of cyclodextrin research. *Pure Appl. Chem.* 76: 1825–1845.

Szejtli, J. and Fenyvesi, E. 2005. Cyclodextrins in active and smart packaging. *Cyclodextrin News* 19: 213–216; 241–245.

Szejtli, J. and Szente, L. 2005. Elimination of bitter, disgusting tastes of drugs and foods by cyclodextrins. *Eur. J. Pharm. Biopharm.* 61: 115–125.

Szente, L. and Szejtli, J. 2004. Cyclodextrins as food ingredients. *Trends Food Sci. Technol.* 15: 137–142.

Tamamoto, L. C., Schmidt, S. J., and Lee, S. Y. 2010. Sensory properties of ginseng solutions modified by masking agents. *J. Food Sci.* 75: S341–S347.

Tee, O. S., Gadosy, T. A., and Giorgi, J. B. 1997. Effect of β-cyclodextrin on the reaction of α-amino acid anions with p-nitrophenyl acetate and p-nitrophenyl hexanoate. *Can. J. Chem.* 75: 83–91.

Tian-Xiang, X. and Anderson, B. D. 1990. Inclusion complexes of purine nucleosides with cyclodextrins: II. Investigation of inclusion complex geometry and cavity microenvironment. *Int. J. Pharmaceut.* 28: 45–55.

Tobitsuka, K., Miura, M., and Kobayashi, S. 2005. Interaction of cyclodextrins with aliphatic acetate esters and aroma components of La France pear. *J. Agric. Food Chem.* 53: 5402–5406.

Tobitsuka, K., Miura, M., and Kobayashi, S. 2006. Retention of a European pear aroma model mixture using different types of saccharides. *J. Agric. Food Chem.* 54: 5069–5076.

Touil, S., Palmeri, J., Tingry, S., Bouchtalla, S., and Deratani, A. 2008. Generalized dual-mode modelling of xylene isomer sorption in polyvinylalcohol membranes containing cyclodextrins. *J. Membr. Sci.* 317: 2–13.

Truong, V. Y. T., Boyer, R. R., McKinney, J. M., O'Keefe, S. F., and Williams, R. C. 2010. Effect of α-cyclodextrin-cinnamic acid inclusion complexes on populations of *Escherichia coli* O157:H7 and *Salmonella enterica* in fruit juices. *J. Food Protect* 73: 92–96.

United Nations Food and Agriculture Organization/World Health Organization (FAO/WHO), Joint Expert Committee on Food Additives (JECFA). 1995. Evaluation of certain food additives and contaminants: 44th report of JECFA. WHO Technical Report Series nr 859. Geneva, Switzerland: WHO, p. 54.

United Nations Food and Agriculture Organization/World Health Organization (FAO/WHO), Joint Expert Committee on Food Additives (JECFA). 2000. Evaluation of certain food additives and contaminants: 51st report of JECFA. WHO Technical Report Series nr 891. Geneva, Switzerland: WHO, p. 168.

United Nations Food and Agriculture Organization/World Health Organization (FAO/WHO), Joint Expert Committee on Food Additives (JECFA). 2002. Evaluation of certain food additives and contaminants: 57th report of JECFA. WHO Technical Report Series nr 909. Geneva, Switzerland: WHO, p. 171.

Xiao, Y. and Chung, T. S. 2007. Functionalization of cellulose dialysis membranes for chiral separation using beta-cyclodextrin. *J. Membr. Sci.* 290: 78–85.

Yang, Y., Zhengbiao, G., Hui, X., Fengwei, L., and Zhang, G. 2010. Interaction between amylose and β-cyclodextrin investigated by complexing with conjugated linoleic acid. *J. Agric. Food Chem.* 58: 5620–5624.

Yuen, F. and Tam, K. C. 2010. Cyclodextrin-assisted assembly of stimuli-responsive polymers in aqueous media. *Soft Matter* 6: 4613–4630.

CHAPTER 9

Functional and Physicochemical Properties of Cyclodextrins

Jun Zhao and Shan-Jing Yao

CONTENTS

9.1 INTRODUCTION

Cyclodextrins (CDs) are a series of oligosaccharides first found by Villiers in 1891 and thereafter were described in detail by Schardinger. CDs are produced by acting on starch with a specific CD glucanotransferase (CGTase, EC 2.4.1.19) from *Bacillus macerans* [1]. In 1930s, Freudenberg and his coworkers revealed that the CD molecules were built from maltose units containing only α-1,4-glycosidic linkages, and they also postulated CDs as cyclic structure. Since 1950s, many scientists, especially the research group led by Szejtli, have paid great attention to the preparation methods, structure, and physicochemical properties of CDs; and their works showed CDs as promising molecules for industrial possibilities [1,2].

9.2 FUNDAMENTALS OF CYCLODEXTRINS

The CD family comprises several cyclic oligosaccharides, among which α-(**1**), β-(**2**), and γ-(**3**) CDs are the well known and most commonly used, containing six, seven, and eight D-glucopyranose units, respectively. In addition, many large CDs have been produced and identified up to now, including δ-, ε-, ζ-, η-, θ-CDs, or even CDs comprised of dozens of D-glucopyranose units [3–7]. They show more variable and flexible structures compared with α-, β-, and γ-CD. CDs having fewer than six glucose residues are not known to exist, probably due to steric hindrance and ring strain [8].

1   2   3

9.2.1 Characteristics of the Macrocyclic Structure of Cyclodextrins

CDs have doughnut-shapes (Figure 9.1) with all the glucopyranoside units in the ⁴C₁ chair conformation. In a CD molecule, all primary hydroxyl groups locate on the narrow edge of the ring

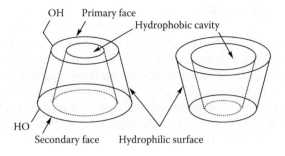

Figure 9.1 Scheme of the hydrophobic and hydrophilic regions of CDs.

(primary face), while all the secondary ones are placed on the other edge (secondary face). The apolar C_3 and C_5 hydrogens and ether-like oxygens are at the inside of the torus-like molecules [9,10]. Thanks to this structure, CDs have the hydrophilic outer surface and the hydrophobic hollow cavity (shown in Figure 9.1).

The CD rings are fairly rigid due to intramolecular hydrogen bonds formed between the O_2–H and the O_3–H of adjacent glucopyranose units [11]. The secondary hydroxyl groups in β-CD molecule can form the complete hydrogen-bond belt which imposes restriction on the inclination of the glucose units. This feature is probably the reason why β-CD is rather rigid than other CDs. Unlike β-CD, the hydrogen-bond belt is incomplete in the α-CD molecule for one glucopyranose unit is in a distorted position; thus, only four hydrogen bonds are fully established actually. The γ-CD ring is noncoplanar with a bit distortion; it shows more flexible conformation as a result [1,11–13]. CDs with more than eight glucopyranose units are more warped and flexible in their conformation, so that the rings exhibit various dramatic shapes, such as the elliptical structure formed by δ-CD (CD9) [3], the "8"-shaped feature twisted by CD26 [6], etc.

The main characteristics of α-, β-, and γ-CDs are summarized in Table 9.1 [1,14].

Table 9.1   Main Characteristics of α-, β-, and γ-CD

Property	α-CD	β-CD	γ-CD
No. of glucose units	6	7	8
Empirical formula (anhydrous)	$C_{36}H_{60}O_{30}$	$C_{42}H_{70}O_{35}$	$C_{48}H_{80}O_{40}$
Mol. weight (anhydrous)	972.85	1134.99	1297.14
Cavity diameter, nm	0.47~0.53	0.60~0.65	0.75~0.83
Height of cavity, nm	0.79±0.01	0.79±0.01	0.79±0.01
Diameter of outer periphery, nm	1.46±0.04	1.54±0.04	1.75±0.04
Approx. volume of cavity, nm³	0.174	0.262	0.427
Approx. cavity volume in 1 mol CD, mL	104	157	256
Approx. cavity volume in 1 g CD, mL	0.10	0.14	0.20
Crystal forms from water	Hexagonal plates	Monoclinic parallelograms	Quadratic prisms
Crystal water, wt%	10.2	13.2~14.5	8.13~17.7
$[\alpha]_D$ (25°C)	150±0.5	162.5±0.5	177.4±0.5
pK_a (25°C)	12.33	12.20	12.08
$\Delta H°$ (solution), kJ mol⁻¹	32.10	34.78	32.35
$\Delta S°$ (solution), J mol⁻¹ K⁻¹	57.75	48.96	61.52

Sources: Data from Szejtli, J., *Chem. Rev.*, 98, 1743, 1998; Connors, K.A., *Chem. Rev.*, 97, 1325, 1997; Jozwiakowskia, M.J. and Connors, K.A., *Carbohydr. Res.*, 143, 51, 1985.

9.2.2 Solubility of Cyclodextrins

CDs are easily dissolved in water due to the plenty of hydroxyl groups existing on the surface of the molecules. However, the three common CDs show remarkable difference in their solubility in water. β-CD has the lowest solubility among the three, probably because of its complete hydrogen-bond belt around the secondary face. The thermodynamics of solution show that the relatively low solubility of β-CD is associated with both less favorable $\Delta H°$ and $\Delta S°$ (listed in Table 9.1 [14,15]). Whereas α-CD shows a bit distortion in its molecule resulting in the increase of both $\Delta H°$ and $\Delta S°$, so it has much larger solubility than β-CD, and thus, the noncoplanar γ-CD is even more soluble.

CDs are insoluble in most of the common organic solvents, even in some binary aqueous-organic solvent mixtures, although CDs are prone to solving in pyridine, *N,N*-dimethylformamide (DMF), dimethyl sulfoxide (DMSO), and several polyhydric alcohols. Some of the organic solvents form inclusion complexes with CDs and crystallize; thus, they can be used as precipitants in CD purification process.

9.2.3 Reactivity of Cyclodextrins

9.2.3.1 Chemical Modifications

Nearly all modifications of CDs take place at the hydroxyl groups. The three types of hydroxyl groups presented at the 2-, 3-, and 6-positions of CDs show distinct reactivity in chemical modification. They usually compete for the reagent, making selective modification extremely difficult [16]. Generally, those hydroxyl groups at the 6-position are the most nucleophilic and the minimum steric hindrance; thus, many reagents especially with bulky groups easily approach and attack this position. The hydroxyl groups at the 2-position are the most acidic ($pK_a = 12.2$ [17]), and they will be the first to get deprotonated in basic condition, while those at the 3-position are the most inaccessible, probably due to the formed hydrogen bonds and more crowded space at the secondary side than the primary side [18–20].

Khan et al. have summarized the methods for modification of CDs as follows (Figure 9.2): (1) under normal circumstances (usually in weak base), an electrophilic reagent attacks the 6-position; (2) in strong base (usually containing NaOH in organic solvents), the oxyanion formed

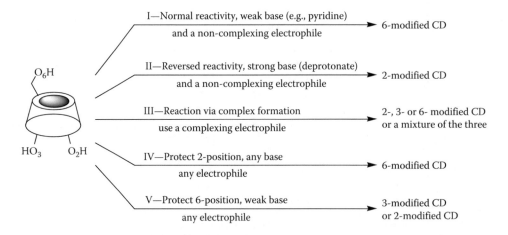

Figure 9.2 Overview of the methods for modification of CDs. (Modified from Khan, A.R. et al., *Chem. Rev.*, 98, 1977, 1998.)

at 2-position is more nucleophilic than other nondeprotonated hydroxyl groups, and thus, the 2-substituted product is the predominant product; (3) the electrophilic reagent forms a complex with CD, and then the orientation of the reagent within the complex may determine the nature of the product; (4) the 2-position of CD is protected, and thus, one can direct the incoming electrophile to the 6-position; and (5) the protection of the primary side enables one to direct the incoming electrophile exclusively to hydroxyl group at the 2-positions.

It is noted that the reagent activity plays an important role in determining the nature of the product. More reactive reagents such as trimethylsilyl chloride (TMSCl, **4**) [21] will attack the hydroxyl groups less selectively, whereas less reactive reagents like *tert*-butyldimethylsilyl chloride (TBDMSCl, **5**) [22] or trityl chloride **6** [23] will react more selectively with the 6-position hydroxyl groups. CD cavity size and solvents also significantly affect the type of the product. There are some evidences showing that tosyl chloride reacts with α-CD to give the 2-substituted product in aqueous solutions, but give the 6-substituted product in pyridine; while it reacts with β-CD, the product is 6-substituted product either in aqueous solutions or in pyridine [24,25]. Moreover, the selective modification is quite distinct for different CDs in the same reaction. The direct activation of primary hydroxyl groups via bulky reagent (such as triphenylphosphonium salt **7**) is more effective for β-CD than α-CD. The α-CD, however, often suffered an indirect strategy involving protection of all hydroxyl groups and the following selective deprotection of the primary ones [26].

9.2.3.2 *Hydrolysis of Cyclodextrins*

CDs are stable enough in neutral and basic aqueous solutions, but in solutions of low pH, CDs are inclined to degrade to glucose or maltose. The hydrolysis constant of β-CD is much lower than that of linear dextrin in hydrochloric acid, probably because of the lack of reductive ends in CD's cyclic molecule [27,28]. The hydrolysis notably accelerates as temperature rises, whereas CDs are rather stable in diluted acids at room temperature.

Several α-amylases have the ability to degrade CDs, and the hydrolysis rate is dependent on the size of CD ring and the source of amylase. Generally, γ-CD is more sensitive to α-amylase (such as human salivary and pancreatic α-amylases [29]) than α- or β-CD, and it is found to be hydrolyzed at appreciable rate. While α-CD has comparatively good tolerance for some of α-amylases mainly because it has such a small ring with large steric hindrance, thus, the α-amylase may not approach to its 1,4-α-glycosidic bond.

9.2.4 Bioadaptability of Cyclodextrins

The studies toward bioadaptability of CDs came from pharmacy researches. It nearly reaches a consensus that the CD molecules will permeate biological membranes with considerable difficulty, since the relatively lipophilic membrane has a low affinity for the hydrophilic and relatively large CD molecules [30]. They usually remain in the aqueous membrane exterior such as saliva or the tear fluid, as observed in the experiment on animals [9]. Their adsorption and metabolism are rather difficult, and most of CDs are excreted with unchanged form. For example, β-CD is recovered almost completely in the intact form after intravenous injection to rats [31]. Oral feeding of α- or β-CD to rats does not cause any toxic reactions due to lack of absorption from the gastrointestinal tract [32–34], and their intravenous LD_{50} for rats is given as 0.788 g·kg^{-1} for β-CD and 1.00 g·kg^{-1} for α-CD [35]. Also γ-CD is considered nontoxic when given intravenously according to animal tests [36]. However, β-CD at higher doses shows somewhat notable nephrotoxicity to rat, probably because of its low solubility and the resultant crystallization in the lysosomes [35], and also β-CD is not suitable for parenteral formulations [34,37].

The cytotoxicity of CDs and hydroxypropylated CD (HP-CD) derivatives was evaluated in vitro by Leroy-Lechat et al. [38]. It showed their cytotoxicity was not specific to the cell type, which verified the hypothesis that they destructed of plasma membranes by forming inclusion complexation of membrane components (cholesterol or phospholipids) [39–41]. In comparison with respective natural parent CD, HP-CDs showed the cytotoxicity reduction, especially for HP-α-CD and HP-γ-CD. The cytotoxicity of other CD derivatives also completed for dimethyl-β-CD (DM-β-CD) [42], sulfobutylether-β-CD (SBE-β-CD) [43], etc.; and 2-HP-β-CD, 6-O-maltosyl-β-CD, and β-CD sulfate can be safely used in parenteral formulations [31].

9.3 CYCLODEXTRIN-BASED SUPRAMOLECULAR STRUCTURES

Supramolecular chemistry, often described as "chemistry beyond the molecules," is proposed by Lehn in 1978. According to Lehn, a supermolecule is an organized, complex entity that is created from the association of two or more chemical species held together and organized by intermolecular (noncovalent) forces [44–48]. It aims at constructing highly complex, functional chemical systems from components, and the construction of any supramolecular systems involves molecular recognition and selective combination, which are fundamentals to all supramolecular chemistry (Figure 9.3) [44,48–50]. The CD molecule is a good cyclic host that possesses molecular recognition capabilities based on its dramatic hydrophobic cavity.

9.3.1 Inclusion Complexes: A Study of Complexation with Flavonoids

A CD molecule possesses a cavity that is suitable for the inclusion of various organic molecules (guest molecule), with a requirement that the guests should fit into the cavity, even if

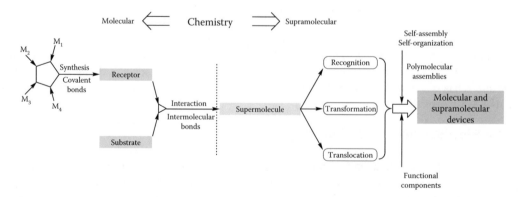

Figure 9.3 From molecular to supramolecular chemistry: molecules, supermolecules, and molecular and supramolecular devices. (Modified from Lehn, J.-M., *Pure Appl. Chem.*, 50, 871, 1978.)

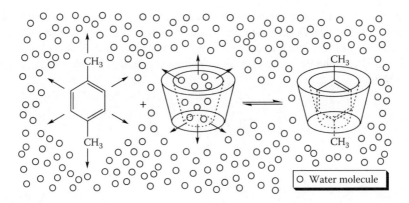

Figure 9.4 A scheme showing the complexation process of *p*-xylene with CD cavity. (Modified from Szejtli, J., *Chem. Rev.*, 98, 1743, 1998.)

only partially [17]. These CD complexes provide important usages in pharmacy, for example, they improve the bad solubility of drugs, or reduce their strong volatilization and pungent odors [9,51,52]. Generally, the interactions of CD and guest molecules involve weak interactions, including hydrogen bonds, electrostatic interactions, van der Waals forces, hydrophobic, π-π stacking interactions, steric effects, etc., instead of covalent bonding [11]. In most cases, the inclusion and complexation of guest molecules is directly driven with the hydrophobic interaction between the cavity and the less polar part of the guest molecule. However, the CD cavity is often occupied by water molecules in an aqueous solution [53]; thus, the complexation process should involve the removal of the enthalpy-rich water molecules from hydrophobic CD cavity into bulk water phase and meanwhile the transport of low-enthalpy guest molecule into the vacated site, as described in Figure 9.4 [1,54]. It is a spontaneous process in thermodynamic, and the favorable reduction of free energy (ΔG) observed is mostly attributable to a large negative ΔH [55,56].

It is a fundamental that the geometry of the guests especially the huge ones may greatly affect the structure and thus the stability of inclusion complexes. When a guest molecule is larger than the CD cavity space, it will be only partially included in the cavity. Take the complexes formed with quercetin **8** and its analogues **8–11**, for example. All these molecules form 1:1 complexes with β-CD in stoichiometry. The aglycone quercetin **8** is relatively small, so a large portion of its skeleton can be included in the β-CD cavity. It is revealed by Xiu's group using the PM3 and ONIOM2 quantum-mechanical semiempirical method that the bond connected B with C ring is inclined to the molecular axis of β-CD, and the B ring will locate near the secondary hydroxyls of the β-CD cavity [57,58]. One intermolecular hydrogen bond is formed between 7-hydroxy group of quercetin and a 6-OH of β-CD, which will stabilize the quercetin/β-CD complex (the stability constant given as 402 L·mol⁻¹ at 25°C [59]). 3-*O*-methyl-quercetin **9** also places the B ring into β-CD cavity and presents molecular structure to that of quercetin [60]. The intermolecular hydrogen bond, however, is not reported to exist, so that it is supposed that the B ring inserts the β-CD cavity less deeply than quercetin **8**. When a glycosyl group is substituted at 3-*O*-position, the guest will not fit the β-CD cavity well owing to the hindrance of the large glycosyl group, which is confirmed by the complex of rutin **10** with β-CD. The huge 3-*O*-[6″-*O*-(α-L-rhamnosyl)-β-D-glucosyl] group in rutin molecule makes the aglycone deviate from the molecular axis of β-CD; as a result, the binding site for this molecule is only the single B ring penetrating into the β-CD cavity with the shallow position [61]. The stability constant for rutin/β-CD complex is 265 ± 12 L·mol⁻¹ at 20°C and pH 7.5 [62], which is lower than that of the quercetin/β-CD complex mentioned earlier. Out of the ordinary, another quercetin glycoside isoquercitrin **11** presents a unique configuration toward β-CD cavity. An analysis of ^{1}H NMR spectroscopy suggests that isoquercitrin prefers to insert the chromone A ring into β-CD cavity rather than the phenyl B ring, and most hydroxyl groups of isoquercitrin orientate outside the hydrophobic cavity [55]. This binding orientation is also consistent with energy calculation results.

8 quercetin	R = H	
9 3-O-methyl-quercetin	R = CH$_3$	

Some CD derivatives may present better binding capacity to guest molecules compared with their parent CDs. This conclusion has been supported by researches on the inclusion properties of CD derivatives with some flavonoids. According to Zheng et al., the quercetin/2-HP-β-CD complex possesses much higher binding constant (1.10×10^4 L·mol^{-1} at 297 K) than native β-CD does (1.03×10^3 L·mol^{-1}), while the value of quercetin/SBE-β-CD complex is even higher (2.53×10^4 L·mol^{-1}) [63] (other researches showed this regularity as well, but the stability constants' values are greatly disparate [59,64]). Coincidentally, the solubility of quercetin is most enhanced by SBE-β-CD, suggesting the solubility enhancement presents positive correlation with the complex stability. DM-β-CD, a third good host compound, usually forms the flavonol complexes with high stability constants [55,65]. It means the substituent groups outside the CD cavity assist in the guest binding. They enlarge the primary diameters of β-CD from 0.65 [1] to 0.91 nm for DM-β-CD and 1.03 nm for 2-HP-β-CD [55], resulting in a notable shape change to the hydrophobic cavity, thus improving the matching degree with guest molecule. Furthermore, the attached sulfobutylether and hydroxypropyl groups may provide a more hydrophilic microenvironment for the whole complex, which is the probable reason of solubility enhancement [64].

9.3.2 Cyclodextrin-Based Artificial Enzymes

Enzymes are usually proteins of high molecule weight that can selectively catalyze various bioreactions. Enzymes reduce the activation energy of the reaction, but do not affect the free-energy change or the equilibrium constant. They are specific, versatile, and quite effective in bioreactions under ambient conditions. It is recognized that the enzyme–substrate complex is formed by weak forces, and the particular substrate binds to a specific site inside the enzyme molecule, like the "lock" and the "key" which is central to much of supramolecular chemistry [66]. It is quite similar to the complexation of guest molecules with CDs. In fact, CDs appear to be ideal molecules as a basis for an enzyme mimic, and there have been thousands of artificial enzymes synthesized for various applications indeed. They cover a wide range of reactions and show miraculous characteristics [67].

9.3.2.1 Mimics of Hydrolase

In 1970, Breslow's group reported a CD enzyme mimic **12** that pyridine-2,5-dicarboxylic acid reacted with α-CD through esterification of one secondary hydroxyl group [68]. This compound was further chelated with nickel(II) chloride and then treated with pyridine-2-carboxaldoxime, giving a hydrolase mimic which had the ability to catalyze the hydrolysis of *p*-nitrophenyl acetate **13** bound in the CD cavity (Figure 9.5). The rate accelerated almost 10^3 times over the uncatalyzed rate when 0.01 M **12** was added.

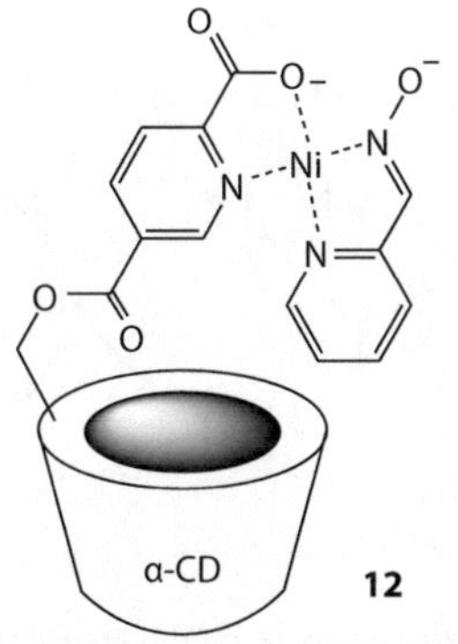

Imidazole ring is a useful functional group included in many enzyme mimics because it usually appears in the catalytic center of serine proteases in the form of histidine. A famous miniature

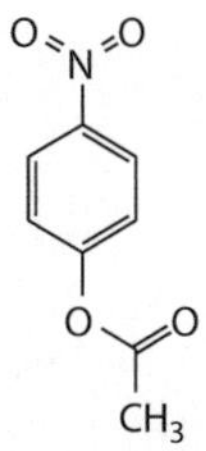

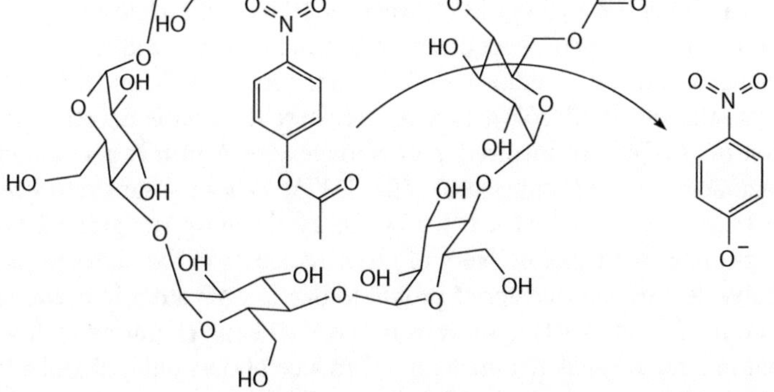

Figure 9.5 A scheme of nickel-chelated hydrolase mimic reported by Breslow, and its hydrolysis of *p*-nitrophenyl acetate bound in the CD cavity. (Reprinted with permission from Cragg, P.J., *Supramolecular Chemistry—From Biological Inspiration to Biomedical Applications*, Springer Press, Dordrecht, the Netherlands, 2010; Jullian, C. et al., *Spectrochim. Acta Part A*, 67, 230, 2007.)

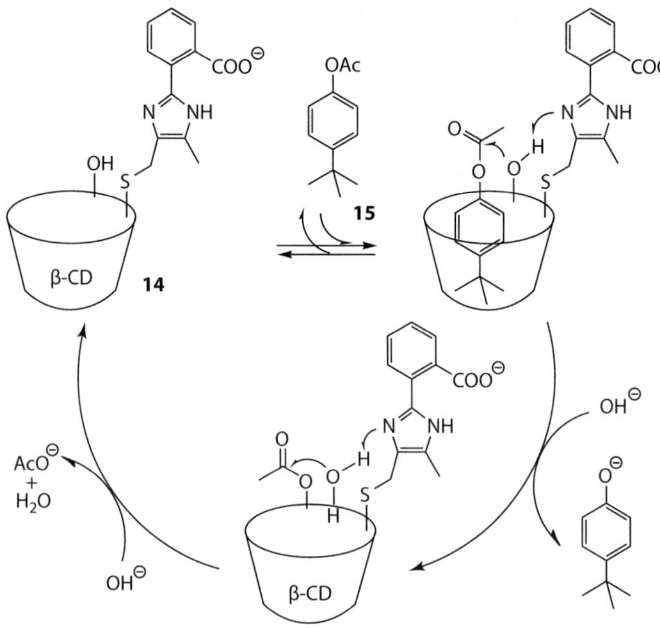

Figure 9.6 A scheme of catalytic mechanism of hydrolyzing **15** by the chymotrypsin mimic **14**. (Modified from D'Souza, V.T. and Bender, M.L., *Acc. Chem. Res.*, 20, 146, 1987; Breslow, R. and Overman, L.E., *J. Am. Chem. Soc.*, 92, 1075, 1970.)

model of serine proteases containing imidazole is compound **14** synthesized by Bender's group [69]. An imidazole carrying benzoate group is attached to the secondary face of a β-CD ring in **14**. The CD cavity is registered as the substrate binding site and the imidazole group as the catalytic center (Figure 9.6). When catalyzing the hydrolysis of *p-tert*-butylphenyl acetate **15**, this enzyme mimic shows comparable value of k_{cat}/K_m to the hydrolysis of **13** catalyzed by the native chymotrypsin [70].

9.3.2.2 Mimics of Ribonuclease

Ribonuclease catalyzes the hydrolytic cleavage of RNA. This cleavage occurs in two steps, involving the forming of cyclic phosphate diester group in the first and later its ring-opening hydrolysis [71]. In ribonuclease A, two imidazoles of histidine units (His[12] and His[119]) play an important role of such catalysis that one imidazole acts as the free base (Im) while the other protonated (ImH[+]) as an acid catalyst [67,72–74]. An inspiration of ribonuclease A mimic is compound **16** with two imidazole groups attached to the primary face of β-CD [75]. When the substrate 4-*tert*-butylcatechol cyclic phosphate **18** is bound into the β-CD cavity, the cyclic phosphate group is accessible to both the imidazole rings of **16**. A further delivery of water molecule to the phosphorus occurs with the action of one imidazole, and the transfer of proton to one oxygen atom in cyclic phosphate by the other *N*-protonated imidazole assists hydrolysis of the P–O bond (Figure 9.7). It is also found that **16** carries this out in a region-selective manner to produce almost only **19** and a trace of **20**. As a comparison, another ribonuclease mimic **17** is synthesized with its catalytic groups further from the cavity. This new catalyst can be completely selective to produce only **20** when **18** is cleaved (Figure 9.7) [71].

The corresponding derivatives of α- and γ-CD are synthesized as well. It is found that all these catalysts are able to hydrolyze the substrate derived from 4-methylcatechol, but only the β-CD- and

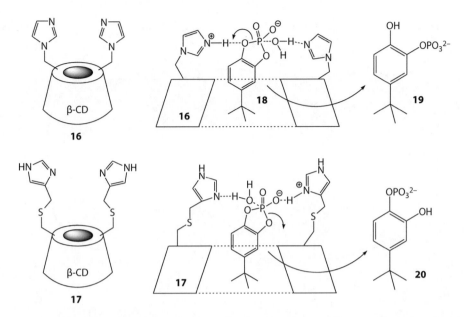

Figure 9.7 A scheme of two ribonuclease mimicries **16** and **17** and their different selectivity when hydrolyzing compound **18**. (Modified from Breslow, R., *Science*, 218, 532, 1982; D'Souza, V.T. et al., *Biochem. Biophys. Res. Commun.*, 129, 727, 1985.)

γ-CD-based catalysts can hydrolyze the substrate based on 4-*tert*-butylcatechol [76]. This selectivity comes from the goodness of fit between the substrate and the CD cavity. The β-CD-based catalyst binds most strongly with the 4-*tert*-butylcatechol derivative among the substrates studied, and this combination has the highest k_{cat} as a result.

Some ribonuclease mimics are "metalloenzyme." One example is a β-CD-based compound **21** containing one substituent of Zn^{2+} chelated tren group and the other substituent of imidazole [77]. In order not to bind each other, the two catalytic groups are attached to opposite sides of β-CD. When the **18** is bound to this catalyst and the following hydrolysis happens, it leads to the selective formation of product **19** rather than **20** at good rate.

9.3.2.3 Mimics of Vitamin B6 Enzymes

Pyridoxal phosphate **22** and pyridoxamine phosphate **23**, the active forms of vitamin B_6, function as coenzymes for transaminase that catalyzes the transamination between an α-amino

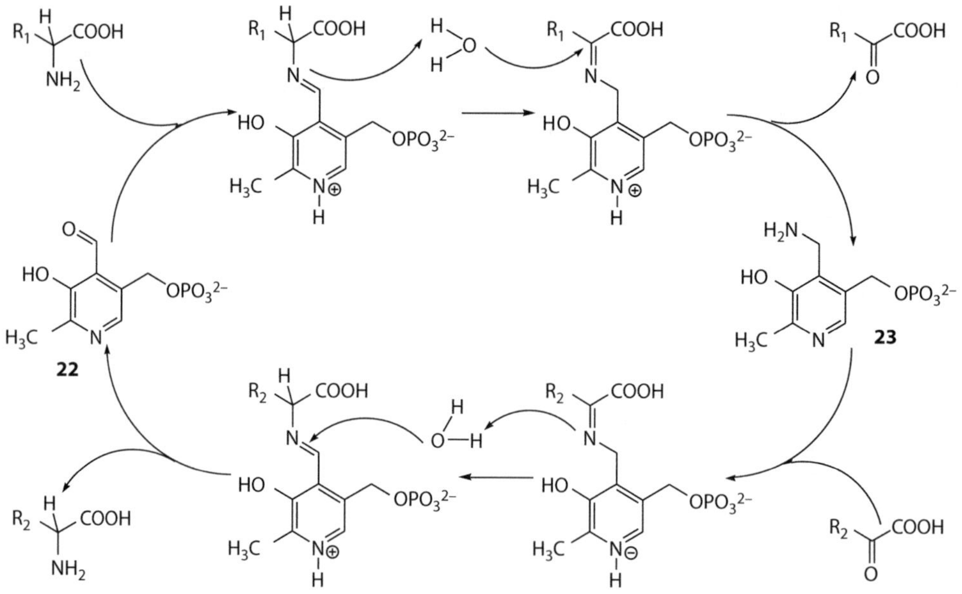

Figure 9.8 A scheme of transamination mechanism, involving the coenzyme pyridoxamine phosphate. (Modified from Breslow, R., *Artificial Enzymes*, WILEY-VCH, Weinheim, Germany, 2005; Jullian, C. et al., *Spectrochim. Acta Part A*, 71, 269, 2008.)

acid and an α-keto acid (the mechanism shown in Figure 9.8 [67]). The enzyme–coenzyme combination can be simply mimicked by linking pyridoxamine to the primary face of β-CD through a sulfur atom. The compound **24** is the first typical example of transaminase mimic that is able to transform indolepyruvic acid **25** to tryptophan **26** (Figure 9.9) [78]. The selectivity for aromatic α-keto acids is usually higher than the aliphatic ones, probably due to the geometry of the

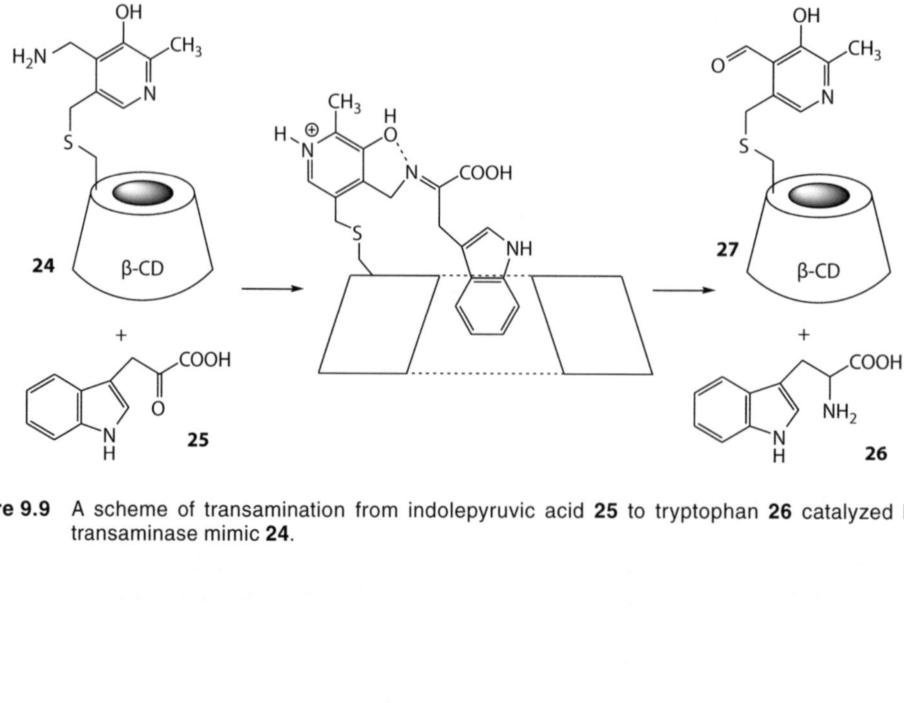

Figure 9.9 A scheme of transamination from indolepyruvic acid **25** to tryptophan **26** catalyzed by the transaminase mimic **24**.

substrates. A convincing example is that phenylpyruvic acid **28** is transaminated ca. 100 times as rapidly as is pyruvic acid **29** by **24**, while the better binding substrate (*p-tert*-butylphenyl) pyruvic acid **30** shows higher selectivity relative to pyruvic acid over 15,000 [72,79]. The enzyme mimic with the pyridoxamine attached to the O_3–H of β-CD exhibits similar properties to catalyst **24** [80].

Higher selectivity is also found to a set of four compounds **31(1–4)** when they bind the substrates with different geometry. The discrimination of the four mimics is observed that the endo pairs of isomers **31(1)** and **31(3)** prefer *p*-substituted phenylpyruvic acids, while the exo pairs **31(2)** and **31(4)** show more affinity to *m*-substituted substrates. Compared with **24**, however, neither the endo pair nor the exo pair of isomers leads to a decrease in the selectivity for transamination of phenylpyruvic acid relative to pyruvic acid **29** [79].

When the catalysts **24** and **31(1–4)** catalyze the synthesis of amino acids, they show some selectivity because of the chirality of the CD unit. However, without a chirally mounted basic group to deliver the proton, the stereoselective is not satisfactory [72]. A bifunctional transaminase mimic **32** is synthesized with both a pyridoxamine and an ethylenediamine unit substituted to the adjacent primary hydroxyl groups of β-CD [81]. The introduction of amino group provides flexible chains to promote proton transfer and exhibits remarkable L-chiral induction from keto-prochiral groupings. The transaminating reaction is carried out with the substrates **25**, **28**, and **29** catalyzed by **32**. The result is exciting that the enantiomeric excess (*e.e.* value) observed for L-tryptophan, L-phenylalanine, and L-alanine is 90%, 96%, and 96%, respectively.

Vitamin B_6 is a cofactor taking part not only in transamination but also in many other amino acid metabolic reactions. Some other enzyme mimics also use them as cofactors. Compound **27** is a mimic of tryptophan synthase which catalyzes the β-replacement reaction of serine with indole (Figure 9.10) [82]. This pyridoxal-bound β-CD catalyst produces tryptophan **26** when incubated with indole, β-chloroalanine, and $Al_2(SO_4)_3$ at pH 5.2 and 100°C for 30 min. However, tryptophan yield is low although it produces several times more tryptophan than the reaction catalyzed by simple pyridoxal. In addition, L-tryptophan is produced just in ca. 10% excess relative to the D-enantiomer, indicating bad stereoselectivity for this catalyst.

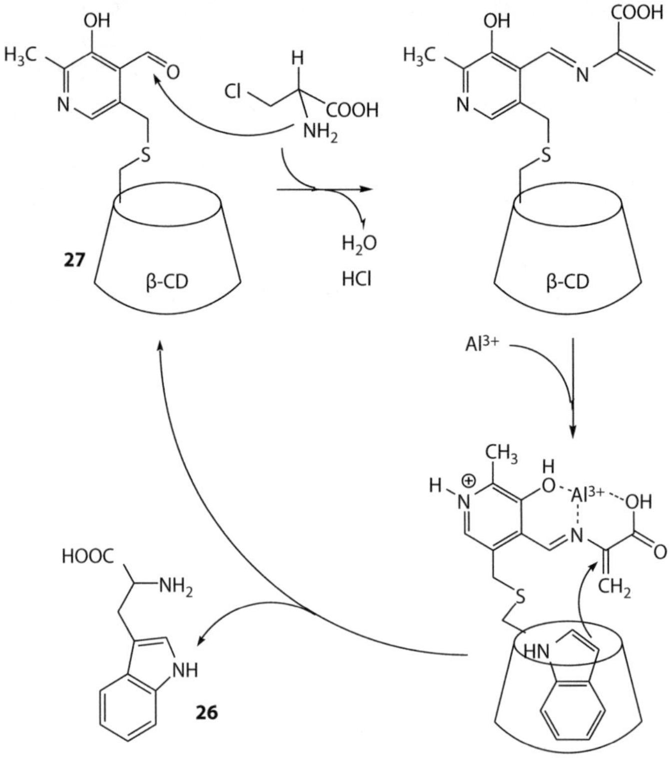

Figure 9.10 A scheme of the synthesis of tryptophan **26** with β-chloroalanine and indole catalyzed by the tryptophan synthase mimic **27**. (Modified from Breslow, R., *Artificial Enzymes*, WILEY-VCH, Weinheim, Germany, 2005; Jullian, C. et al., *Spectrochim. Acta Part A*, 71, 269, 2008.)

9.3.2.4 Mimics of Cytochrome P450 Enzymes

Cytochrome P450 enzymes catalyze the oxidation of organic substances including metabolic intermediates (such as lipids and steroids) as well as xenobiotic substances (drugs and other toxic chemicals), usually in the form of monooxygenase reactions as

$$R\text{---}H + O_2 + 2H^+ + 2e^- \longrightarrow R\text{---}OH + H_2O \tag{9.1}$$

Cytochrome P450 enzymes are a family of hemoproteins containing heme group (a porphyrin molecule chelated with iron) in their catalytic site as the cofactors participating in electron transfer in oxidations. For this reason, the artificial models for P450 enzymes generally consist of metalloporphyrins. However, the oxidation catalyzed by metalloporphyrins itself presents nearly no special positional selectivity initially, presumably due to the lack of the steric restriction [83]. A set of Mn(III) porphyrin derivatives **33**, **34**, and **35** carrying several β-CD units have been synthesized by Breslow's group [84]. They can epoxidize the double bonds of long-chain stilbene substrates in aqueous solution with the aid of an oxidant iodosobenzene. Among them, **33** and **34** show positional selectivity to a certain extent, since the long-chain substrate prefers to bind into the two CD cavities in the diagonal position and the substrate spans the porphyrin ring. As a result, the oxygen carried by Mn(III) will be added to the double bond only at one face of the porphyrin plane, when the opposite face is shielded by adamantanecarboxylate **36**. In contrast, the catalyst **35** does not bind the substrate in this effective binding geometry, and it gives very poor selectivity consequently.

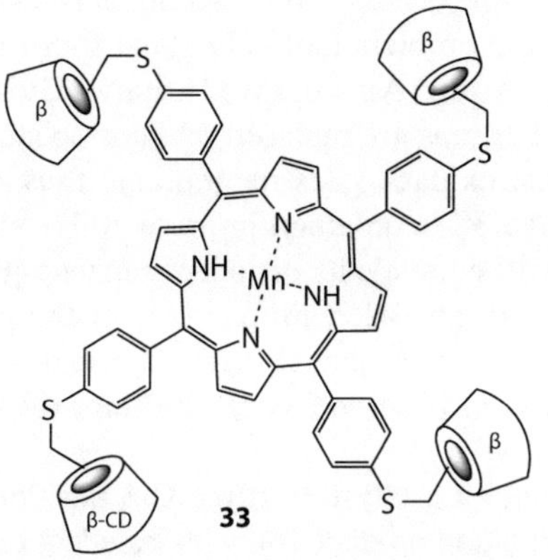

33

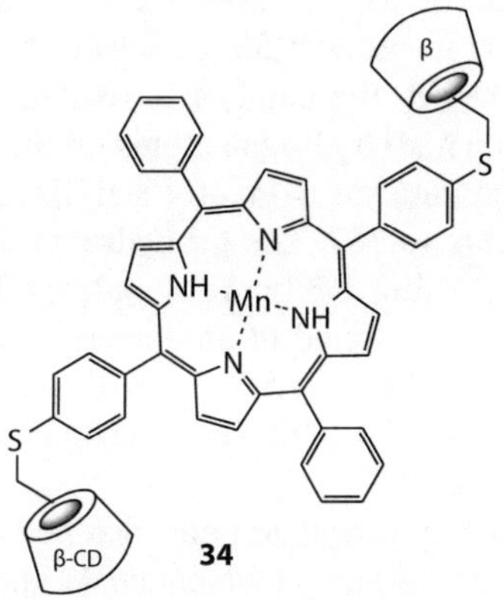

34

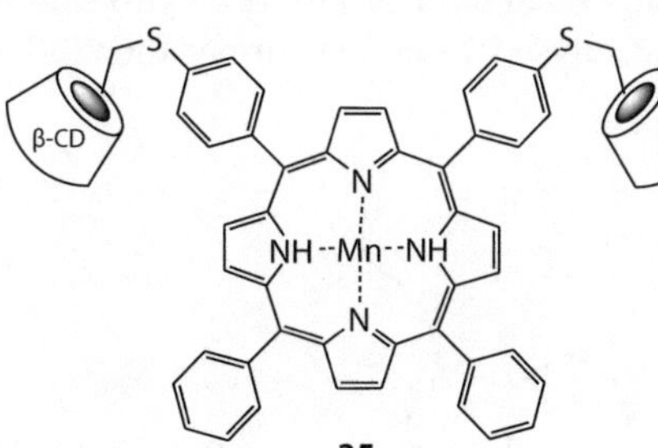

35

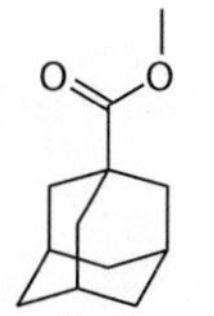

36

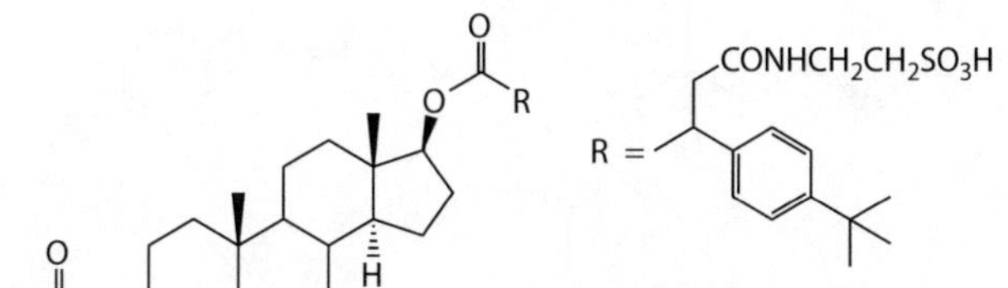

37 X = H
38 X = OH

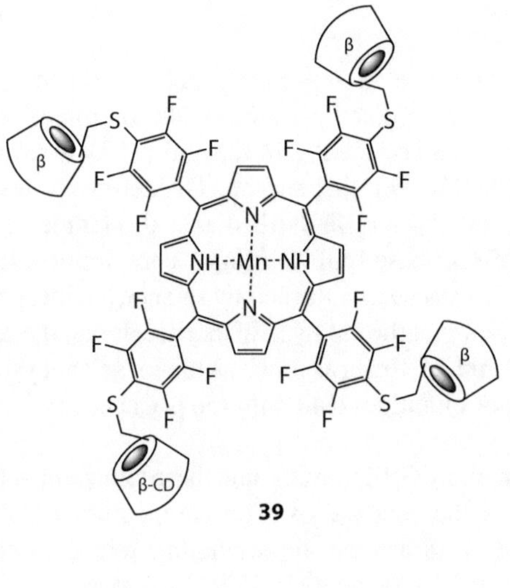

39

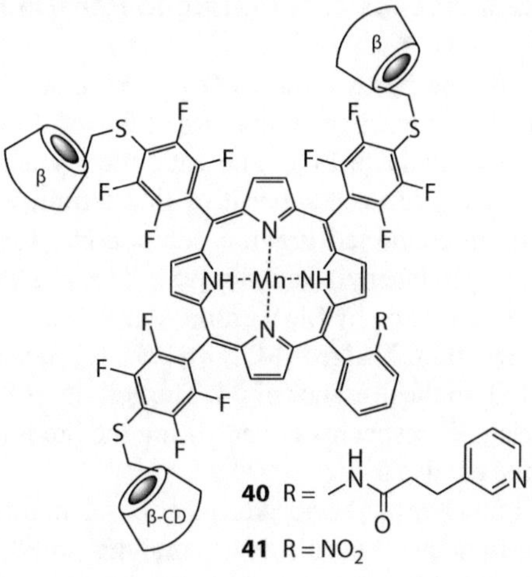

40 R =
41 R = NO₂

A further study has been carried out on hydroxylating the steroid substrate **37** with the catalyst **33**, using iodosobenzene as the oxidant as well [85,86]. It succeeds in introducing α-hydroxyl group at C-6 position in product **38**. However, it suffers the problem that only about four turnovers are realized before the catalyst is oxidatively destroyed. An improved catalyst **39** has been prepared in which all hydrogen atoms on the phenyl groups are replaced by fluorine atoms [87]. The fluorinated catalyst is greatly stabilized against oxidative destruction, and thus, the turnover number soars to 187. The 6α-hydroxyl derivative **38** is obtained in up to 100% yield with 95% conversion using 1% (mol) of catalyst **39** and 10 equivalents of iodosobenzene [88]. More turnover number is found to another two catalysts **40** and **41**, with one face of the porphyrin ring substituted by a pyridine derivative **40** or an *o*-nitrophenyl group **41** instead of tetrafluorophenyl-thio-β-CD [89]. The amazing turnovers are observed as ca. 2000 for **40** and 3000 for **41**, respectively.

Moreover, the C-9 hydroxylation has also been realized by catalyst **39** if the C-6 position is blocked by a third binding group (compound **42** and its oxidation product **43**), with excellent turnovers [88,90]. Other well-designed similar P450 mimics have been also studied by Breslow's group in detail and show delightful results in catalysis of steroid hydroxylation [90–92].

42 X = H
43 X = OH

9.3.2.5 *Mimics of Glutathione Peroxidase*

Glutathione peroxidase (GPX) plays a significant role in scavenging active oxygen in human body. It catalyzes the reduction of harmful hydroperoxide by glutathione **44** to protect cells from oxidative damage [93]. The catalytically active center of GPX is recognized as a selenocysteine residue in each subunit of this tetramer [94,95]. For this reason, GPX mimics are usually selenium-contained organic compounds, such as ebselen **45** [96], diaryl diselenides [97], and α-(phenylselenenyl) ketones [98]. The study of CD-based GPX mimics containing selenium is being carried on by Liu's group. The 6-iodo-β-CD has been modified by selenocysteine, producing 6,6′-selenium-bridged-β-CD **46** [99]. It is observed that the catalyst **46** can accelerate the reduction of H_2O_2 in the presence of **44**. The activity is 82 and 4.2 times as much as that of selenocysteine and ebselen **45**, respectively, indicating the binding of glutathione **44** with the β-CD cavity is essential for the catalysis.

Tellurium has been taken place of selenium in many GPX mimics, and these tellurium-substituted CDs may show even higher catalytic ability as the catalyst **46**. The compounds 2,2′-ditellurobridged-β-CD **47** and 6,6′-ditelluro-bridged-β-CD **48** are two representative tellurium-contained GPX mimics prepared by Liu's group [100,101]. It is confirmed by ¹H NMR that the complexation between **47** and **44** may be dominated by hydrogen bonding between carbonyl groups and the

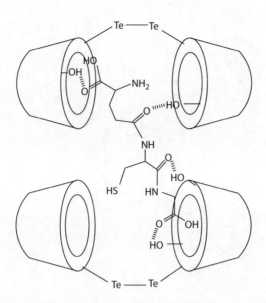

Figure 9.11 A possible structure of the complex formed with **44** and **47**. (Modified from Hao, Y.-Q. et al., *J. Incl. Phenom. Macrocyclic Chem.*, 54, 171, 2006; Ren, X.-J. et al., *ChemBioChem*, 3, 356, 2002.)

hydroxyl groups outside the β-CD, as illustrated in Figure 9.11. Thanks to the matched structure, **47** catalyzes the reduction of some hydroperoxides about 3.4×10^5 times more efficiently than diphenyl diselenide **49**, or even 46 times more efficiently than ebselen **45** [102,103]. Recently, a novel bifunctional artificial enzyme **50** has been developed by self-assembly of an adamantyl-modified porphyrin ring and CD-based telluronic acid [104]. This antioxidative enzyme has both superoxide dismutase (SOD) and GPX activities; the latter is attributed to the tellurol moiety substituted to the C-2 position of all four β-CDs. Compared with the well-known ebselen **45**, the catalyst **50** accelerates the reduction rate of 27-fold when it uses 4-nitrobenzenethiol and cumene hydroperoxide as substrates.

50

9.3.3 Rotaxanes Containing Cyclodextrins

The CD-based molecular machines are the "machines at the molecular level." They are attractive supramolecular systems, appealing to chemists both for their aesthetics and for the challenge of designing their synthesis [105,106]. Among these dramatic artworks created by chemists, rotaxane is a classic superstructure of topological importance. Generally, the fabrication of rotaxane is divided into two steps. The first synthetic step is that one linear axle penetrates through macrocycles to form a pseudorotaxane, and the next step is to block the axle with bulky stopper groups to form a rotaxane (Figure 9.12) [107,108].

9.3.3.1 Pseudorotaxanes and Polypseudorotaxanes with Cyclodextrins

Due to the property of forming inclusion complexes with many lipophilic guests, CDs and their derivatives are widely used for these self-assembled suprastructures. When a threadlike long-chain molecule penetrates the CD's cavity, a pseudorotaxane will be formed [108]. These pseudorotaxanes are usually unstable, since they are maintained by weak interactions, depending on the nature of the guest entrapped in the CD cavity. The CD molecules may slide through the axle or even fall off. However, their stability will be raised if polymers are acted as the axle instead (usually called polypseudorotaxane), probably thanks to the strong overall van der Waals interactions between CD cavities and the hydrophobic polymer chain. The well-known pseudorotaxane **51** formed with α-CD and polyethylene glycol (PEG) prepared by Harada's group is a representative example of that (Figure 9.13a) [109,110]. When PEG (molecular weight more than 200) is mixed with α-CD in an aqueous solution at room

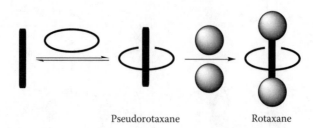

Pseudorotaxane   Rotaxane

Figure 9.12   A scheme of rotaxane and catenane formed from a pseudorotaxane.

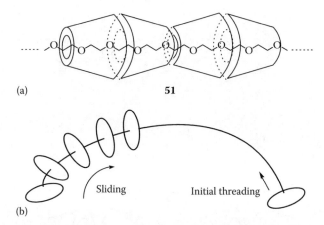

(a)   51

Sliding   Initial threading

(b)

Figure 9.13 (a) A scheme of pseudorotaxane **51** formed with α-CD and PEG (Modified from Harada, A. et al., *Macromolecules*, 26, 5698, 1993; Nepogodiev, S.A. and Stoddart, J.F., *Chem. Rev.*, 98, 1959, 1998.) and (b) the threading of α-CD along the PEG chain (Modified from Ceccato, M. et al., *Langmuir*, 13, 2436, 1997; Harada, A. and Kamachi, M., *Macromolecules*, 23, 2821, 1990.)

temperature, the solution becomes turbid in minutes, and the final complex is obtained by freeze drying. The threading is a spontaneous process with α-CD molecules sliding along the linear PEG chain (Figure 9.13b), and this process is affected by the nature of the solvent and the temperature, according to Ceccato et al. [111,112]. The molecular dynamics simulating results and x-ray diffraction studies show that the α-CDs adopt head-to-head and tail-to-tail sequences instead of head-to-tail sequences, owing to the stronger hydrogen bonds formed between adjacent α-CDs [113,114].

Unlike α-CD, β-CD does not form stable pseudorotaxane with PEG since its cavity is a bit large for this slim polymer chain, but it does thread into polypropylene glycol (PPG) to generate another stable pseudorotaxane **52** [115,116]. In this complex, β-CDs are not stacked in a slightly inclined column, and all of their secondary hydroxyl groups take part in intermolecular hydrogen bonding to an adjacent molecule [117]. The γ-CD, due to its larger cavity than that of α- and β-CD, threads the polymers with larger side chains like polyisobutylene (pseudorotaxane **53**) [118], or even forms double-stranded inclusion complexes with some slim polymer chains, such as PEG [119] and polyethylene adipate (PEA) [120], forming the pseudorotaxanes **54** and **55**, respectively.

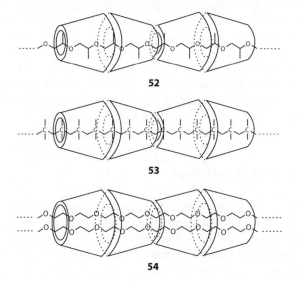

52

53

54

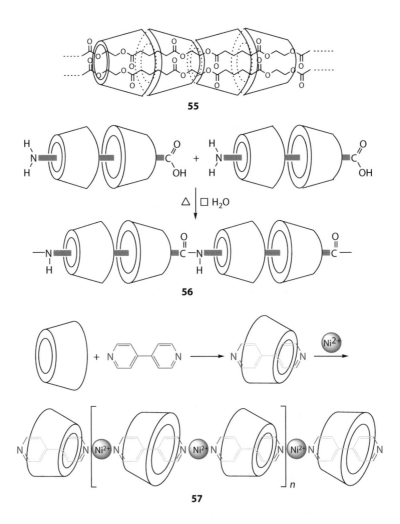

55

56

57

Another kind of pseudorotaxane can be prepared by the polymerization of the guest molecules in CD complexes. It is reported that one α,ω-amino acid forms inclusion complex with two β-CDs, with its amino and carboxyl group exposed outside the CD cavity [121]. When annealing the microcrystalline inclusion compounds at 150°C–250°C under vacuum, the polycondensation happens between α-amino group and adjacent ω-carboxyl group; a pseudorotaxane **56** is obtained as a result. Liu et al. have developed a novel method to prepare a metal-chelated pseudorotaxane **57** [122]. The ligand 4,4′-dipyridine that is included in a β-CD cavity with 1:1 stoichiometry can coordinate with nickel(II) ions, and thus, the axle "polymerizes" only by the coordination bond. The pseudorotaxane **57** reaches to 450 nm in length; moreover, it will be easily disassembled by adding metal chelators.

9.3.3.2 Cyclodextrin-Based Rotaxanes

When both ends of the "axle" are blocked, a pseudorotaxane becomes a rotaxane. Generally, the stoppers should have enough steric hindrance to prevent the CD ring from falling off. The first CD-based rotaxane **58** was prepared by Ogino and Ohata by using α,ω-diaminoalkanes as the axle and cobalt(III)-ethylenediamine complexes as the stopper [123,124]. It was found that the yield of rotaxane was greatly influenced by the length of axle, and α,ω-diaminoalkanes with 12 methylene units gave the highest yield. Some other CD-based rotaxanes appearing later were prepared with similar method based on the coordination of transition metals with the end groups of axle [125–127]. A part of stoppers can block the pseudorotaxane with electrostatic interactions; as an instance, the rotaxane **59** designed by Rao and Lawrence is prepared with biphenyl diammonium salt penetrating the cavity of heptakis (2,6-*O*-methyl) -β-CD and then blocked with tetraphenylboron anion by electrostatic attraction [128]. However, these rotaxanes are blocked by noncovalent bonding; they are prone to dissociating under certain circumstance as a result. The problem can be solved by taking covalent bonding between the end group of axle and the stopper. Harada et al. chose 2,4,6-trinitrobenzenesulfonic acid sodium salt (TNBS) as a stopping agent and diaminohexaethylene dihydrochloride (DAHE-2HCl) as an axle, developing stable nonionic rotaxanes **60** and **61** based on 2,6-di-*O*-methyl-α-CD and 2,3,6-tri-*O*-methyl-α-CD, respectively [129]. They also developed a quaint rotaxane **62** with CD molecule entrapped in a rotaxane by an electric trap, where the multiple cations in the axle prevented the movement of the CD ring [130].

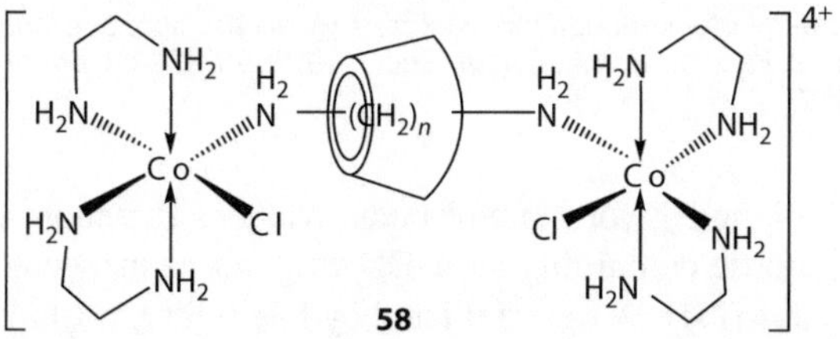

58

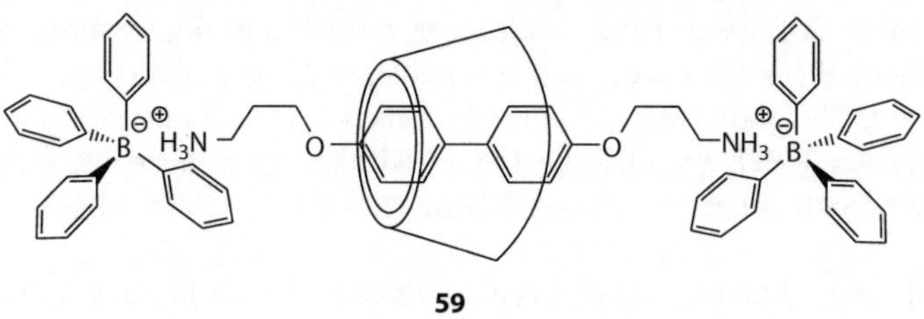

59

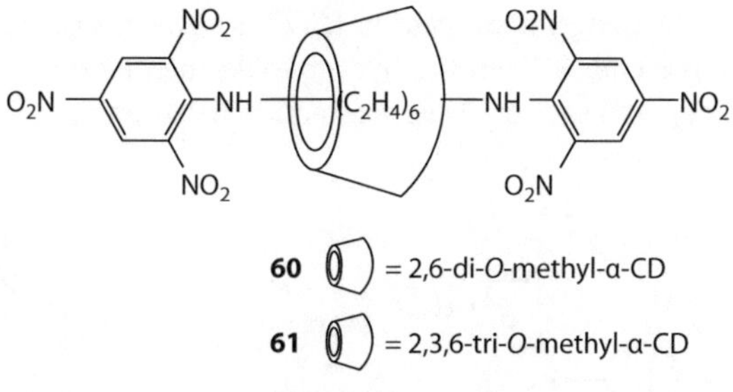

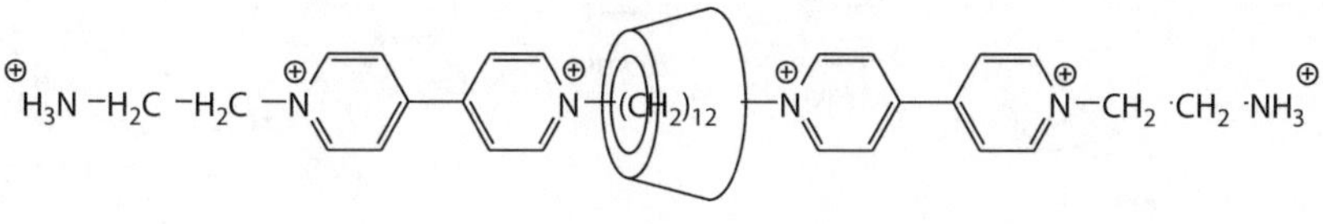

60 ⬭ = 2,6-di-*O*-methyl-α-CD

61 ⬭ = 2,3,6-tri-*O*-methyl-α-CD

62

Some CD-based rotaxanes show photoinduced activity. They act as the molecular shuttle that the CD molecules will slide and locate at other positions as the wavelength changes. These smart

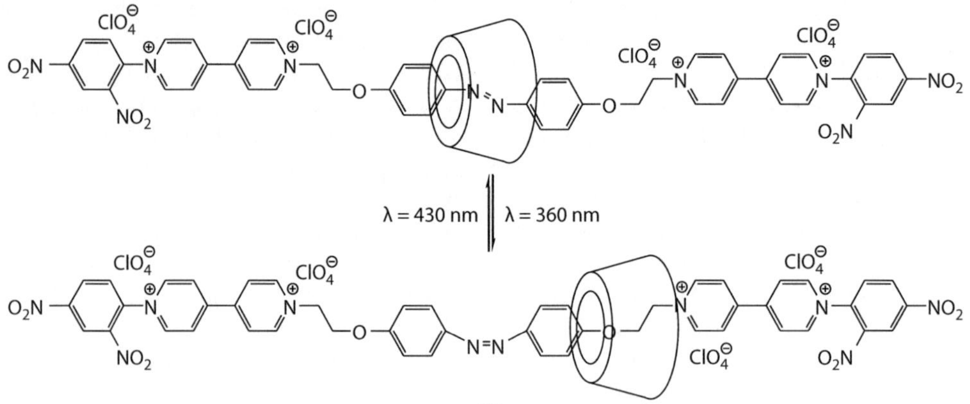

Figure 9.14 The *trans-/cis-* photoisomerization occurred to an azobenzene-based rotaxane **63**. (Modified from Murakami, H. et al., *J. Am. Chem. Soc.*, 119, 7605, 1997; Harada, A. et al., *Chem. Commun.*, 18, 1413, 1997.)

devices may provide good models for the molecular switches in nanoscale. The rotaxane **63** is a light-driven molecular shuttle containing an α-CD ring, an azobenzene chain, and 2,4-dinitrophenyl bulky terminal caps [131]. When irradiated by UV light (360 nm), the photoisomerization occurs from the *trans-* to the *cis-*configuration of the azobenzene unit in **63**; as a result, the α-CD molecule is driven to shift away from azobenzene moiety and finally exists at the methylene spacer moiety (Figure 9.14). This process is reversible, and it can revert to the initial state when irradiated at 430 nm. The photoisomerization phenomenon is also observed for **64**, a symmetric stilbene-based rotaxane [132]. The stilbene axle of **64** undergoes reversible *E/Z* photoisomerization when the wavelength converts between 340 nm ($E \rightarrow Z$) and 265 nm ($Z \rightarrow E$), as described in Figure 9.15.

A solvent- and temperature-sensitive molecular shuttle **65** contains an α-CD ring, two dodecamethylene units, three 4,4′-bipyridinium groups, and capped with two 2,4-dinitrophenyl groups [133]. Attributed to the energy barrier of cationic bipyridinium group, the movement of α-CD is restricted. However, site exchange occurs as the temperature rises to coalescence temperature (T_c, ca. 130°C). The free energy of activation ($\Delta G°$) for this process can be calculated as ca. 84 kJ·mol^{-1}. The driven force comes from the hydrophobic interaction between the CD ring and the dodecamethylene moiety, and the repulsive interaction between the CD ring and bipyridinium

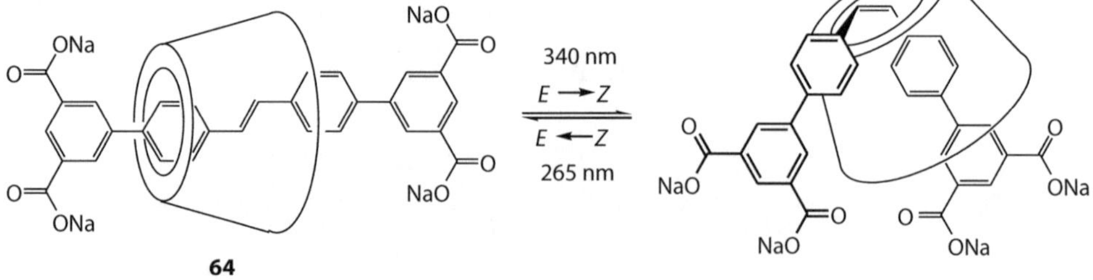

Figure 9.15 The reversible *E/Z* photoisomerization of stilbene-based rotaxane **64**. (Modified from Stanier, C.A. et al., *Angew. Chem. Int. Ed.*, 41, 1769, 2002; Kawaguchi, Y. and Harada, A., *J. Am. Chem. Soc.*, 122, 3797, 2000.)

cations. It is also found that the rate of site exchange appears faster in DMSO-d_6 than in D_2O due to the additional hydrophobic interaction in DMSO-d_6.

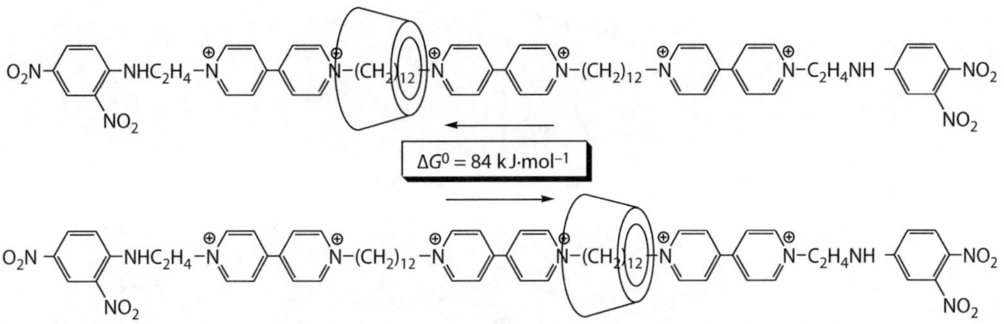

65

9.3.3.3 Polyrotaxanes Constructed with Cyclodextrins

A polyrotaxane is prepared by blocking the polymer axle penetrated through many CD cavities, as similar as the rotaxane preparation. Harada et al. used PEG-bisamine as the axle to prepare α-CD/PEG-bisamine polypseudorotaxane **66** and then endcapped **66** by 2,4-dinitrofluorobenzene in DMF, giving a typical polyrotaxane **67**, called "molecular necklace" (Figure 9.16)

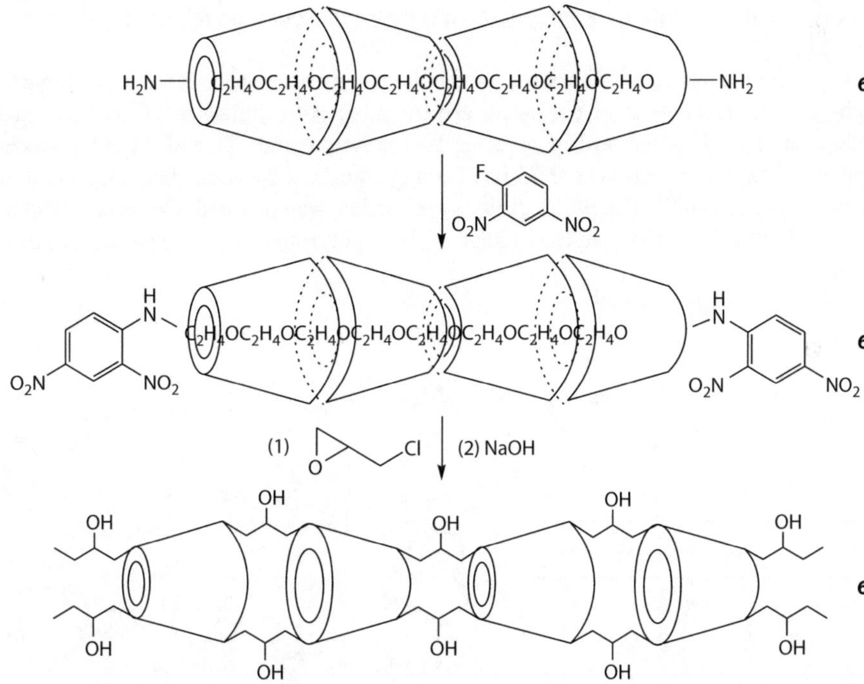

Figure 9.16 Synthesis of polyrotaxane **67** and molecular tube **68** from the polypseudorotaxane **66**. (Modified from Harada, A. et al., *Chem. Soc. Rev.*, 38, 875, 2009; Harada, A. et al., *J. Am. Chem. Soc.*, 116, 3192, 1994.)

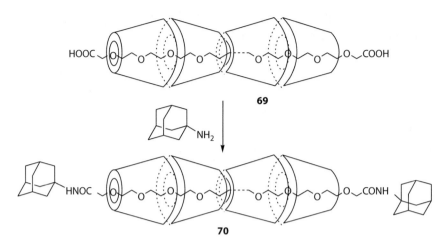

Figure 9.17 Synthesis of polyrotaxane **70** from the polypseudorotaxane **69** formed with PEG-dicarboxylic acid and α-CDs. (Modified from Araki, J. et al., *Macromolecules*, 38, 7524, 2005; Harada, A. et al., *Nature*, 364, 516, 1993.)

[134–136]. They also synthesized monodisperse α,ω-PEG-bisamine (28 mer, MW = 1248) as the axle [137]. This molecular necklace may provide a template for CD polymerization. When adjacent CD rings within the necklace **67** are cross-linked by reaction with epichlorohydrin in alkaline solution, the molecular tube **68** will be separated from the axle after cleavage of the stoppers [138,139]. The large cyclic compound adamantanamine is also a good stopper that is used to cap the α-CD/PEG-dicarboxylic-acid polypseudorotaxane **69**, giving the product **70** (Figure 9.17) [140].

In addition, the macrocyclic CD ring is an excellent stopper as well. It has been reported that oligothiophene diboric ethylene glycol esters are threaded with dimethyl-β-CDs and react with 6-*O*-(4-iodophenyl)-β-CD afterward, generating two polyrotaxanes **71** and **72** [141]. Recently, a β-CD-capped α-CD/PEG polyrotaxane **73** has been synthesized by click chemistry via a one-pot strategy with high yield [142]. The β-CD cavities are further included with the terminal adamantyl groups of modified PEG, resulting in the linkage of these polyrotaxanes to construct a nanowire **74** (Figure 9.18).

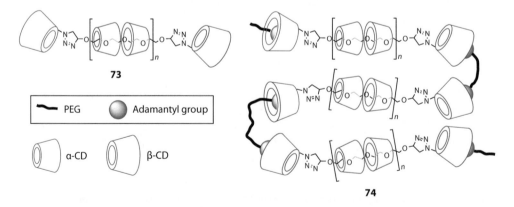

Figure 9.18 A β-CD-capped α-CD/PEG polyrotaxane **73** and the nanowire **74** constructed by polyrotaxane **73** and the adamantyl-modified PEG. (Modified from Wu, J.-Y. et al., *Macromolecules*, 43, 2252, 2010; Araki, J. et al., *Macromolecules*, 38, 7524, 2005.)

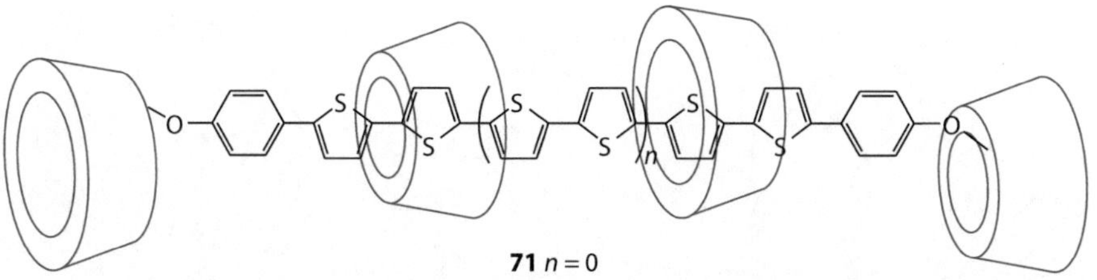

71 *n* = 0
72 *n* = 1

The degradable polyrotaxanes have been developed to provide a potential for biomedical applications. A controlled biodegradable polyrotaxane **75** is synthesized based on polypseudo-rotaxane **51** terminated with a short peptide chain of L-phenylalanyl-glycyl-glycine [143]. It is observed that **75** will be degraded by aminopeptidase M in vitro, though it has high molecular weight of 16,000. Two pH-triggerable degradable polyrotaxanes have been designed by Loethen et al. [144]. They select an azidated PEG 1500 as the axle to thread with α-CD capped with different substituents of vinyl ether and ester. A study of degradation kinetics reveals that the vinyl ether cholesterol-endcapped polyrotaxane **76** is sensitive to acidic condition, while the *tert*-butoxycarbonyl-tryptophan ester-endcapped polyrotaxane **77** will be solubilized in alkaline solution. These polyrotaxanes can be readily and efficiently prepared and disassembled, which offer a convenient way to fabricate intelligent molecular machines.

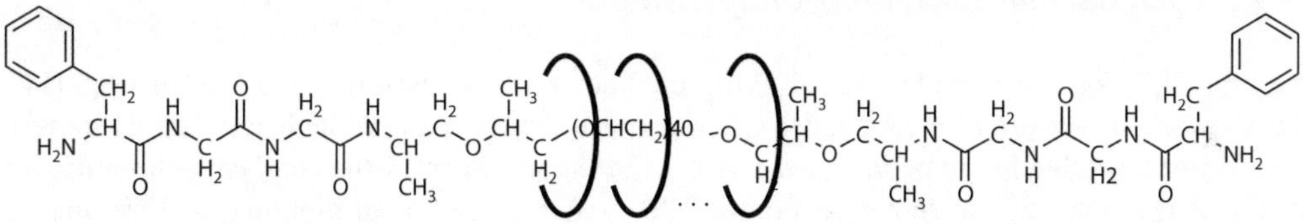

75

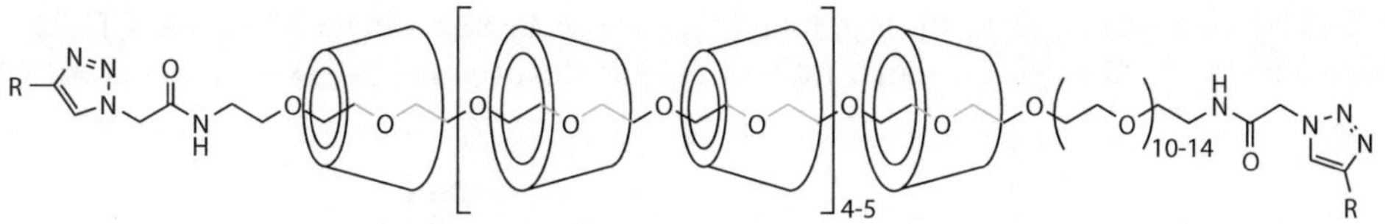

76 R =

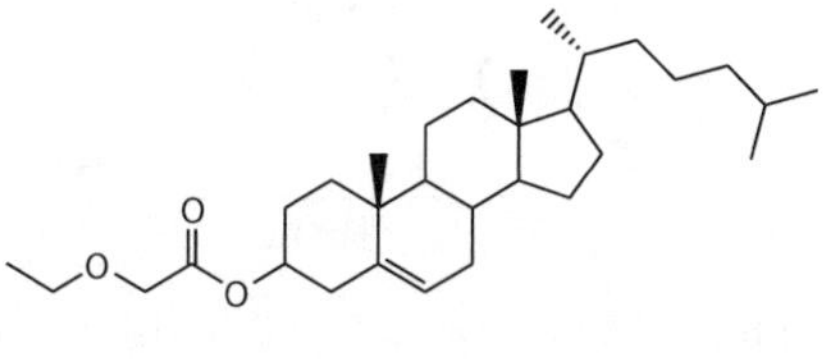

77 R =

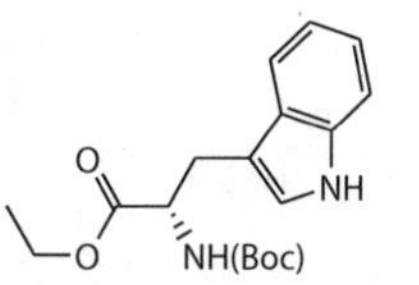

9.4 CYCLODEXTRIN POLYMERS

CD polymers (CDPs) are the polymers containing many CD units. In CDPs, CDs are either covalently immobilized onto a matrix or cross-linked to form linear chains or networks (refer to Figure 9.19) [145]. Generally, CDPs are prepared in three ways as follows [146]: (1) cross-linking CDs with bi- or multifunctional reagents, such as epichlorohydrin, bisepoxides, and diisocyanates (the products usually have the structure of linear chains or networks, as shown in Figure 9.19b and c); (2) polymerization happening to the double bonds in the side chains of CD derivatives (like acrylic substituted CDs), producing the polymers containing pendant CD units; and (3) linking

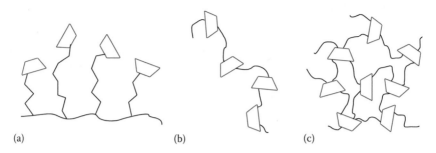

Figure 9.19 A scheme of CDPs: (a) CD immobilized onto a matrix, (b) linear CDP obtained by cross-linking, and (c) CDP cross-linked networks. (Modified from Crini, G. and Morcellet, M., *J. Sep. Sci.*, 25, 789, 2002; Solms, J. and Egli, R.H., *Helv. Chim. Acta*, 48, 1225, 1965.)

the CD rings to a polymeric backbone with covalent bonding, that is, modifying the polymeric backbone with CDs, as shown in Figure 9.19a. The CDPs preserve the property to include organic molecules that belongs to the CD cavity; consequently, they may find specific applications in separation process, catalysis, and pharmacy.

9.4.1 Cross-Linked Cyclodextrin Polymers

9.4.1.1 Cyclodextrin-Epichlorohydrin Polymers

The first CDP was prepared by cross-linking method reported by Solms and Egli in 1965 [147]. In their work, a mixture of α-, β-, and γ-CD was reacted with epichlorohydrin in NaOH solution, and a water-insoluble block polymer was formed in the presence of the catalyst sodium borohydride (the reaction scheme shown in Figure 9.20). After drying, comminuting, and sieving, the compressible powders in an amorphous form were obtained as a good adsorbent for several dyes and phenols. The cross-linking reaction between β-CD and epichlorohydrin has been particularly described by Fenyvesi et al. [148]. It is found that the molecular weight of soluble CD-EP varies from 1400 to 5300 when the molar ratio of epichlorohydrin and β-CD increases from 2:1 to

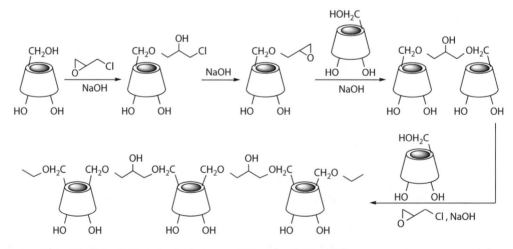

Figure 9.20 The cross-linking reaction scheme of CD with epichlorohydrin.

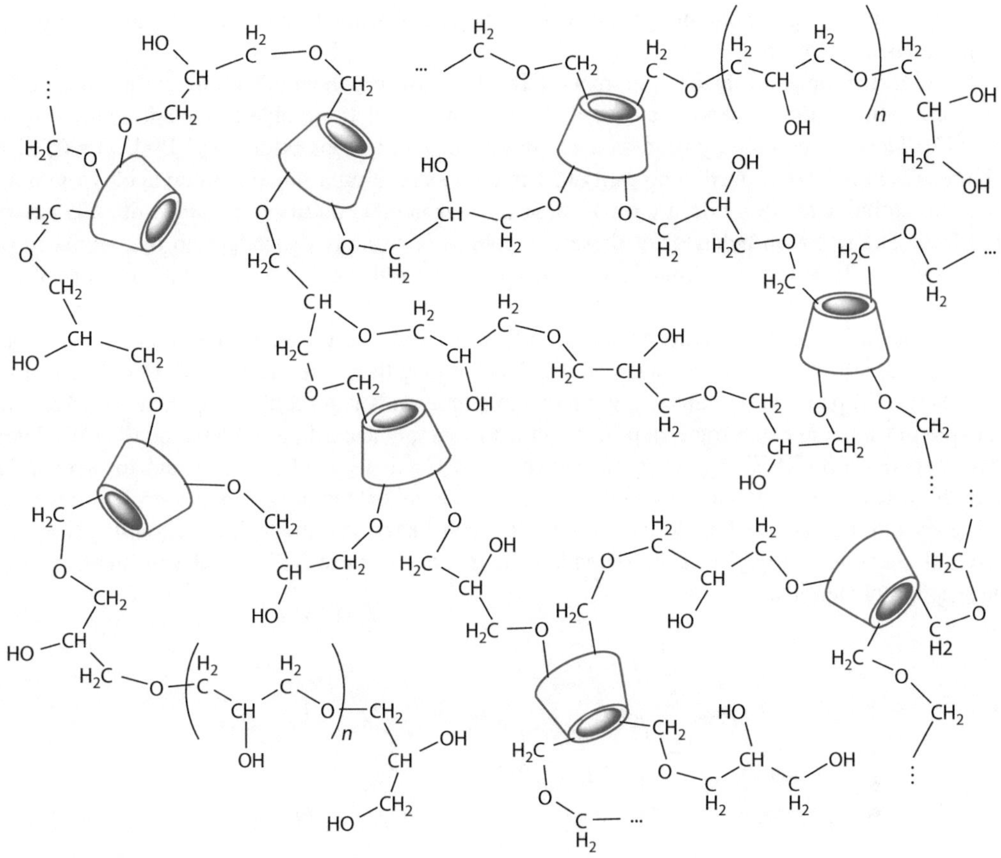

Figure 9.21 A possible network structure of CD-EP polymer. (Modified from Crini, G. and Morcellet, M., *J. Sep. Sci.*, 25, 789, 2002; Solms, J. and Egli, R.H., *Helv. Chim. Acta*, 48, 1225, 1965.)

20:1, and the further increase of this ratio may result in the formation of insoluble gel-like beads. A possible structure of CD-EP is depicted in Figure 9.21, in which CD molecules act as the nodes of the three-dimensional cross-linked network, linked by interconnected glycidyl arms [149,150]. This highly hydrophilic polymer swells in water with high water uptake capacity, but insoluble in most organic solvents.

Wiedenhof et al. developed the preparation method and produced spherical CD-EP microbeads by the reversed-phase suspension cross-linking technique [151]. They chose o-xylene or methyl isobutyl ketone as the dispersed phase containing the dispersant Nonidet P-40 (a nonionic detergent), mixed them with a solution of NaOH and α- or β-CD, stirred vigorously and added epichlorohydrin, and finally obtained the CD-EP beads after washing with acetone. The beads were packed in a column to separate a mixture of amino acids and several aromatic organics, showing distinct retention behavior from Sephadex gels [152,153]. However, due to their flexibility and compressibility, the CD-EP beads prepared earlier cannot suffer the high pressure drop in packed column [17,154]. Some efforts have been made to overcome the shortcoming of poor rigidity. Preparing copolymers of CD-EP with polyvinyl alcohol (PVA) is an effective method adopted by Szejtli's group [155]. A second method developed by Nussstein et al. is to cross-link the CD derivatives (such as hydroxyalkyl-β-CD) instead of unsubstituted CDs [154]. The third one is to coat or deposit a layer of CD-EP onto the rigid inorganic granules or fill the granules in CD-EP to form organic-inorganic composite beads [156,157].

Cyclodextrin-epichlorohydrin polymers (CD-EPs) show excellent adsorbing ability to specific organics owing to the preservation of hydrophobic cavities that are able to include many organics [149]. They are commonly used as a supporting matrix in separations [152,158]. The CD-EP microbeads have been proved to be well utilizable for separating aromatic amino acids from nonaromatic amino acids in a mixture on account of their notable ability to include with CD cavity [152,159]. The retention behavior of these compounds is strongly dependent on the geometry of these compounds and the resulting fit with the CD cavity, reflected by the stability constant of the complex. Therefore, it is not surprising that tryptophan 26 has the maximum retention value among all amino acids in β-CD-EP stationary phase [54]. As a similar result, adenosine-3′-monophosphate (3′-AMP, 78) may retain most strongly in β-CD-EP among the nucleotides [160]. CD-EP can also work as a solid-phase extractant to capture some compounds from multicomponent samples. An example of that is to catch trace steroids from urine sample according to Moon et al. [161]. They have prepared a β-CD-EP by entrapment with 0.3 M $CaCl_2$ to yield an extractant to recover 77 steroids in urine. The polymer exhibits particularly excellent binding capacity to hydroxylated estrogens with the recovery not less than 96%. It is considered that the hydrogen bonding between steroidal phenolic hydroxyl and the secondary hydroxyl groups of β-CD plays significant role in the sorption of steroids.

As the same to CD monomer, CD-EPs can be applied in enantiomer recognization and resolution. An attempt of resolving mandelic acid derivatives with β-CD-EP has been completed by Harada et al., revealing a descending resolution order of methyl mandelate 79, O-methyl mandelic acid 80, ethyl mandelate 81, and mandelic acid 82 [162]. β-CD-EP binds L-(+)-isomers preferentially over D-(−)-isomers, and surprisingly, D-(−)-methyl mandelate can be obtained with 100% optical purity in the primary fraction from chromatographic column. A successful resolution of two indole alkaloids, vincadifformine 83 and quebrachamine 84, has been achieved with the polymer CDP-25 in good yield [163]. The selectivity for the enantiomer pairs is sensitive to the pH of mobile phase, the size of CD cavity, and the shape of enantiomer molecules [164].

The CD-EPs (usually oligomer) can be immobilized to other matrices for separations and purifications, in which CD-EPs serve as the ligands to capture specific molecules in mobile phase. There have been several examples of separating natural products with these materials. The isoflavonoid puerarin **85** in the extracts of *Radix puerariae* has been separated by Tan's group with the aid of various oligo-β-CD-coupled matrices, including Sepharose HP [165], polyacrylate beads [166], and polystyrene matrix [167]. In addition, the purification of some other natural products such as epigallocatechin gallate **86** from green tea extracts [168] and paclitaxel **87** from *Taxus cuspidata* [169] has also been studied. They find that oligo-β-CD as an excellent ligand displays high affinity to these natural products. Further investigations on the adsorption mechanism by NMR spectroscopy confirm the significance of the inclusion complex formation between these molecules and the β-CD cavity in such oligomer, via the same mechanism as monomeric β-CD [170,171].

It is noticed that CD-EP can also be utilized as the matrices for expanded bed adsorption (EBA). A beneficial exploration is that a CD-EP POLY RINGDEX B has been applied in expanded bed for protein refolding by Mannen et al. [172]. With CD as an artificial chaperone, the monomeric enzyme α-glucosidase can be refolded with a yield of twofolds at a protein concentration higher than fivefolds. Unlike the conventional liquid-phase artificial chaperone system, the reusable CD-EP may be suitable for scale-up for industrial protein production. A set of novel matrices named CroCD-TuC special for EBA have been developed by our group [157]. CroCD-TuC matrices are the β-CD-EP microbeads filled with tungsten carbide powder, prepared by the reversed-phase suspension cross-linking technique. The densifier tungsten carbide increases the density of the β-CD-EP as high as 2.09 g·cm^{-3} and also improves their rigidity. Their hydrodynamic properties have been studied, and the residence time distributions (RTDs) analysis is introduced to estimate the liquid mixing in the expanded bed [173]. It reveals that two of these matrices are superior to others in hydrodynamic properties, and they will be promising matrices for further EBA use.

9.4.1.2 Cyclodextrin-Polyurethane Polymers

Diisocyanates are a series of cross-linker among which 1,6-hexamethylene diisocyanate **88** (HMDI), toluene-2,4-diisocyanate **89** (2,4-TDI), 1,4-phenylene diisocyanate **90** (PDI), and 4,4′-diphenylmethane diisocyanate **91** (DPMDI) are the mostly used. The isocyanate groups in these cross-linkers are so active that they react with the CDs' hydroxyl groups to form CD-polyurethane polymers (CD-PUs), with pyridine or DMF as the solvent (described in Figure 9.22).

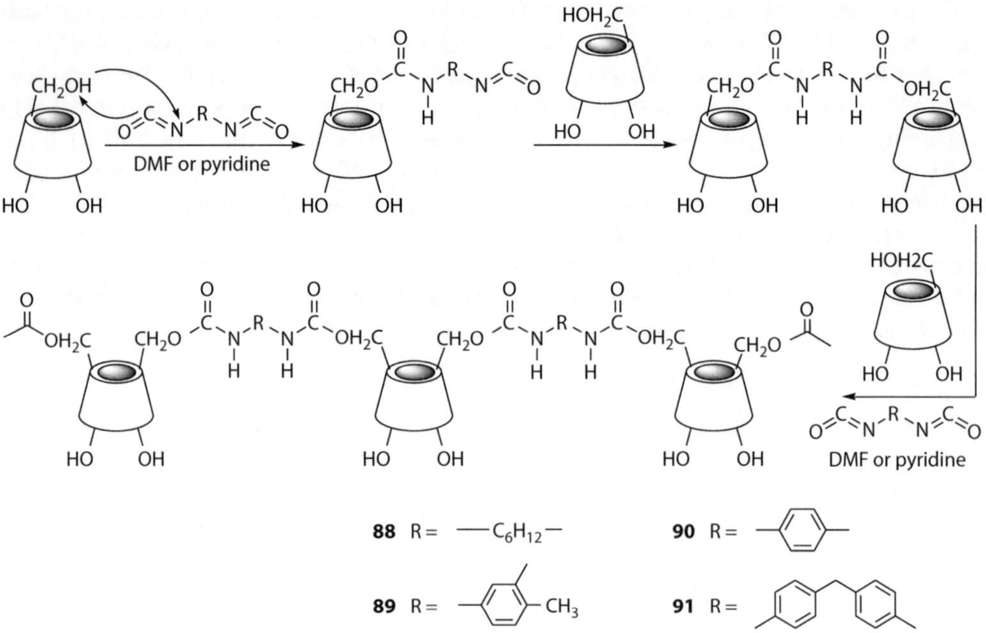

Figure 9.22 The cross-linking reaction scheme of CD with diisocyanates.

CD-PUs are good adsorbents for chromatographic usage, as the same to CD-EPs. CD-PUs exhibit strong interaction with guest molecules containing π-electrons or heteroatoms and even hydrogen bonds [174]. A separation of six aromatic amino acids was tried by CD-PUs according to Mizobuchi et al., among which four amino acids (phenylglycine **92**, tyrosine **93**, tryptophan **26**, and phenylalanine **94**) were separated completely, and the retention on CD-PU became more strongly at higher pH value [175]. The adsorption behavior of aromatic amino acids on a β-CD-HMDI (cross-linked with β-CD and **88**) resin was studied by Tang et al. [176]. It exhibited the adsorption efficiencies of three aromatic amino acids as the descending order of L-tryptophan > L-phenylalanine > L-tyrosine under the same conditions, for the reason of hydrophobicity of amino acid molecules and the resultant strength of interactions between amino acid molecule and the polymer. Moreover, CD-PUs have been coated onto the surface of rigid matrices by in situ cross-linking method to prepare the matrices for HPLC applications [177], and the matrices can be used for the separation of many organics, for example, 2,4-dinitrophenyl amino acids [178].

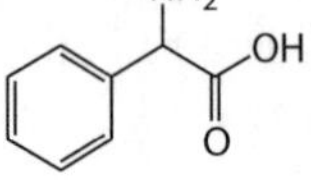

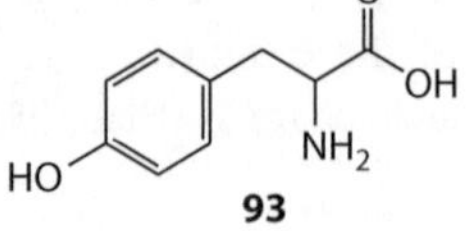

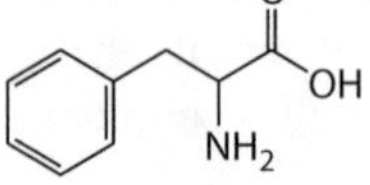

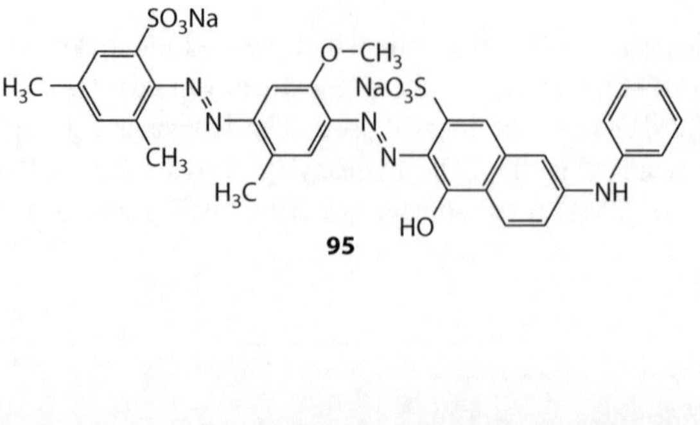

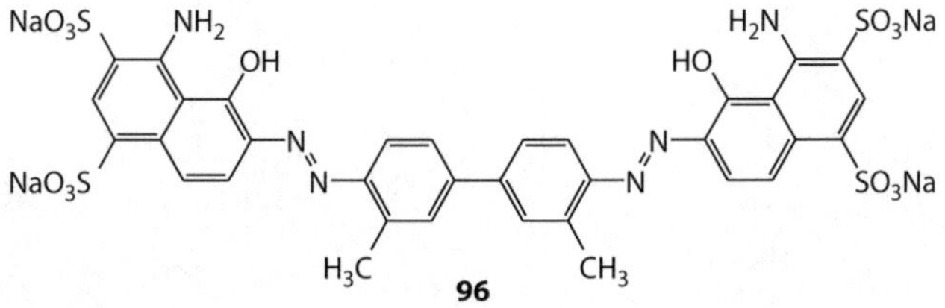

96

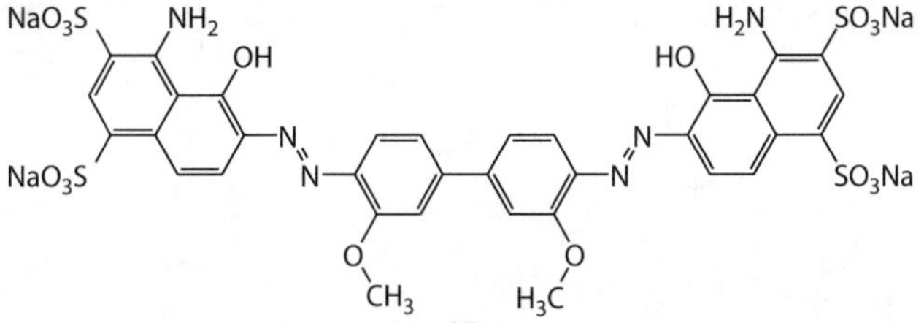

97

One of the most important applications of CD-PU is to remove specific organic compounds from aqueous solution. It is of great significance in the treatment of wastewater with the aim of reducing the emission of hazardous substances to the environment. Bhaskar et al. prepared a β-CD-HMDI resin acting as a solid-phase extractant for the preconcentration of several carcinogenic aromatic amines from water in the optimum pH 8.5 [179]. The recovery of aromatic amines for this CD-contained resin is found to be higher compared with the common C_{18} materials. It is exciting that CD-PU has extraordinary capability to phenols, with the phenol molecules both entrapped into the CD cavities and inserted in the polyurethane skeleton attributing to the hydrogen bonding between phenolic hydroxyl and N–H groups (Figure 9.23) [180]. The copolymer adsorbents cross-linked of CDs (β-CD or mixed CDs) with HMDI **88** in a 1:8 molar ratio show best removal efficiency to phenol, *m*- or *p*-cresol, and xylenol not less than 70%. The adsorbents can be easily regenerated by extraction with methanol without notable change of adsorbing capability, exhibiting a potential for treatment of a raw industrial wastewater from phenolic resin processing [181]. Additionally, CD-PUs are convenient for the adsorption and removal of azo dyes from aqueous solutions. It has been published that the β-CD-HMDI resins exhibit high sorption capacities toward Direct Violet 51 (**95**) of 69% at least, and the adsorption is highly dependent on the host-guest interactions and the presence of sulfonate groups of the anionic dyes [182]. The β-CD-DPMDI (cross-linked by **91**) resins adsorbing two azo dyes, Evans Blue **96** and Chicago Sky Blue **97**, follow the Langmuir isotherm model, and they present better adsorption capability as compared with β-CD-HMDI resin [183].

Since 1990s, molecular imprinting technique has been introduced in CDP synthesis to produce molecularly imprinted CDPs (MI-CDPs). These specially designed CDPs exhibit specific adsorbing capability toward the template molecule due to the increase of binding affinity and substrate selectivity, and thus, they provide a family of new receptors for the detection of organics. Komiyama's group has made significant contributions to MI-CDPs [184]. They have synthesized various MI-CDPs by cross-linking CDs or their derivatives with HMDI **88** or TDI **89**, aiming to the recognition of steroids such as cholesterol **98** [185,186] and stigmasterol **99** [187,188]. Of course, these MI-CDPs are used for complexation of steroids (it is interesting that many types of CDPs did not adsorb cholesterol **98** [184]). Besides steroids, the MI-CDPs can be designed as the receptors for many other biomolecules including amino acid derivatives [189], oligopeptides [190], bilirubin **100** [191], and even lysozyme [192]. Recently, Muk and Narayanaswamy et al. have cross-linked β-CD with TDI **89** in the presence of the template molecule *N*-phenyl-1-naphthylamine (NPN, **101**), deriving the MIβ-CDP-TDI for NPN detection [193]. The high binding affinity is

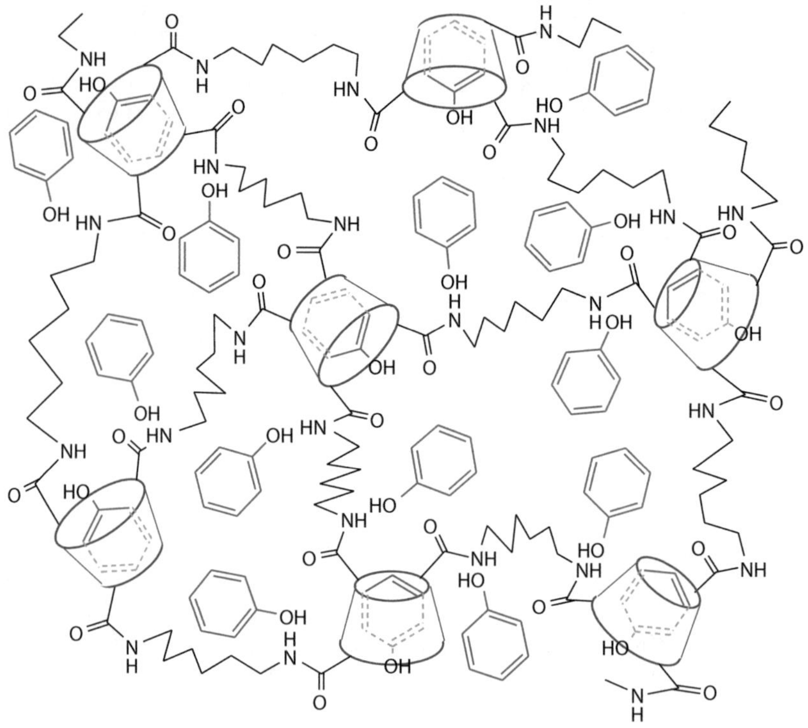

Figure 9.23 A scheme of the adsorbed phenol onto β-CD-PU beads in aqueous solution. (Modified from Yamasaki, H. et al., *J. Chem. Technol. Biotechnol.*, 83, 991, 2008; Kim, I.W. et al., *Anal. Chim. Acta*, 569, 151, 2006.)

observed, and the fluorescence signal is linear corresponding to the analyte content. Moreover, molecular imprinting promotes a better sensing signal by about 16% to the analyte NPN due to the increase of binding affinity and substrate selectivity toward the template molecule. A similar MIβ-CD-TDI has been also synthesized by them to develop a surface plasmon resonance sensor for dextromethorphan **102** [194]. The sensor will be expected to track the consumption and metabolism of drugs by enzymes.

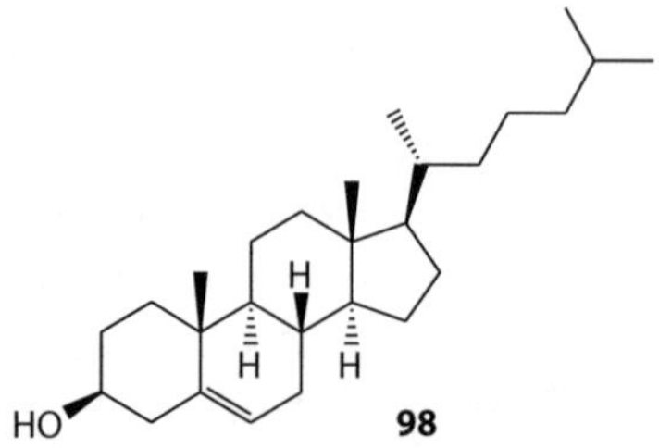

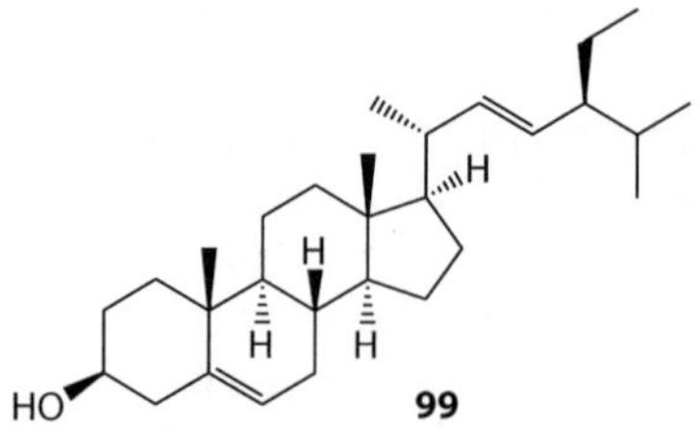

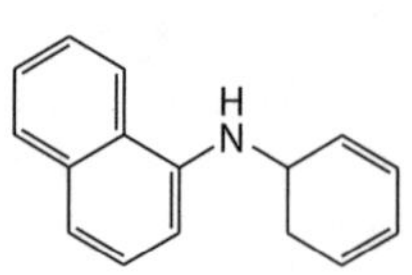

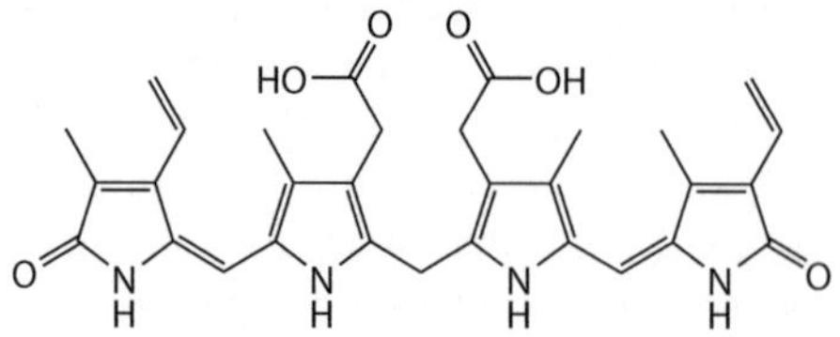

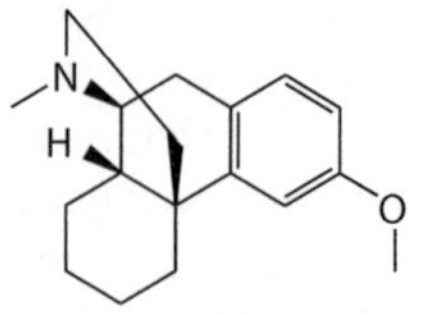

9.4.1.3 Other Cross-Linked Cyclodextrin Polymers

The syntheses of cross-linked CDPs were reported a lot with different bi- or multifunctional reagents. Pariot et al. prepared microcapsules by interfacial cross-linking β-CD with terephthaloyl chloride **103** (5%) in 1 mol·L^{-1} NaOH solution, producing the microcapsules with mean size range of 10–35 µm [195]. They selected *p*-nitrophenol as the guest molecule to evaluate their binding ability to organics, showing a rapid complexation achieved with the maximal binding capacity of 97.8 µmol·g^{-1}. Glutaraldehyde was used as the cross-linker to synthesize a copolymer of PVA and CDs for adsorbing drugs and some other organics, such as diclofenac **104**, indometacin **105**, congo red **106**, etc. [196]. It was considered that the nontoxic microspheres show good biocompatibility for drug delivery systems. The synthesis of succinic anhydride cross-linked β-CDP has been achieved in anhydrous DMF with the aid of NaH, forming a complexation agent in ion flotation process [197].

103   **104**   **105**

106

Even polycarboxylic acids (such as citric acid) can work as the cross-linker as well. An interesting example is that β-CD, NaH$_2$PO$_4$, PVA 1799, citric acid monohydrate, and deionized water are mixed in a wide open vessel and heated at 140°C for a few hours; after the water generated instantly driven away, a copolymer will be formed as a result [198]. This citric acid cross-linked copolymer exhibits excellent adsorption sensitivity toward methylene blue. Another copolymer of β-CD and PEG 400 is cross-linked by citric acid with the same method [199]. The polymer is a good adsorbent for alkaline organics, for example, aniline, and it shows larger affinity toward aniline compared with other CDPs owing to the large amount of acidic carboxyl groups in the polymer skeleton.

9.4.2 Polymerization of Monomers Containing Cyclodextrins

Acryloyl CDs possess carbon-carbon double bonds in the substituent, so that it can be polymerized when initiated, just as the polymerization of acrylic acid. The CD rings are the residues on the side chain of the polymer, like pendants hanging from a necklace. Harada et al. obtained acryloyl **107** and **108** and *N*-acrylyl-6-aminocaproyl CDs (α-**109** and β-**110**) by reacting the respective nitrophenyl esters with α- or β-CD [200]. The two monomers were further polymerized using 2,2′-azobis(isobutyronitrile) (AIBN, **111**) as an initiator at 60°C for 20–43 h, respectively, generating water-soluble polymers **112** and **113** with the molecular weight range of 10^4–10^5. Also acryloyl α- and β-CD have been copolymerized with acrylamide, acrylic acid, and vinyl pyrrolidone to give water soluble polymers in high yields. Surprisingly, polyacryloyl-β-CD **112** shows enzymatic activity in the hydrolysis of *p*-nitrophenyl esters [201]. The polymer is found to exhibit a

Michaelis–Menten-type kinetics as β-CD does, but it forms a more stable complex than β-CD with *p*-nitrophenyl-*p*-nitrobenzoate **114**. The enhanced catalytic ability of **112** is probably attributed to the "cooperative effect" between two neighboring β-CD moieties on a polymeric chain, since two CD residues are required to bind such a large substrate molecule.

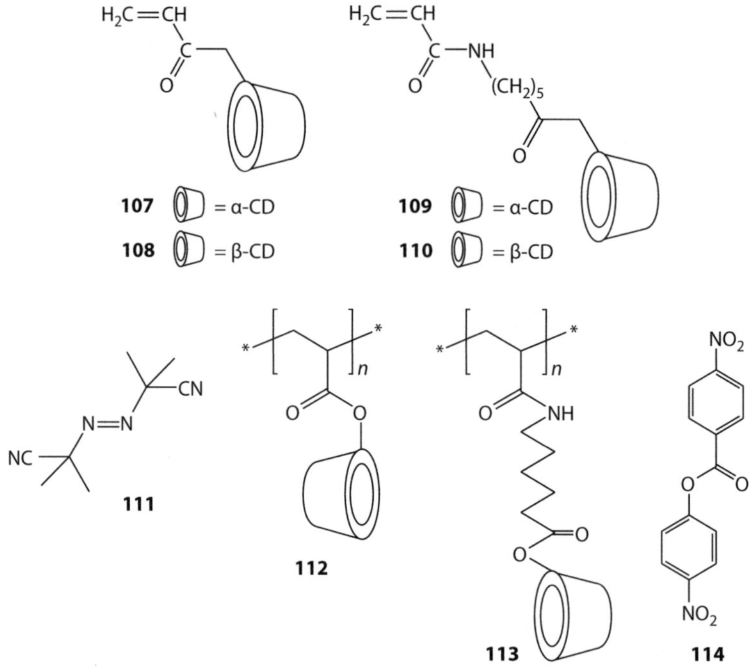

A CD derivative 2-hydroxy-3-methacryloyloxy-propyl-β-CD (BETAW7MAHP, **115**) can be used as a monomer copolymerized with 2-hydroxyethyl methacrylate (HEMA, **116**) in water to form hydrophilic insoluble polymer **117** according to Janus et al. (Figure 9.24) [202]. Similar with other CDPs, the polymer **117** exhibits high sorption capacities toward substituted benzene derivatives, and thus, it may be used for the recovery of organic pollutants from aqueous solutions.

Acrylamido-methyl β-CD **118** is an important monomer for the preparation of CDPs. A mixture of **118** and 1-vinyl-2-pyrrolidone **119** will be copolymerized when adding the initiator AIBN **111**, and the copolymer **120** will be derived as a result (Figure 9.25a) [203]. An interesting phenomenon

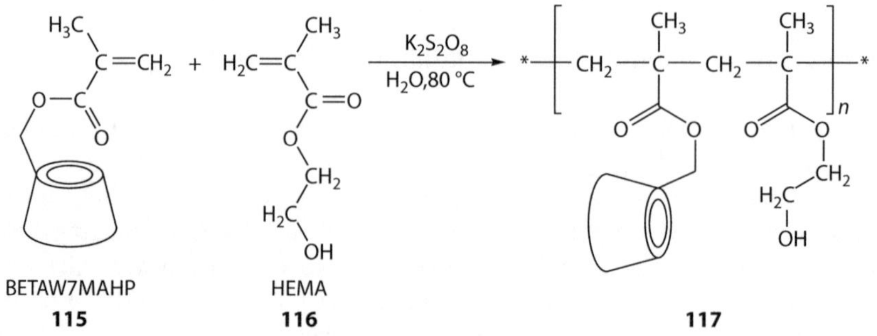

Figure 9.24 Synthesis of the copolymers **117**. (Modified from Janus, L. et al., *React. Funct. Polym.*, 42, 173, 1999; Harada, A. et al., *Macromolecules*, 9, 701, 1999.)

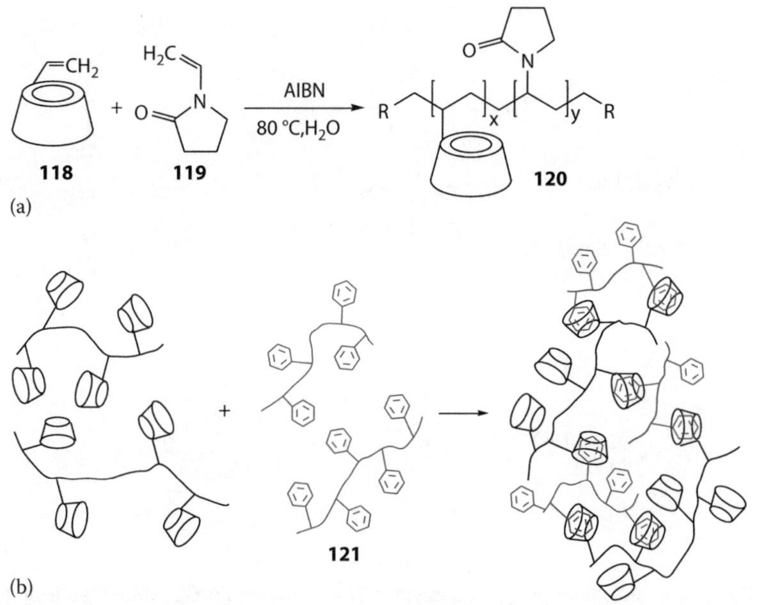

Figure 9.25 (a) Copolymerization of acrylamido-methyl β-CD **118** with 1-vinyl-2-pyrrolidone **119** and (b) a scheme of self-assembly of **120** with dextran-benzoate **121** in aqueous solutions. (Modified from Nielsen, A.L. et al., *Colloids Surf. B*, 73, 267, 2009; Harada, A. et al., *Macromolecules*, 9, 705, 1976.)

is observed for **120** that it can be self-assembled with another polymer dextran-benzoate **121** to form particles by the complexation interactions of benzoate and CD cavities (Figure 9.25b). The particles with diameter in the nano-/micrometer range may serve as a vector for drug delivery purposes.

9.4.3 Cyclodextrins Linking to Polymeric Backbones

9.4.3.1 Cyclodextrin Bonded to Synthesized Support

CDs are bonded to water-insoluble matrices as functional ligands to interact freely with the solutes by inclusion. These materials are so-called immobilized CD resins. The first example is an adsorbent with α-CD bonded to the commercialized support Sepharose 6B that is activated by a bifunctional reagent 1,4-bis(2,3-epoxypropoxy)-butane [204]. When worked as a stationary phase, it exhibits better affinity to β-amylase than albumin, and consequently, it retains β-amylase on the column. Epoxides are excellent activator for Sepharose 6B, and these gel-like matrices show high affinity to starch-degrading enzymes [205]. The polyacrylamide gel beads Bio-Gel P-2 is also activated and then coupled with substituted CDs to prepare stationary phases for chromatographic separations of various positional isomers of phenols, aromatic amines, and nitrophenyl derivatives [206–208].

A biocompatible polymeric β-CD resin **122** is prepared by grafting β-CD to the partially aminated polyacrylamide [209]. The synthesis process involves the polymerization of porous cross-linked polyacrylamide beads, activation of amino groups by epichlorohydrin, and linking β-CDs to the epoxy groups. An adsorption investigation of bilirubin **100** by this resin reveals that **122** is a good adsorbent for unconjugated bilirubin with high capacity. The copolymerization

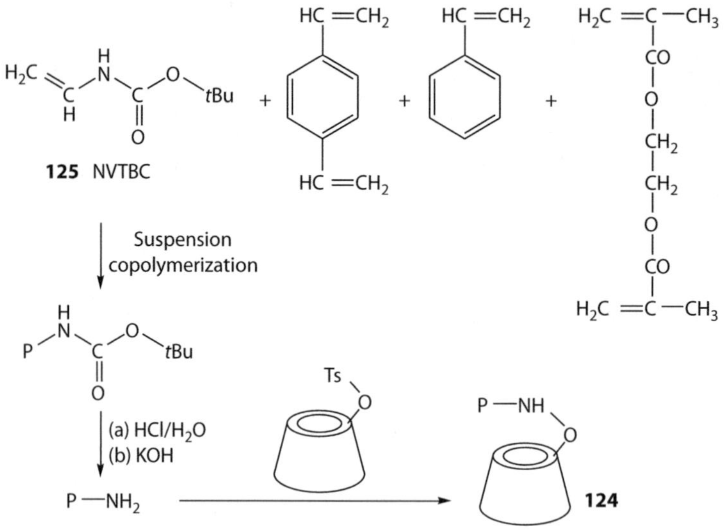

Figure 9.26 Synthesis of a macroporous copolymer **124** containing β-CD. (Modified from Crini, G. et al., *J. Appl. Polym. Sci.*, 69, 1419, 1998; Sreenivasan, K., *J. Appl. Polym. Sci.*, 69, 1419, 1998.)

of HMDI **88** and HEMA **116** has been studied by Sreenivasan with β-CD grafted, for the purpose of adsorbing cholesterol **98** and testosterone **123** [210,211]. He has developed a molecularly imprinted copolymer based on β-CD-coupled HEMA **116** as well, by selecting cholesterol **98** as the print molecule [212].

Crini et al. have synthesized macroporous polymer beads **124** containing *N*-vinyl-tertio-butyl carbamate **125** (NVTBC) in the presence of cross-linker divinylbenzene by suspension copolymerization [213]. The tosyl-β-CD is reacted with vinylamine group formed after solvolysis and then attaches to the polymer backbones (Figure 9.26). The high sorption capacities toward substituted benzene derivatives are found to these functionalized polymers, and they are a candidate for treatment of wastewater.

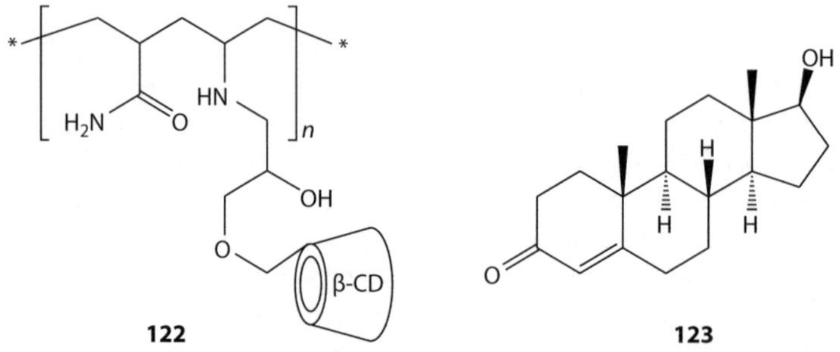

9.4.3.2 Cyclodextrin Grafted onto Chitosan (Chitosan-g-CD)

Chitosan (compound **126**), a polyaminosaccharide normally obtained by deacetylation of chitin, is an attractive biomaterial for its unique chemical and biological properties. The chitosan-*g*-CD possesses the cumulative effects of inclusion, size specificity, and transport properties of CD cavities, as well as the controlled release ability of the polymeric matrix [214]. The preparation and properties of these functionalized polymers have been summarized by Prabaharan and Mano [215]. The CDs have been grafted onto chitosan by using bi- or multifunctional reagents such as HMDI **88** [216–218], 1-ethyl-3-(3-dimethylaminopropyl) carbodiimide (EDC, **127**) [219,220], epichlorohydrin [221], etc. Sometimes the modified CDs including tosylated CDs [222,223], monochlorotriazinyl substituted CDs **128** [224,225], and CD citrate [226] are directly attached to amino groups of chitosan to form functionalized polymers.

In recent years, many chitosan-*g*-CD polymers have been developed for various usages, especially for the drug delivery systems or the carriers for controlled release. Alonso's group has derived a new type of nanoparticles made of chitosan and carboxymethyl-β-CD (CM-β-CD), finding them stable in simulated intestinal fluid pH 6.8 at 37°C for at least 4 h [227]. The association of insulin and heparin with them has been investigated, showing that insulin is very fast released while heparin is much slower. They have also prepared several fluorescein-modified nanoparticles composed of low-molecular-weight chitosan and β-CD derivatives [228]. Plasmid DNA encoding secreted alkaline phosphatase (pSEAP) has been associated to the nanoparticles later for gene delivery to the airway epithelium. The in vitro transfection study shows that a significantly higher gene expression can be observed for the nanoparticle formulations than naked pDNA. The system composed of chitosan and CM-β-CD is a promising vector among these nanocarriers. Other nanocarriers [229–231] have been developed as well by their group, and all these nanocarriers show possibilities for drug delivery.

Many other research groups focus on developing the carriers for pharmacy as well. A thiolated carboxymethyl chitosan (CMC) graft β-CD drug delivery carrier has been exploited by linking CM-β-CD onto CMC, using the condensing agents EDC **127** and *N*-hydroxysuccinimide (NHS) **129**, and further grafted with cysteine methyl ester hydrochloride **130** [232]. The in vitro experiment reveals that a fivefold increase in the adhesion time is found in this polymer than in chitosan, and the release of the entrapped ketoprofen **131** is slower than chitosan. Zhang et al. have investigated the controlled protein release nanocomplexes by attaching mono(6-(2-aminoethyl) amino-6-deoxy)-β-CD **132** to *N*-succinylated chitosan **133** with the aid of carbodiimide, deriving the chitosan-*g*-CDen **134** (Figure 9.27) [233]. They further assemble **134** with insulin to compose nanocomplexes with the size range of 190–328 nm. Insulin released from interpolymeric networks is controlled by a drug diffusion process in vitro, and it will be accelerated in acidic medium because of the electrostatic repulsion between both positive insulin and the polymer. As a carrier for drug loading and controlled release, chitosan-*g*-CD has been already touched on tissue engineering. Prabaharan and Jayakumar have promoted a type of chitosan-*g*-CD scaffolds (with glutaraldehyde treated and then freeze dried) used as synthetic extracellular matrices to fill the gap during the healing process [234]. It is found that the loading of drug ketoprofen **131** increases by increasing the chitosan-*g*-β-CD concentration or by decreasing the cross-linking densities. The ketoprofen

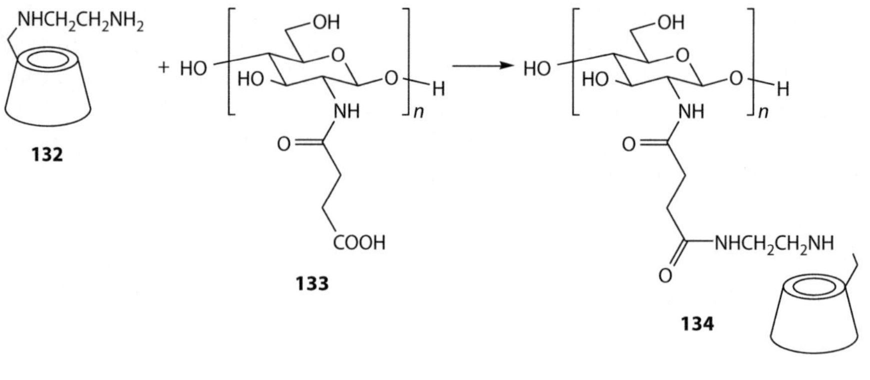

Figure 9.27 The synthesis of chitosan-*g*-CDen **134** by attaching **132** to *N*-succinylated chitosan **133**. (Modified from Zhang, X.-G. et al., *Carbohydr. Polym.*, 77, 394, 2009; Trapani, A. et al., *Eur. J. Pharm. Biopharm.*, 75, 26, 2010.)

entrapped in the chitosan-*g*-β-CD scaffolds will present a slower and steadier release into the medium than the unalloyed chitosan scaffolds after an initially abrupt release. The property of sustained drug delivery may enhance both the inductive and cell transplantation approaches to tissue engineering. Other similar works have been also reported by Ji et al., with methotrexate **135** and calcium folinate **136** as the model drugs [235].

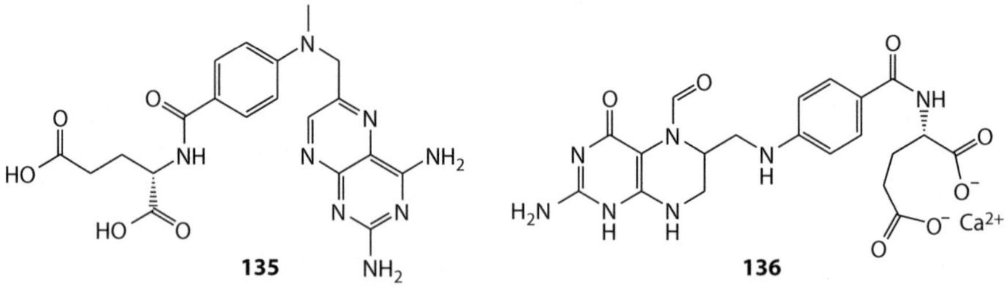

9.5 CONCLUSIONS

The seductive CD molecules have been appealing to thousands of researchers in production, physicochemical properties, and applications. CDs are derived from starch, but exhibit distinct properties from starch. As representative hosts, these macrocyclic compounds possess the magic hydrophobic cavities that show the ability to complex various guest compounds and also to construct captivating and exquisite supramolecular devices.

In this chapter, we have described the basic characteristics belonging to CDs and their derivatives, the main CD-based supramolecular structures, and the widely applied CDPs. Recently, progressive creations have been developed to expand the field of application, which have been partially included in this chapter. Many types of CD derivatives and CD-based materials have been available or even commercialized, and thus, further researches are facilitated. It is a pleasure that researchers have been still keeping high enthusiasm for the magic cyclic molecules, both in pure research and in applied technologies. In future, CDs will expectedly provide ways and means for producing more functionalized materials for industrial, analytical, environmental, and medical use, or serve as the models for self-assembled molecular machines and bioorganizations.

REFERENCES

1. Szejtli, J. 1998. Introduction and general overview of cyclodextrin chemistry. *Chem. Rev.* 98: 1743–1753.
2. Hedges, A. R. 1998. Industrial applications of cyclodextrins. *Chem. Rev.* 98: 2035–2044.
3. Fujiwara, T., Tanaka, N., and Kobayashi, S. 1990. Structure of δ-cyclodextrin 13.75H$_2$O. *Chem. Lett.* 19: 739–742.
4. Takaha, T., Yanase, M., Takata, H., Okada, S., and Smith, S. M. 1996. Potato D-enzyme catalyzes the cyclization of amylose to produce cycloamylose, a novel cyclic glucan. *J. Biol. Chem.* 271: 2902–2908.
5. Jacob, J., Gessler, K., Hoffmann, D., Sanbe, H., Koizumi, K., Smith, S. M., Takaha, T., and Saenger, W. 1998. Strain-induced "band flips" in cyclodecaamylose and higher homologues. *Angew. Chem. Int. Ed.* 37: 605–609.
6. Larsen, K. L. 2002. Large cyclodextrins. *J. Inclusion Phenom. Macrocyclic Chem.* 43: 1–13.
7. Taira, H., Nagase, H., Endo, T., and Ueda, H. 2006. Isolation, purification and characterization of large-ring cyclodextrins (CD36~CD39). *J. Inclusion Phenom. Macrocyclic Chem.* 56: 23–28.
8. Sundararajan, P. R. and Rao, V. S. R. 1970. Conformational studies on cycloamyloses. *Carbohydr. Res.* 13: 351–358.
9. Valle, M. D. E. M. 2004. Cyclodextrins and their uses: A review. *Process Biochem.* 39: 1033–1046.
10. Saenger, W., Jacob, J., Gessler, K., Steiner, T., Hoffmann, D., Sanbe, H., Koizumi, K., Smith, S. M., and Takaha, T. 1998. Structures of the common cyclodextrins and their larger analogues—Beyond the doughnut. *Chem. Rev.* 98: 1787–1802.
11. Harata, K. 1998. Structural aspects of stereodifferentiation in the solid state. *Chem. Rev.* 98: 1803–1827.
12. Lindner, K. and Saenger, W. 1982. Crystal and molecular structure of cyclohepta-amylose dodecahydrate. *Carbohydr. Res.* 99: 103–115.
13. Harata, K. 1987. The structure of the cyclodextrin complex. XX. Crystal structure of uncomplexed hydrated γ-cyclodextrin. *Bull. Chem. Soc. Jpn.* 60: 2763–2767.
14. Connors, K. A. 1997. The stability of cyclodextrin complexes in solution. *Chem. Rev.* 97: 1325–1357.
15. Jozwiakowskia, M. J. and Connors, K. A. 1985. Aqueous solubility behavior of three cyclodextrins. *Carbohydr. Res.* 143: 51–59.
16. Khan, A. R., Forgo, P., Stine, K. J., and D'Souza, V. T. 1998. Methods for selective modifications of cyclodextrins. *Chem. Rev.* 98: 1977–1996.
17. Li, S. and Purdy, W. C. 1992. Cyclodextrins and their applications in analytical chemistry. *Chem. Rev.* 92: 1457–1470.
18. Hybl, A., Rundle, R. E., and Williams, D. E. 1965. The crystal and molecular structure of the cyclohexaamylose-potassium acetate complex. *J. Am. Chem. Soc.* 87: 2779–2788.
19. Saenger, W., Noltemeyer, M., Manor, P. C., Hingerty, B., and Klar, B. 1976. "Induced-fit"-type complex formation of the model enzyme α-cyclodextrin. *Bioorg. Chem.* 5: 187–195.
20. Rong, D. and D'Souza, V. T. 1990. A convenient method for functionalization of the 2-position of cyclodextrins. *Tetrahedron Lett.* 31: 4275–4278.
21. Cramer, F., Mackensen, G., and Kensse, K. 1969. Über einschlußverbindungen, XX. ORD-spektren und konformation der glucose-einheiten in cyclodextrinen. *Chem. Ber.* 102: 494–508.
22. Takeo, K., Ueraura, K., and Mitoh, H. 1988. Derivatives of α-cyclodextrin and the synthesis of 6-*O*-α-D-glucopyranosyl-α-cyclodextrin. *J. Carbohydr. Chem.* 7: 293–308.
23. Boger, J., Brenner, D. G., and Knowles, J. R. 1979. Symmetrical triamino-per-*O*-methyl- α-cyclodextrin: Preparation and characterization of primary trisubstituted α-cyclodextrins. *J. Am. Chem. Soc.* 101: 7630–7631.
24. Takahashi, K., Hattori, K., and Toda, F. 1984. Monotosylated α- and β-cyclodextrins prepared in an alkaline aqueous solution. *Tetrahedron Lett.* 25: 3331–3334.
25. Fujita, K., Nagamura, S., and Imoto, T. 1984. Convenient preparation and effective separation of the C-2 and C-3 tosylates of α-cyclodextrin. *Tetrahedron Lett.* 25: 5673–5676.
26. Boger, J., Corcoran, R. J., and Lehn, J.-M. 1978. Cyclodextrin chemistry. Selective modification of all primary hydroxyl groups of α- and β-cyclodextrins. *Helv. Chim. Acta* 61: 2190–2218.
27. Szejtli, J. 1988. *Cyclodextrin Technology*. Dordrecht, the Netherlands: Kluwer.
28. Szejtli, J. 1977. Interaction of hydrochloric acid with cyclodextrin. *Stärke* 29: 410–413.

29. Marshall, J. J. and Miwa, I. 1981. Kinetic difference between hydrolyses of gamma-cyclodextrin by human salivary and pancreatic alpha-amylases. *Biochim. Biophys. Acta* 661: 142–147.

30. Rajewski, R. A. and Stella, V. J. 1996. Pharmaceutical applications of cyclodextrins. 2. In vivo drug delivery. *J. Pharm. Sci.* 85: 1142–1169.

31. Uekama, K., Hirayama, F., and Irie, T. 1998. Cyclodextrin drug carrier systems. *Chem. Rev.* 98: 2045–2076.

32. Andersen, G. H., Robbins, F. M., Domingues, F. J., Moores, R. G., and Long, C. L. 1963. The utilization of Schardinger dextrins by the rat. *Toxicol. Appl. Pharmacol.* 5: 257–266.

33. Szejtli, J. and Sebestyen, G. 1979. Resorption, metabolism and toxicity studies on the peroral application of β-cyclodextrin. *Stärke* 31: 385–389.

34. Irie, T. and Uekama, K. 1997. Pharmaceutical applications of cyclodextrins. III. Toxicological issues and safety evaluation. *J. Pharm. Sci.* 86: 147–162.

35. Frank, D. W., Gray, J. E., and Weaver, R. N. 1976. Cyclodextrin nephrosis in the rat. *Am. J. Pathol.* 83: 367–382.

36. Munro, I. C., Newberne, P. M., Young, R. R., and Bär, A. 2004. Safety assessment of γ-cyclodextrin. *Regul. Toxicol. Pharm.* 39: 3–13.

37. Loftsson, T. and Duchêne, D. 2007. Cyclodextrins and their pharmaceutical applications. *Int. J. Pharm.* 329: 1–11.

38. Leroy-Lechat, F., Wouessidjewe, D., Andreux, J.-P., Puisieux, F., and Duchêne, D. 1994. Evaluation of the cytotoxicity of cyclodextrins and hydroxypropylated derivatives. *Int. J. Pharm.* 101: 97–103.

39. Frijlink, H. W., Eissens, A. C., Hefting, N. R., Poelstra, K., Lerk, C. F., and Meijer, D. K. F. 1991. The effect of parenterally administered cyclodextrins on cholesterol levels in the rat. *Pharm. Res.* 8: 9–16.

40. Szejtli, J., Cserhati, T., and Szogyi, M. 1986. Interactions between cyclodextrins and cell membrane phospholipids. *Carbohydr. Polym.* 6: 35–49.

41. Gould, S. and Scott, R. C. 2005. 2-Hydroxypropyl-β-cyclodextrin (HP-β-CD): A toxicology review. *Food Chem. Toxicol.* 43: 1451–1459.

42. Yoshida, A., Arima, H., Uekama, K., and Pitha, J. 1988. Pharmaceutical evaluation of hydroxyalkyl ethers of β-cyclodextrins. *Int. J. Pharm.* 46: 217–222.

43. Rajewski, R. A., Traiger, G., Bresnahan, J., Jaberaboansari, P., and Stella, V. J. 1995. Preliminary safety evaluation of parenterally administered sulfoalkyl ether β-cyclodextrin derivatives. *J. Pharm. Sci.* 84: 927–932.

44. Lehn, J.-M. 1978. Cryptates: Inclusion complexes of macropolycyclic receptor molecules. *Pure Appl. Chem.* 50: 871–892.

45. Lehn, J.-M. 1988. Supramolecular chemistry—Scope and perspectives molecules, supermolecules, and molecular devices. *Angew. Chem. Int. Ed.* 27: 89–112.

46. Lehn, J.-M. 1993. Supramolecular chemistry. *Science* 260: 1762–1763.

47. Lehn, J.-M. 1994. Supramolecular chemistry. *J. Chem. Sci.* 106: 915–922.

48. Lehn, J.-M. 1995. *Supramolecular Chemistry—Concepts and Perspectives*. Weinheim, Germany: VCH.

49. Ariga, K. and Kunitake, T. 2006. *Supramolecular Chemistry—Fundamentals and Applications*. Berlin, Heidelberg, New York: Springer-Verlag.

50. Lehn, J.-M. 2007. From supramolecular chemistry towards constitutional dynamic chemistry and adaptive chemistry. *Chem. Soc. Rev.* 36: 151–160.

51. Brewster, M. E. and Loftsson, T. 2007. Cyclodextrins as pharmaceutical solubilizers. *Adv. Drug Deliv. Rev.* 59: 645–666.

52. Yuan, H.-N., Yao, S.-J., Shen, L.-Q., and Mao, J.-W. 2009. Preparation and characterization of inclusion complexes of β-cyclodextrin-BITC and β-cyclodextrin-PEITC. *Ind. Eng. Chem. Res.* 48: 5070–5078.

53. Manor, P. C. and Saenger, W. 1974. Topography of cyclodextrin inclusion complexes. III. Crystal and molecular structure of cyclohexaamylose hexahydrate, the $(H_2O)_2$ inclusion complex. *J. Am. Chem. Soc.* 96: 3630–3639.

54. Rekharsky, M. V. and Inoue, Y. 1998. Complexation thermodynamics of cyclodextrins. *Chem. Rev.* 98: 1875–1917.

55. Wang, Y.-L., Qiao, X.-N., Li, W.-C., Zhou, Y.-H., Jiao, Y., Yang, C., Dong, C., Inoue, Y., and Shuang, S.-M. 2009. Study on the complexation of isoquercitrin with β-cyclodextrin and its derivatives by spectroscopy. *Anal. Chim. Acta* 650: 124–130.

56. Tabushi, I., Kiyosuke, Y., Sugimoto, T., and Yamamura, K. 1978. Approach to the aspects of driving force of inclusion by α-cyclodextrin. *J. Am. Chem. Soc.* 100: 916–919.

57. Yan, C.-L., Li, X.-H., Xiu, Z.-L., and Hao, C. 2006. A quantum-mechanical study on the complexation of β-cyclodextrin with quercetin. *J. Mol. Struct. (Theochem)* 764: 95–100.

58. Yan, C.-L., Xiu, Z.-L., Li, X.-H., Teng, H., and Hao, C. 2007. Theoretical study for quercetin/β-cyclodextrin complexes: Quantum chemical calculations based on the PM3 and ONIOM2 method. *J. Inclusion Phenom. Macrocyclic Chem.* 58: 337–344.

59. Pralhad, T. and Rajendrakumar, K. 2004. Study of freeze-dried quercetin-cyclodextrin binary systems by DSC, FT-IR, X-ray diffraction and SEM analysis. *J. Pharm. Biomed. Anal.* 34: 333–339.

60. Schwingel, L., Fasolo, D., Holzschuh, M., Lula, I., Sinisterra, R., Koester, L., Teixeira, H., and Bassani, V. L. 2008. Association of 3-*O*-methylquercetin with β-cyclodextrin: complex preparation, characterization and *ex vivo* skin permeation studies. *J. Inclusion Phenom. Macrocyclic Chem.* 62: 149–159.

61. Ding, H.-Y., Chao, J.-B., Zhang, G.-M., Shuang, S.-M., and Pan, J.-H. 2003. Preparation and spectral investigation on inclusion complex of β-cyclodextrin with rutin. *Spectrochim. Acta Part A* 59: 3421–3429.

62. Shuang, S.-M., Pan, J.-H., Guo, S.-Y., Cai, M.-Y., and Liu, C.-S. 1997. Fluorescence study on the inclusion complexes of rutin with β-cyclodextrin, hydroxypropyl-β-cyclodextrin and γ-cyclodextrin. *Anal. Lett.* 30: 2261–2270.

63. Zheng, Y., Haworth, I. S., Zuo, Z., Chow, M. S. S., and Chow, A. H. L. 2005. Physicochemical and structural characterization of quercetin-β-cyclodextrin complexes. *J. Pharm. Sci.* 94: 1079–1089.

64. Jullian, C., Moyano, L., Yanez, C., and Olea-Azar, C. 2007. Complexation of quercetin with three kinds of cyclodextrins: An antioxidant study. *Spectrochim. Acta Part A* 67: 230–234.

65. Jullian, C., Orosteguis, T., Pérez-Cruz, F., Sánchez, P., Mendizabal, F., and Olea-Azar, C. 2008. Complexation of morin with three kinds of cyclodextrin: A thermodynamic and reactivity study. *Spectrochim. Acta Part A* 71: 269–275.

66. Cragg, P. J. 2010. *Supramolecular Chemistry—From Biological Inspiration to Biomedical Applications.* Dordrecht, the Netherlands; London, U.K.: Springer.

67. Breslow, R. 2005. *Artificial Enzymes.* Weinheim, Germany: WILEY-VCH.

68. Breslow, R. and Overman, L. E. 1970. An "artificial enzyme" combining a metal catalytic group and a hydrophobic binding cavity. *J. Am. Chem. Soc.* 92: 1075–1077.

69. D'Souza, V. T., Hanabusa, K., O'Leary, T., Gadwood, R. C., and Bender, M. L. 1985. Synthesis and evaluation of a miniature organic model of chymotrypsin. *Biochem. Biophys. Res. Commun.* 129: 727–732.

70. D'Souza, V. T. and Bender, M. L. 1987. Miniature organic models of enzymes. *Acc. Chem. Res.* 20: 146–152.

71. Breslow, R. 1982. Artificial enzymes. *Science* 218: 532–537.

72. Breslow, R. and Dong, S. D. 1998. Biomimetic reactions catalyzed by cyclodextrins and their derivatives. *Chem. Rev.* 98: 1997–2011.

73. Breslow, R. 1991. How do imidazole groups catalyze the cleavage of RNA in enzyme models and in enzymes? Evidence from "negative catalysis." *Acc. Chem. Res.* 24: 317–324.

74. Anslyn, E. and Breslow, R. 1989. Proton inventory of a bifunctional ribonuclease model. *J. Am. Chem. Soc.* 111: 8931–8932.

75. Breslow, R., Doherty, J. B., Guillot, G., and Lipsey, C. 1978. β-Cyclodextrinylbisimidazole, a model for ribonuclease. *J. Am. Chem. Soc.* 100: 3227–3229.

76. Breslow, R. and Schmuck, C. 1996. Goodness of fit in complexes between substrates and ribonuclease mimics: Effects on binding, catalytic rate constants, and regiochemistry. *J. Am. Chem. Soc.* 118: 6601–6605.

77. Dong, S. D. and Breslow, R. 1998. Bifunctional cyclodextrin metalloenzyme mimics. *Tetrahedron Lett.* 39: 9343–9346.

78. Breslow, R., Hammond, M., and Lauer, M. 1980. Selective transamination and optical induction by a β-cyclodextrin-pyridoxamine artificial enzyme. *J. Am. Chem. Soc.* 102: 421–422.

79. Breslow, R., Canary, J. W., Varney, M., Waddell, S. T., and Yang, D. 1990. Artificial transaminases linking pyridoxamine to binding cavities: Controlling the geometry. *J. Am. Chem. Soc.* 112: 5212–5219.

80. Breslow, R. and Czarnik, A. W. 1983. Transaminations by pyridoxamine selectively attached at C-3 in β-cyclodextrin. *J. Am. Chem. Soc.* 105: 1390–1391.

81. Tabushi, I., Kuroda, Y., Yamada, M., Higashimura, H., and Breslow, R. 1985. A-(modified B_6)-B-[ω-amino(ethylamino)]-β-cyclodextrin as an artificial B_6 enzyme for chiral aminotransfer reaction. *J. Am. Chem. Soc.* 107: 5545–5546.

82. Weiner, W., Winkler, J., Zimmerman, S. C., Czarnik, A. W., and Breslow, R. 1985. Mimics of tryptophan synthetase and of biochemical dehydroalanine formation. *J. Am. Chem. Soc.* 107: 4093–4094.

83. Breslow, R., Brown, A. B., McCullough, R. D., and White, P. W. 1989. Substrate selectivity in epoxidation by metalloporphyrin and metallosalen catalysts carrying binding groups. *J. Am. Chem. Soc.* 111: 4517–4518.

84. Breslow, R., Zhang, X.-J., Xu, R., Maletic, M., and Merger, R. 1996. Selective catalytic oxidation of substrates that bind to metalloporphyrin enzyme mimics carrying two or four cyclodextrin groups and related metallosalens. *J. Am. Chem. Soc.* 118: 11678–11679.

85. Breslow, R., Zhang, X.-J., and Huang, Y. 1997. Selective catalytic hydroxylation of a steroid by an artificial cytochrome P-450 enzyme. *J. Am. Chem. Soc.* 119: 4535–4536.

86. Breslow, R., Huang, Y., Zhang, X.-J., and Yang, J. 1997. An artificial cytochrome P450 that hydroxylates unactivated carbons with regio- and stereoselectivity and useful catalytic turnovers. *Proc. Natl. Acad. Sci. USA* 94: 11156–11158.

87. Breslow, R., Gabriele, B., and Yang, J. 1998. Geometrically directed selective steroid hydroxylation with high turnover by a fluorinated artificial cytochrome P-450. *Tetrahedron Lett.* 39: 2887–2890.

88. Yang, J. and Breslow, R. 2000. Selective hydroxylation of a steroid at C-9 by an artificial cytochrome P-450. *Angew. Chem. Int. Ed.* 39: 2692–2694.

89. Breslow, R., Yang, J., and Yan, J.-M. 2002. Biomimetic hydroxylation of saturated carbons with artificial cytochrome P-450 enzymes—Liberating chemistry from the tyranny of functional groups. *Tetrahedron* 58: 653–659.

90. Yang, J., Gabriele, B., Belvedere, S., Huang, Y., and Breslow, R. 2002. Catalytic oxidations of steroid substrates by artificial cytochrome P-450 enzymes. *J. Org. Chem.* 67: 5057–5067.

91. Breslow, R. 2001. Biomimetic selectivity. *Chem. Rec.* 1: 3–11.

92. Breslow, R., Yan, J.-M., and Belvedere, S. 2002. Catalytic hydroxylation of steroids by cytochrome P-450 mimics. Hydroxylation at C-9 with novel catalysts and steroid substrates. *Tetrahedron Lett.* 43: 363–365.

93. Rotruck, J. T., Pope, A. L., Ganther, H. E., Swanson, A. B., Hafeman, D. G., and Hoekstra, W. G. 1973. Selenium: Biochemical role as a component of glutathione peroxidase. *Science* 179: 588–590.

94. Epp, O., Ladenstein, R., and Wendel, A. 1983. The refined structure of the selenoenzyme glutathione peroxidase at 0.2-nm resolution. *Eur. J. Biochem.* 133: 51–69.

95. Ren, B., Huang, W.-H., Åkesson, B., and Ladenstein, R. 1997. The crystal structure of seleno-glutathione peroxidase from human plasma at 2.9 Å resolution. *J. Mol. Biol.* 268: 869–885.

96. Müller, A., Cadenas, E., Graf, P., and Sies, H. 1984. A novel biologically active seleno-organic compound—1: Glutathione peroxidase-like activity *in vitro* and antioxidant capacity of PZ 51 (Ebselen). *Biochem. Pharmacol.* 33: 3235–3239.

97. Wilson, S. R., Zucker, P. A., Huang, R. R. C., and Spector, A. 1989. Development of synthetic compounds with glutathione peroxidase activity. *J. Am. Chem. Soc.* 111: 5936–5939.

98. Cotgreave, I. A., Moldeus, P., Brattsand, R., Hallberg, A., Andersson, C. M., and Engman, L. 1992. α-(Phenylselenenyl)acetophenone derivatives with glutathione peroxidase-like activity: A comparison with ebselen. *Biochem. Pharmacol.* 43: 793–802.

99. Ren, X.-J., Liu, J.-Q., Luo, G.-M., Zhang, Y., Luo, Y.-M., Yan, G.-L., and Shen, J.-C. 2000. A novel selenocysteine-β-cyclodextrin conjugate that acts as a glutathione peroxidase mimic. *Bioconjug. Chem.* 11: 682–687.

100. Ren, X.-J., Xue, Y., Liu, J.-Q., Zhang, K., Zheng, J., Luo, G.-M., Guo, C.-H., Mu, Y., and Shen, J.-C. 2002. A novel cyclodextrin-derived tellurium compound with glutathione peroxidase activity. *ChemBioChem* 3: 356–363.

101. Dong, Z.-Y., Liu, J.-Q., Mao, S.-Z., Huang, X., Luo, G.-M., and Shen, J.-C. 2006. A glutathione peroxidase mimic 6,6′-ditellurobis (6-deoxy-β-cyclodextrin) with high substrate specificity. *J. Inclusion Phenom. Macrocyclic Chem.* 56: 179–182.

102. Hao, Y.-Q., Wu, Y.-Q., Liu, J.-Q., Luo, G.-M., and Yang, G.-D. 2006. ^{1}H NMR study on the inclusion complex of glutathione with a glutathione peroxidase mimic, 2,2′-ditelluro-bridged β-cyclodextrins. *J. Inclusion Phenom. Macrocyclic Chem.* 54: 171–175.

103. Dong, Z.-Y., Huang, X., Mao, S.-Z., Liang, K., Liu, J.-Q., Luo, G.-M., and Shen, J.-C. 2006. Cyclodextrin-derived mimic of glutathione peroxidase exhibiting enzymatic specificity and high catalytic efficiency. *Chem. Eur. J.* 12: 3575–3579.

104. Yu, S.-J., Huang, X., Miao, L., Zhu, J.-Y., Yin, Y.-Z., Luo, Q., Xu, J.-Y., Shen, J.-C., and Liu, J.-Q. 2010. A supramolecular bifunctional artificial enzyme with superoxide dismutase and glutathione peroxidase activities. *Bioorg. Chem.* 38: 159–164.

105. Dodziuk, H. 2006. *Cyclodextrins and Their Complexes—Chemistry, Analytical Methods, Applications.* Weinheim, Germany: WILEY-VCH.

106. Harada, A. 2001. Cyclodextrin-based molecular machines. *Acc. Chem. Res.* 34: 456–464.

107. Amabilino, D. B. and Stoddart, J. F. 1995. Interlocked and intertwined structures and superstructures. *Chem. Rev.* 95: 2725–2828.

108. Nepogodiev, S. A. and Stoddart, J. F. 1998. Cyclodextrin-based catenanes and rotaxanes. *Chem. Rev.* 98: 1959–1976.

109. Harada, A. and Kamachi, M. 1990. Complex formation between poly(ethylene glycol) and α-cyclodextrin. *Macromolecules* 23: 2821–2823.

110. Harada, A., Li, J., and Kamachi, M. 1993. Preparation and properties of inclusion complexes of poly(ethylene glycol) with α-cyclodextrin. *Macromolecules* 26: 5698–5703.

111. Ceccato, M., Lo Nostro, P., and Baglioni, P. 1997. α-Cyclodextrin/polyethylene glycol polyrotaxane: A study of the threading process. *Langmuir* 13: 2436–2439.

112. Lo Nostro, P., Lopes, J. R., and Cardelli, C. 2001. Formation of cyclodextrin-based polypseudorotaxanes: Solvent effect and kinetic study. *Langmuir* 17: 4610–4615.

113. Pozuelo, J., Mendicuti, F., and Mattice, W. L. 1997. Inclusion complexes of chain molecules with cyclo-amyloses. 2. Molecular dynamics simulations of polyrotaxanes formed by poly(ethylene glycol) and α-cyclodextrins. *Macromolecules* 30: 3685–3690.

114. Harada, A., Li, J., Kamachi, M., Kitagawa, Y., and Katsube, Y. 1998. Structures of polyrotaxane models. *Carbohydr. Res.* 305: 127–129.

115. Harada, A., Okada, M., Li, J., and Kamachi, M. 1995. Preparation and characterization of inclusion complexes of poly(propylene glycol) with cyclodextrins. *Macromolecules* 28: 8406–8411.

116. Harada, A. 1996. Preparation and structures of supramolecules between cyclodextrins and polymers. *Coord. Chem. Rev.* 148: 115–133.

117. Kamitori, S., Matsuzaka, O., Kondo, S., Muraoka, S., Okuyama, K., Noguchi, K., Okada, M., and Harada, A. 2000. A novel pseudo-polyrotaxane structure composed of cyclodextrins and a straight-chain polymer: Crystal structures of inclusion complexes of β-cyclodextrin with poly(trimethylene oxide) and poly(propylene glycol). *Macromolecules* 33: 1500–1502.

118. Harada, A., Suzuki, S., Okada, M., and Kamachi, M. 1996. Preparation and characterization of inclusion complexes of polyisobutylene with cyclodextrins. *Macromolecules* 29: 5611–5614.

119. Harada, A., Li, J., and Kamachi, M. 1994. Double-stranded inclusion complexes of cyclodextrin threaded on poly(ethylene glycol). *Nature* 370: 126–128.

120. Harada, A., Nishiyama, T., Kawaguchi, Y., Okada, M., and Kamachi, M. 1997. Preparation and characterization of inclusion complexes of aliphatic polyesters with cyclodextrins. *Macromolecules* 30: 7115–7118.

121. Steinbrunn, M. B. and Wenz, G. 1996. Synthesis of water-soluble inclusion compounds from polyamides and cyclodextrins by solid-state polycondensation. *Angew. Chem. Int. Ed.* 35: 2139–2141.

122. Liu, Y., Zhao, Y.-L., Zhang, H.-Y., and Song, H.-B. 2003. Polymeric rotaxane constructed from the inclusion complex of β-cyclodextrin and 4,4′-dipyridine by coordination with nickel(II) ions. *Angew. Chem. Int. Ed.* 42: 3260–3263.

123. Ogino, H. 1981. Relatively high-yield syntheses of rotaxanes. Syntheses and properties of compounds consisting of cyclodextrins threaded by α,ω-diaminoalkanes coordinated to cobalt(III) complexes. *J. Am. Chem. Soc.* 103: 1303–1304.

124. Ogino, H. and Ohta, K. 1984. Syntheses and properties of rotaxane complexes. 2. Rotaxane consisting of α- or β-cyclodextrin threaded by (μ-α,ω-diaminoalkane)bis[chlorobis-(ethylenediamine)-cobalt(III)] complexes. *Inorg. Chem.* 23: 3312–3316.

125. Yamanari, K. and Shimura, Y. 1983. Stereoselective formation of rotaxanes composed of poly-methylene-bridged dinuclear cobalt(III) complexes and α- or β-cyclodextrin. *Bull. Chem. Soc. Jpn.* 56: 2283–2289.

126. Yamanari, K. and Shimura, Y. 1984. Stereoselective formation of rotaxanes composed of polymethylene-bridged dinuclear cobalt(III) complexes and α-cyclodextrin. II. *Bull. Chem. Soc. Jpn.* 57: 1596–1603.

127. Wylie, R. S. and Macartney, D. H. 1992. Self-assembling metal rotaxane complexes of α-cyclodextrin. *J. Am. Chem. Soc.* 114: 3136–3138.

128. Rao, T. V. S. and Lawrence, D. S. 1990. Self-assembly of a threaded molecular loop. *J. Am. Chem. Soc.* 112: 3614–3615.

129. Harada, A., Li, J., and Kamachi, M. 1997. Non-ionic [2]rotaxanes containing methylated α-cyclodextrins. *Chem. Commun.* 18: 1413–1414.

130. Kawaguchi, Y. and Harada, A. 2000. An electric trap: A new method for entrapping cyclodextrin in a rotaxane structure. *J. Am. Chem. Soc.* 122: 3797–3798.

131. Murakami, H., Kawabuchi, A., Kotoo, K., Kunitake, M., and Nakashima, N. 1997. A light-driven molecular shuttle based on a rotaxane. *J. Am. Chem. Soc.* 119: 7605–7606.

132. Stanier, C. A., Alderman, S. J., Claridge, T. D. W., and Anderson, H. L. 2002. Unidirectional photoinduced shuttling in a rotaxane with a symmetric stilbene dumbbell. *Angew. Chem. Int. Ed.* 41: 1769–1772.

133. Kawaguchi, Y. and Harada, A. 2000. A cyclodextrin-based molecular shuttle containing energetically favored and disfavored portions in its dumbbell component. *Org. Lett.* 2: 1353–1356.

134. Harada, A., Li, J., and Kamachi, M. 1992. The molecular necklace: A rotaxane containing many threaded α-cyclodextrins. *Nature* 356: 325–327.

135. Harada, A., Li, J., Nakamitsu, T., and Kamachi, M. 1993. Preparation and characterization of polyrotaxanes containing many threaded α-cyclodextrins. *J. Org. Chem.* 58: 7524–7528.

136. Shigekawa, H., Miyake, K., Sumaoka, J., Harada, A., and Komiyama, M. 2000. The molecular abacus: STM manipulation of cyclodextrin necklace. *J. Am. Chem. Soc.* 122: 5411–5412.

137. Harada, A., Li, J., and Kamachi, M. 1994. Preparation and characterization of a polyrotaxane consisting of monodisperse poly(ethylene glycol) and α-cyclodextrins. *J. Am. Chem. Soc.* 116: 3192–3196.

138. Harada, A., Li, J., and Kamachi, M. 1993. Synthesis of a tubular polymer from threaded cyclodextrins. *Nature* 364: 516–518.

139. Harada, A., Takashima, Y., and Yamaguchi, H. 2009. Cyclodextrin-based supramolecular polymers. *Chem. Soc. Rev.* 38: 875–882.

140. Araki, J., Zhao, C.-M., and Ito, K. 2005. Efficient production of polyrotaxanes from α-cyclodextrin and poly(ethylene glycol). *Macromolecules* 38: 7524–7527.

141. Sakamoto, K., Takashima, Y., Yamaguchi, H., and Harada, A. 2007. Preparation and properties of rotaxanes formed by dimethyl-β-cyclodextrin and oligo(thiophene)s with β-cyclodextrin stoppers. *J. Org. Chem.* 72: 459–465.

142. Wu, J.-Y., He, H.-K., and Gao, C. β-Cyclodextrin-capped polyrotaxanes: One-pot facile synthesis via click chemistry and use as templates for platinum nanowires. *Macromolecules* 43: 2252–2260.

143. Ooya, T., Eguchi, M., and Yui, N. 2001. Enhanced accessibility of peptide substrate toward membrane-bound metalloexopeptidase by supramolecular structure of polyrotaxane. *Biomacromolecules* 2: 200–203.

144. Loethen, S., Ooya, T., Choi, H. S., Yui, N., and Thompson, D. H. 2006. Synthesis, characterization, and pH-triggered dethreading of α-cyclodextrin-poly(ethylene glycol) polyrotaxanes bearing cleavable endcaps. *Biomacromolecules* 7: 2501–2506.

145. Frömming, K.-H. and Szejtli, J. 1994. *Cyclodextrins in Pharmacy.* Dordrecht, the Netherlands: Kluwer.

146. Mocanu, G., Vizitiu, D., and Carpov, A. 2001. Cyclodextrin polymers. *J. Bioact. Compat. Pol.* 16: 315–342.

147. Solms, J. and Egli, R. H. 1965. Harze mit Einschlusshohlräumen von Cyclodextrin—Strucktur, *Helv. Chim. Acta* 48: 1225–1228.

148. Fenyvesi, E., Szilasi, M., Zsadon, B., Szejtli, J., and Tudos, F. 1981. Water-soluble cyclodextrin polymers and their complexing properties. In *Proceedings of the 1st International Symposium on Cyclodextrins*, Budapest, Hungary, 1981, pp. 345–356.

149. Crini, G. and Morcellet, M. 2002. Synthesis and applications of adsorbents containing cyclodextrins. *J. Sep. Sci.* 25: 789–813.

150. Kobayashi, N., Shirai, H., and Hojo, N. 1989. Virtues of a poly-cyclodextrin for the de-aggregation of organic molecules in water. *J. Polym. Sci., Part C: Polym. Lett.* 27: 191–195.

151. Wiedenhof, N., Lammers, J. N. J. J., and van Panthaleon van Eck, C. L. 1969. Properties of cyclodextrins. III. Cyclodextrin-epichlorohydrin resins. Preparation and analysis. *Stärke* 21: 119–123.

152. Wiedenhof, N. 1969. Properties of cyclodextrins. IV. Features and use of insoluble cyclodextrin-epichlorohydrin resins. *Stärke* 21: 163–166.
153. Wiedenhof, N. and Trieling, R. G. 1971. Properties of cyclodextrins. V. Inclusion isotherm and kinetics of inclusion of benzoic acid and *m*-chlorobenzoic acid on β-E25 cyclodextrin-epichlorohydrin resin. *Stärke* 23: 129–132.
154. Nussstein, P., Staudinger, G., and Kreuzer, F.-H. 1994. Cyclodextrin polymers and process for their preparation. US patent 5360899.
155. Szejtli, J., Fenyvesi, E., Zoltan, S., Zsadon, B., and Tudos, F. 1980. Cyclodextrin-poly(vinyl alcohol) polymeric compositions in film, fiber, bead, and block polymers. DE patent 2 927 733.
156. Thuaud, N., Sebille, B., Deratani, A., and Lelievre, G. 1991. Retention behavior and chiral recognition of β-cyclodextrin-derivative polymer adsorbed on silica for warfarin, structurally related compounds and Dns-amino acids. *J. Chromatogr.* 555: 53–64.
157. Zhao, J., Lin, D.-Q., Wang, Y.-C., and Yao, S.-J. 2010. A novel β-cyclodextrin polymer/tungsten carbide composite matrix for expanded bed adsorption: Preparation and characterization of physical properties. *Carbohydr. Polym.* 80: 1085–1090.
158. Smolková-Keulemansová, E. 1982. Cyclodextrins as stationary phases in chromatography. *J. Chromatogr.* 251: 17–34.
159. Zsadon, B., Szilasi, M., Tudos, F., Fenyvesi, E., and Szejtli, J. 1979. Inclusion chromatography of amino acids on beta-cyclodextrin-polymer gel beds. *Stärke* 31: 11–12.
160. Hoffman, J. L. 1973. Chromatography of nucleic acids on cross-linked cyclodextrin gels having inclusion-forming capacity. *J. Macromol. Sci. Chem.* A7: 1147–1157.
161. Moon, J.-Y., Jung, H.-J., Moon, M. H., Chung, B. C., and Choi, M. H. 2008. Inclusion complex-based solid-phase extraction of steroidal compounds with entrapped β-cyclodextrin polymer. *Steroids* 73: 1090–1097.
162. Harada, A., Furue, M., and Nozakura, S.-I. 1978. Optical resolution of mandelic acid derivatives by column chromatography on crosslinked cyclodextrin gels. *J. Polym. Sci. Polym. Chem. Ed.* 16: 189–196.
163. Zsadon, B., Decsei, L., Szilasi, M., Tudos, F., and Szejtli, J. 1983. Inclusion chromatography of enantiomers of indole alkaloids on a cyclodextrin polymer stationary phase. *J. Chromatogr.* 270: 127–134.
164. Zsadon, B., Szilasi, M., Decsei, L., Ujhazy, A., and Szejtli, J. 1986. Variation of the selectivity in the resolution of alkaloid enantiomers on cross-linked cyclodextrin polymer stationary phases. *J. Chromatogr.* 356: 428–432.
165. He, X.-L., Tan, T.-W., Xu, B.-Z., and Janson, J.-C. 2004. Separation and purification of puerarin using β-cyclodextrin-coupled agarose gel media. *J. Chromatogr. A* 1022: 77–82.
166. Yang, L., Zhu, Y., Yan, T.-W., and Janson, J.-C. 2007. Coupling oligo-β-cyclodextrin on polyacrylate beads media for separation of puerarin. *Process Biochem.* 42: 1075–1083.
167. Yang, L. and Tan, T.-W. 2008. Enhancement of the isolation selectivity of isoflavonoid puerarin using oligo-β-cyclodextrin coupled polystyrene-based media. *Biochem. Eng. J.* 40: 189–198.
168. Xu, J., Zhang, G.-F., Tan, T.-W., and Janson, J.-C. 2005. One-step purification of epigallocatechin gallate from crude green tea extracts by isocratic hydrogen bond adsorption chromatography on β-cyclodextrin substituted agarose gel media. *J. Chromatogr. B* 824: 323–326.
169. Yang, L., Tan, T.-W., and Zhang, L.-Q. 2009. Process utilized oligo-β-cyclodextrin substituted agarose gel medium for efficient purification of paclitaxel from *Taxus cuspidata*. *Sep. Purif. Technol.* 68: 119–124.
170. Xu, J., Sandström, C., Janson, J.-C., and Tan, T. W. 2006. Chromatographic retention of epigallocatechin gallate on oligo-β-cyclodextrin coupled Sepharose media investigated using NMR. *Chromatographia* 64:7–11.
171. Yang, L., Zhang, H.-Y., Tan, T.-W., and Rahman, A. U. 2009. Thermodynamic and NMR investigations on the adsorption mechanism of puerarin with oligo-β-cyclodextrin-coupled polystyrene-based matrix. *J. Chem. Technol. Biotechnol.* 84: 611–617.
172. Mannen, T., Yamaguchi, S., Honda, J., Sugimoto, S., and Nagamune, T. 2001. Expanded-bed protein refolding using a solid-phase artificial chaperone. *J. Biosci. Bioeng.* 91: 403–408.
173. Zhao, J., Lin, D.-Q., and Yao, S.-J. 2009. Expansion and hydrodynamic properties of β-cyclodextrin polymer/tungsten carbide composite matrix in an expanded bed. *J. Chromatogr. A* 1216: 7840–7845.
174. Mizobuchi, Y., Tanaka, M., and Shono, T. 1980. Preparation and sorption behaviour of cyclodextrin polyurethane resins. *J. Chromatogr.* 194: 153–161.

175. Mizobuchi, Y., Tanaka, M., and Shono, T. 1981. Separation of aromatic amino acids on β-cyclodextrin polyurethane resins. *J. Chromatogr.* 208: 35–40.
176. Tang, S.-W., Kong, L., Ou, J.-J., Liu, Y.-Q., Li, X., and Zou, H.-F. 2006. Application of cross-linked β-cyclodextrin polymer for adsorption of aromatic amino acids. *J. Mol. Recognit.* 19: 39–48.
177. Carbonnier, B., Janus, L., Lekchiri, Y., and Morcellet, M. 2004. Coating of porous silica beads by in situ polymerization/crosslinking of 2-hydroxypropyl β-cyclodextrin for reversed-phase high performance liquid chromatography applications. *J. Appl. Polym. Sci.* 91: 1419–1426.
178. Kim, I. W., Choi, H. M., Yoon, H. J., and Park, J. H. 2006. β-Cyclodextrin-hexamethylene diisocyanate copolymer-coated zirconia for separation of racemic 2,4-dinitrophenyl amino acids in reversed-phase liquid chromatography. *Anal. Chim. Acta* 569: 151–156.
179. Bhaskar, M., Aruna, P., Jeevan, R. J. G., and Radhakrishnan, G. 2004. β-Cyclodextrin-polyurethane polymer as solid phase extraction material for the analysis of carcinogenic aromatic amines. *Anal. Chim. Acta* 509: 39–45.
180. Yamasaki, H., Makihata, Y., and Fukunaga, K. 2008. Preparation of crosslinked β-cyclodextrin polymer beads and their application as a sorbent for removal of phenol from wastewater. *J. Chem. Technol. Biotechnol.* 83: 991–997.
181. Yamasaki, H., Makihata, Y., and Fukunaga, K. 2006. Efficient phenol removal of wastewater from phenolic resin plants using crosslinked cyclodextrin particles. *J. Chem. Technol. Biotechnol.* 81: 1271–1276.
182. Ozmen, E. Y., Sezgin, M., Yilmaz, A., and Yilmaz, M. 2008. Synthesis of β-cyclodextrin and starch based polymers for sorption of azo dyes from aqueous solutions. *Bioresour. Technol.* 99: 526–531.
183. Yilmaz, E., Memon, S., and Yilmaz, M. 2010. Removal of direct azo dyes and aromatic amines from aqueous solutions using two β-cyclodextrin-based polymers. *J. Hazard. Mater.* 174: 592–597.
184. Asanuma, H., Hishiya, T., and Komiyama, M. 2000. Tailor-made receptors by molecular imprinting. *Adv. Mater.* 12: 1019–1030.
185. Asanuma, H., Kakazu, M., Shibata, M., and Hishiya, T. 1997. Molecularly imprinted polymer of β-cyclodextrin for the efficient recognition of cholesterol. *Chem. Commun.* 1971–1972.
186. Asanuma, H., Kakazu, M., Shibata, M., Hishiya, T., and Komiyama, M. 1998. Synthesis of molecularly imprinted polymer of β-cyclodextrin for the efficient recognition of cholesterol. *Supramol. Sci.* 5: 417–421.
187. Hishiya, T., Shibata, M., Kakazu, M., Asanuma, H., and Komiyama, M. 1999. Molecularly imprinted cyclodextrins as selective receptors for steroids. *Macromolecules* 32: 2265–2269.
188. Hishiya, T., Asanuma, H., and Komiyama, M. 2002. Spectroscopic anatomy of molecular-imprinting of cyclodextrin. Evidence for preferential formation of ordered cyclodextrin assemblies. *J. Am. Chem. Soc.* 124: 570–575.
189. Osawa, T., Shirasaka, K., Matsui, T., Yoshihara, S., Akiyama, T., Hishiya, T., Asanuma, H., and Komiyama, M. 2006. Importance of the position of vinyl group on β-cyclodextrin for the effective imprinting of amino acid derivatives and oligopeptides in water. *Macromolecules* 39: 2460–2466.
190. Song, S.-H., Shirasaka, K., Katayama, M., Nagaoka, S., Yoshihara, S., Osawa, T., Sumaoka, J., Asanuma, H., and Komiyama, M. 2007. Recognition of solution structures of peptides by molecularly imprinted cyclodextrin polymers. *Macromolecules* 40: 3530–3532.
191. Yang, Y., Long, Y.-Y., Cao, Q., Li, K.-A., and Liu, F. 2008. Molecularly imprinted polymer using β-cyclodextrin as functional monomer for the efficient recognition of bilirubin. *Anal. Chim. Acta* 606: 92–97.
192. Zhang, W., Qin, L., He, X.-W., Li, W.-Y., and Zhang, Y.-K. 2009. Novel surface modified molecularly imprinted polymer using acryloyl-β-cyclodextrin and acrylamide as monomers for selective recognition of lysozyme in aqueous solution. *J. Chromatogr. A* 1216: 4560–4567.
193. Muk Ng, S. and Narayanaswamy, R. 2009. Molecularly imprinted β-cyclodextrin polymer as potential optical receptor for the detection of organic compound. *Sens. Actuators B* 139: 156–165.
194. Roche, P. J. R., Muk Ng, S., Narayanaswamy, R., Goddard, N., and Page, K. M. 2009. Multiple surface plasmon resonance quantification of dextromethorphan using a molecularly imprinted β-cyclodextrin polymer: A potential probe for drug-drug interactions. *Sens. Actuators B* 139: 22–29.
195. Pariot, N., Edwards-Lévy, F., Andry, M.-C., and Lévy, M.-C. 2000. Cross-linked β-cyclodextrin microcapsules: Preparation and properties. *Int. J. Pharm.* 211: 19–27.

196. Constantin, M., Fundueanu, G., Bortolotti, F., Cortesi, R., Ascenzi, P., and Menegatti, E. 2004. Preparation and characterisation of poly(vinyl alcohol)/cyclodextrin microspheres as matrix for inclusion and separation of drugs. *Int. J. Pharm.* 285: 87–96.

197. Girek, T., Kozlowski, C. A., Koziol, J. J., Walkowiak, W., and Korus, I. 2005. Polymerisation of β-cyclodextrin with succinic anhydride. Synthesis, characterisation, and ion flotation of transition metals. *Carbohydr. Polym.* 59: 211–215.

198. Zhao, D., Zhao, L., Zhu, C.-S., Huang, W.-Q., and Hu, J.-L. 2009. Water-insoluble β-cyclodextrin polymer crosslinked by citric acid: Synthesis and adsorption properties toward phenol and methylene blue. *J. Inclusion Phenom. Macrocyclic Chem.* 63: 195–201.

199. Zhao, D., Zhao, L., Zhu, C.-S., Tian, Z.-B., and Shen, X.-Y. 2009. Synthesis and properties of water-insoluble β-cyclodextrin polymer crosslinked by citric acid with PEG-400 as modifier. *Carbohydr. Polym.* 78: 125–130.

200. Harada, A., Furue, M., and Nozakura, S.-I. 1976. Cyclodextrin-containing polymers. 1. Preparation of polymers. *Macromolecules* 9: 701–704.

201. Harada, A., Furue, M., and Nozakura, S.-I. 1976. Cyclodextrin-containing polymers. 2. Cooperative effects in catalysis and binding. *Macromolecules* 9: 705–710.

202. Janus, L., Crini, G., El-Rezzi, V., Morcellet, M., Cambiaghi, A., Torri, G., Naggi, A., and Vecchi, C. 1999. New sorbents containing beta-cyclodextrin. Synthesis, characterization, and sorption properties. *React. Funct. Polym.* 42: 173–180.

203. Nielsen, A. L., Steffensen, K., and Larsen, K. L. 2009. Self-assembling microparticles with controllable disruption properties based on cyclodextrin interactions. *Colloids Surf. B* 73: 267–275.

204. Vretblad, P. 1974. Immobilization of ligands for biospecific affinity chromatography *via* their hydroxyl groups. The cyclohexaamylose-β-amylase system. *FEBS Lett.* 47: 86–89.

205. Subbaramaiah, K. and Sharma, R. 1984. Polarimetric determination of substitution of oxirane groups on Epoxy-activated Sepharose 6B gel by cyclodextrins. *J. Biochem. Biophys. Methods* 10: 65–68.

206. Tanaka, M., Mizobuchi, Y., Sonoda, T., and Shono, T. 1981. Retention behaviour of benzene derivatives on chemically bonded β-cyclodextrin phases. *Analy. Lett.* 14: 281–290.

207. Tanaka, M., Kawaguchi, Y., Nakae, M., Mizobuchi, Y., and Shono, T. 1982. Separation of disubstituted benzene isomers on chemically bonded cyclodextrin stationary phases. *J. Chromatogr.* 246: 207–214.

208. Hattori, K., Takahashi, K., Mikami, M., and Watanabe, H. 1986. Novel high-performance liquid chromatographic adsorbents prepared by immobilization of modified cyclodextrins. *J. Chromatogr.* 355: 383–391.

209. Wang, H.-J., Ma, J.-B., Zhang, Y.-H., and He, B.-L. 1997. Adsorption of bilirubin on the polymeric β-cyclodextrin supported by partially aminated polyacrylamide gel. *React. Funct. Polym.* 32: 1–7.

210. Sreenivasan, K. 1996. Grafting of β-cyclodextrin-modified 2-hydroxyethyl methacrylate onto polyurethane. *J. Appl. Polym. Sci.* 60: 2245–2249.

211. Sreenivasan, K. 1998. Solvent effect on the interaction of steroids with a novel methyl β-cyclodextrin polymer. *J. Appl. Polym. Sci.* 69: 1419–1427.

212. Sreenivasan, K. 1998. Synthesis and evaluation of a beta cyclodextrin-based molecularly imprinted copolymer. *J. Appl. Polym. Sci.* 70: 15–18.

213. Crini, G., Janus, L., Morcellet, M., Torri, G., Naggi, A., Bertini, S., and Vecchi, C. 1998. Macroporous polyamines containing cyclodextrin: Synthesis, characterization, and sorption properties. *J. Appl. Polym. Sci.* 69: 1419–1427.

214. Auzely, R. and Rinaudo, M. 2001. Chitosan derivatives bearing pendant cyclodextrin cavities: Synthesis and inclusion performance. *Macromolecules* 34: 3574–3580.

215. Prabaharan, M. and Mano, J. F. 2006. Chitosan derivatives bearing cyclodextrin cavities as novel adsorbent matrices. *Carbohydr. Polym.* 63: 153–166.

216. Sreenivasan, K. 1998. Synthesis and preliminary studies on a β-cyclodextrin-coupled chitosan as a novel adsorbent matrix. *J. Appl. Polym. Sci.* 69: 1051–1055.

217. Chiu, S. H., Chung, T. W., Giridhar, R., and Wu, W. T. 2004. Immobilization of β-cyclodextrin in chitosan beads for separation of cholesterol from egg yolk. *Food Res. Int.* 37: 217–223.

218. Zha, F., Li, S.-G., and Chang, Y. 2008. Preparation and adsorption property of chitosan beads bearing β-cyclodextrin cross-linked by 1,6-hexamethylene diisocyanate. *Carbohydr. Polym.* 72: 456–461.

219. Furusaki, E., Ueno, Y., Sakairi, N., Nishi, N., and Tokura, S. 1996. Facile preparation and inclusion ability of a chitosan derivative bearing carboxymethyl-β-cyclodextrin. *Carbohydr. Polym.* 29: 29–34.

220. Aoki, N., Nishikawa, M., and Hattori, K. 2003. Synthesis of chitosan derivatives bearing cyclodextrin and adsorption of *p*-nonylphenol and bisphenol A. *Carbohydr. Polym.* 52: 219–223.
221. Zhang, X., Wang, Y., and Yi, Y. 2004. Synthesis and characterization of grafting α-cyclodextrin with chitosan. *J. Appl. Polym. Sci.* 94: 860–864.
222. Chen, S. and Wang, Y. 2001. Study on β-cyclodextrin grafting with chitosan and slow release of its inclusion complex with radioactive iodine. *J. Appl. Polym. Sci.* 82: 2414–2421.
223. Wang, H., Fang, Y., Ding, L., Gao, L., and Hu, D. 2003. Preparation and nitromethane sensing properties of chitosan thin films containing pyrene and β-cyclodextrin units. *Thin Solid Films* 440: 255–260.
224. Martel, B., Devassine, M., Crini, G., Weltrowski, M., Bourdonneau, M., and Morcellet, M. 2001. Preparation and sorption properties of a β-cyclodextrin-linked chitosan derivative. *J. Polym. Sci. Part A: Polym. Chem.* 39: 169–176.
225. Abdel-Halim, E. S., Abdel-Mohdy, F. A., Al-Deyab, S. S., and El-Newehy, M. H. 2010. Chitosan and monochlorotriazinyl-β-cyclodextrin finishes improve antistatic properties of cotton/polyester blend and polyester fabrics. *Carbohydr. Polym.* 82: 202–208.
226. El-Tahlawy, K., Gaffar, M. A., and El-Rafie, S. 2006. Novel method for preparation of β-cyclodextrin/ grafted chitosan and its application. *Carbohydr. Polym.* 63: 385–392.
227. Krauland, A. H. and Alonso, M. J. 2007. Chitosan/cyclodextrin nanoparticles as macromolecular drug delivery system. *Int. J. Pharm.* 340: 134–142.
228. Teijeiro-Osorio, D., Remuñán-López, C., and Alonso, M. J. 2009. Chitosan/cyclodextrin nanoparticles can efficiently transfect the airway epithelium *in vitro*. *Eur. J. Pharm. Biopharm.* 71: 257–263.
229. Maestrelli, F., Garcia-Fuentes, M., Mura, P., and Alonso, M. J. 2006. A new drug nanocarrier consisting of chitosan and hydroxypropylcyclodextrin. *Eur. J. Pharm. Biopharm.* 63: 79–86.
230. Teijeiro-Osorio, D., Remunan-Lopez, C., and Alonso, M. J. 2009. New generation of hybrid poly/ oligosaccharide nanoparticles as carriers for the nasal delivery of macromolecules. *Biomacromolecules* 10: 243–249.
231. Trapani, A., Lopedota, A., Franco, M., Cioffi, N., Ieva, E., Garcia-Fuentes, M., and Alonso, M. J. 2010. A comparative study of chitosan and chitosan/cyclodextrin nanoparticles as potential carriers for the oral delivery of small peptides. *Eur. J. Pharm. Biopharm.* 75: 26–32.
232. Prabaharan, M. and Gong, S.-Q. 2008. Novel thiolated carboxymethyl chitosan-*g*-β-cyclodextrin as mucoadhesive hydrophobic drug delivery carriers. *Carbohydr. Polym.* 73: 117–125.
233. Zhang, X.-G., Wu, Z.-M., Gao, X.-J., Shu, S.-J., Zhang, H.-J., Wang, Z., and Li, C.-X. 2009. Chitosan bearing pendant cyclodextrin as a carrier for controlled protein release. *Carbohydr. Polym.* 77: 394–401.
234. Prabaharan, M. and Jayakumar, R. 2009. Chitosan-graft-β-cyclodextrin scaffolds with controlled drug release capability for tissue engineering applications. *Int. J. Biol. Macromol.* 44: 320–325.
235. Ji, J.-G., Hao, S.-L., Liu, W.-Q., Wu, D.-J., Wang, T.-F., and Xu, Y. 2011. Preparation, characterization of hydrophilic and hydrophobic drug in combine loaded chitosan/cyclodextrin nanoparticles and in vitro release study. *Colloids Surf. B* 83: 103–107.

CHAPTER **10**

Introduction of Starch into Plastic Films
Advent of Starch in Plastics

R.E. Harry-O'kuru and Sherald H. Gordon

CONTENTS

10.1 INTRODUCTION

For years, plastics manufacturers labored to imbue plastics with qualities that make them resistant to the environment in which they are used. Naturally, such inertness and durability are highly desirable in durable goods. Unfortunately, this longevity has spilled over into materials meant only for transient utility. The ubiquitous results of the spillover are litter and pollution in lakes and marine environments as well as a disproportionate occupancy of landfill space. Exacerbating the pollution problem is the slow or imperceptible rate of degradation of most plastics wherever they are found. There is no question as to the necessity to replace wood and some metals with appropriate plastics in order to free the former for specialized use only. So it would be essential to develop new breeds of plastics that would be environmentally friendly and yet meet their utilitarian purpose—in other words, a generation of plastics with a useful life span after which they should degrade at a measurable rate when discarded in a favorable environment.

Actually, the first idea of incorporating a biodegradable substance into plastics did not arise from any concern regarding the environment. Rather, Griffin's[1] idea was to improve writability on plastic bags by creating a matted surface through incorporation of starch granules during formulation of polyethylene blown films. Even the Northern Regional Research Center's (NRRC) Otey's[2] first priority seemed to have been more concerned with replacing foreign oil components used in plastics with domestically produced cornstarch. This was a legitimate response to the need at the time period. Today, however, realization of the urgency for solutions to the problems posed by current plastics technology and its unforeseen effects on the environment demands more insightful approach. The solutions to these problems will require greater appreciation of the scope of the issues involved and bold new thought processes that would lead to novel formulation techniques that incorporate more naturally biodegradable materials into low-density polyethylene (LDPE) or other low-cost synthetics. In fact, it may even be feasible to modify starch, so it becomes the sole constituent of the plastic. At the NRRC (now NCAUR), much work and experience had been gained in formulation and manufacture of ternary plastic blown films consisting 10%–50% starch with LDPE and EAA (ethylene-co-acrylic acid).[3,4]

Although acceptable physical properties (percent elongation, tensile strength) as well as biodegradation information have been obtained for these plastics,[5,6] the technology was not adopted by industry because the cost was not competitive. For starch to play a larger role in plastics, it was imperative to have a clearer understanding of the interaction between the components constituting the plastic film including bonding types and morphology. In this regard, a study of bonding characteristics of the NRRC films was undertaken via Fourier transform infrared spectroscopy (FT-IR), solid-state ^{13}C-NMR spectrometry, and differential scanning calorimetry (DSC). A follow-up approach was to chemically modify the starch component of the film in order to glean information of its morphology spectroscopically (FT-IR, NMR) and through scanning light microscopy. From the results of this probe, one would gain some insight into the possible biodegradative pathway of the ternary film components. A final objective of the study was to investigate how chemically modified starch could be used to replace EAA which was the most expensive component in the formulation of the ternary starch plastic film with LDPE.

Since the formulation process of the plastic films involved alkali and heat treatment of starch, EAA, urea, LDPE, and water, it was conceivable that new covalent bonding as well as van der Waal's interaction may result as the system was heated to lower water content and consequent loss of ammonia. To verify this possibility, FT-IR spectra of the blown film were analyzed for carboxylate ions, ester groupings, and carboxylic acid stretching frequencies. The spectral data obtained showed no presence of ester vibrational modes (1730–1750 cm^{-1}) region. But observed were vibrational modes corresponding to hydrogen-bonded carboxylic acids (1704 cm^{-1}) and carboxylate ions (1550–1480 cm^{-1}). The carboxylate absorption band disappeared when the plastic was treated with dilute mineral acid, whereas the 1704 cm^{-1} band was enhanced. From the FT-IR spectral data, it was concluded that no new covalent bonds were formed during processing.

For structure–function relationship in the plastic blown film, it was necessary to obtain more information from additional spectroscopic techniques that could correlate and or expand on the data already obtained. The FT-IR technique showed the presence of inter- and intramolecular H bonding between starch –OH's and carboxylic acid groupings of EAA as indicated by the broad O–H bands observed at 3600–3100 cm^{-1} and the 1704 cm^{-1} of the carboxylic acid as compared to 1720–1710 cm^{-1} for free CO$_2$H group. The solid-state ^{13}C-NMR spectra showed higher field resonances for the aliphatic moieties (14.00 ppm [–CH$_3$], 30.00 ppm [–CH$_2$-]) and the expected carbohydrate frequencies for starch at (62.57 ppm [C-6 α-D-glucopyranosyl], 72.55 ppm [C-2, C-3, C-5 α-D-glucopyranosyl], 84.00 ppm [C-4 α-D-glucopyranosyl], and 96–101.98 ppm [C-1 α-D-glucopyranosyl]) 183.00 ppm for the –CO$_2$H of EAA.

Thermal analysis: DSC is a technique used to evaluate the purity and crystallinity of polymeric materials.[7] The expectation in applying this technique was to gain information regarding homogeneity of component distribution in the plastic film. An intimate mixture would lead to a lowering of melting point of the blend, whereas a mixing dominated by segregated domains would have little or no effect on the melting temperature. A partially homogeneous mixture, on the other hand, would give a more complex pattern of melting transitions. The observed thermograms for the ternary plastic films were deceptively simple as though the incompatibles had become miscible forming a single compound. A closer inspection and analysis of the thermograms, however, showed that the only melting transition (T_m) was at a higher temperature (113.4°C) compared to that of the starting LDPE (111.5°C). This seeming improvement in melting point of LDPE in the blown film could be interpreted in one of two ways. Firstly, it is a known phenomenon that when two pure distinct materials are mixed in equal proportions, the melting point of the mixture is generally depressed as each component is acting as an impurity to the other. There is, however, an occasional mixture that gives an elevated melting point depending on component ratio. The second interpretation to this observation is based on the incompatibility of the components that have been forced to coexist in the plastic film. It is possible that the components have been partitioned into more ordered domains, thus allowing for greater degree of crystallinity of the LDPE fraction and hence the observed higher melting point (113.4°C). To investigate the latter hypothesis, a combination of techniques was used. Strips of film 1.0 cm wide were mounted in paraffin wax and sectioned along the length, that is, cross sections by microtome techniques. The paraffin was then removed with xylene and the micro-sections mounted on microscope slides. The micro-sections were then treated with aqueous periodic acid followed by basic Fuschin reaction to generate aldehyde groups in the carbohydrate moieties; the generated aldehydes groups then reacted with Fuschin reagent to give a blue-violet stain. The samples were then rinsed with water and alcohol and dried and sealed with glass cover plates. Examination of the specimens under scanning light microscope equipped with a 35 mm SLR camera at high power (10×100 in oil immersion) and high contrast showed a layered structure from edge to edge of blue-violet laminas separated by red "furrows." The blue-violet color being the starch-rosaniline complex, and the red regions were the unstained PE/EAA bands. These bands terminate at the edges of the film. It was therefore concluded that the alternating bands of starch and LDPE–EAA were a result of partition and cohesive forces at work in the plastic film, forcing greater ordering of the partitioned components and hence the higher T_m observed in the DSC thermograms. From the regularity of the alternating bands, one could surmise why the film behaved like a homogeneous material as indicated by the single endotherm of the ternary blown film.

10.1.1 Modification of Cornstarch for Formulation with LDPE in Extrusion Blown Films

Although USDA research initiated the introduction of larger amounts of starch into plastic films (20%–50% w/w), "Otey[4] technology" as an improvement over the earlier Griffin process (6%–15% starch),[3] cost of the product turned out was not competitive with existing plastics in the market. The higher cost of the product was due in large part to the polyethylene-*co*-acrylic acid, EAA used as compatibilizer for starch-LDPE blown film. Added to this disadvantage, earlier work had also indicated that while EAA facilitated formulation of starch-LDPE/EAA into film, a starch-EAA complex that is resistant to biodegradation was also produced. To make the earlier process cost-effective, an investigation into derivatization of cornstarch with substituent groups that may render the starch more lipophilic and hence more compatible with LDPE without need for EAA in the process was undertaken. Substituent group candidates selected for this study were the triphenylmethyl (trityl), palmitoyl, *O*-butyl, *O*-(2-hydroxybutyl), and longer-chain hydroxyalkyl groups.

Trityl starch: Trityl starches of degrees of substitution (D.S.) 0.33, 0.50, and 0.80 were synthesized using a modification of the method of Roberts.[8] In a typical reaction, 40.0 g of cornstarch (Buffalo 3401) was added to 400 mL of dry pyridine in a 1.0 L reaction flask that had been purged with N_2. The contents were stirred and warmed to 60°C followed by rapid addition of trityl chloride (210.0 g, 750 mmol). Stirring of the slurry at 60°C was continued overnight. Higher temperatures (90°C–100°C) tended to lead to decomposition of trityl chloride resulting in a dark brown product. The product mixture was then poured slowly into stirring methanol, giving a light yellow to a light brown precipitate (ppt). The ppt was filtered, triturated with toluene and washed in sequence with 50% methanol–water and methanol, and dried in vacuo to yield 102 g. The calculated D.S. was 0.8:

$$D.S. = \frac{162\,T}{26,000 - 243\,T}$$

where $T = \%$ trityl.

Analysis of the tritylated starch by FT-IR (KBr) and solid-state ^{13}C-NMR showed the following absorbances in the IR (cm^{-1}): 3424 ms (OH stretch-starch); 3059 m (C–H aromatic stretch); 2930 m (C–H stretch starch); 1599, 1491 m (C=C breathing mode of the trityl grouping); 1164, 1089, and 1033 vs. (CCO starch); and 704 s (aromatic C–H out-of-plane bending). The ^{13}C-NMR chemical shift resonances (ppm) were observed at 142.74 (ipso Cs of the phenyl rings), 128.13 (aromatic), 102.61 (C-1 α-D-glucopyranosyl), 87.19 (tertiary aliphatic C of the trityl group), 82.27 (C-4 α-D-glucopyranosyl), 73.72 (C-2, C-3, C-5 α-D-glucopyranosyl), and 61.06 (C-6 α-D-glucopyranosyl).

The blown plastic film made from a formulation of the trityl starch of D.S. 0.33 with LDPE showed poor physical properties (tensile strength and elongation). Although this derivatized starch was lipophilic, it was incompatible with LDPE. The lack of compatibility might have arisen from the bulkiness of the substituent group on the starch.

Starch palmitates: Palmitates having D.S. 0.09, 0.48, and 0.71 were synthesized from palmitoyl chloride. A typical reaction procedure was carried out as given in the following text. In a dry 1 L three-necked RBF equipped with a reflux condenser, addition funnel, a magnetic stirrer, and nitrogen inlet, dry cornstarch (119.5 g, A. E. Staley Mfg. Co.) was added. The system was purged with N$_2$ ca. 15.0 min and then dry toluene (140.0 mL) and dry pyridine (48.0 mL) were added and the mixture stirred. Meanwhile, dry toluene (60.0 mL) and palmitoyl chloride (170 mL, 0.61 mol) were placed in the addition funnel and slowly added to the stirring reaction mixture. After all the acid chloride had been added, the mixture was heated to reflux ca. 2 h. The heat was removed and the reaction mixture allowed to cool to room temperature overnight. Reaction progress was monitored through small aliquots that were checked by FT-IR spectroscopy. Reheating was sometimes necessary until IR spectrum showed complete reaction. The reaction flask was cooled, and the brown to straw-yellow product mixture was poured slowly into 1500 mL of vigorously stirred MeOH. The resulting solid was filtered, washed with more MeOH followed by hot hexane, and dried to give 174 g. Trituration of the crude solid with hot hexane and drying gave a product, with m.p. 120°C–130°C (D.S. 0.48); it showed the following FT-IR absorbances (ν_{KBr} cm^{-1}): 3400 (OH), 2923 s (–CH$_3$, –CH$_2^-$ asym. stretch), 2851 (–CH$_2^-$ sym. stretch), and 1744 (–CO$_2$R stretch), and for ^{13}C-NMR (ppm), 14.55 (–CH$_3$), 23.5 (–CH$_2^-$ contraction), 32.82 (–CH$_2^-$ contiguous to C=O), 72.88 (C-2, C-3, C-5 α-D-glucopyranosyl), 62.08 (C–6 α-D-glucopyranosyl), 84.24 (C–4 glucopyranosyl), 96–102 (–C–1 α-D-glucopyranosyl), and 172 (O=C–OR).

10.1.1.1 O-(2-Hydroxybutyl) and O-Butyl Starches

Hydroxyethyl and hydroxypropyl starches had been reported[9] to impart improved physical characteristics to ternary starch-based films. To expand study of the hydroxyalkyl starch ethers, our main goal was to investigate the effects of increasing alkyl chain length of substituent groups on the derivatized starch and the consequent physical properties of the starch-containing blown film especially if the derivatized starch would become compatible enough with LDPE as to eliminate EAA in the formulation.

Procedure:

Typically, the synthetic procedure for the *O*-(2-hydroxybutyl) starch was by a modification of the method by Kesler and Hjermstad[9] Cornstarch (100.0 g, 10% moisture, A. E. Staley Mfg. Co.) was placed in a dry resin flask (1 L) provided with a mechanical stirrer, reflux condenser, and a nitrogen inlet. Isopropanol (130.0 mL), Na_2SO_4 (15.00 g), and NaOH (5.86 g dissolved in 9.0 mL H_2O) were added. The resin flask was clamped in a 45°C–50°C oil bath and purged with N_2 (15 min) while the contents were being stirred. Butylene oxide (45.0 mL, 36.91 g, that is, 29% relative to starch, Aldrich Chemical Co.) was added and N_2 flow cut off after 5 min. The reaction was allowed to run for ca. 40 h at the end of which interval an aliquot taken from the reaction mixture showed that only slight amount of product had been formed, and the reaction time was extended to 60 h. After this interval, the oil bath was removed, the mixture was allowed to cool to room temperature, and glacial acetic acid was added to the stirring slurry until pH 5 was reached. The solid was filtered and washed in sequence with water, 80% EtOH, absolute EtOH, and finally with acetone and dried in vacuo to give 112.0 g of product which looked very much like unmodified starch. The molar substitution ratio (MS) for these derivatives was determined using a modification of the procedure of Jones et al.[10,11]:

$$MS = \frac{\% \text{ substituent}}{100 - \% \text{ substituent}} \times \frac{162.14}{MW \text{ substituent}}$$

where

MS represents molar substitution of *O*-(2-hydroxybutyl) starch
162.14 = molecular mass of AGU (anhydroglucose unit)
substituent is the 2-hydroxybutyl group

10.1.1.2 Characterization of O-Butyl and O-(2-Hydroxybutyl) Starches

FT-IR v_{KBr} cm^{-1}: 2966, 2931 (–CH3, –CH$_2$– asym. stretch), 2883 (–C–H stretch alkyl), and 1152 (–C–O–C– ether). ^{13}C-NMR (ppm): 100.86 (C–1 of α-D-glucopyranosyl), 82.0 (C–4), 72.57 (C–2, C–3, C–5 of α-D-glucopyranosyl and C–2′ of butyl moiety), 62.23 (C–6 α-D-glucopyranosyl and C–1′ primary butyl carbon), 25.32 (–CH$_2$–), and 10.23 (–CH$_3$).

10.1.2 Characterization of O-(2-Hydroxybutyl) and O-Butyl Starch/LDPE-Containing Blown Films

Analyses of the DSC thermograms of the two blown films gave results that were hardly distinguishable from that of NRRC's starch/LDPE/EAA films even though the former had a lower (5%) amount of EAA content in the formulation. The thermograms of the blown films containing *O*-(2-hydroxybutyl) and *O*-butyl starch components showed slightly broader endotherms with a corresponding exotherm, respectively. There was a slight shift, however, in the outset temperature of the exotherm which could be attributable to hysteresis or super cooling resulting from inefficient heat transfer mechanism in the material as it cooled to the point of crystallization.

Intrinsic Viscosity of *O*-(2-Hydroxybutyl) and *O*-Butyl Starches: Brookfield Viscometer

Sample	Spindle #	rpm	% Slurry	Ave. Reading	Temp., °C	η
HBS	3	6	3	9×200	50	1800
HBS	3	6	5	30.5×200	50	6100
OBS	3	6	5	4.0×200	60	800
OBS	3	6	5	10.0×200	60	2000
Starch	3	6	5	3.0×200	70	600

10.1.3 Adhesion Behavior of Modified Cornstarches to *Lactobacillus amylovorus*

Adhesion studies are used to screen potential substrates or microorganisms that will exhibit competitive advantage over others found in the environment. For example, an organism that binds to a particular substrate would have a head start over one that does not if the substrate is used as a food source. Thus, to evaluate the effect that different substituent groups on starch would have on its adhesion behavior for microorganisms, *Lactobacillus amylovorus*[12] was studied using each derivatized starch. The data obtained showed that starch ethers tend to enhance adhesion and hence may be readily biodegraded in the environment, whereas the esters used tended to inhibit adhesion as shown in the following table.

Comparative Adhesion Behavior of Derivatized Starches to *L. amylovorus*[12]

Derivative	% Microbial Adhesion
O-Butyl starch [10%, dialyzed]	82±0.5 (2)
O-Butyl starch [8%, soluble fraction in dialysis bag]	83±0.1 (2)
Whole *O*-butyl starch [10%, undialyzed, cold water swelling]	57±16 (2)
O-(2-Hydroxybutyl) starch [10%]	58±1.3 (2)
Control [10%, regular cornstarch]	60±5.0
6-*O*-Trityl starch	58.6±13.7
Starch palmitate [D.S.=0.48]	32.7±17
Starch palmitate [D.S.=0.09]	21.0±3.9
Starch-EAA ester	48.6±4.9
Starch-EAA anhydride	29.0±1.3

From the spectroscopic, thermal, and textural analyses of the ternary film, one can infer that the components were not homogeneously distributed but were rather partitioned into domains that allowed the LDPE component to improve its crystallinity as compared to the starting material. Since the LDPE/starch/EAA film was destined for use as an agricultural mulch material, it was important that its cost be affordable for the high volume needed by end users. The study, therefore, was to understand the intrinsic characteristics of the blown film in relation to its components. In exploring substrates that would react with starch to form a platform for replacing EAA in the LDPE/starch/EAA composite, the most promising substituent alkyl groups were the *O*-(2-hydroxybutyl) and *O*-butyl groups. Butyl starch ethers and LDPE formulated with 5% EAA resulted in films that exhibited physical characteristics which compared favorably to NRRC's ternary films having higher levels (20%) of EAA. This ushered in the possibility that EAA can actually be eliminated from the formulation without adverse effects on the performance of the product. The obvious advantages in the use of these starch derivatives included lowering of the gelatinization temperature of the derivatized starch, thus ruling out the use of urea and most especially the level of EAA usually needed in earlier formulations. This lowered the cost of production. Another advantage was the decreased tendency for retrogradation of amylose in the film, thus ensuring flexibility. Although the use of these starch ether derivatives in the studies were novel and preliminary in nature, the obtained data clearly indicated that starch ethers with the appropriate alkyl substituent chain length could effectively replace EAA from the formulation of starch-containing plastic blown films and thereby make the process economical while ensuring biodegradability of the film after its useful life span.

REFERENCES

1. Griffin, G. L. Biodegradable synthetic resin sheet material containing starch and a fatty material. U.S. Patent No. 4016117 (1977).

2. Otey, F. H., Westoff, R. P., and Doane, W. M. Starch-based blown films, *Ind. Eng. Chem. Prod. Res. Dev.* 19, 592 (1980).

3. Otey, F. H., Westoff, R. P., and Doane, W. M. Starch-based blown films II, *Ind. Eng. Chem. Res.* 26, 1659 (1987).

4. Otey, F. H. and Doane, W. M. Starch-based degradable plastic film, *Proceedings of Symposium on Degradable Plastics Society of the Plastics Industry*, Washington, DC, June 1987.

5. Gould, J. M. Microbial degradation of plastics containing starch, *Conference of Solid Waste Management and Materials Policy*, New York, January 25–27, 1989.

6. Gould, J. M., Gordon, S. H., Dexter, L. B., and Swanson, C. L. Microbial degradation of plastics containing starch, *Corn Utilization Conference*, Columbus, OH, November 17–18, 1988.

7. Richardson, M. J. Thermal analysis of polymers. In *Comprehensive Polymer Science: The Synthesis, Characterization, Reactions, and Applications of Polymers*, G. Allen and J. C. Bevington, eds., Vol. 1, p. 890, Pergamon Press, Oxford , U.K. (1989).

8. Roberts, H. J. *Methods in Carbohydrate Chemistry IV*, p. 311, R. L. Whistler, ed., Academic Press, New York (1964 and references therein).

9. Kesler, C. C. and Hjermstad, E. T. *Methods in Carbohydrate Chemistry IV*, (71), p. 304, R. L. Whistler, ed., Academic Press, New York (1964).

10. Jones, L. R. and Riddick, J. R. Colorimetric determination of propionaldehyde, *Anal. Chem.* 26(6), 1035 (1954).

11. Harry-O'kuru, R. E., Moser, K. B., and Gordon, S. H. Colorimetric determination of the molar substitution ratio of 2-hydroxybutyl and related starch ethers, *Carbohydr. Polym.* 17, 313–318 (1992).

12. Imam, S. H. and Harry-O'kuru, R. E. Adhesion of *Lactobacillus amylovorus* to insoluble and derivatized cornstarch granules, *Appl. Environ. Microbiol.* 57(4), 1128–133 (1991).

CHAPTER 11

Starch-Based Edible Films and Coatings

Mahesh Gupta, Charles Brennan, and Brijesh K. Tiwari

CONTENTS

11.1 INTRODUCTION

Starch is a polymeric carbohydrate composed of anhydroglucose units. This is not a uniform material, and most starches contain two types of glucose polymers: a linear chain molecule termed "amylase" and a branched polymer of glucose termed "amylopectin" (Rodriguez et al., 2006). Starches are often used in industrial foods as fillers or structurizing agents. During recent years, starch has been used to produce biodegradable films to partially or entirely replace plastic polymers. This is mainly due to its low cost as a raw material, its renewability in terms of the wheat crop, and its good mechanical properties (Xu et al., 2005). High-amylose starch such as cornstarch is a good source for film formation; for instance, free-standing films can be produced from aqueous solution of gelatinized amylose and then drying of the material in a controlled manner. Normal cornstarch consists of approximately 25% amylose and 75% amylopectin. Mutant varieties of corn are produced which contain starch with up to 85% amylose (Whistler and Daniel, 1985). Mark et al. (1966) reported that films produced from high-amylose cornstarch (71% amylose) had no detectable oxygen permeability at RH levels less than 100%. This was true for both unplasticized and plasticized (16% glycerol) films. This result is surprising in light of the fact that addition of plasticizers and absorption of water molecules by hydrophilic polymers generally lead to increased polymer chain mobility and subsequent to increased gas permeability (Banker et al., 2000). Partial etherification of high-amylose starch with propylene oxide, to yield the hydroxypropylated derivative, improves water solubility.

Edible films are defined as a thin layer of material which can be consumed and provides a barrier to moisture, oxygen, and solute movement for the food. The material can be a complete food coating or can be disposed as a continuous layer between food components (Guilbert, 2000). Edible films can be formed as food coatings and free-standing films and have the potential to be used with food as gas aroma barrier (Kester and Fennema, 1986). However, technical information is still needed to develop films for food application (Donhowe and Fennema, 1993). Edible films and coatings have received considerable attention in recent years because of their advantages over synthetic films. The main advantage of edible films over traditional synthetics is that they can be consumed with the packaged products. That being the fact, there is no package to dispose even if the films are not consumed. Thus, they could still contribute to the reduction of environmental pollution. The films are produced exclusively from renewable, edible ingredients and, therefore, are anticipated to degrade more readily than polymeric materials.

The films can enhance the organoleptic properties of packaged foods provided they contain various components such as flavorings, colorings, and sweeteners (Bourtoom, 2008). The films can also be used for individual packaging of small portions of food, particularly products that currently are not individually packaged for practical reasons such as pears, beans, nuts, and strawberries. An interesting application of the films in the food industry is to apply them inside heterogeneous foods at the interfaces between different layers of components. In this way, they can be tailored to prevent deteriorative intercomponent moisture and solute migration in foods such as pizzas, pies, and candies. In a similar application, they can also be used on the surface of food to control the diffusion rate of preservative substances from the surface to the interior of the food. Recent research has also focused on their functionality as carriers for antimicrobial and antioxidant agents. Another possible application for edible films could be their use in multilayer food packaging materials together with nonedible films. In this case, the edible films would be the internal layers in direct contact with food materials. As mentioned previously, production of edible films causes less waste and pollution; however, their permeability and mechanical properties are generally poorer than synthetic films (Kester and Fennema, 1986). Extensive research is needed for the development of new starch-based materials, methods of film formation, methods to improve film properties, and the potential applications.

11.2 HISTORY OF EDIBLE FILMS AND COATINGS

Edible films and coatings have been used for hundreds of years. For example, wax has been applied to citrus fruits to delay their dehydration since the twelfth and thirteenth centuries in China (Debeaufort et al., 1998). Yuba obtained from the skin of boiled soy milk, essentially a protein film, was used to preserve the appearance of some foodstuffs in Asia in the fifteenth century (Debeaufort et al., 1998; Han and Gennadios, 2005). In the sixteenth century, fats were used to coat meat cuts to prevent shrinkage. Lard or wax was used to enrobe fruit and other foodstuffs in England (Miller and Krochta, 1997). Later, in the nineteenth century, gelatin films were used to cover meat stuffs, while sucrose was chosen as an edible protective coating on nuts such as almonds and hazelnuts to prevent oxidation and rancidness (Debeaufort et al., 1998). In the last 30 years, petrochemical polymers, commonly called plastic, have been the most widely used materials for packaging because of their high performance and low cost (Callegarin and Quezada-Gallo, 1997). However, the serious environmental problems associated with their nonbiodegradability have urged scientists to search for new alternative materials (Petersson and Stading, 2005). Thus, edible or biodegradable packagings made from various biological resources and their applications have recently been investigated. Shellac and wax coatings on fruits and vegetables, zein coatings on candies, and sugar coatings on nuts are the most common commercial examples of edible coatings. Cellulose ethers (carboxymethyl cellulose, hydroxypropyl, and methylcellulose) have been used as ingredients in coatings for fruits, vegetables, meats, nuts, confectionery, bakery, grains, and other agricultural products (Han and Gennadios, 2005).

11.3 CLASSIFICATION OF EDIBLE FILMS AND COATINGS

Edible films can be produced from materials with film–forming ability. During manufacturing, film materials must be dispersed and dissolved in a solvent such as water, alcohol, or mixture of water and alcohol or a mixture of other solvents. Plasticizers, antimicrobial agents, colors, or flavors can be added in this process. Adjusting the pH and/or heating the solutions may be done for the specific polymer to facilitate dispersion. The film solution is then casted and dried at a desired temperature and relative humidity to obtain free-standing films. In food applications, film solutions could be applied to food by several methods such as dipping, spraying, brushing, and panning followed by drying. Components used for the preparation of edible films can be classified into three categories: hydrocolloids (such as proteins, polysaccharides, and alginate), lipids (such as fatty acids, acylglycerol, waxes), and composites (Donhowe and Fennema, 1993).

11.4 EDIBLE FILM MATERIALS AND THEIR PREVIOUS APPLICATIONS

Environmental concern about the use of synthetic plastics for food packaging has led to increased interest in biodegradable and edible film research (McHugh et al., 1993; Lai and Padua, 1997). Both dehydration and growth of microbial organisms in food products have been delayed by using edible and biodegradable films and coatings. Moreover, the flavor, odor, and overall organoleptic characteristics were not modified. Many materials from biological resources have been used for edible or biodegradable film and coating formulations, such as polysaccharides, proteins, lipids, or their mixtures (Debeaufort et al., 1998). Waxes and oils (mineral oils, paraffin, beeswax, shellac, etc.) were used largely as coatings on fruits, such as oranges, lemons, apples, and pears; they create a really efficient barrier to water and can prevent weight loss. Polysaccharides used in edible or biodegradable films and coatings include cellulose, starch, pectin, and algal gum. Proteins from various plant and animal sources—including wheat gluten (Kayserilioglu et al., 2001), soy protein, zein (Lai and Padua, 1997; Lai et al., 1997), and milk protein (Letender et al., 2002)—have also been used in edible films. Lipids and their derivatives are mainly used in films or coatings to improve their moisture barrier properties. The properties of edible films depend on the type of film–forming materials and especially on their structural cohesion. Additives—such as plasticizers, cross-linking agents, antimicrobial agents, antioxidants, and texture agents—are used to alter the functional properties of the films. Among the natural polymers, starch has been considered as one of the most promising candidates for future materials because of the attractive combination of price, availability, and thermoplasticity (Lai and Padua, 1997; Mali et al., 2005). Starch-based resins have been made into compost bags, disposable food-service items (cutlery, plates, cups, etc.), packaging materials (loose, fill, and films), coatings, and other specialty items (Lai et al., 1997). Edible films and coatings from starch mainly find applications in the meat, poultry, seafood, fruit, vegetable, grains, and candies industries (Debeaufort et al., 1998).

Edible films and coatings can be used as a host for carrying basic nutrients and/or nutraceuticals that are lacking or are present in only low quantity in fruits and vegetables. The development of nutritionally fortified edible films and coatings strongly depends on the type of carriers (film-forming materials) and the type and concentration of nutraceuticals added into the film–forming solutions.

11.4.1 Plasticizers

Edible films and coatings need to have good elasticity and flexibility, a low brittleness, and a high toughness to prevent cracking during handling and storage (Barreto et al., 2003). Therefore, plasticizers of low molecular weight (nonvolatile) are typically added to hydrocolloid film–forming solutions to modify the flexibility of edible films. Plasticizers with characteristics such as small size, high polarity, more polar groups per molecule, and greater distance between polar groups within a molecule generally impart greater plasticizing effects on a polymeric system. Indeed, they act by increasing the free volume or in other words by decreasing intermolecular attractions between adjacent polymeric chains by reducing hydrogen bonding between polymer chains (Myllarinen et al., 2002). Generally, plasticizers are required for polysaccharide-based edible films. Their amount added into hydrocolloid film–forming preparations varies between 10% and 60% by weight of the hydrocolloid. The most commonly used plasticizers are polyols (propylene glycol [Jagannath et al., 2006], glycerol [Myllarinen et al., 2002], sorbitol [Cerqueira et al., 2009], polyethylene glycol [Bourtoom et al., 2006]), oligosaccharides (sucrose [Veiga-Santos et al., 2007]), and water. In fact, water is a ubiquitous natural diluent, which plasticizes and/or antiplasticizes some films depending on the amount sorbed onto the films' matrix (Chang et al., 2000). Thus, film moisture content, as affected by the relative humidity of the surrounding environment, largely affects film properties (Krochta, 2002).

11.4.2 Antimicrobial, Antioxidant, and Other Functional Agents

The new generation of edible films and coatings is being especially designed to increase their functionalities by incorporating natural or chemical antimicrobial agents, antioxidants, enzymes, or functional ingredients such as probiotics, minerals, and vitamins (Bifani et al., 2007; Lin and Zhao, 2007; Vargas et al., 2008). Edible coating can enhance the nutritional value of foods by carrying basic nutrients and/or nutraceuticals in its matrix. The sensory quality of coated products can be also improved if the matrix flavor and pigment are added.

11.4.3 Antimicrobial Agents

Common chemical antimicrobial agents used in food systems, such as benzoic acid, propionic acid, sodium benzoate, sorbic acid, and potassium sorbate, may be incorporated into edible films and coatings to inhibit the outgrowth of both bacterial and fungal cells (Cha and Chinnan, 2004). However, due to the health concerns of consumers related to chemical preservatives, the demand for natural foods has spurred the search for natural biopreservatives in edible film–forming preparations. Common biopreservatives that may be used in edible films and coatings are bacteriocins, such as lacticin and pediocin (Santiago-Silva et al., 2009), and antimicrobial enzymes, such as chitinase and glucose oxidase (Suppakul et al., 2003). Recently, different studies have also been conducted on the use of the enzyme lactoperoxidase (LPS) in edible films and coatings (Mecitoğlu and Yemenicioğlu, 2007). LPS that has antimicrobial and antioxidant properties is found in the mammary, salivary, and lachrymal glands of mammals and in their respective secretions (Seifu et al., 2005). LPS shows bactericidal effect on Gram-negative bacteria and bacteriostatic effect on Gram-positive bacteria (Kussendrager and van Hooijdonk, 2000). LPS also presents antifungal and antiviral activities (Seifu et al., 2005). Natural antimicrobial compounds have also been incorporated into protein or polysaccharide-based matrices: rosemary, garlic essential oils (Seydim and Sarikus, 2006), oregano, lemon grass, or cinnamon oil at different concentrations.

11.4.4 Antioxidant Agents

Generally, edible films and coatings have lipids added to them to reduce water vapor transfer due to its hydrophobic character. The incorporation of antioxidants in edible film–forming preparations to increase product shelf life by protecting foods against oxidative rancidity, degradation, and discoloration is becoming very popular (Baldwin et al., 1995). Natural antioxidants such as phenolic compounds and vitamins E and C in place of synthetic antioxidants are extensively used in edible films. Essential oils also exhibit a wide range of biological effects, including antioxidant and antimicrobial properties. The phenolic components, such as carvacrol, camphor, eugenol, linalool, and thymol, are most active and appear to act principally as membrane permeabilizers (Burt, 2004). Maizura et al. (2007) reported that starch–alginate edible films containing lemon grass oil are effective in inhibiting the growth of *Escherichia coli* O157:H7 at all levels. However, generally, all these studies showed their efficacy in vitro against various microorganisms, but they were not tested with real foods.

11.5 STARCH FILM–FORMING MECHANISMS: GELATINIZATION AND RECRYSTALLIZATION

Often, the first step in the production of starch films is heating starch in water. When heated at high water content, starch is gelatinized and transformed from a semicrystalline granular material into a system containing granular remnants, or to an amorphous paste with no structure at all (Smits et al., 2003). The process is termed "gelatinization," which corresponds to an irreversible swelling and breakage of starch granules, and leaching of amylose and amylopectin into the solutions. A gradual dissolution of starch granules allows a further hydration up to a point where the whole structure of the starch granules is completely disintegrated (Endres et al., 1994). Often, noncrystalline swollen granular remnants named "ghosts" remain even after a long period of gelatinization (Figure 11.1) (Smits et al., 2003; Mehyar and Han, 2004; Mathew et al., 2006; Zhang and Han, 2006). Studies have shown that the presence of sugars in the water increases the gelatinization temperature of starch. Various mechanisms have been proposed to explain this,

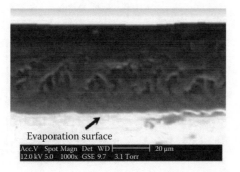

Figure 11.1 SEM of starch film with swollen granules. (From Phan, T.D. et al., *J. Agric. Food Chem.*, 53, 973, 2005.)

including the ability of sugar to compete for water against starch and the ability of sugars to reduce water activity in the system, resulting in a reduction in the plasticizing effect of the sucrose–water solvent (Maaurf et al., 2001). At temperatures higher than the glass transition temperature (T_g), starch materials are in a rubbery state, and retrogradation (or recrystallization) occurs easily when gelatinized starch is stored at high RH or high plasticizer contents (Delville et al., 2003). In the rubbery state, high RH or high plasticizer content favors starch macromolecular mobility, which facilitates the development of crystallinity (Delville et al., 2003). Recently, the mechanism of starch retrogradation has been extensively investigated. Starch retrogradation occurs as a result of intermolecular hydrogen bonding between O-6 of D-glucosyl residues of amylose molecules and OH-2 of D-glucosyl residues of short side chains of amylopectin molecules (Figure 11.2) (Tako and Hizukuri, 2002). It can be attributed to intermolecular hydrogen bonding between OH-2 of D-glucosyl residues of amylose molecules and O-6 of D-glucosyl residues of short side chains of amylopectin molecules (Figure 11.3). In addition to intermolecular hydrogen bonding between amylose and amylopectin, hydrogen bonding between O-3 and OH-3 of D-glucosyl residues on different amylopectin molecules may also occur (Tako and Hizukuri, 2002). Intramolecular association of amylopectin molecules was not suggested to exist, while intramolecular hydrogen bonding might take place between OH-6 and adjacent hemiacetal oxygen atoms of the D-glucosyl residues within the amylose molecules (Tako and Hizukuri, 2002). The cluster formation begins with the formation of crystalline lamellae composed of double helices of amylopectin short chains (represented by rectangular boxes). Then the packing of double helices forms crystalline clusters (Delville et al., 2003). Increased water content increases the degree of crystallinity and the kinetics of crystallization, while increased glycerol content slows the crystallization kinetics in starch amorphous rubbery amylopectin systems (Delville et al., 2003). Crystallites may act as physical cross-linking points that generate internal stresses or cracks which lead to damage of the starch products (Delville et al., 2003). Therefore, while crystallinity increases, elongation (E) decreases drastically, and tensile strength (TS) and modulus of elasticity (EM) increase. Reportedly, a B-type polymorph of the crystallite develops in the aged gel of all starches, irrespective of their pattern of

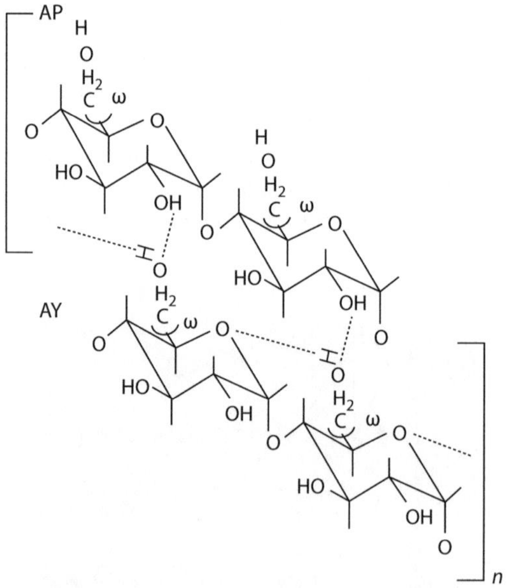

Figure 11.2 Hydrogen bonding between amylose and amylopectin molecules (dashed lines represent hydrogen bonds; AY, amylose; AP, short side chains of amylopectin molecules). (From Tako, M. and Hizukuri, S., *Carbohydr. Polym.*, 48, 397, 2002.)

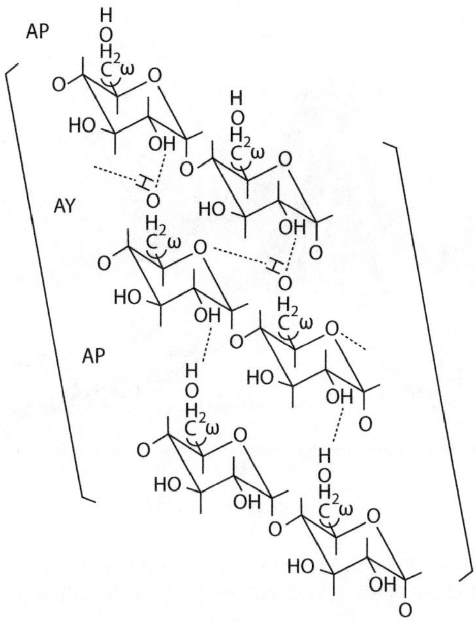

Figure 11.3 Retrogradation mechanism of starch (dashed lines represent hydrogen bonds; AY, amylose; AP, short side chains of amylopectin molecules). (From Tako, M. and Hizukuri, S., *Carbohydr. Polym.*, 48, 397, 2002.)

crystallinity in the natural state (Abd-Karim et al., 2000). However, the type of polymorph developed in aged cereal starch gel may also depend on water content. Samples containing more than 43% moisture develop the B-pattern on aging, whereas those containing less than 29% moisture develop the A-pattern (Abd-Karim et al., 2000).

The crystallite structure of starch films is often analyzed using an X-ray diffraction (XRD) technique, equipped with a 1° divergence slit and a 0.1 mm receiving slit, between $2\theta = 3°$ and $2\theta = 40°$ or 60° (where θ is the angle between the incident radiation and the diffracting plane) with a step size of $2\theta = 0.02°$ or 0.05° when using CuKα1 radiation (the wavelength of the x-ray, $\lambda = 0.154\ 10$ nm), 40 kV and 40 or 50 mA (Mali et al., 2002, 2006; Myllarinen et al., 2002; Zimeri and Kokini, 2002; Romero-Bastida et al., 2005; Mathew et al., 2006). Film samples are first cut into rectangular shapes and clamped onto a quartz monochromator. The typical XRD patterns of starch films (Figure 11.4) are characterized by sharp peaks associated with the crystalline diffraction and an amorphous zone. The amorphous fraction of the sample can be estimated by the area between the smooth curve drawn following the scattering hump and the baseline joining the background within the low- and high-angle points. The crystalline fraction can be estimated by the upper region above the smooth curve (Mali et al., 2006). Therefore, the crystallinity of the starch films can be calculated using the following equation:

$$Crystallinity = \frac{A_c}{(A_c + A_a)} \times 100\%$$

where

A_c is the crystalline area on the XRD

A_a is the amorphous area on the XRD (Yoo and Jane, 2002)

The crystallinity of starch films is dependent on the processing conditions, such as the completeness of amylose dissolution in water, drying conditions (rate and temperature), plant origin of the starch,

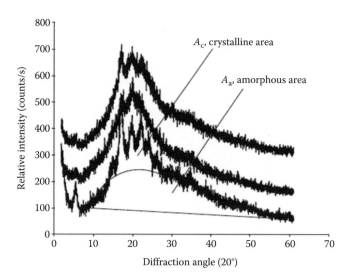

Figure 11.4 XRD pattern (B-type) of starch films made by thermal gelatinization at 60 days of storage at 25°C. (From Romero-Bastida et al., *Carbohydr. Polym.*, 60, 235, 2005.)

moisture content of the films, and the temperature of storage (Mali et al., 2002). For example, when starch films are stored at temperatures below the T_g, the starch polymers are in a stable glassy state, and crystallization does not occur or is extremely slow. However, recrystallization of starch can occur at temperatures above T_g at a rate depending on the difference between T_g and the storage temperature (Mali et al., 2006). The crystallinity of the starch films increased with storage time. As storage time increases, the width of the XRD peak decreases, but its peak intensity increases, showing an increase in crystallinity of the starch (Mali et al., 2002). Plasticizers were also found to affect the crystallinity of starch. According to Mali et al. (2006), glycerol limited crystal growth and recrystallization. Glycerol could interact with the polymeric chains and interfere with polymer chain alignment due to steric hindrances. However, Garcia et al. (2000) reported that plasticizers (glycerol and water) favored polymer chain mobility and allowed the development of more stable crystalline structure during shorter periods of storage. In contrast to the conclusions of Mali et al. (2006), Smits et al. (2003) found that starch films without plasticizers showed less recrystallinity than the plasticized starch films. They attributed this phenomenon to the mobility of starch polymer chains. Plasticized starch polymers could easily vibrate and align to form crystallites, while the unplasticized starch polymers interact with each other strongly and lose mobility.

11.6 APPEARANCE AND PHYSICAL PROPERTIES OF STARCH FILMS

The appearance of starch films depends on the other ingredients added into the starch dispersion. Pure starch films without any additives are usually colorless and transparent, but brittle. Films with polyols—such as glycerol, ethylene glycol, or sorbitol—are also colorless, but flexible. Films containing monosaccharides—such as fructose, mannose, and glucose—as plasticizers are yellowish, with the extent of the color depending on the concentration of the plasticizers used in the films (Zhang and Han, 2006). When the microstructure of starch films is observed under a light microscope, starch films reveal a characteristic surface pattern with representative withered "ghost" granules (Mehyar and Han, 2004; Mathew et al., 2006). Starch films can also be observed by scanning electron microscopy (SEM). Under SEM observation, the surface of starch films is smooth and homogeneous (Figure 11.1).

11.7 MECHANICAL PROPERTIES OF STARCH FILMS

Starch films are often characterized by tensile tests from which three mechanical properties are obtained: TS, EM, and E. TS is a measurement of the strength of the film. It is calculated by dividing the force needed to break the film by the cross-sectional area of the initial specimen. The value of TS should not be affected by film thickness (Phan et al., 2005). The E value represents the flexibility of the film. It is defined as the percentage of the change in the length of the specimen relative to the original length. The EM value, also known as Young's modulus or the elastic modulus, is the fundamental measurement of the film stiffness. It is calculated from the initial linear slope of the stress–strain curve (Figure 11.5). The higher the EM values of the films, the stiffer the films are (Mali et al., 2005). The test methods follow the procedure of ASTM D882-91 (ASTM, 1991). The universal testing machine is widely chosen to test the film mechanical properties. According to the method, the initial grip distance is set at 5 cm. The choice of crosshead speed depends on the E value. Because of the hydrophilicity of starch films, it is necessary to condition the films in a certain RH prior to the tests. Usually, the conditioning RH is 50%, accomplished using saturated calcium sulfate solution in a well-sealed chamber at room temperature. Typical force–deformation curves are shown in Figure 11.5.

Apart from the original source of the starch, four other factors were found to affect the mechanical properties of the films: plasticizer content, T_g, crystallinity, and ratio of amylose to amylopectin. During the last few years, the effect of plasticizers on the mechanical properties of films prepared from starch, amylose, amylopectin, and mixtures of starch and other biopolymers have been widely studied (Myllarinen et al., 2002). Normally, plasticizers are used to increase E values and to decrease TS and EM.

This is because plasticizers can increase the free volume in the amorphous phase and reduce interaction between the starch polymer chains. However, an antiplasticization effect of plasticizers was found when the plasticizer concentration was below a critical level (Godbillot et al., 2006). At a temperature above T_g, the starch films are in a rubbery state and are flexible and extendible because more free volume is available in the starch film matrix. In contrast, at temperatures below T_g, the films are in a glassy state and are brittle. As crystallinity increases, the TS and EM of starch films increase, but E decreases because crystallites behave like hard particles or physical cross-linkers (Liu, 2005). Amylose and amylopectin films are mechanically different (Figure 11.5)

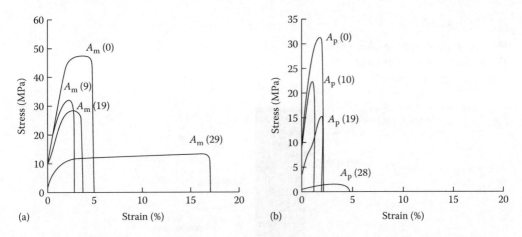

Figure 11.5 Tensile stress–strain curves of (a) amylose (A_m) and (b) amylopectin (A_p) films with various glycerol contents, measured at RH of 50% and at 20°C. *Abbreviations*: A_m (9), amylose films containing 9% glycerol; A_p (10), amylopectin films containing 10% glycerol. (From Myllarinen, P. et al., *Carbohydr. Polym.*, 50, 355, 2002.)

(Myllarinen et al., 2002). Pure amylose films are stronger, whereas pure amylopectin films are more brittle. Films made of a mixture of amylose and amylopectin were studied by Lourdin et al. (1995); the results showed that a preponderance of amylose in starch films leads to higher TS, whereas a preponderance of amylopectin leads to lower TS. This is presumably due to the higher degree of crystallinity in starch films containing more amylase (Liu, 2005).

11.8 BARRIER PROPERTIES OF STARCH FILMS

When an edible film is applied on the food, the surface gas concentration may change during storage. The chemical composition of the food surface is dynamic, and these changes occur mainly due to food metabolism, microbial respiration, gas solubility, and permeability of the edible film. Food and microbial metabolism are responsible for consumption of oxygen and CO_2 production. The microbial activity can affect the composition of oxygen and CO_2; however, this is only significant when it has reached the end of shelf life and deterioration is evident (Pfeiffer and Menner, 1999). Therefore, the efficiency of an edible coating strongly depends on its barrier properties to gas, water vapor, aroma, and oil which in turn depend on the chemical composition and structure of the coating-forming polymers, the characteristics of the product, and the storage conditions.

11.9 GAS BARRIER

Coatings are used to create a controlled or modified atmosphere inside the fruits and vegetables that will delay ripening and senescence in a manner similar to the most costly controlled or modified atmosphere storage. The transport of gas molecules in an edible film consists of three steps (Figure 11.6): adsorption of the permeant onto the edible film surface, diffusion of the permeant from one side of the edible film to the other side, and then the desorption of the permeant from the edible film.

Diffusion of simple gases (e.g., O_2, N_2, CO_2) through a film generally obeys Fick's law, i.e. (Crank, 1975),

$$J = -D\frac{\partial C}{\partial x} = D\frac{C_1 - C_2}{l} \tag{11.1}$$

where
  l is the thickness of the edible film
  D is the diffusion coefficient
  C is the concentration

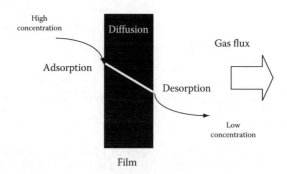

Figure 11.6 Transport of gas molecules in an edible film.

Table 11.1   Solubility, Diffusivity, and Kinetic Diameter Data of N_2, O_2, and CO_2 Gases in Water at 25°C

Gas	Solubility Coefficient mol $(L\ atm)^{-1}$	Diffusivity 10^{-9} (m^2/s)	Kinetic Diameter $(10^{-10}\ m)$
N_2	6.1×10^{-4}	2	3.65
O_2	1.3×10^{-3}	2.42	3.46
CO_2	3.4×10^{-2}	1.91	3.3

Source: From Lide, D.R., *Handbook of Chemistry and Physics*, CRC Press/Taylor & Francis, Boca Raton, FL, 2007.

Moreover, Henry's law says that the concentration of the gas can be expressed as

$$C = \Sigma\, p \tag{11.2}$$

where
  Σ is the solubility coefficient of the gas in the membrane
  p is the pressure of the gas

The solubility, diffusivity, and kinetic diameter data of N_2, O_2, and CO_2 gases in water at 25°C are presented in Table 11.1. The kinetic diameters of CO_2, O_2, and N_2 are very close. However, CO_2 is much more permeable (product of diffusivity by the solubility coefficient) than the two other gases because of its much higher solubility in moist films due to presence of water in the films. Furthermore, the solubility and diffusivity of O_2 are larger than N_2. Therefore, in the case of a polar film matrix, these characteristics attribute to the permeability in the order of $CO_2 > O_2 > N_2$.

Furthermore, what happens inside the edible film is not relevant to the design; therefore, the concept of permeability can be obtained by combining Fick's first law (Equation 11.1) with Henry's law (Equation 11.2):

$$J = D\Sigma(p_1 - p_2)/1 \tag{11.3}$$

The product of diffusion coefficient and solubility coefficient is equal to the permeability coefficient, $\Pi = D\,\Sigma$, which characterizes the intrinsic permeability of the edible film. For an i-multilayer edible film, the flux J will be described by

$$J = \frac{p_1 - p_2}{\sum_i \dfrac{l_i}{\Pi_i}} \tag{11.4}$$

It is important to point out that in respiring foods, the permeability of the edible film is extremely relevant since foods consume O_2 to produce an equilibrium condition. Oxygen uptake by a food often results in deleterious reactions which affect its flavor, nutritional quality, and acceptability. The oxygen uptake for the food (R) is usually estimated by a Michaelis–Menten model (Talasila and Cameron, 1997):

$$R = R_{max}\left(\frac{P_{O_2}}{K + P_{O_2}}\right) \tag{11.5}$$

Thus, it is possible to predict easily the equilibrium partial pressure of oxygen when $R=J$ with the following equation:

$$V_F R_{max}\left(\frac{P_{O_2}}{K+P_{O_2}}\right)=\frac{A_F}{\sum_i \frac{l_i}{\Pi_{O_2,i}}}\left(p_{O_2,AIR}-p_{O_2}\right)$$

(11.6)

where V_F and A_F are volume of food and area of food, respectively.

11.10 WATER VAPOR PERMEABILITY

In the case of water vapor diffusing through a polar film matrix, J will increase as Δp is moved up in the vapor pressure spectrum. This occurs because the adsorption isotherms for hydrophilic polymers are sigmoidal, causing Σ to increase significantly when the vapor pressure rises. The greater the amount of adsorbed water plasticizes the more the film causes the diffusion coefficient D to increase. Fick's law is not strictly obeyed, and D is not a constant for water vapor migration through hydrophilic film matrices (Kester and Fennema, 1986). The WVP is calculated by the following equation (Gontard et al., 1993):

$$WVP=\frac{\Delta m}{\Delta t}\times A\times\Delta P$$

(11.7)

The WVP corresponds to the amount of water vapor (Δm) transferred through a film of area (A), of thickness l during a finite time (Δt). The ΔP is the vapor pressure difference across the film. Published WVP values for edible films should be cautiously compared because of the difference in plasticizers, temperature, relative humidity gradient, etc. The increase of the film permeability with thickness indicates a water affinity of the film that could be attributed to hydrophilic compounds. Then it seems that the influence of thickness varies with the lipid nature or the film composition (Morillon et al., 2002).

11.11 AROMA BARRIER

An understanding of sorption between flavor compounds and edible film matrix requires knowledge of the chemical and physical structures of both the flavor compound and the polymer (Voilley and Souchon, 2006). Studies on the aroma permeability of hydrocolloid edible films are very scarce. The flavor diffusion coefficient D in edible films is in the order of $10-13\,m^2\,s^{-1}$ (Debeaufort et al., 1998; Voilley and Souchon, 2006). Flavor absorption may affect the flavor of a food as well as the mechanical properties of the edible film, such as tensile and heat seal strength and permeability, or cause delamination of the polymeric structure (Hirose et al., 1998). Finally, an example study of aroma permeability was reported by Miller and Krochta (Miller et al., 1998); a whey protein film was better than vinylidene chloride copolymer and was comparable to ethylene vinyl alcohol copolymer as barrier to limonene (citrus aroma).

11.12 OIL BARRIER

Film coatings may become a good alternative to reduce oil uptake during frying. Several hydrocolloids with thermal gelling or thickening properties, such as proteins or carbohydrates, were tested to reduce oil migration (Williams and Mittal, 1999; Han and Krochta, 2001; Garcia et al., 2002).

Hydrocolloid films are expected to be oil resistant due to their substantial hydrophilicity. Garcia et al. (2002) reported that, in deep-fat frying potato strips and dough disks, methylcellulose coatings are more effective in reducing oil uptake than hydroxyl-propyl-methyl cellulose. Sorbitol addition was necessary to maintain coating integrity and improve barrier properties. Some potentiality and applications of edible coatings as an oil barrier are presented by Garcia et al. (2002).

11.13 OPTICAL PROPERTIES: GLOSS, TRANSPARENCY, AND COLOR

In recent years, research efforts have been made to improve mechanical and barrier properties of edible films and coatings, but few studies have focused on optical properties such as color, gloss, and transparency. Optical properties are characteristics of surfaces which are detected by human vision affecting some crucial aspects of food quality. The internal and surface film microstructure plays an important role in optical properties of the film. Indeed, the intensity of light reflected by the coated food is determined both by the light directly reflected at the interface between air and the coated food surface (specular reflection), and by the light reemitted out of the surface in all directions after penetrating into the coating of the food and scattering internally (indirect reflection). Specular reflection is related to the gloss of the coating. According to the ASTM standard D-523, it is measured at 20°, 60°, and 85° angles from the normal to the coating surface using a flat surface gloss meter. Polished black glass with a refractive index of 1.567 possesses a specular gloss of 100%. The gloss of edible coatings is affected by coating microstructure, in particular, by the type and level of surfactant, the distribution and the size particle diameter of dispersed phase, the relative humidity, the storage time, and the surface roughness (Trezza and Krochta, 2000; Nikolova et al., 2005). The film transparency is measured through the surface reflectance spectra in a spectro-colorimeter. A good method to determine the transparency is based on the Kubelka–Munk theory (Hutchings, 1999). This theory models the reflected and transmitted spectrum of a colored layer based on a material-dependent scattering and absorption function. The assumptions of this theoretical model are the following:

- A translucent colorant layer on the top of an opaque background.
- Within the colorant layer, both absorption and scattering occur.
- Light within the colorant layer is completely diffuse.
- Geometry should be diffuse illumination and diffuse collection.

The colors of edible films are evaluated by a colorimeter or a spectrophotometer. The parameters L (luminosity), a (red–green), and b (yellow–blue) are the edible film color values in the CIELAB scale. A white disk (L_0, a_0, and b_0) was used like standard color. From these values, chrome ($C=[a^2+b^2]^{1/2}$), hue ($H=\arctan b/a$), whiteness index ($WI=100-[(100-L)^2+a^2+b^2])^{1/2}$, and difference of color ($\Delta E=[(\Delta L)^2+(\Delta a)^2+(\Delta b)^2]^{1/2}$) can be calculated, where $\Delta L=L-L_0$, $\Delta a=a-a_0$, and $\Delta b=b-b_0$.

11.14 SUMMARY

Starch-based edible films and coatings represent a stimulating route for creating new packaging materials mainly because these are available with a wide range of physicochemical properties that can help to ease many problems encountered with the packaging of food and food products. Edible films and coatings can be used where plastic packaging cannot be applied, and many additives—such as antioxidants and antimicrobials—can be added into the films or coatings to provide new functionality. The main advantage is that they can be eaten, and hence, no waste is created or left in the environment; they are totally nonpolluting, creating environmentally friendly packaging material. Perceived benefits and potentials of edible films and coating have attracted industries for their use. Components used for the preparation of edible films can be classified into three categories: hydrocolloids, lipids, and composites. Hydrocolloid films possess good barrier properties to oxygen, carbon dioxide, and lipids but not to water vapor. Most hydrocolloid films also possess superb mechanical properties, which are quite useful for fragile food products. Extensive research is still needed on the methods of films formation and methods to improve film properties for potential industrial applications. The microstructural observations and thickness measurements helped explain the differences in water vapor and gas permeabilities of starch-based films and coatings. High-amylose-content starch films with a thicker structure had the lowest values for water vapor and gas permeabilities, and films without plasticizer with a multilayer structure exhibited the largest values.

REFERENCES

Abd-Karim, A.; Norziah, M. H.; and Seow, C. C. Methods for the study of starch retrogradation, *Food Chem.*, 2000, 71(1), 9–36.

ASTM. Standard test method for tensile properties of thin plastic sheeting. D882. In: *Annual Book of American Society for Testing Methods*. Philadelphia, PA: ASTM, 1991, pp. 313–321.

Baldwin, E. A.; Nisperos-Carriedo, M. O.; and Baker, R. A. *Crit. Rev. Food Sci.*, 1995, 35, 509–524.

Banker, G. S.; Gore, A. Y.; and Swarbrick, J. *J. Pharm. Pharmacol.*, 2000, 18, 173–176.

Barreto, P. L. M.; Pires, A. T. N.; and Soldi, V. *Polym. Degrad. Stabil.*, 2003, 79, 147–152.

Bifani, V.; Ramirez, C.; Ihl, M.; Rubilar, M.; Garcia, A.; and Zaritzky, N. *LWT-Food Sci. Technol.*, 2007, 40, 1473–1481.

Bourtoom, T. Edible films and coating: Characteristics and properties, *Int. Food Res. J.*, 2008, 15(3), 237–248.

Bourtoom, T.; Chinnan, M. S.; Jantawat, P.; and Sanguandeekul, R. *Food Sci. Technol. Int.* 2006, 12, 119–126.

Burt, S. A. *Int. J. Food Microbiol.*, 2004, 94, 223–253.

Callegarin, F. and Quezada–Gallo, J. A. Lipid and biopackaging, *JAOCS*, 1997, 74, 1183–1192.

Cerqueira, M. A.; Lima, A. M.; Souza, B. W. S.; Teixeira, J. A.; Moreira, R. A.; and Vicente, A. A. *J. Agric. Food Chem.*, 2009, 57, 1456–1462.

Cha, D. S. and Chinnan, M. S. *Crit. Rev. Food Sci.*, 2004, 44, 223–237.

Chang, Y. P.; Cheah, P. B.; and Seow, C. C. *J. Food Sci.*, 2000, 65, 1–7.

Crank, J. *The Mathematics of Diffusion*, 2nd edn., Oxford, U.K.: Clarendon Press, 1975.

Debeaufort, F.; Quezada-Gallo, J. A.; and Voilley, A. *Crit. Rev. Food Sci.*, 1998, 38, 299–313.

Delville, J.; Joly, C.; Dole, P.; and Bliard, C. *Carbohydr. Polym.*, 2003, 53, 373–381.

Donhowe, Y. G. and Fennema, O. R. *J. Food Process Preserv.*, 1993, 17, 247–257.

Endres, H. J.; Kammer-Stetter, H.; Hobelsberger, M.; and Reichenhall, B. *Starch*, 1994, 46(12), 474–480.

Garcia, M. A.; Ferrero, C.; Bertola, N.; Martino, M.; and Zaritzky, N. *Innovat. Food Sci. Emerg. Technol.*, 2002, 3, 391–397.

Garcia, M. A.; Martino, M. N.; Zaritzky, N. E.; and Plata, L. *Starch*, 2000, 52, 118–124.

Godbillot, L.; Dole, P.; Joly, C.; Roge, B.; and Mathlouthi, M. *Food Chem.*, 2006, 96, 380–386.

Gontard, N.; Guilbert, S.; and Cuq, J. L. *J. Food Sci.*, 1993, 58, 206–211.

Guilbert, S. *Bull. Int. Dairy Fed.*, 2000, 346, 10–16.

Han, J. H. and Gennadios, A. In *Innovations in Food Packaging*; Han, J. H.; Ed.; Oxford, U.K.: Elsevier Academic Press, 2005; pp. 239–262.

Han, C. and Krochta, J. M. *J. Food Sci.*, 2001, 66, 294–299.

Hirose, K.; Harte, B. R.; Giacin, J. R.; Miltz, J.; and Stine, C. In *Food and Packaging Interactions*; Hotchkiss, J. H.; Ed.; ACS Symposium Series 365, Washington, DC: American Chemical Society, 1998; pp. 28–41.

Hutchings, J. B. *Food Color and Appearance*, Gaithersburg, MD: Aspen Publisher, 1999.

Jagannath, J. H.; Nanjappa, C.; Das Gupta, D. K.; and Bawa, A. S. *Int. J. Food Sci. Technol.*, 2006, 41, 498–506.

Kayserilioglu, D. S.; Stevels, W. M.; Mulder, W. J.; and Akkas, N. *Starch*, 2001, 53, 381–386.

Kester, J. J. and Fennema, O. R. *Food Technol.*, 1986, 40, 47–59.

Krochta, J. M. In *Protein-Based Films and Coatings*; Gennadios, A.; Ed.; Boca Raton, FL: CRC Press, 2002; pp. 1–41.

Kussendrager, K. D. and van Hooijdonk, A. C. M. *Br. J. Nutr.*, 2000, 84, 19–25.

Lai, H. M. and Padua, G. W. *Cereal Chem.*, 1997, 74(6), 771–775.

Lai, H. M.; Padua, G. W.; and Wei, L. S. *Cereal Chem.*, 1997, 74(1), 83–90.

Letender, M.; D'aprano, G.; Lacroix, M.; Salmieri, S.; and St-Gelais, D. *J. Agric. Food Chem.*, 2002, 50, 6017–6022.

Lin, D. and Zhao, Y. *Comp. Rev. Food Sci. Food Saf.*, 2007, 6, 60–75.

Liu, Z. In *Innovations in Food Packaging*; Han, J. H.; Ed.; Oxford, U.K: Elsevier Academic Press, 2005; pp. 318–337.

Lourdin, D.; Della Valle, G.; and Colonna, P. *Carbohydr. Polym.*, 1995, 27, 261–270.

Maaurf, A. G.; Man, Y. B.; Asbi, B. A.; Junainah, A. H.; and Kennedy, J. F. *Carbohydr. Polym.*, 2001, 45, 335–345.

Maizura, M.; Fazilah, A.; Norziah, M. H.; and Karim, A. A. *J. Food Sci.*, 2007, 72, 324–330.

Mali, S.; Grossmann, M. V. E.; Garcia, M. A.; Martino, M. N.; and Zaritzky, N. E. Microstructural characterization of yam starch films, *Carbohydr. Polym.*, 2002, 50, 379–386.

Mali, S.; Grossmann, M. V. E.; Garcia, M. A.; Martino, M. N.; and Zaritzky N. E. Mechanical and thermal properties of yam starch films, *Food Hydrocoll.*, 2005, 19, 157–164.

Mali, S.; Grossmann, M. V. E.; Garcia, M. A.; Martino, M. N.; and Zaritzky, N. E. *J. Food Eng.*, 2006, 75, 453–460.

Mark, A. M.; Roth, W. B.; Mehltretter, C. L.; and Rist, C. E. *Food Technol.*, 1966, 20, 75–80.

Mathew, S.; Brahmakumar, M.; and Abraham, T E. Microstructural imaging and characterization of the mechanical, chemical, thermal, and swelling properties of starch–chitosan blend films, *Biopolym.*, 2006, 82(2), 176–187.

Mchugh, T. H.; Avena-Bustillos, R.; and Krochta, J. M. *J. Food Sci.*, 1993, 58(4), 899–903.

Mecitoğlu, C. and Yemenicioğlu, A. *Food Chem.*, 2007, 104, 726–733.

Mehyar, G. F. and Han, J. H. *J. Food Sci.*, 2004, 69(9), E449–E454.

Miller, K. S. and Krochta, J. M. *J. Appl. Polym. Sci.*, 1997, 88, 1095–1099.

Miller, K. S.; Upadhyaya, S. K.; and Krochta, J. M. *J. Food Sci.*, 1998, 63, 244–247.

Morillon, V.; Debeaufort, F.; Bond, G.; Capelle, M.; and Voilley, A. *Crit. Rev. Food Sci.* 2002, 42, 67–89.

Myllarinen, P.; Partanen, R.; Seppala, J.; and Forssell, P. *Carbohydr. Polym.*, 2002, 50, 355–361.

Nikolova, K.; Panchev, I.; and Sainov, S. *J. Optoelectron. Adv. M.*, 2005, 7, 1439–1444.

Petersson, M. and Stading, M. *Food Hydrocoll.*, 2005, 19, 123–132.

Pfeiffer, T. and Menner, M. *Fleischwirtschaft*, 1999, 79, 79–84.

Phan, T. D.; Debeaufort, F.; Luu, D.; and Volley, A. *J. Agric. Food Chem.*, 2005, 53, 973–981.

Rodriguez, M.; Oses, J.; Ziani, K.; and Mate, J. I. *Food Res. Int.*, 2006, 39, 840–846.

Romero-Bastida, C. A.; Bello-Perez, L. A.; Garcia, M. A.; Martino, M. N.; Solorza-Feria, J.; and Zaritzky, N. E. Physicochemical and microstructural characterization of films prepared by thermal and cold gelatinization from non-conventional sources of starches, *Carbohydr. Polym.*, 2005, 60, 235–244.

Santiago-Silva, P.; Soares, N. F. F.; Nobrega, J. E.; Junior, M. A. W.; Barbosa, K. B. F.; Volp, A. C. P.; Zerdas, E. R. M. A.; and Wurlitzer, N. *J. Food Control*, 2009, 20, 85–89.

Seifu, E.; Buys, E. M.; and Donkin, E. F. *Trends Food Sci. Technol.*, 2005, 16, 137–154.

Seydim, A. C. and Sarikus, G. *Food Res. Int.*, 2006, 39, 639–644.

Smits, A. L. M.; Kruiskamp, P. H.; Van Soest, J. J. G.; and Vliegenthart, J. F. G. *Carbohydr. Polym.*, 2003, 51, 417–424.

Suppakul, P.; Miltz, J.; Sonneveld, K.; and Bigger, S.W. *J. Food Sci.*, 2003, 68, 408–420.

Tako, M. and Hizukuri, S. *Carbohydr. Polym.*, 2002, 48, 397–401.

Talasila, P. C. and Cameron, A. C. *J. Food Sci.*, 1997, 62, 926–930.

Trezza, T. A. and Krochta, J. M. *J. Food Sci.*, 2000, 65, 658–662.

Vargas, M.; Pastor, C.; Chiralt, A.; McClements, D. J.; and Gonzalez-Martinez, C. *Crit. Rev. Food Sci.*, 2008, 48, 496–511.

Veiga-Santos, P.; Oliveira, L. M.; Cereda, M. P.; and Scamparini, A. R. P. *Food Chem.*, 2007, 103, 255–262.

Voilley, A. and Souchon, I. In *Flavour in Food*; Voilley, A. and Etievant, P.; Eds.; Cambridge, U.K.: Woodhead Publishing Limited and CRC Press LLC, 2006; pp. 117–132.

Whistler, R. L. and Daniel, J. R. In *Food Chemistry*; Fennema, O. R.; Ed.; New York: Marcel Dekker, 1985; p. 69.

Williams, R. and Mittal, G. S. *LWT-Food Sci. Technol.*, 1999, 32, 440–445.

Xu, X. Y.; Kim, K. M.; Hanna, M. A.; and Nag, D. *Ind. Crop. Prod.*, 2005, 21, 185–192.

Yoo, S. H. and Jane, J. L. *Carbohydr. Polym.*, 2002, 49, 297–305.

Zhang, Y. and Han, J. H. *J. Food Sci.*, 2006, 71(6), E253–E261.

Zimeri, J. E. and Kokini, J. L. *Carbohydr. Polym.*, 2002, 48, 299–304.

CHAPTER 12

Starch as a Feedstock for Bioproducts and Packaging

Gregory M. Glenn, Syed H. Imam, William J. Orts, and Kevin Holtman

CONTENTS

12.1 INTRODUCTION

The importance of developing sustainable products and practices is being recognized more so today than ever because of concerns over a global dependence on petroleum feedstocks. Crude oil has proven to be a very versatile, abundant, and valuable feedstock. While fuel continues to be the primary product made from crude oil, approximately 16%–20% is used for other products including plastics, fertilizers, chemicals, detergents, lubricants, pesticides, paints, etc. [1]. In fact, the value of petroleum in nonfuel products has led some to lament that petroleum is much too valuable a resource to simply burn as a fuel source. Reducing our dependency on petroleum feedstocks will be extremely challenging not only because of the vast amounts of petroleum consumed but also because of its historically low cost and versatility in making a host of products.

No single resource will be able to completely displace petroleum and all of the petroleum-based products. Notwithstanding, agriculture is one of the resources that is gaining importance in providing feedstocks for many industrial products. Historically, agriculture was developed primarily to provide for the food and feed requirements of society. However, as a feedstock, agriculture meets many of the criteria sought for as a replacement for petroleum. For example, agricultural feedstocks are a bountiful, renewable, and sustainable resource. Agricultural starches, oils, proteins, and fibers are currently used for a wide array of industrial products. This chapter will focus on efforts to utilize starch as a feedstock for replacing some petroleum-based products including food packaging.

12.2 FEEDSTOCKS FOR PLASTICS

12.2.1 Petroleum

Petroleum-based plastics have grown to play an integral role in the global marketplace. About 4.6% of the petroleum consumed in the United States is made into plastics [1]. Plastics have many outstanding properties that make them functional for a wide array of products. Plastics generally are low cost, may be processed into films or fibers, or may be molded into intricate parts [2]. Plastics have excellent chemical and moisture resistance and good thermal and electrical insulation, and they are lightweight yet strong [2]. It is no wonder that plastics have gained market share in the construction, furniture, automotive, textile, and packaging industries.

Global production of all types of plastics currently hovers at 245 MMT (million metric tons) which represents just under 6% of the world's crude oil production (Table 12.1). The plastic materials include thermoplastics, polyurethanes, thermosets, elastomers, adhesives, coatings, and sealants [3]. Much of the public concern over plastics is centered on single-use plastic food and beverage containers as well as polyethylene bags. The functional life of single-use containers may be less than the time required for a hot food or beverage to cool, and yet, the container may persist in the environment for decades or centuries depending on its composition. Perhaps nothing has come to represent the problem of plastics in the environment more than the use of polystyrene-foam food containers and packaging. In the United States, 0.82 MMT of plastic food service items, primarily made of polystyrene, were discarded in landfills in 2009 [4]. While recycling efforts have been very successful for paper and metal containers, there is virtually no significant recycling of polystyrene-foam food service packaging in the United States [4]. This is troublesome because polystyrene degrades particularly slowly, and because of its low density, it tends to be easily carried by water runoff into storm drains and accumulate in rivers, streams, or oceans [5]. A study conducted by a coastal U.S. state found that 15% of the total volume of litter collected from storm drains was comprised of polystyrene foam [5]. These concerns as well as concerns of litter in the city landscape and the amount of food packaging going into landfills have led scores of municipalities to enact legislation to ban the use of polystyrene for food service containers within their jurisdictions [5].

Table 12.1   Global and U.S. Consumption of Petroleum and Global Production of Starch-Rich Crops

Global consumption of petroleum	4242 [12]
Global production of plastics	245 [3]
Global production of grains (2010)	2243 [84,85]
Global production of potatoes and cassava (2010)	558 [85]
Potential global starch supply	1552[a]
U.S. consumption of petroleum (2009)	968 [12]
U.S. petroleum used for plastics (2006)	46.3 [1]

2009 Data (MMT)
[a] Starch supply data based on 65% starch extraction from grains and 17% extraction from fresh potato and cassava. Values in MMT.

One of the alternatives to plastic containers is paper. The amount of paper used in packaging and containers is approximately three times that of plastics [4]. Even though the production of paper pulp has been touted as damaging to the environment as plastics, there are factors associated with plastic packaging that should be taken into consideration such as the recycling rate [6]. Nearly 72% of paper packaging is recycled while less than 14% of plastic packaging is recycled [4]. Approximately 32% of the thermoplastics produced in the United States were converted into packaging and containers [4]. Plastic packaging and containers comprised 5.2% of municipal solid waste or approximately 11.4 MMT [4]. Paper litter tends to break down when it gets wet and disintegrate into fibers in aquatic environments. Plastics are generally very hydrophobic and do not disintegrate or degrade in the landscape or aquatic environments. In some beach areas, 55% of the litter on beaches is from plastics [7].

Another concern with the use of plastics is its environmental impact on the marine environment. Great masses of waste have been dumped in the world's oceans. Some of the debris is from "ocean dumping" by ships or barges from municipalities. However, a significant amount of debris is also thought to wash out to sea from storm drains and rivers [8]. While some waste materials slowly biodegrade or sink in aquatic environments, plastics tend to float at or near the water surface where they may persist for generations. The North Pacific Central Gyre (NPCG) [9], also known as the "Great Pacific Garbage Patch," is an expansive area in the Pacific Ocean where ocean currents gather floating waste from the North Pacific. Roughly 80% of the floating ocean debris consists of plastic [8].

The impact of plastic debris in the marine environment is believed to be greater than once thought. Birds, mammals, and fish may become entangled in plastic debris or ingest plastic particles [9]. More recent studies show that some seaborne plastics slowly disintegrate over time into microscopic debris due to ultraviolet radiation exposure and the mechanical agitation of waves. These particles float in a zone just beneath the water. There is evidence to support that the microscopic debris can absorb chemical toxins from seawater [10]. Recent studies have found microscopic plastic debris in plankton, thus raising fears of the accumulation of toxins and plastic debris in sea life that ultimately thrive on plankton [10,11].

Aside from the environmental and functional aspects, cost plays a key role in the development of new feedstocks to replace petroleum. The escalating cost of petroleum compared to agricultural commodities such as corn is a motivating factor for diverting more agricultural feedstocks to industrial products. During the period from 1946 to 1975, the price for a barrel of oil closely matched the price for a bushel of corn (Figure 12.1). The average price per barrel of oil increased sharply in the mid-1970s but remained below $50 until recently when the price surged to over $100 per barrel. The higher price of oil is attributed to an ever-increasing demand, especially from large developing countries such as India and China [12]. Production of nonrenewable resources such as petroleum is thought to follow a model proposed by Mario Hubbert in 1956. The production rate of a resource follows a bell curve where production initially increases and then tails off after peaking [13]. The term "Hubbert's Peak" was coined to describe the peak oil production for individual production sites as well as the collective global production. There are some who claim that the global oil supply has peaked and that there is insufficient supply to meet growing demand [13]. This assessment is supported by recent production data that reveal a decline in global oil production (Figure 12.2). Whether the cause of the decline is due to diminishing oil reserves or to a short-term fall in demand, it is clear that the need to develop alternative feedstocks to make fuel and products is becoming more urgent [13].

12.3 STARCH FEEDSTOCKS

Starch is the major component of important grain crops such as wheat, corn, and rice [14,15]. It is also the major solids component of root crops including potato and cassava. While starch is synthesized in small amounts in the chloroplasts of leaves, most of the plant starch supply is

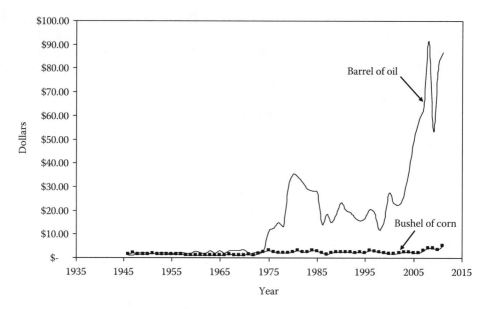

Figure 12.1 Price comparison between a barrel of crude oil and a bushel of corn. (From Oil Production, *The World Factbook*, Central Intelligence Agency, Skyhorse Publishing, New York, 2011; United States Department of Agriculture, Economic Research Service, *Grains Database: Yearbook Tables*. 2010. http://www.ers.usda.gov/data/FeedGrains/FeedYearbook.aspx)

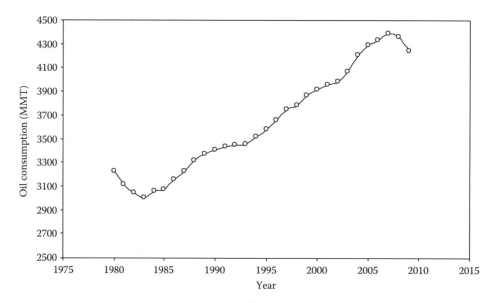

Figure 12.2 Global oil consumption. (From Oil Production, *The World Factbook*, Central Intelligence Agency, Skyhorse Publishing, New York, 2011.)

synthesized in amyloplasts where granules that range 2–100 μm in diameter are formed. The granules are prevalent in grains and root crops where they serve as a form of energy storage for the plant. Over 80% of commercial starch supplies are extracted from corn, while most of the remaining starch is extracted from wheat, potato, and cassava [14]. Starch is the lowest-priced and most abundant food commodity in the world (Table 12.2) [16].

Table 12.2 Increase in Price of Grain Commodities over a 5 Year Period

Commodity	$/Bu (2005)	$/Bu (2010)
Corn	2.00	5.05–5.75
Wheat	3.40	5.60–5.80
Sorghum	1.86	5.15–5.85
Oats	1.63	2.30–2.50
Barley	2.53	3.80–4.0

Source: From United States Department of Agriculture, Economic Research Service, *Grains Database: Yearbook Tables.* 2010. http://www.ers.usda.gov/data/FeedGrains/FeedYearbook.aspx

The use of starch as a feedstock for industrial products is not new. Millions of metric tons of starch already are used annually as an ingredient for paper sizing, adhesives, gypsum wallboard, and sizing for textiles [16]. Using starch as a feedstock to displace petroleum is complicated by the fact that agriculture itself is heavily dependent on petroleum feedstocks. Petroleum-based fertilizers, pesticides, and fuels to operate equipment for pumping water and for crop cultivation, harvest, and distribution are critical to agricultural production. A further complication is that starch is an important food source for the world's population. In fact, most of the production of starchy crops is used to help feed the world. The total global production of starch-rich crops including the amount used for food is less than 53% of the global consumption of petroleum oil (Table 12.1). It is clear that whatever surplus starch that can be spared for making industrial products will not be nearly enough to meet the current demand for petroleum feedstocks. Nevertheless, there is a sufficient starch supply to meet some of the growing needs for bioproducts.

The quantity of starch that is actually available for industrial products without impacting critical food supplies is a matter of debate. The use of starch and other bioproducts for industrial uses reduces the amount of food available to sustain the world's population and can generally lead to higher food costs [17]. On the other hand, the benefit of higher food prices is that it provides an incentive for farmers to increase production and provides an economic stimulus for rural and agricultural communities. Nowhere is this more apparent than with the U.S. corn industry. The corn industry has been a model of efficiency and productivity for many years. The yield per acre of corn was relatively constant from the late 1860s to the mid to late 1950s. However, since that time, yields have steadily climbed to a several-fold increase (Figure 12.3).

During the nearly 30-year period from 1946 through the early 1970s, the price per bushel of corn closely tracked the price for a barrel of oil (Figure 12.1). The price per bushel of corn fluctuated between $1 and $2 per bushel for nearly 30 years. Historically, the increases in the price per bushel of corn have coincided with dramatic increases in the price of oil such as during the mid-1970s (Figure 12.2). A recent increase in corn price has coincided with another large increase in oil price (Figure 12.2). Until recently, the real value U.S. farmers received per bushel of corn consistently trended downward (Figure 12.4). During this time, farming operations that were most efficient remained profitable by increasing production per acre and consolidating smaller family farms to reach efficiencies of scale. In contrast, the price for oil has increased dramatically relative to corn prices (Figure 12.1). The trends underscore the problem with continuing to rely heavily on petroleum and the relative value of utilizing agricultural feedstocks as an alternative resource.

State and federal research institutions have been forward thinking in utilizing agricultural resources in place of petroleum feedstocks. Their research has helped develop new markets for

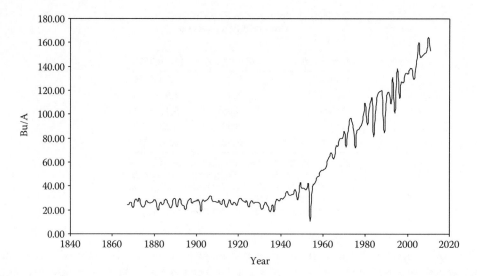

Figure 12.3 Changes in U.S. corn yields (bushels per acre) over the last 146 years. (From United States Department of Agriculture, Economic Research Service, *Grains Database: Yearbook Tables*. 2010. http://www.ers.usda.gov/data/FeedGrains/FeedYearbook.aspx)

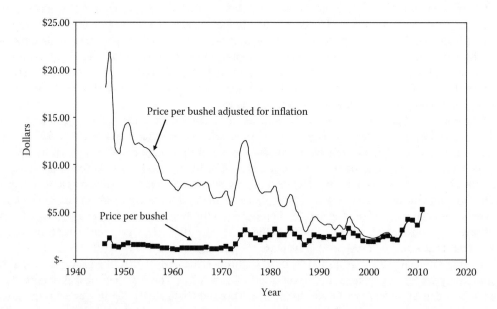

Figure 12.4 Comparison of price (dollars) per bushel and inflation adjusted prices since 1946. (From United States Department of Agriculture, Economic Research Service, *Grains Database: Yearbook Tables*. 2010. http://www.ers.usda.gov/data/FeedGrains/FeedYearbook.aspx)

agricultural commodities and helped bolster rural economies. Increasing the value of agricultural commodities has, in turn, resulted in increased production and greater efficiency. The most successful new market to date is the corn-based fuel ethanol industry. The amount of corn used to make ethanol has increased dramatically in the last few years (Figure 12.5). In 2010, nearly 40% of the U.S. corn production was diverted to ethanol production (Table 12.3). In spite of the amount of corn that is being diverted into industrial products, farmers have been able to produce enough grain to increase the supply for feed and export markets compared to earlier years (Table 12.4). The amount

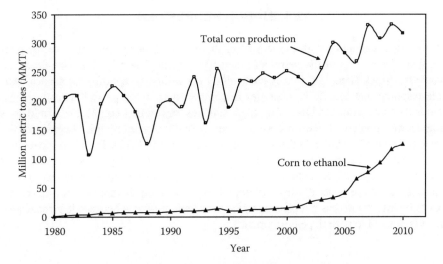

Figure 12.5 Total corn production and the amount of corn diverted into making ethanol. (From United States Department of Agriculture, Economic Research Service, Grains Database: *Yearbook Tables*. 2010. http://www.ers.usda.gov/data/FeedGrains/FeedYearbook.aspx)

Table 12.3 U.S. Consumption of Petroleum and Production of Starch-Rich Crops and Ethanol Production

U.S. gasoline consumption	414 [86]
U.S. grain production (including corn)	400 [84]
U.S. corn production	316 [84]
U.S. potato production	20 [83]
Potential U.S. starch supply	263[a]
Corn used for ethanol production	126 [83]
Corn ethanol production	39.8 [87]

2009 Data (MMT).
[a] Starch supply data based on 65% starch extraction from grains and 17% extraction from fresh potato. Values in MMT.

Table 12.4 Production and Use of U.S. Corn in MMT

Year	Production	Industrial	Feed	Domestic Use	Export
1975	148	13	91	104	42
2010	316	162	131	293	49
Increase (%)	113	1174	44	182	17

Source: From United States Department of Agriculture, Economic Research Service, *Grains Database: Yearbook Tables*. 2010. http://www.ers.usda.gov/data/FeedGrains/FeedYearbook.aspx

of U.S. corn that can ultimately be diverted from food and feed to meet the demand for industrial products may be debatable, but current data indicate that it must be greater than the 160 MMT currently used. The development of lignocellulosic ethanol could displace some or all of corn ethanol in the future, making available vast quantities of corn starch available for other industrial products such as bioplastics.

12.4 STARCH PROPERTIES

Starch is not abundant enough to meet all of the feedstock needs currently met by petroleum. However, there are sectors of the petroleum industry such as plastic food service products where a renewable feedstock such as starch is particularly appealing. In contrast to plastics, starch is a renewable resource that readily biodegrades in composting conditions and is easily metabolized by human and other mammals. Diverting starch supplies toward making plastics would not significantly impact the supply or prices of corn or other agricultural commodities because of the small market size (11.4 MMT) relative to the total supply of corn (316 MMT). The challenge is in making starch-based plastics that have the functionality and cost points comparable to those of plastics. While starch granules are insoluble in cold water, starch is hydrophilic as opposed to petroleum-based plastics which are generally hydrophobic. In spite of their differences in properties, starch continues to be investigated as a replacement for petroleum products because it is renewable, inexpensive, and compatible with the environment.

12.5 STARCH AS FILLER

Two general approaches have been undertaken to develop starch-based plastics. The first approach is to simply add starch as a filler for plastic resins. Early work with starch and plastic showed that up to 30% starch from, corn, rice, or cassava could be added to polyethylene while maintaining acceptable thermal stability and melt-flow properties needed for most processing needs [18]. Adding starch as a filler in plastics does not improve the biodegradation of the plastic component and it generally decreases the strength and elasticity of plastic film and makes them more moisture sensitive [18,19]. However, incorporating starch granules of relatively small diameter has less of a negative effect on the mechanical properties of plastic composites [19]. This may be due to the larger surface area of the smaller granules. Some manufacturers continue to use starch as a cheap filler in plastic food packaging primarily to increase the renewable content of food packaging.

Interest in nanoparticle fillers has grown in recent years because of their ability to improve the properties of composite materials. Nanoparticle fillers have a very high surface area which increases the surface interaction between the nanoparticles and the matrix material [20]. Nanoparticles from inorganic clay minerals, such as montmorillonite and hectorite, have been used for making polymer nanocomposites [21]. These clays consist of platelets that become intercalated under proper mixing conditions which facilitates their dispersion in a plastic matrix. Plastic composites made with clay nanoparticles have superior mechanical strength and gas-barrier properties [20–22]. Plastic nanocomposites made with cellulose microfibrils have been prepared [23]. As with starch, cellulose has the advantage of being a renewable resource. Cellulose microfibrils are long crystalline fibers ranging from 10 to 20 nm in diameter with an average aspect ratio of 20–100 [24–26]. Due to their high surface area per weight ratio, cellulose microfibrils are thought to work similar to clay nanoparticles as a reinforcing particle in plastic matrix materials [24–29]. Cellulose microfibrils are effective in increasing the strength of matrices due to their high axial Young's moduli (137 GPa) which are similar to Kevlar and stronger than steel [30,31]. Cellulose microfibril nanocomposites have improved strength, lower density, less abrasiveness, and other properties deemed useful for many different applications [23].

As mentioned previously, starch granules are in the micrometer size range (10–100 μm). The granule structure must be disrupted in order to form nanoparticles. Starch nanocrystals have been made by breaking down the starch granule structure with either HCl or H_2SO_4 [14]. Angellier et al. [32] partially hydrolyzed waxy starch in H_2SO_4 to make starch nanocrystals 15–30 nm in diameter. They were able to add up to 20% of the starch nanoparticles without reducing elongation to break

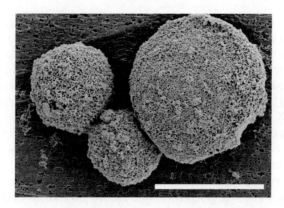

Figure 12.6 Porous microspheres made from atomized, air-classified droplets of starch collected in ethanol and dried. Scale bar = 10 μm. (From Glenn, G.M. et al., *J. Agric. Food Chem.*, 58, 7, 4180, 2010.)

of the matrix material. Ma et al. [33] made starch particles 50–100 nm by precipitating a starch paste with ethanol while subjecting the solution to constant stirring. They did not report on the fate of starch granule remnants. Nevertheless, the starch nanoparticles prepared in this manner increased the tensile yield strength and Young's modulus of composite materials [33]. Glenn et al. [34] completely destructurized starch granules of high-amylose corn starch granules by heating an aqueous suspension to 140°C with moderate shear. The resulting starch paste was atomized through a nozzle, air classified, and collected in ethanol [35]. Although the starch particles collected from the ethanol bath were in the micrometer range (1–12 μm), the particles had a very high surface area due to the porous nature of the particles (Figure 12.6). These high surface area nanoporous starch microspheres could be useful in making composites with properties similar to those of nanocomposites.

12.6 STARCH AS A FERMENTATION FEEDSTOCK

Polyhydroxyalkanoate (PHA) and polylactic acid (PLA) are the two common biopolymers derived from microbial fermentation of starch [36,37]. PLA is a transparent thermoplastic with a glass transition temperature around 60°C and a melt temperature of 170°C–180°C [36,38]. Its high modulus compares to that of PET or cellophane [39]. The use of PLA in films for packaging applications is occurring on a commercial scale. However, PLA must be plasticized in order to obtain films with adequate flexibility [40]. Plasticizers that have been studied for PLA include poly(3-methyl-1,4-dioxan-2-one), 6 poly(ethylene oxide), citrate esters, triacetin, and poly(ethylene glycol)s (PEGs) [38,41].

PLA is made from lactic acid produced via bacterial fermentation typically using starch as a feedstock [42]. PLA is either produced directly from the condensation of lactic acid or through the ring opening polymerization of lactide [36,43,44]. Lactic acid exists as two stereoisomers, or enantiomers, either L or D [39,44]. The ratio of L to D enantiomers determines the structure and, thus, the properties of the resultant polymer [36,44].

The D contamination in the PLA tends to lower the melting point and the degree of crystallization as well as lower the tensile strength of the material [39,44]. The ring-opening process tends to produce a polymer that is mostly L, which may be preferred commercially for fibers, thermoforming, injection molding, or foaming depending on the specific amount of D present.

PHAs are natural biopolymers that are synthesized and catabolized by microorganisms [45,46]. The interest in PHAs is because they process similarly to many important petroleum-based plastics, and yet they are completely renewable and biodegradable in many environments [47]. Unlike starches, PHAs are hydrophobic. The properties of PHAs depend on their chemical structure which varies with feedstock composition, processing conditions, and selection of microorganisms [48]. PHAs are polymers containing 10^3–10^4 3-hydroxy fatty acid monomers [48]. They accumulate as small granules (0.2–0.5 μm) made of 2-hydroxy fatty acid monomers [49]. PHAs accumulate in microorganisms as a storage molecule in response to an external stress such as a nutrient imbalance [50].

Interest in PHAs has grown because of their useful functional properties. For instance, films made from PHAs may have gas-barrier properties similar to poly(vinyl chloride) and poly(ethylene terephthalate) which make them suitable for packaging applications [51]. Although PHAs show great promise, they have yet to develop a significant market due to their high production costs [47]. Starch has been used as a source of glucose for fermentative production of PHAs to reduce the costs [47]. Huang et al. [47] used amylases to produce glucose from extruded bran and cornstarch which, in turn, was used as a feedstock for PHA production from *Haloferax mediterranei*. Further research to develop ways of reducing the production costs of PHAs could eventually increase the viability of PHAs as a replacement for petroleum-based plastics.

Ethanol is a renewable product that has gained interest in recent years as a feedstock for plastic. Whether the ethanol is based on corn starch as in the United States or sugar as in Brazil, there is growing interest in making "green" plastics that are derived from renewable resources. The Brazilian chemical company Braskem opened a 0.2 MMT capacity plant to produce high-density polyethylene in 2010. The company has also announced plans to open another smaller plant (0.032 MMT) in 2013 to produce polypropylene [52]. As oil prices continue to climb and petroleum resources diminish, green plastics from renewable resources continue to grow and become more of a viable alternative to petroleum-based plastics.

12.7 STARCH AS A MATRIX

12.7.1 Microcellular Foam

In addition to starch being used as a filler and a feedstock for plastics, starch is also used as a matrix material. The benefit of using starch as a matrix material is that the final product can have a much higher starch content. One interesting material that has been studied as a replacement for plastic foam products is starch-based microcellular foam (MCF) [53]. Starch-based MCF is made by dehydrating aqueous gels of starch with a desiccating solvent such as ethanol. The ethanol is removed by air-drying or critical point drying resulting in a dry, porous foam that can be made into blocks, panels, beads, or powders depending on the desired application [54]. The open cell, porous structure of MCF provides excellent thermal insulation [53]. While the MCF has good compressive strength, it is rather weak and brittle when tensile or flexural stress is applied [53]. The MCF is not moisture resistant and would require a moisture-resistant film or barrier to meet the functional requirements of many commercial applications.

MCF has a large surface area and a very small pore size which give it properties that could be useful for other applications. MCF is able to adsorb small amounts of a number of volatile compounds and lower the partial pressure in the headspace above the MCF manifold [54,55]. This property could be useful for encapsulating flavors for use in dry foods such as cake mixes or for biodegradable controlled-release devices [54,55]. The small pore size and porous matrix structure help the MCF to absorb and immobilize liquids which makes the MCF an effective reservoir for controlled-release devices [54]. The MCF has been tested as a substitute for titanium dioxide as a whitening pigment in paper production [56,57]. MCF has also been tested in chemical encapsulation and column chromatography [58]. More recently, the foam was tested as a means of protecting honeybees from parasitic mites. Small spheres of MCF foam were prepared by atomization and air classification techniques (Figure 12.6). Once loaded with a control agent for mites, the spheres are small enough in diameter to be ingested by honeybees and small enough to attach to honeybees similar to pollen grains [35]. The foam has also been tested in controlled-release devices where the device has no need to be recovered from the field because they can be safely ingested or will readily degrade in the field [35,59].

12.7.2 Baking

Single-use food packaging can be made from starch composites using a baking process [60]. The composites are made by combining starch with fiber and other minor components such as a mold-release agent and a filler. The composites are cost competitive with their plastic counterpart and are functional as food service containers.

The baking technology is an adaptation of a baking technology used in the food industry for making ice-cream cones and wafer cookies [61]. The process involves preparing an aqueous starch slurry that is deposited in a heated mold and baked for approximately 45 s. Moisture is baked out of the slurry as steam that produces a foam core as it forges its way through the viscous, molten slurry. The fiber component adds strength and flexibility of the baked item as the mold is opened and demolding occurs [62,63].

One of the functional problems of the starch containers was the moisture sensitivity. Moisture resistance was achieved by laminating the containers with a vapor barrier film. Coincidentally, the containers had greater strength and flexibility because of the film laminate added to the baked starch substrate. Starch was blended with natural rubber latex or with poly(vinyl alcohol) to provide better moisture resistance in the baked product [62,64,65]. However, to date, the laminate film has been the preferred method of providing moisture resistance. Starch-based, baked products are produced commercially as single-use, disposable plates and bowls. These items degrade in composting conditions in approximately 35 days.

12.7.3 Extrusion

Starch can be heated in a closed vessel with predetermined quantities of water to form homogeneous melts [66]. Extrusion is one process that can successfully melt starch using low water concentrations and has been used extensively in the food industry for many years to create puffed snack foods [67,68]. It is an energy efficient cooking system able break down the starch granule structure through a combination of high shear, temperature, and pressure. When processed by extrusion, the thermoplastic starch (TPS) behaves similar to thermoplastic resin-like products and can be further processed by injection molding or another extrusion step [66,69].

Although TSP may be injection molded or extruded similarly to thermoplastics, it remains moisture sensitive and unstable. The main use of TSP is for loose-fill packing materials and sheets of starch foam for cushioning applications [70–73]. Moisture resistance can be improved by chemically modifying starch through grafting or acetylation, for example [74]. However, chemical modifications of starch can be cost prohibitive. Blending TSP with other miscible polymers can be an effective means of producing a blend that can substitute for thermoplastic resins [69,75,76]. Starch has been blended with a number of aliphatic polyesters to improve biodegradability. These blends include polycaprolactone [77], PLA [78], polyhydroxybutyrate-covalerate (PHBV) [79], and vinyl alcohol copolymer [80] among others.

The use of TSP as a replacement for single-use food packaging has been commercialized by at least two European companies, Novamont in Italy [75,80] and Bio-plast in Germany [81]. These starch-based materials are currently used to make film for disposable bags, cutlery, and food service products. However, production capacity is still low (0.02 MMT) compared to the global market. Production of TSP-based plastics will likely target niche markets until they become more price competitive with petroleum-based commodity plastics.

12.8 CONCLUSIONS

Much progress has been achieved in developing starch-based feedstocks as a partial replacement for petroleum-based feedstocks. Although starch remains a poor direct substitute for plastics, composite starch-based materials have useful functional properties and are in commercial production as a replacement for single-use plastic food service containers. In recent years, starch has developed an even greater role as a fermentation feedstock. Ethanol and PLA are the two most important starch fermentation products developed for replacing petroleum fuel and plastics. Further growth of starch-based fermentation products is likely as new fermentation processes and products are identified and commercially developed. The utilization of other agricultural fermentation feedstocks such as plant biomass will be important to meet ever-growing demands for fuels and products and help prevent diverting too much starch or other food resources into industrial products and will also help reduce overdependence on petroleum feedstocks.

REFERENCES

1. U.S. Energy Information Administration. Oil uses. 2011. http://www.eia.doe.gov/tools/faqs/faq. cfm?id=34&t=6 (accessed June 9, 2011).
2. American Chemistry Council. Life cycle of a plastic product. 2011. http://www.americanchemistry. com/s_plastics/doc.asp?cid=1571&did=5972 (accessed June 9, 2011).
3. Johansson, J.-E., Compelling facts about plastics; PlasticsEurope Market Research Group (PEMRG): 2009. http://www.plasticsrecyclers.eu/uploads/media/eupr/Compelling/PlasticsEurope.pdf (accessed June 9, 2011).
4. Solid Waste and Emergency Response (5306P). United States Environmental Protection Agency: (EPA-530-F-010-012). 2010. http://www.epa.gov/wastes/ (accessed June 9, 2011).
5. Gorden, M., Ban on expanded polystyrene foam takeout food packaging. In *California Water Action*. 2006. http://www.cleanwateraction.org/ca (accessed June 9, 2011).
6. Hunt, R. G., LCA considerations of solid waste management systems-case studies using a combination of life cycle assessment (LCA) and life cycle costing (LCC). *Resources, Conservation, and Recycling* 1995, 14(33), 225–231.
7. Nakashima, E.; Isobe, A.; Magome, S.; Kako, S.; Deki, N., Using aerial photography and in situ measurements to estimate the quantity of macro-litter on beaches. *Marine Pollution Bulletin* 2011, 62(4), 762–769.
8. Walker, T. R.; Reid, K.; Arnould, J. P. Y.; Croxall, J. P., Marine debris surveys at Bird Island, South Georgia 1990–1995. *Marine Pollution Bulletin* 1997, 34(1), 61–65.
9. Boerger, C. M.; Lattin, G. L.; Moore, S. L.; Moore, C. J., Plastic ingestion by planktivorous fishes in the North Pacific Central Gyre. *Marine Pollution Bulletin* 2010, 60(12), 2275–2278.
10. Kaiser, J., The dirt on ocean garbage patches. *Science* 2010, 328(5985), 1506.
11. Zhang, Y.; Zhang, Y.-B.; Feng, Y.; Yang, Z.-J., Reduce the plastic debris: A model research on the great Pacific Ocean garbage patch. *Advanced Materials Research* 2010, 113–114, 59–63.
12. U.S. Department of Energy, International petroleum consumption, 2009.
13. Bardi, U., Peak oil: The four stages of a new idea. *Energy* 2009, 34(3), 323–326.
14. Le Corre, D.; Bras, J.; Dufresne, A., Starch nanoparticles: A review. *Biomacromolecules* 2010, 11(5), 1139–1153.
15. Ellis, R. P.; Cochrane, M. P.; Dale, M. F. B.; Duffus, C. M.; Lynn, A.; Morrison, I. M.; Prentice, R. D. M.; Swanston, J. S.; Tiller, S. A., Starch production and industrial use. *Journal of the Science of Food and Agriculture* 1998, 77(3), 289–311.

16. French, D., Organization of starch granules. In *Starch: Chemistry and Technology*, 2nd edn.; Whistler, R. L.; Bemiller, J. N.; Paschall, E. F., Eds. Academic Press Inc.: Orlando, FL, 1984; pp. 183–247.

17. Cassman, K. G.; Liska, A. J., Food and fuel for all: Realistic or foolish? *Biofuels, Bioproducts and Biorefining* 2007, 1(1), 18–23.

18. Griffin, G. J. L., Biodegradable fillers in thermoplastics. In *Fillers and Reinforcements for Plastics*, Vol. 134. American Chemical Society: Washington, DC, 1974; pp. 159–170.

19. Lim, S.; Jane, J. L.; Rajagopalan, S.; Seib, P. A., Effect of starch granule size on physical properties of starch-filled polyethylene film. *Biotechnology Progress* 1992, 8(1), 51–57.

20. Chiou, B.-S.; Yee, E.; Glenn, G. M.; Orts, W. J., Rheology of starch-clay nanocomposites. *Carbohydrate Polymers* 2005, 59(4), 467–475.

21. Liao, H.-T.; Wu, C.-S., Synthesis and characterization of polyethylene-octene elastomer/clay/biodegradable starch nanocomposites. *Journal of Applied Polymer Science* 2005, 97(1), 397–404.

22. Maiti, P.; Nam, P. H.; Okamoto, M.; Hasegawa, N.; Usuki, A., Influence of crystallization on intercalation, morphology, and mechanical properties of polypropylene/clay nanocomposites. *Macromolecules* 2002, 35(6), 2042–2049.

23. Panaitescu, D. M.; Donescu, D.; Bercu, C.; Vuluga, D. M.; Iorga, M.; Ghiurea, M., Polymer composites with cellulose microfibrils. *Polymer Engineering & Science* 2007, 47(8), 1228–1234.

24. Favier, V.; Chanzy, H.; Cavaille, J. Y., Polymer nanocomposites reinforced by cellulose whiskers. *Macromolecules* 2002, 28(18), 6365–6367.

25. Dubief, D.; Samain, E.; Dufresne, A., Polysaccharide microcrystals reinforced amorphous poly(Î²-hydroxyoctanoate) nanocomposite materials. *Macromolecules* 1999, 32(18), 5765–5771.

26. Dufresne, A.; Dupeyre, D.; Vignon, M. R., Cellulose microfibrils from potato tuber cells: Processing and characterization of starch-cellulose microfibril composites. *Journal of Applied Polymer Science* 2000, 76(14), 2080–2092.

27. Angles, M. N.; Dufresne, A., Plasticized starch/tunicin whiskers nanocomposite materials. 2. Mechanical behavior. *Macromolecules* 2001, 34(9), 2921–2931.

28. Paillet, M.; Dufresne, A., Chitin whisker reinforced thermoplastic nanocomposites. *Macromolecules* 2001, 34(19), 6527–6530.

29. Dufresne, A.; Vignon, M. R., Improvement of starch film performances using cellulose microfibrils. *Macromolecules* 1998, 31(8), 2693–2696.

30. Wu, Q.; Henriksson, M.; Liu, X.; Berglund, L. A., A high strength nanocomposite based on microcrystalline cellulose and polyurethane. *Biomacromolecules* 2007, 8(12), 3687–3692.

31. Leitner, J.; Hinterstoisser, B.; Wastyn, M.; Keckes, J.; Gindl, W., Sugar beet cellulose nanofibril-reinforced composites. *Cellulose* 2007, 14(5), 419–425.

32. Angellier, H. L. N.; Molina-Boisseau, S.; Dufresne, A., Mechanical properties of waxy maize starch nanocrystal reinforced natural rubber. *Macromolecules* 2005, 38(22), 9161–9170.

33. Ma, X.; Jian, R.; Chang, P. R.; Yu, J., Fabrication and characterization of citric acid-modified starch nanoparticles/plasticized-starch composites. *Biomacromolecules* 2008, 9(11), 3314–3320.

34. Glenn, G. M.; Klamczynski, A.; Chiou, B.-S.; Orts, W. J.; Imam, S. H.; Wood, D. F., Temperature related structural changes in wheat and corn starch granules and their effects on gels and dry foam. *Starch—Stärke* 2008, 60(9), 476–484.

35. Glenn, G. M.; Klamczynski, A. P.; Woods, D. F.; Chiou, B.; Orts, W. J.; Imam, S. H., Encapsulation of plant oils in porous starch microspheres. *Journal of Agricultural and Food Chemistry* 2010, 58(7), 4180–4184.

36. Mecking, S., Nature or petrochemistry?—Biologically degradable materials. *Angewandte Chemie International Edition* 2004, 43(9), 1078–1085.

37. Poirier, Y., Green chemistry yields a better plastic. *Nature Biotechnology* 1999, 17, 960–961.

38. Martin, O.; Avérous, L., Poly(lactic acid): Plasticization and properties of biodegradable multiphase systems. *Polymer* 2001, 42(14), 6209–6219.

39. Garlotta, D., A literature review of poly(lactic acid). *Journal of Polymers and the Environment* 2001, 9(2), 63–84.

40. Ljungberg, N.; Wesslén, B., Tributyl citrate oligomers as plasticizers for poly (lactic acid): Thermo-mechanical film properties and aging. *Polymer* 2003, 44(25), 7679–7688.

41. Kulinski, Z.; Piorkowska, E.; Gadzinowska, K.; Stasiak, M., Plasticization of poly(l-lactide) with poly(propylene glycol). *Biomacromolecules* 2006, 7(7), 2128–2135.

42. John, R. P.; Nampoothiri, K. M.; Pandey, A., Fermentative production of lactic acid from biomass: An overview on process developments and future perspectives. *Applied Microbiology Biotechnology* 2007, 74, 524–534.

43. Cheng, M.; Attygalle, A. B.; Lobkovsky, E. B.; Coates, G. W., Single-site catalysts for ring-opening polymerization: Synthesis of heterotactic poly(lactic acid) from rac-lactide. *Journal of the American Chemical Society* 1999, 121(49), 11583–11584.

44. Siparsky, G.; Voorhees, K.; Dorgan, J.; Schilling, K., Water transport in polylactic acid (PLA), PLA/polycaprolactone copolymers, and PLA/polyethylene glycol blends. *Journal of Polymers and the Environment* 1997, 5(3), 125–136.

45. Akar, A.; Akkaya, E.; Yesiladali, S.; Çelikyilmaz, G.; Çokgor, E.; Tamerler, C.; Orhon, D.; Çakar, Z., Accumulation of polyhydroxyalkanoates by Microlunatus phosphovorus under various growth conditions. *Journal of Industrial Microbiology and Biotechnology* 2006, 33(3), 215–220.

46. Babel, W.; Steinbüchel, A.; Kim, Y.; Lenz, R., Polyesters from microorganisms. In *Biopolyesters*, Vol. 71. Springer: Berlin/Heidelberg, Germany, 2001; pp. 51–79.

47. Huang, T.-Y.; Duan, K.-J.; Huang, S.-Y.; Chen, C., Production of polyhydroxyalkanoates from inexpensive extruded rice bran and starch by Haloferax mediterranei. *Journal of Industrial Microbiology and Biotechnology* 2006, 33(8), 701–706.

48. Suriyamongkol, P.; Weselake, R.; Narine, S.; Moloney, M.; Shah, S., Biotechnological approaches for the production of polyhydroxyalkanoates in microorganisms and plants—A review. *Biotechnology Advances* 2007, 25(2), 148–175.

49. Luengo, J. M.; García, B.; Sandoval, A.; Naharro, G.; Olivera, E. R., Bioplastics from microorganisms. *Current Opinion in Microbiology* 2003, 6(3), 251–260.

50. Reddy, C. S. K.; Ghai, R.; Rashmi; Kalia, V. C., Polyhydroxyalkanoates: An overview. *Bioresource Technology* 2003, 87(2), 137–146.

51. Choi, H. J.; Kim, J.; Jhon, M. S., Viscoelastic characterization of biodegradable poly(3-hydroxybutyrate-co-3-hydroxyvalerate). *Polymer* 1999, 40(14), 4135–4138.

52. Esposito, F., Braskem planning to add PVC and PE capacity. *Plastic News*, April 8, 2011. http://plastics-news.com/headlines2.html?id=21655 (accessed June 9, 2011).

53. Glenn, G. M.; Irving, D. W., Starch-based microcellular foams. *Cereal Chemistry* 1995, 72(2), 155–161.

54. Glenn, G. M.; Klamczynski, A. P.; Takeoka, G.; Orts, W. J.; Wood, D.; Widmaier, R., Sorption and vapor transmission properties of uncompressed and compressed microcellular starch foam. *Journal of Agricultural and Food Chemistry* 2002, 50(24), 7100–7104.

55. Buttery, R. G.; Glenn, G. M.; Stern, D. J., Sorption of volatile flavor compounds by microcellular cereal starch. *Journal of Agricultural and Food Chemistry* 1999, 47(12), 5206–5208.

56. El-Tahlawy, K.; Venditti, R. A.; Pawlak, J. J., Aspects of the preparation of starch microcellular foam particles crosslinked with glutaraldehyde using a solvent exchange technique. *Carbohydrate Polymers* 2007, 67(3), 319–331.

57. Bolivar, A. I.; Venditti, R. A.; Pawlak, J. J.; El-Tahlawy, K., Development and characterization of novel starch and alkyl ketene dimer microcellular foam particles. *Carbohydrate Polymers* 2007, 69(2), 262–271.

58. Glenn, G. M.; Klamczynski, A. P.; Ludvik, C.; Shey, J.; Imam, S. H.; Chiou, B.; McHugh, T.; Orts, W. J.; Wood, D.; DeGrandi-Hoffman, G.; Offeman, R., Permeability of starch gel matrices and select films to solvent vapors. *Journal of Agricultural and Food Chemistry* 2006, 54(9), 3297–3304.

59. Glenn, G. M.; Klamczynski, A. P.; Shey, J.; Chiou, B.-S.; Holtman, K. M.; Wood, D. F.; Ludvik, C.; DeGrandi-Hoffman, G.; Orts, W.; Imam, S., Controlled release of 2-heptanone using starch gel and polycaprolactone matrices and polymeric films. *Polymers for Advanced Technologies* 2007, 18(8), 636–642.

60. Andersen, P. J.; Hodson, S. K. Molded starch-bound containers and other articles having natural and/or synthetic polymer coatings. U.S. Patent# 6030673, 2000.

61. Tiefenbacher, K. F., Starch-based foamed materials-use and degradation properties. *Journal of Macromolecular Science, Part A: Pure and Applied Chemistry* 1993, 30(9), 727–731.

62. Glenn, G. M.; Orts, W. J.; Nobes, G. A. R., Starch, fiber and $CaCO_3$ effects on the physical properties of foams made by a baking process. *Industrial Crops and Products* 2001, 14(3), 201–212.

63. Shogren, R. L.; Lawton, J. W.; Doane, W. M.; Tiefenbacher, K. F., Structure and morphology of baked starch foams. *Polymer* 1998, 39(25), 6649–6655.

64. Cinelli, P.; Chiellini, E.; Lawton, J. W.; Imam, S. H., Foamed articles based on potato starch, corn fibers and poly(vinyl alcohol). *Polymer Degradation and Stability* 2006, 91(5), 1147–1155.
65. Shey, J.; Imam, S. H.; Glenn, G. M.; Orts, W. J., Properties of baked starch foam with natural rubber latex. *Industrial Crops and Products* 2006, 24(1), 34–40.
66. Stepto, R. F. T., Thermoplastic starch. *Macromolecular Symposia* 2009, 279(1), 163–168.
67. Lai, L. S.; Kokini, J. L., Physicochemical changes and rheological properties of starch during extrusion (a review). *Biotechnology Progress* 1991, 7(3), 251–266.
68. Wiedmann, W.; Strobel, E., Compounding of thermoplastic starch with twin-screw extruders. *Starch—Stärke* 1991, 43(4), 138–145.
69. Swanson, C. L.; Shogren, R. L.; Fanta, G. F.; Imam, S. H., Starch-plastic materials—Preparation, physical properties, and biodegradability (a review of recent USDA research). *Journal of Polymers and the Environment* 1993, 1(2), 155–166.
70. Willett, J. L.; Shogren, R. L., Processing and properties of extruded starch/polymer foams. *Polymer* 2002, 43(22), 5935–5947.
71. Lacourse, N. L.; Altieri, P. A., Biodegradable packing material and the method of preparation. U.S. Patent# 4863655, 1989.
72. Nabar, Y.; Raquez, J. M.; Dubois, P.; Narayan, R., Production of starch foams by twin-screw extrusion: Effect of maleated poly(butylene adipate-co-terephthalate) as a compatibilizer. *Biomacromolecules* 2005, 6(2), 807–817.
73. Nabar, Y.; Ramani, N.; Schindler, M., Twin-screw extrusion production and characterization of starch foam products for use in cushioning and insulation applications. *Polymer Engineering & Science* 2006, 46(4), 438–451.
74. Shogren, R. L.; Fanta, G. F.; Doane, W. M., Development of starch based plastics—A reexamination of selected polymer systems in historical perspective. *Starch—Stärke* 1993, 45(8), 276–280.
75. Bastioli, C.; Cerutti, A.; Guanella, I.; Romano, G. C.; Tosin, M., Physical state and biodegradation behavior of starch-polycaprolactone systems. *Journal of Polymers and the Environment* 1995, 3(2), 81–95.
76. Averous, L.; Boquillon, N., Biocomposites based on plasticized starch: Thermal and mechanical behaviours. *Carbohydrate Polymers* 2004, 56(2), 111–122.
77. Averous, L.; Moro, L.; Dole, P.; Fringant, C., Properties of thermoplastic blends: Starch-polycaprolactone. *Polymer* 2000, 41(11), 4157–4167.
78. Huneault, M. A.; Li, H., Morphology and properties of compatibilized polylactide/thermoplastic starch blends. *Polymer* 2007, 48(1), 270–280.
79. Innocentini-Mei, L. H.; Bartoli, J. R.; Baltieri, R. C., Mechanical and thermal properties of poly(3-hydroxybutyrate) blends with starch and starch derivatives. *Macromolecular Symposia* 2003, 197(1), 77–88.
80. Bastioli, C., Properties and applications of Mater-Bi starch-based materials. *Polymer Degradation and Stability* 1998, 59(1–3), 263–272.
81. Angellier, H. L. N.; Molina-Boisseau, S.; Dole, P.; Dufresne, A., Thermoplastic starch-waxy maize starch nanocrystals nanocomposites. *Biomacromolecules* 2006, 7(2), 531–539.
82. Oil Production. *The World Factbook*. Central Intelligence Agency, Skyhorse Publishing, New York, 2011.
83. United States Department of Agriculture, Economic Research Service, *Grains Database: Yearbook Tables*. 2010. http://www.ers.usda.gov/data/FeedGrains/FeedYearbook.aspx (accessed June 9, 2011).
84. United States Department of Agriculture, *World Agricultural Supply and Demand Estimates*. 2011; pp. WASDE-491-8.
85. Food and Agriculture Organization of the United Nations. *Statistical Yearbook*. 2010. http://faostat.fao.org/DesktopDefault.aspx?PageID=567&lang=en#ancor (accessed June 9, 2011).
86. American Fuels, 2010 Gasoline Consumption. 2011. http://americanfuels.blogspot.com/2011/02/2010-gasoline-consumption.html (accessed June 9, 2011).
87. Renewable Fuels Association. *Historic U.S. Fuel Ethanol Production*. 2011. http://www.ethanolrfa.org/pages/statistics (accessed June 9, 2011).

CHAPTER 13

Chemometric Analysis of Multicomponent Biodegradable Plastics by Fourier Transform Infrared Spectrometry
The R-Matrix Method

Sherald H. Gordon

CONTENTS

13.1 INTRODUCTION

Research is being conducted in a number of private, academic, and government laboratories worldwide to develop biodegradable materials that neither persist in the environment nor create the undesirable ecological effects for which ubiquitous synthetic plastics have become notorious. The intent is to produce consumer products that biodegrade much more rapidly than synthetic petroleum-based plastics. Examples are agricultural mulches for water retention in crops and films used for weed control, as well as rigid materials such as eating utensils, packaging materials, bottles, and other containers that are disposable and can biodegrade after use.

The kinds of biodegradable plastics that are being developed recently at USDA and other laboratories include combinations of renewable plant polymers with microbial polyesters blended by injection molding or extrusion. These composites are then tested to measure the rate and extent of biodegradation of each component over time under environmental conditions. Methods used for testing biodegradation generally consist of weight loss measurements of the plastics in situ. However, with multicomponent plastics, when weight loss is the only criterion, it is not possible to determine the relative amounts of each component remaining in a biodegraded specimen. It cannot be assumed, for example, that weight loss is due to degradation of any one component in the plastic while all of the other components remain intact. Therefore, the testing method must be capable of measuring the amount of degradation in each and every component.

The problem of measuring degradation of each individual component in a multicomponent biodegradable plastic is not a trivial one. In the years since biodegradable plastics were first invented, this perplexing problem appeared to be an insurmountable obstacle for developers seeking to optimize their performance and biodegradability. The history of research to understand the structure–property relationships in multicomponent plastics is replete with attempts to quantify specific components in biodegraded samples. A number of wet chemical extraction and biochemical assays[1-6] have been used in studies of polymer blends having only two components, such as blends of a plant polymer and a synthetic polymer. Several quantitative and semiquantitative instrumental tests have been developed and used in USDA to study biodegradation of plastics. The instrumental tests include such advanced techniques as fluorescence microscopy[4,6] and cross-polarization/magic angle spinning ^{13}C solid-state NMR spectroscopy (CP/MAS ^{13}C-NMR), which was used to estimate the extent of hydrolysis of starch in enzymatically degraded starch-PE-EAA injection-molded plastics,[5] in addition to carbon-13 analysis[7] and Fourier transform infrared (FT-IR) spectrometry.[8-13]

CP/MAS ^{13}C-NMR is a novel technique that compares spectra of biodegraded and unbiodegraded plastics. By measuring the diminution of the ^{13}C peaks of the carbohydrate portion, it was possible to estimate the residual starch content in three-component starch-EAA-PE blends after biodegradation.[5] Results agreed reasonably well with reducing sugar assays. CP/MAS ^{13}C-NMR can probe tertiary structure, as well as interactions between individual components. However, the data are usually only semiquantitative, and, even worse, they say nothing about the relative concentrations of the noncarbohydrate components.

Carbon-13 analysis is another novel approach that has been applied to biodegradation of starch-containing plastic and other polymer blends.[7] The method uses the fact that ^{13}C, the naturally occurring stable isotope of carbon, is enriched in starch compared to petroleum-based plastic. Coupled with weight loss data, the ^{13}C measurements before and after biodegradation permit definition of a near linear combination of the ^{13}C signatures of the components in each blend, which makes one blend isotopically distinct from another. Thus, the relative extent of biodegradation of all components in a blend can be computed. However, it can be proved mathematically that since a weight loss function is the sole constraint on the linear mathematical model used,[14] the system is unfortunately indeterminate for a blend with more than two components.

13.2 FT-IR SPECTROMETRY

FT-IR spectrometry is a well-established and widely used instrumental technique for quantitative analyses of components in solid materials. Although it is routinely used to measure relative concentrations of components in multicomponent polymeric materials,[15] in analyses of biodegraded plastics, FT-IR spectrometry extracts only information about the relative concentrations of components, not the relative weight losses. To determine the relative weight losses of components in a multicomponent plastic, when more than one component biodegrades, the total weight loss from the test specimen must first be determined gravimetrically.[8,16] Given this total weight loss value, FT-IR spectrometry can be used with several chemometric methods[9-11] for the multicomponent analysis.

In the simplest cases from a chemometric point of view, such as the starch-PE-EAA blend shown in Figure 13.1, starch is the only one of the three components that shows significant absorbance in the carbohydrate (960–1190 cm^{-1}) region of the infrared spectrum. Quantitation of the starch relative to PE and EAA is possible and straightforward simply by measuring the absorbance over this spectral range. But, it is the more complex cases, in which all components show infrared bands that overlap each other, that are by far the most common in biodegradable plastics research.

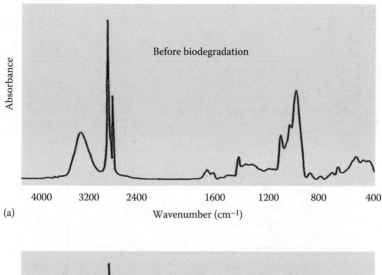

(a)

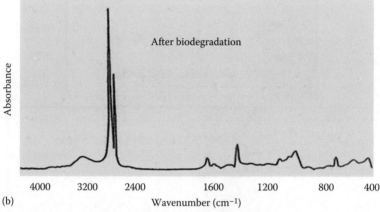

(b)

Figure 13.1 FT-IR spectra of starch-PE-EAA plastic (a) before and (b) after 72 h enzymatic biodegradation.

Depending upon whether all components are known or whether their pure component spectra are available, these complex cases often present difficult chemometric problems to the analyst. If, in these cases, calibration data are available from spectra of a series of known composite plastics, the component concentrations can be determined as the solution to a system of simultaneous equations by an approach known as multiple linear regression (MLR).[17]

13.3 MULTIVARIATE FT-IR CALIBRATION AND ANALYSIS

MLR is the standard approach for calibration and analysis of FT-IR spectra when two or more components are present. Simultaneous determination of multiple components in plastic is possible because no two components have identical infrared spectra.[18] For the FT-IR spectra of biodegradable plastics consisting of three different mixtures of three component polymers, as depicted in Figure 13.2, the Beer–Lambert law[19–21] relates the absorbance ($\mathbf{A}$) of a sample to the concentrations ($\mathbf{c}$) of its components:

$$A_{jl} = \sum_{i=1}^{n} k_{ji} c_{il} \tag{13.1}$$

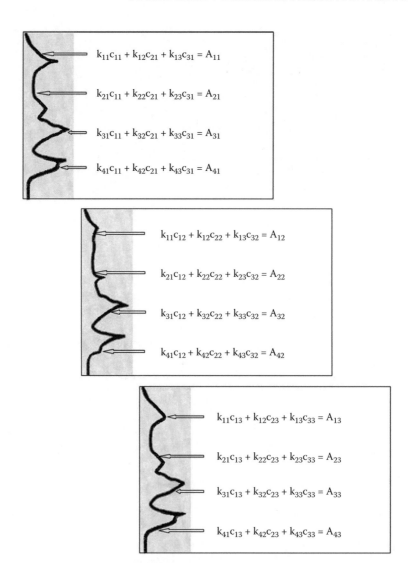

Figure 13.2 FT-IR spectra of three mixtures of three components with Beer–Lambert law equations for absorbances at four wavenumbers.

for wavenumbers $j = 1, 2, 3,\ldots, w$; components $i = 1, 2, 3,\ldots, n$; and mixtures $l = 1, 2, 3,\ldots, m$, where

$\mathbf{k}_{ji}$ is the absorption coefficient (absorptivity) of component i at wavenumber j

$\mathbf{c}_{il}$ is the concentration of component i in mixture l

This is the system of linear simultaneous equations shown in Figure 13.2. The FT-IR spectral data form a $w \times m$ matrix of absorbances ($\mathbf{A}$) which contains the spectra at w wavenumbers for each of the m mixtures in the calibration set. A matrix of $w \times n$ absorption coefficients ($\mathbf{K}$) of the n components in each mixture at the w wavenumbers is formed from the known mixtures. A matrix of $n \times m$ concentrations ($\mathbf{C}$) of the n components in each of the m mixtures in the calibration set is also formed from known concentrations of components in the mixtures.

Expressed in matrix notation, the system of equations for the Beer–Lambert law is

$$\mathbf{A} = \mathbf{KC} \tag{13.2}$$

where
   A is the dependent variable
   C is the independent variable
   K is a $w \times n$ matrix of absorption coefficients at w wavenumbers for each of the n components

   Thus, the example in Figure 13.2 has $m = 3$, $n = 3$, and $w = 4$. To solve this system, the three mixtures are first calibrated for the elements of the matrix **K** using a maximum likelihood criterion that gives the least squares regression estimate

$$\mathbf{K} = \mathbf{AC}^{\mathrm{T}}\left(\mathbf{CC}^{\mathrm{T}}\right)^{-1} \tag{13.3}$$

After calibration, in the validation step, computed **K** values are used to predict the unknown concentrations of components in a test plastic by

$$\mathbf{C} = (\mathbf{K}^{\mathrm{T}}\mathbf{K})^{-1}\mathbf{K}^{\mathrm{T}}\mathbf{A} \tag{13.4}$$

where **A** are the absorbances from the spectra of samples of the unknown test plastics. These equations are well known as the classical **K**-matrix method.[11,21] In chemometrics, this **K**-matrix method is regarded as the best MLR method for well-conditioned systems (linear responses, low noise, no collinearities, etc.).[22] The **K**-matrix approach was preferred for analyses of substrates in solid-state fermentations. In those preliminary experiments using the **K**-matrix method with biodegradable solids composed of starch, protein, and lipid, the concentration results were accurate to less than 5% standard error of prediction.[23]

   Despite the advantage for well-conditioned systems, the **K**-matrix method has experimental disadvantages that limit its widespread application to multicomponent plastics. Primarily, the concentrations of all components in the calibration standards must be known. Also, the number of infrared wavenumbers used must equal or exceed the number of components in the calibration standards. These are significant disadvantages of the **K**-matrix method for practical use.

   A second chemometric method, known as the **P**-matrix method, is MLR based on the inverse of the classical **K**-matrix model.[21,24] As with the classical model, the theory assumes the matrix form of the Beer–Lambert law (Equation 13.2) but expresses the law in its inverse form,

$$\mathbf{C} = \mathbf{AK}^{-1} \tag{13.5}$$

This equation is written in terms of a new matrix **P** which is defined as the inverse of **K**; hence,

$$\mathbf{C} = \mathbf{PA} \tag{13.6}$$

where, in contrast with the **K**-matrix model,
   C is now the dependent variable
   A is the independent variable

In MLR, the least squares solution for **P** is

$$\mathbf{P} = \mathbf{CA}^{\mathrm{T}}\left(\mathbf{AA}^{\mathrm{T}}\right)^{-1} \tag{13.7}$$

Unknown concentrations **C** are then computed by simply multiplying **P** by **A** as in Equation 13.6.

The **P**-matrix method has a distinct advantage of not requiring knowledge of the concentrations of the components in the calibration standards. However, the number of calibration standards must equal or exceed the number of wavenumbers, which greatly limits the number of wavenumbers that can be used. This reverse of the situation with the **K**-matrix method is a significant disadvantage because unless an excessively large set of calibration standards is tested, it is unlikely that the **P**-matrix method will include the optimum set of wavenumbers needed.

A third chemometric method, known as the **Q**-matrix method, is based on the fundamental hypothesis that the spectrum of an unknown mixture can be "synthesized" or simulated as a linear combination of the spectra of the pure components known to be included in the mixture.[24] This is analogous to the infrared regression technique called spectral curve fitting. The approach is the reverse of the disproportionation process used in multiple spectral subtraction. The scale factors used in such disproportionation are related to the elements of the **Q**-matrix. The **Q**-matrix method has been described by a preeminent chemometrician[13] as a "classical" MLR method despite the fact that the algorithm is somewhat different from the **K**-matrix approach. The simulation is expressed in matrix-vector notation as the product

$$\mathbf{a}_{w1} = \mathbf{A}_{wm}\mathbf{Q}_{m1} \qquad (13.8)$$

where
  $\mathbf{a}_{w1}$ is a column vector of absorbances across the wavenumber range in the spectrum of the unknown mixture
  $\mathbf{A}_{wm}$ is the matrix of absorbances from spectra of the known pure components used as calibration standards
  $\mathbf{Q}_{m1}$ is a column vector of coefficients which gives the least squares fit of the pure component spectra to the unknown spectrum

If $\mathbf{c}_{n1}$ is the column vector of the n concentrations to be determined and $\mathbf{C}_{nm}$ is an identity matrix of unity concentrations of the known pure components, then

$$\mathbf{c}_{n1} = \mathbf{C}_{nm}\mathbf{Q}_{m1} \qquad (13.9)$$

from which the system of equations can be solved directly, without discrete computation of the elements of the **Q**-matrix, by

$$\mathbf{c}_{n1} = \mathbf{C}_{nm}\left(\mathbf{A}_{mw}{}^{\mathrm{T}}\mathbf{A}_{wm}\right)^{-1}\mathbf{A}_{mw}{}^{\mathrm{T}}\mathbf{a}_{w1} \qquad (13.10)$$

where the subscripts are integer indices of the w wavenumbers, n components, and m mixtures, as defined earlier in the **K**-matrix method, and where the constraints $m \geq n$ and $w \geq n$ are applied for physically meaningful results.

Since these constraints are not too restrictive, Equation 13.10 appears to be suitable for practical applications, including biodegradable plastics. However, it can be shown that, in practice if not in principle, this **Q**-matrix approach is inherently unlikely to produce the true solution due to the experimental improbability of obtaining and combining a set of pure component spectra which have the same relative absorbances that exist in the unknown mixture spectrum.

Both the **K**-matrix and **P**-matrix methods require prior calibration from a set of spectra of a series of known mixtures, while the **Q**-matrix method requires prior calibration from a set of spectra of only the pure components. For analyses of biodegradable plastics, all of these methods necessitate preparation of calibration standards by extruding or injection molding of either an extensive series of known blends or the pure components. Also, these conventional MLR methods are often beset with difficulties including sample preparation, nonlinear absorbances, and irrelevant information in the data (interfering signals, noise, analyte–analyte interactions, experimental error, collinearities) in addition to other problems, not the least of which is the selection of analytical wavelengths. Advances in the science of FT-IR spectrometry and chemometrics[21–23] have overcome some of these difficulties, but even more advanced chemometric methods are needed to surmount the increasingly challenging multivariate analytical problems.

Accordingly, in this chapter, a new chemometric model is presented that uses absorbance ratios from FT-IR spectra of pure components to quantify degradation of all individual components in biodegradable multicomponent plastics. This new chemometric MLR method is called the **R**atio-matrix method, or dubbed **R**-matrix method, because it uses a matrix of ratios as the basis of the mathematical model. The **R**-matrix method enables multivariate quantitation without external calibration, and it potentially obviates the need for wavelength selection by the analyst.

13.4 THEORY AND DERIVATION OF THE R-MATRIX MODEL

In the **K**- and **P**-matrix methods, if the number of components is known and infrared spectra of different known multicomponent plastics are available, it is possible to compute the relative concentrations of the components in the test plastic. Concentrations can be simultaneously computed even where there are no unique absorbances in the pure component spectra. These conventional MLR methods require, first, computation of calibration parameters in equations relating spectral absorbances at given analytical wavenumbers to the concentration of each component in known mixtures of the components, and then, solving the equations for the unknown concentrations of the components from absorbances of the unknown mixtures.[22] As with the **K**- and **P**-matrix models, the **R**-matrix model also starts with the Beer–Lambert law[19–21] which assumes that the spectra of the components are additive and the infrared spectral absorbance is a linear function of the component concentrations. Again, the Beer–Lambert law is represented by a system of multivariate linear equations each having the form

$$\sum_{i=1}^{n} k_{ji} c_{il} = A_{jl} \tag{13.11}$$

or in matrix-vector form

$$\sum \mathbf{k}_{wn} \mathbf{c}_{nm} = \mathbf{A}_{wm} \tag{13.12}$$

where $\mathbf{k}_{wn}$ refers to absorption coefficients for each of the component concentrations $\mathbf{c}_{nm}$ in mixtures having absorbances $\mathbf{A}_{wm}$ and where the subscripts are integer indices of the w wavenumbers, n components, and m mixtures, as defined earlier in the **K**-matrix method. For example, the absorbance for the first mixture in a set of different mixtures of three components with concentrations c_{11}, c_{21}, and c_{31} at the first wavenumber is

$$k_{11}c_{11} + k_{12}c_{21} + k_{13}c_{31} = A_{11} \tag{13.13}$$

while the absorbance for the second mixture of the same three components at the first wavenumber is

$$k_{11}c_{12} + k_{12}c_{22} + k_{13}c_{32} = A_{12} \tag{13.14}$$

When concentrations of the components in all of the mixtures are known, the absorption coefficients $\mathbf{k}_{wn}$ for each wavenumber can be computed by MLR using the $\mathbf{K}$-matrix method.[25,26] In the analysis of an unknown mixture, the absorbances $\mathbf{A}_{wm}$ at the same wavenumbers are measured, and the desired concentrations $\mathbf{c}_{nm}$ of the components in the mixture are directly calculated.

In the case of multicomponent biodegradable plastics, where it is desired to determine the weight losses of individual polymers, the component concentrations are known before biodegradation but unknown after biodegradation. Consider, for example, a biodegradable plastic composed of three components. The infrared spectra before and after biodegradation will have absorbances at three wavenumbers ($w \geq n$ and $m \geq n$ required for solution) defined by the Beer–Lambert law as follows:

Before biodegradation	After biodegradation

$$
\begin{aligned}
k_{11}c_{11} + k_{12}c_{21} + k_{13}c_{31} &= A_{11} & k_{11}c_{12} + k_{12}c_{22} + k_{13}c_{32} &= A_{12} \\
k_{21}c_{11} + k_{22}c_{21} + k_{23}c_{31} &= A_{21} & k_{21}c_{12} + k_{22}c_{22} + k_{23}c_{32} &= A_{22} \\
k_{31}c_{11} + k_{32}c_{21} + k_{33}c_{31} &= A_{31} & k_{31}c_{12} + k_{32}c_{22} + k_{33}c_{32} &= A_{32}
\end{aligned}
\tag{13.15}
$$

where c_{11}, c_{21}, and c_{31} are known concentrations in the original mixture, but c_{12}, c_{22}, and c_{32} are unknown concentrations in the biodegraded mixture to be determined. If these are the only components present, then their concentrations, expressed as weight fractions, in each mixture sum to unity:

$$
c_{11} + c_{21} + c_{31} = 1 \quad c_{12} + c_{22} + c_{32} = 1
\tag{13.16}
$$

When spectra of the pure component polymers are available, the ratios, r_{wn}, of the absorbance bands at each wavenumber in a given spectrum are known (measurable), even though the absorption coefficients $\mathbf{k}_{wn}$ are unknown. And, because the concentration in a pure component (the neat polymer) is unity, the known absorbance ratios are equal to the ratios of the unknown absorption coefficients as follows:

$$
r_{21} = \frac{k_{21}}{k_{11}} \quad r_{22} = \frac{k_{22}}{k_{12}} \quad r_{23} = \frac{k_{23}}{k_{13}}
$$

$$
r_{31} = \frac{k_{31}}{k_{11}} \quad r_{32} = \frac{k_{32}}{k_{12}} \quad r_{33} = \frac{k_{33}}{k_{13}}
\tag{13.17}
$$

Rewriting Equations 13.15 in terms of these absorbance ratios, r_{wn}, gives the new set of absorbance equations:

$$
\begin{aligned}
k_{11}c_{11} + k_{12}c_{21} + k_{13}c_{31} &= A_{11} & k_{11}c_{12} + k_{12}c_{22} + k_{13}c_{32} &= A_{12} \\
r_{21}k_{11}c_{11} + r_{22}k_{12}c_{21} + r_{23}k_{13}c_{31} &= A_{21} & r_{21}k_{11}c_{12} + r_{22}k_{12}c_{22} + r_{23}k_{13}c_{32} &= A_{22} \\
r_{31}k_{11}c_{11} + r_{32}k_{12}c_{21} + r_{33}k_{13}c_{31} &= A_{31} & r_{31}k_{11}c_{12} + r_{32}k_{12}c_{22} + r_{33}k_{13}c_{32} &= A_{32}
\end{aligned}
\tag{13.18}
$$

which in matrix notation is

$$
\begin{bmatrix} 1 & 1 & 1 \\ r_{21} & r_{22} & r_{23} \\ r_{31} & r_{32} & r_{33} \end{bmatrix}
\begin{bmatrix} k_{11}c_{11} \\ k_{12}c_{21} \\ k_{13}c_{31} \end{bmatrix}
=
\begin{bmatrix} A_{11} \\ A_{21} \\ A_{31} \end{bmatrix}
\quad
\begin{bmatrix} 1 & 1 & 1 \\ r_{21} & r_{22} & r_{23} \\ r_{31} & r_{32} & r_{33} \end{bmatrix}
\begin{bmatrix} k_{11}c_{12} \\ k_{12}c_{22} \\ k_{13}c_{32} \end{bmatrix}
=
\begin{bmatrix} A_{12} \\ A_{22} \\ A_{32} \end{bmatrix}
\tag{13.19}
$$

where the two systems of equations, before and after biodegradation, contain the same 3×3 matrix of absorbance ratios. This ratio matrix, or $\mathbf{R}$-matrix as it is called, and its determinant are designated by different brackets. Thus,

$$\mathbf{R} = \begin{bmatrix} 1 & 1 & 1 \\ r_{21} & r_{22} & r_{23} \\ r_{31} & r_{32} & r_{33} \end{bmatrix} \qquad \det \mathbf{R} = \begin{vmatrix} 1 & 1 & 1 \\ r_{21} & r_{22} & r_{23} \\ r_{31} & r_{32} & r_{33} \end{vmatrix} \tag{13.20}$$

If analytical wavenumbers are selected that have absorbance ratios such that $\mathbf{R}$ is nonsingular (det $\mathbf{R} \neq 0$), the two systems of three equations in three unknowns (Equations 13.19) are determinate and can therefore be solved using Cramer's rule[27] as follows:

$$k_{11}c_{11} = \frac{\begin{vmatrix} A_{11} & 1 & 1 \\ A_{21} & r_{22} & r_{23} \\ A_{31} & r_{32} & r_{33} \end{vmatrix}}{\det \mathbf{R}} \underset{=}{let} b_{11} \qquad k_{11}c_{12} = \frac{\begin{vmatrix} A_{12} & 1 & 1 \\ A_{22} & r_{22} & r_{23} \\ A_{32} & r_{32} & r_{33} \end{vmatrix}}{\det \mathbf{R}} \underset{=}{let} b_{12}$$

$$k_{12}c_{21} = \frac{\begin{vmatrix} 1 & A_{11} & 1 \\ r_{21} & A_{21} & r_{23} \\ r_{31} & A_{31} & r_{33} \end{vmatrix}}{\det \mathbf{R}} \underset{=}{let} b_{21} \qquad k_{12}c_{22} = \frac{\begin{vmatrix} 1 & A_{12} & 1 \\ r_{21} & A_{22} & r_{23} \\ r_{31} & A_{32} & r_{33} \end{vmatrix}}{\det \mathbf{R}} \underset{=}{let} b_{22} \tag{13.21}$$

$$k_{13}c_{31} = \frac{\begin{vmatrix} 1 & 1 & A_{11} \\ r_{21} & r_{22} & A_{21} \\ r_{31} & r_{32} & A_{31} \end{vmatrix}}{\det \mathbf{R}} \underset{=}{let} b_{31} \qquad k_{13}c_{32} = \frac{\begin{vmatrix} 1 & 1 & A_{12} \\ r_{21} & r_{22} & A_{22} \\ r_{31} & r_{32} & A_{32} \end{vmatrix}}{\det \mathbf{R}} \underset{=}{let} b_{32}$$

If each solution (absorbance b_{nm}) before biodegradation is divided by the corresponding solution after biodegradation, the unknown absorption coefficients k_{wn} are eliminated, leaving the desired ratios of the concentrations before and after biodegradation in terms of known absorbances A_{wm} and known peak ratios r_{wn} contained in the b_{nm}.

Thus,

$$\frac{c_{11}}{c_{12}} = \frac{b_{11}}{b_{12}} \underset{=}{let} d_1 \qquad \frac{c_{21}}{c_{22}} = \frac{b_{21}}{b_{22}} \underset{=}{let} d_2 \qquad \frac{c_{31}}{c_{32}} = \frac{b_{31}}{b_{32}} \underset{=}{let} d_3 \tag{13.22}$$

where b_{nm} and d_m as defined earlier are introduced only to simplify the notation.

The unknown concentrations c_{12}, c_{22}, and c_{32} are then obtained by equating Equations 13.22 to unity,

$$\frac{c_{12}}{c_{11}} d_1 = \frac{c_{22}}{c_{21}} d_2 = \frac{c_{32}}{c_{31}} d_3 = 1 \tag{13.23}$$

then imposing the constraints of Equations 13.16, substituting for c_{32} (after biodegradation),

$$\frac{c_{12}}{c_{11}} d_1 = \frac{(1 - c_{12} - c_{22})}{c_{31}} d_3 \qquad \frac{c_{22}}{c_{21}} d_2 = \frac{(1 - c_{12} - c_{22})}{c_{31}} d_3 \tag{13.24}$$

and rearranging to give the resulting system of $2(n - 1)$ equations in two unknowns for c_{12} and c_{22}:

$$\left(d_1c_{31} + d_3c_{11}\right)c_{12} + d_3c_{11}c_{22} = d_3c_{11}$$

$$d_3c_{21}c_{12} + \left(d_2c_{31} + d_3c_{21}\right)c_{22} = d_3c_{21}$$

(13.25)

Finally, the matrix form of Equations 13.25 is solved, again by the method of determinants using Cramer's rule as follows:

$$\begin{bmatrix} \left(d_1c_{31} + d_3c_{11}\right) & d_3c_{11} \\ d_3c_{21} & \left(d_2c_{31} + d_3c_{21}\right) \end{bmatrix} \begin{bmatrix} c_{12} \\ c_{22} \end{bmatrix} = \begin{bmatrix} d_3c_{11} \\ d_3c_{21} \end{bmatrix}$$

$$c_{12} = \frac{\begin{vmatrix} d_3c_{11} & d_3c_{11} \\ d_3c_{21} & \left(d_2c_{31} + d_3c_{21}\right) \end{vmatrix}}{\begin{vmatrix} \left(d_1c_{31} + d_3c_{11}\right) & d_3c_{11} \\ d_3c_{21} & \left(d_2c_{31} + d_3c_{21}\right) \end{vmatrix}} \qquad c_{22} = \frac{\begin{vmatrix} \left(d_1c_{31} + d_3c_{11}\right) & d_3c_{11} \\ d_3c_{21} & d_3c_{21} \end{vmatrix}}{\begin{vmatrix} \left(d_1c_{31} + d_3c_{11}\right) & d_3c_{11} \\ d_3c_{21} & \left(d_2c_{31} + d_3c_{21}\right) \end{vmatrix}}$$

(13.26)

$$c_{32} = 1 - c_{12} - c_{22}$$

Thus, in an analytical solution, the desired concentrations of the components in the mixture after biodegradation, c_{12}, c_{22}, and c_{32}, are determined.

As is the case with MLR methods in general, for materials that obey the Beer–Lambert law precisely, the solutions produced by the **R**-matrix method are mathematically unique and exact. However, as with all chemometric methods, the inevitable presence of experimental error causes accuracy to decrease as the number of components increases.[28]

Furthermore, it is well known[14] that quantitative FT-IR spectrometry extracts only concentration information from a sample. The relative concentrations of components in a biodegraded mixture as determined from Equations 13.26 give no information whatsoever about the amount of biodegradation that has occurred in each component. When more than one component biodegrades, these relative component concentrations, by themselves, before and after degradation say nothing about weight loss in each component. In fact, the concentration of a given component in a biodegraded mixture may be lower than, equal to, or higher than its concentration in the original mixture, depending on the fate of the other components. Therefore, the analyst must compute the actual weight losses of each individual component in the biodegraded mixture.

To accomplish this, the analyst must first measure the total weight loss of the mixture and then factor this weight loss into the computations. Total weight loss from biodegraded plastic is easily measured gravimetrically from samples before and after biodegradation. The total weight loss can be expressed as the weight fraction, L_2, of the unbiodegraded plastic. Once L_2 is known, the unknown weights, w_{12}, w_{22}, and w_{32}, of each component remaining in the biodegraded plastic can be calculated from the concentrations computed in Equations 13.26 as follows:

$$w_{12} = c_{12}\left(1 - L_2\right)\left(w_{11} + w_{21} + w_{31}\right)$$

$$w_{22} = c_{22}\left(1 - L_2\right)\left(w_{11} + w_{21} + w_{31}\right)$$

(13.27)

$$w_{32} = c_{32}\left(1 - L_2\right)\left(w_{11} + w_{21} + w_{31}\right)$$

where w_{11}, w_{21}, and w_{31} are the known weights of each of the three polymers in the original unbiodegraded plastic. Finally, the desired weight losses of each individual component, l_{12}, l_{22} and l_{32}, can be calculated from the fraction difference between component weights before and after degradation simply as (13.27)

$$l_{n2} = \frac{w_{n1} - w_{n2}}{w_{n1}} \tag{13.28}$$

Although Equations 13.26 apply to three-component mixtures, the mathematical model and solution algorithm just described can be extended to any number of components. The three-component model was described here rather than the simpler two-component model because the three-component derivation is conceptually more instructive of the **R**-matrix method for multicomponent materials. But, for completeness of information to the reader, and for the convenience of potential users of the **R**-matrix method, the two-component model is also included in the following text.

Without repeating the definitions and mathematical description of the algorithm given earlier in the theory and derivation, the system of equations used to solve the two-component **R**-matrix model is as follows:

Before biodegradation   After biodegradation

$$\begin{bmatrix} 1 & 1 \\ r_{21} & r_{22} \end{bmatrix} \begin{bmatrix} k_{11}c_{11} \\ k_{12}c_{21} \end{bmatrix} = \begin{bmatrix} A_{11} \\ A_{21} \end{bmatrix} \qquad \begin{bmatrix} 1 & 1 \\ r_{21} & r_{22} \end{bmatrix} \begin{bmatrix} k_{11}c_{12} \\ k_{12}c_{22} \end{bmatrix} = \begin{bmatrix} A_{12} \\ A_{22} \end{bmatrix}$$

$$R = \begin{bmatrix} 1 & 1 \\ r_{21} & r_{22} \end{bmatrix} \qquad\qquad \det R = \begin{vmatrix} 1 & 1 \\ r_{21} & r_{22} \end{vmatrix}$$

$$k_{11}c_{11} = \frac{\begin{vmatrix} A_{11} & 1 \\ A_{21} & r_{22} \end{vmatrix}}{\det R} \; \underline{let}\, b_{11} \qquad k_{11}c_{12} = \frac{\begin{vmatrix} A_{12} & 1 \\ A_{22} & r_{22} \end{vmatrix}}{\det R} \; \underline{let}\, b_{12}$$

$$k_{12}c_{21} = \frac{\begin{vmatrix} 1 & A_{11} \\ r_{21} & A_{21} \end{vmatrix}}{\det R} \; \underline{let}\, b_{21} \qquad k_{12}c_{22} = \frac{\begin{vmatrix} 1 & A_{12} \\ r_{21} & A_{22} \end{vmatrix}}{\det R} \; \underline{let}\, b_{22} \tag{13.29}$$

$$\frac{c_{11}}{c_{12}} = \frac{b_{11}}{b_{12}} \; \underline{let}\, d_1 \qquad\qquad \frac{c_{21}}{c_{22}} = \frac{b_{21}}{b_{22}} \; \underline{let}\, d_2$$

$$c_{11} + c_{21} = 1 \quad d_1 \frac{c_{12}}{c_{11}} = d_2 \frac{c_{22}}{c_{21}} = 1 \quad c_{12} + c_{22} = 1$$

$$d_1 \frac{c_{12}}{c_{11}} = d_2 \frac{(1 - c_{12})}{c_{21}}$$

$$c_{12} = \frac{d_2 c_{11}}{d_1 c_{21} + d_2 c_{11}} \qquad\qquad c_{22} = 1 - c_{12}$$

$$w_{12} = c_{12}(1 - L_2)(w_{11} + w_{21}) \qquad w_{22} = c_{22}(1 - L_2)(w_{11} + w_{21})$$

The two algorithms earlier, for solution of the two-component and the three-component systems, are equivalent mathematically, but they are not equally robust in a computational sense due to the inherent decrease in numerical stability as the matrix dimensions increase. This condition arises only when experimental error is present in the data, and it becomes even worse in higher dimension systems. However, in purely abstract experiments, where experimental error is absent and no artificial random error is added, all valid multivariate algorithms work equally well and give solutions whose accuracy is limited only by the arithmetic precision of the computer.

Therefore, it is possible to test the validity of the solution algorithm by computer simulation. One has only to generate exact model data by computer so that no error is present, and then apply the algorithm to test whether it produces the solution known to be exact for the model data. Such computer simulation is both necessary and sufficient to prove the validity of the **R**-matrix model. This was done to test both the two-component and three-component algorithms in this work.[8] The results showed that in both cases the solution algorithms reproduced all of the computer-generated data exactly, within round-off error of the computer. This, together with the analytical solution already obtained in the mathematical derivation of the **R**-matrix model, proved that the algorithms are theoretically correct for spectra that conform to the Beer–Lambert law.

As a concept, it can be stated that the **R**-matrix model described earlier represents a new and unobvious solution to a long-standing chemometric problem in the spectrometry of biodegradable plastics. Particularly, the **R**-matrix makes it possible to extract unknown absorption coefficients of multiple components from only two mixtures without calibration against any external mixtures. As a result, some long-sought advancements in such chemometric preprocessing techniques as spectral scaling and wavelength selection have now become feasible.

Despite the fact that this success in purely abstract experiments, where experimental and random error were absent, represents groundbreaking progress in theoretical chemometrics research, the current situation in practical experiments is another, quite different, matter. All real-world research data contain unavoidable experimental error. And, infrared spectra of solids, especially polymeric materials like plastics, are notorious for their deviation from Beer–Lambert law.[15] In applied chemometrics research, much effort is being invested in approaches to overcome these problems.

Therefore, in this work, it was of interest to test the performance of the **R**-matrix method with real-world data from composite plastics prepared in the laboratory. This was done using two- and three-component polymer formulations designed to simulate weight losses that would result from biodegradation. The results of the analyses were in good agreement with the known values. The two-component results agreed with the known values more closely than the three-component results. Although this trend with matrix dimensions was expected, the accuracy was surprisingly good in both cases.[8]

Besides its theoretical and mathematical properties, a likely reason for the better than expected accuracy of the **R**-matrix model is related to the flexibility it allowed the analyst in spectral scaling and wavenumber selection. It is an onerous task to manually search for wavenumber combinations, scale factors, and other conditions that maximize compliance to the Beer–Lambert law. But, it should be relatively easy to exploit this flexibility and implement the search in software.

13.5 ADVANTAGES OF THE R-MATRIX METHOD AS A CHEMOMETRIC METHOD

Although the classical **Q**-matrix and the novel **R**-matrix model are based on analogous mathematical concepts, the **R**-matrix method can be applied to certain multivariate problems where neither the **Q**-matrix method nor the **K**-matrix or **P**-matrix chemometric method can be applied. A good example of this was the recent application of the **R**-matrix method to solve an important problem with the traditional pressed KBr disk technique that has perplexed infrared spectroscopists for over 50 years.[29] For FT-IR analyses of composite solid materials such as biodegradable plastics, the pressed KBr disk technique has been preferred over other techniques and remains the method of choice for good reasons.[8,29] However, KBr is notorious in infrared spectroscopy because it is plagued with interference from water vapor absorbed during sample preparation.[30–37] This interference by strong water bands from minute levels of water in KBr occurs, unfortunately, in the all-important amide and hydroxyl regions of biomaterial spectra. Consequently, true spectra of the solid pure components in KBr are unknown and unavailable. And, since the minute water concentration in KBr is also unknown and impracticable to measure, the **K**-matrix, **P**-matrix, and **Q**-matrix methods become indeterminate for this problem.[29]

The **R**-matrix method, combined with a spectral curve fitting technique, eliminated this problem. It was able to completely free FT-IR spectra from the interfering water bands in KBr, while the water concentrations were unknown. The **R**-matrix approach enabled solution of this problem which could not be solved by any other known chemometric method. The **R**-matrix model not only caused the system of Beer–Lambert law equations to become determinate but also produced the unknown absorption coefficients of the solid component from only two sample mixtures with different levels of absorbed water.

As a chemometric method, the new **R**-matrix approach enables more accurate quantitative infrared spectra of components in solid biomaterial composites in KBr disks than was previously possible. In fact, the **R**-matrix method even gives the KBr disk technique a decided advantage over the highly touted algorithm for correction of attenuated total reflectance (ATR) spectra, widely known as the Advanced ATR Correction Algorithm, introduced by Nunn and Nishikida.[38] Much error is retained in the ATR spectrum of a protein after treatment with the Advanced ATR Correction Algorithm, but no error was retained in the KBr spectrum of the same protein after correction by the new **R**-matrix algorithm.[29] Therefore, compared to the corrected ATR spectra, the pressed KBr disk spectra corrected by the novel **R**-matrix method of the current work are much better measures of the true infrared spectra of solid biomaterials. This was a breakthrough that removes a major barrier to valid quantitative spectrometry of biochemical structures in many areas of science.

13.6 CONCLUSION

Because this chapter was conceived to provide a concise introduction to the mathematics of the **R**-matrix chemometric method and the science behind it for quantifying degradation in biodegradable plastics, the method presented was described without experimental detail. Obviously, details such as instrument settings, sample handling, and other skills of the analyst are crucial to the success or failure of any analytical method. Therefore, some practical aspects and details of the experimental methods and sampling techniques, as well as some critical theoretical issues underlying the mathematical methods involved here, are referred to discussions elsewhere.[8,39] Chemometric methods are constantly evolving as more difficult problems arise, so the theory behind the analytical methods is never static.[40] Since the wide availability of computer software for chemometrics has been a major factor in the growth and acceptance of this field, the computer algorithms and software used in this work are also described elsewhere.[8,29]

In contrast to the classical **K**-matrix and the **P**-matrix chemometric methods, the new **R**-matrix method does not require prior calibration against known composites and it does not require sampling of more different plastics than the number of components in the plastic being analyzed. Only two measurements, one before and one after biodegradation, are needed, particularly where spectra of the pure components are available. The **R**-matrix approach not only renders the multivariate system of Beer–Lambert law equations amenable to solution but also produces the unknown absorbance coefficients from the two sample mixtures without external calibration. Also, in contrast to the classical **Q**-matrix method, which is based on an analogous mathematical concept, the **R**-matrix method can be applied in certain multivariate problems where the **Q**-matrix method and the **K**-matrix and **P**-matrix methods are indeterminate and cannot be applied.

In this chapter, the theoretical basis and derivation of the mathematical model for multicomponent systems and its solution algorithm were described. First, the validity of the **R**-matrix method was proved mathematically and by computer simulation of multicomponent systems as defined in the new **R**-matrix model. Then the method was validated experimentally by comparing test results with known biodegradable composites prepared in the laboratory, without dependence on or knowledge of the model. Use of the **R**-matrix model to analyze data containing inevitable experimental error also revealed a potential for automatic infrared wavelength selection in software that would obviate the need for wavelength selection by the analyst.

Although the focus of this chapter was on the theory and derivation of the **R**-matrix model rather than the laboratory technology, the reader will understand that it is the experimental technique that ultimately makes or breaks any analytical method. As mentioned earlier, further research is needed for this immature laboratory technology to catch up with the advanced mathematics.[8] Meanwhile, the **R**-matrix method represents a significant advance in the science of chemometrics, especially for research on biodegradable plastics. Like past advances in chemometrics, the **R**-matrix method is expected to be a groundbreaking tool that can be readily applied in other fields of research. For scientists developing biodegradable plastics, the future is promising and exciting. They will participate in the development of green technology to benefit humankind, and they will enjoy the intellectual challenge of creating sophisticated methods to analyze and evaluate designer biobased plastics for a cleaner, healthier, environment.

REFERENCES

1. S. H. Imam, J. M. Gould, S. H. Gordon, M. P. Kinney, A. M. Ramsey, and T. R. Tosteson (1992) *Curr. Microbiol.* 25, 1.
2. S. H. Imam, A. Burgess-Cassler, G. L. Cote, S. H. Gordon, and F. L. Baker (1991) *Curr. Microbiol.* 22, 365.
3. R. L. Shogren, A. R. Thompson, R. V. Greene, S. H. Gordon, and G. Cote (1991) *J. Appl. Polym. Sci.* 47, 2279.
4. R. L. Shogren, A. R. Thompson, F. C. Felker, R. E. Harry-O'kuru, S. H. Gordon, R. V. Greene, and J. M. Gould (1992) *J. Appl. Polym. Sci.* 44, 1971.
5. S. H. Imam, S. H. Gordon, A. R. Thompson, R. E. Harry O'kuru, and R. V. Greene (1993) *Biotechnol. Tech.* 7, 791.
6. C. L. Swanson and R. L. Shogren (1993) *J. Environ. Polym. Degrad.* 1, 155.
7. M. L. Sykes, H.-W. Yeh, I. F. West, R. W. Gauldie, and C. F. Helsley (1993) *J. Polym. Environ.* 2(3), 201–209.
8. S. H. Gordon, S. H. Imam, and C. James (2000) *J. Polym. Environ.* 8(3), 125–134.
9. D. M. Haaland (1990) *Multivariate Calibration Methods Applied to Quantitative FT-IR Analyses, Practical Fourier Transform Infrared Spectroscopy*; Chapter 8, Academic Press: San Diego, CA.
10. R. J. O. Torgrip (2003) Chemometric analysis of first order chemical data. Doctoral thesis. Department of analytical chemistry, Stockholm University, Stockholm, Sweden.
11. R. Karoui, G. Downey, and C. Blecker (2010) *Chem. Rev.* 110, 6144–6168.
12. A. L. Smith (1979) *Applied Infrared Spectroscopy: Fundamentals, Techniques and Applied Problem Solving*; Chapter 6, John Wiley & Sons: New York.
13. D. M. Haaland, R. G. Easterling, and D. A. Vopica (1985) *Appl. Spectrosc.* 39, 73.
14. S. H. Gordon, S. H. Imam, and R. V. Greene (1996) *The Polymeric Materials Encyclopedia: Synthesis, Properties and Applications*; Salamone, J. C., Ed., CRC Press: Boca Raton, FL, Vol. 10, p. 7885.
15. J. L. Koenig (1992) *Spectroscopy of Polymers*; Chapter 2, American Chemical Society: Washington, DC.
16. S. H. Gordon, S. H. Imam, R. L. Shogren, N. S. Govind, and R. V. Greene (2000) *J. Appl. Polym. Sci.* 76, 1767–1776.
17. D. C. Montgomery, E. A. Peck, and C. G. Vining (2001) *Introduction to Linear Regression Analysis*; Chapter 3, John Wiley & Sons: New York.
18. C. G. Smith, R. A. Nyquist, A. J. Pasztor, and S. J. Martin (1991) *Anal. Chem.* 63, 11R.
19. D. F. Swinehart (1962) *J. Chem. Educ.* 39(7), 333.
20. R. L. Pecsok and D. L. Shields (1968) *Modern Methods of Chemical Analysis*; John Wiley & Sons: New York.
21. B. C. Smith (2002) *Quantitative Spectroscopy: Theory and Practice*; Chapters 1–3, Elsevier: Amsterdam, the Netherlands.

22. K. R. Beebe and B. R. Kowalski (1987) *Anal. Chem.* 59, 1007A.
23. S. H. Gordon, R. V. Greene, B. C. Wheeler, and C. James (1993) *Biotechnol. Adv.* 11, 665.
24. G. L. McClure, Ed. (1987) *Computerized Quantitative Infrared Analysis*; ASTM: Philadelphia, PA.
25. C. W. Brown, P. F. Lynch, R. J. Obremski, and D. S. Lavery (1982) *Anal. Chem.* 54, 1472.
26. H. J. Kisner, C. W. Brown, and G. J. Kavarnos (1983) *Anal. Chem.* 55, 1703.
27. H. Anton (1987) *Elementary Linear Algebra*; John Wiley & Sons: New York.
28. S. D. Brown, R. S. Bear, and T. B. Blank (1992) *Anal. Chem.* 64, 22R.
29. S. H. Gordon, A. A. Mohamed, R. E. Harry-Okuru, and S. H. Imam (2010) *Appl. Spectrosc.* 64, 448.
30. F. J. Stevenson and K. M. Goh (1974) *Soil Sci.* 117, 34.
31. K. Schwochau and H. J. Eichhoff (1962) *Z. Naturforsch.* 17b, 582.
32. L. Borka (1975) *Anal. Chem.* 47, 1212.
33. V. M. Malhotra, S. Jasty, and R. Mu (1989) *Appl. Spectrosc.* 43, 638.
34. W. C. Price, W. F. Sherman, and G. R. Wilkinson (1958) *Proc. R. Soc. A* 247, 467.
35. W. K. Thompson (1964) *J. Chem. Soc.* (London), Part IV, 3658.
36. H. V. Walter (1971) *Z. Physik. Chem.* 75, 287.
37. W. P. Norris, A. L. Olsen, and R. G. Brophy (1972) *Appl. Spectrosc.* 26, 247.
38. S. Nunn and K. Nishikida (2003) *Advanced ATR Correction Algorithm, Thermo Electron Corporation Application*, Note AN 01153. Thermo Electron Corporation: Madison, WI.
39. E. V. Thomas (1994) *Anal. Chem.* 66, 795A.
40. K. S. Booksh and B. R. Kowalski (1994) *Anal. Chem.* 66, 782A.

CHAPTER **14**

Starch Polymer as Advanced Material for Industrial and Consumer Products

Randal L. Shogren

CONTENTS

14.1 CURRENT TECHNOLOGIES

Starches serve as an energy source for growing seeds such as corn and wheat and for tubers such as potatoes. Corn is by far the largest starch-rich crop in the United States, and the largest use of corn at 40% continues to be an energy source or feed for livestock (USDA/ERS). Second in volume at 33%, the conversion of cornstarch to ethanol by fermentation has increased greatly recently due to subsidies and the increased price of petroleum. Food starches and starches for industrial use utilize a rather small portion (2%) of the total corn crop. Most industrial applications of starch take advantage of its low cost and its ability to be dispersed in water with heating followed by drying to form a coating, film, or adhesive. Common high-volume applications include paper coatings and sizes, textile coatings, cardboard adhesive, drywall adhesive, etc. (Zobel and Stephen 2006). Recently, starches have become useful as biodegradable packaging foams and as components in films and molded articles (Bastioli 2001; Carvalho 2008; Halley et al. 2008; Shen et al. 2010). In most cases, the starch has been gelatinized or melted in the presence of water or other plasticizer and exists in the form of a largely amorphous solid.

Starches are often chemically modified to improve their functionality (Rutenberg and Solarek 1984). For example, cationic starches are commonly utilized to bind together negatively charged cellulose fibers in papermaking. Starches may also be acetylated, hydroxyalkylated, phosphorylated,

287

or derivatized with hydrophobic reagents to improve water solubility, stabilize viscosity, improve emulsifying properties, or reduce the tendency for recrystallization. A wide variety of such modifications are needed in the food industry to achieve desired stability and mouth-feel properties after severe processing and storage conditions. Commercially available chemically modified starches typically have very low degree of substitution (DS < 0.05).

14.2 WHY ADVANCED MATERIALS BASED ON STARCH ARE NEEDED

Starch-based materials with advanced structures and properties are needed to replace petroleum-based products to deliver new and improved properties and to reduce environmental impacts. Most polymers used today are derived from fossil resources and thus are not sustainable. Starches, in contrast, are annually renewable within the limits determined by agricultural capacity and needs for food and fuel. Starches have the capability of forming highly ordered structures which could significantly enhance many properties, thus allowing new applications or using less for existing ones. Starches are safe for people and the environment while many petroleum-based polymers are made from toxic monomers and often contain significant residuals of these.

14.3 ADVANCED PHYSICALLY MODIFIED AND CHEMICALLY MODIFIED STARCHES

14.3.1 Oriented Systems: Fibers and Films

Amorphous, isotropic starch films typically have rather high strength but low flexibility due to hydrogen bonding between hydroxyl groups. Physical aging (Shogren 1992), antiplasticization (Shogren et al. 1992; Lourdin et al. 1997), and short-chain branching of amylopectin also contribute to brittleness such that samples can behave similarly to glass. In addition, amorphous starch is highly water absorbent. This can be an advantage where water solubility or absorbency is desired. However, it is a severe limitation for most materials applications where contact with high humidity or liquid water causes a drastic loss of strength. Obviously, this water sensitivity as well as brittleness severely limits the usefulness of starch materials.

Recently, Shogren (2007) showed that uniaxially oriented high (70%)-amylose cornstarch films have elongations to break which are much larger than unoriented films. This was also found to be the case for pure potato amylose (see Table 14.1). Tensile strengths of amylose films also increased by a factor of about 2 upon orientation. Yamane (2006) obtained similar results for partially oriented, enzymatically synthesized amylose fibers. Presumably, it is much more difficult to initiate a fracture of oriented amylose or starch since that would require breaking covalent bonds of the starch backbone. Increases in tensile strength and elongation to break with orientation were also found for brittle, synthetic polymers such as polystyrene or polylactic acid.

Optical birefringence is a convenient method to estimate the degree of orientation in a polymer as long as the intrinsic birefringence of a perfectly oriented specimen is known. The birefringence given in Table 14.1 for a highly oriented amylose (0.037) could be regarded as the intrinsic birefringence of amylose although this is likely approximate since orientation may be incomplete. An attempt was made to calculate the intrinsic birefringence for the different crystalline forms of amylose by vector summation of bond polarizabilities (Shogren 2009). Birefringence was found to be sensitive to conformation, but calculated values were somewhat less than experimental values. Extension of this work utilizing more accurate quantum mechanical methods would be desirable.

As shown in Table 14.1, inducing crystallization to the B-form of amylose by annealing causes a large increase in tensile strength of water-soaked films to 7 MPa from 0.4 MPa for amorphous

Table 14.1 Mechanical and Optical Properties of Oriented Amylose and High-Amylose Starch Fibers

Material	Orientation Method	Annealing	Test Condition	Birefringence	Tensile Strength (MPa)	Elongation to Break (%)	Reference
70% amylose cornstarch	None	No	50% relative humidity (r.h.)	0	47±4	7±2	Shogren (2007)
70% amylose cornstarch	Draw[a]	No	50% r.h.	0.0035	44±4	27±12	Shogren (2007)
70% amylose cornstarch	Draw	Yes[b]	Wet[c]	n.d.	1.5±0.2	20±8	Shogren (2007)
Potato amylose	None	No	50% r.h.	0	59±7	11±1	Shogren (2007)
Potato amylose	Draw	No	50% r.h.	0.037	101±15	32±7	Shogren (2007)
Potato amylose	Draw	Yes	Wet	n.d.	7.0±0.9	31±3	Shogren (2007)
Enzymatic synthesized amylose	Spinning[d]	No	?	0.008–0.012	80–105	40–60	Yamane (2006)

a Amylose or starch triacetate stretched five to six times in hot glycerol followed by deacetylation.
b 3 days at 80% r.h., 3 days at 100% r.h., and 1 h in 90°C water.
c Samples immersed in water for 0.5 h.
d Fibers spun from aqueous NaOH solution and drawn through 1 M HCL and methanolic HCl baths.

amylose (data not shown). Unlike amorphous amylose, crystalline amylose swells only slightly in water due to strong intra- and intermolecular hydrogen-bonded networks. Although a wet strength of 7 MPa is much lower than dry strength, it should be adequate for many applications. The effect of crystallization on strength of starch under ambient conditions has been studied by other authors (Shogren et al. 1992; Van Soest and Borger 1997) and generally has been found to lead to an increase in strength since crystallites act as physical cross-links.

Another method by which oriented, crystalline amylose might be prepared is epitaxial crystallization. In this method, an inorganic single crystal or highly oriented, crystalline polymer film is immersed in an amylose solution in a poor solvent leading to ordered deposition of the amylose onto the crystal. The feasibility of such an experiment has not been demonstrated yet for amylose except for one study of growth of amylose-butanol crystals on cellulose (Helbert and Chanzy 1994) but has been done for many other polymers (Mauritz et al. 1978; Wittmann and Lotz 1990). Such a method would seem to be appropriate mainly for the production of rather small, thin pieces of ordered amylose.

In order to develop practical applications of oriented starch, large-scale methods of orientation as well as sources of rather pure amylose are needed. Since Yamane (2006) demonstrated the feasibility of wet-spinning amylose fibers from aqueous NaOH solution, this would seem to be a logical route for optimization and scale-up. Further research would be needed to control crystallinity, improve orientation, and minimize solvent use. Electrospinning would also seem to be a good starting point for the small-scale production of amylose fibers (Gordon et al. 2006; Lee et al. 2009). Biaxial orientation of polymer films is a common commercial process although it would be more complicated for amylose since solvent baths would probably be required. Pure amylose is typically prepared by precipitation of amylose–butanol complexes from aqueous starch dispersions, but this is rather expensive so that little is prepared commercially. Amylose can also be prepared enzymatically by polymerization of glucose-1-phosphate with glucan phosphorylase, but this is currently very expensive (Ohdan et al. 2006). Recently, very-high-amylose cornstarch (90% amylose) and potato starch (86% amylose) have become available (Thuwall et al. 2006; Pearlstein et al. 2007), so these may have good potential as economical substitutes for pure amylose.

Some applications for oriented, crystalline amylose fibers and films might include optical polarizer films, sutures, wound coverings, slowly digestible dietary fiber, food packaging, sausage casings, and personal hygiene articles. Cost and properties of competing materials such as cellulose, gelatin, and polylactic acid would have to be weighed to determine if development of amylose products would be warranted.

14.3.2 Spherulitic Systems: Native Starch Granules and Amylose–Lipid Complexes

Native starch granules are semicrystalline, microscopic, spherical particles and have elegant molecular ordering (Donald 2001; Tester et al. 2004; Perez and Bertoft 2010). Granular size ranges from 1 to 60 μm in diameter depending on botanical source (Srichuwong and Jane 2007). The amylopectin component of starch is thought to be oriented radially within the starch granule, giving rise to alternating crystalline and amorphous layers. Degree of crystallinity is typically 20%–40%, and crystal structures termed A, B, C, and V are found depending on type of starch (Blanshard 1987). As a result of this crystallinity, starch granules are water resistant.

There have been, however, very few practical applications for native starch granules. Some examples are dusting powders to replace talcum in surgical gloves (Fitt and McNary 1992) and fillers in plastics (Wang et al. 2008; Willett 2009). The latter, however, do not offer particular advantages over mineral fillers. Native starches can be digested with acids into submicrometer-sized fragments even down to the nanoparticulate range. Starch nanoparticles have been reviewed recently (Le Corre et al. 2010), so they will not be described in detail here. A successful application of starch nanoparticles has been as a replacement for carbon black and silica in tires (Scott 2002). Tires having such starch reinforcement have significantly lower rolling resistance and hence improved fuel economy. A suspension of starch microparticles in water tends to have a lubricious feel similar to fats and hence have been used as fat substitutes in foods (Jane 1992). A suspension of starch nanoparticles in water may also have utility as an industrial lubricant, but no tests of this have been reported in the literature.

Amylose readily forms helical inclusion complexes with hydrophobic molecules such as lipids, and these have been studied in detail (Bengang and Liming 2010; Putseys et al. 2010). Complexes with 6, 7, and 8 glucose residues per turn of the helix are known to form depending on the size of the included molecule (Winter and Sarko 1978; Cardoso et al. 2007; Nishiyama et al. 2010). It has been demonstrated that amylose–fatty acid complexes crystallize into spherulites having well-defined shapes and sizes depending on added ligand, cooling rate, and pH (Davies et al. 1980; Fanta et al. 2008; Bhosale and Ziegler 2010). Such spherulites are stable in water up to the boiling point. Applications for these spherulites have not been developed as yet, but some possibilities might be carriers for fragrances, drug delivery, controlled release, and slowly digestible starch.

The fatty acid component of the spherulites can be extracted with alcohol/water mixtures leaving an array of hollow "nanotubes" (Shogren et al. 2006). After extraction, the spherulites become water soluble, so cross-linking of some type would be needed to stabilize the structure in aqueous environments. Such spherulites could have interesting absorption properties and thus might serve as adsorbents for trace organic pollutants.

14.3.3 Electrically Conductive Systems

Starches are intrinsically insulating materials with conductance values of around 10^{-13} S/cm for extruded films in the dry state (Finkenstadt and Willett 2005). Therefore, to confer some degree of electrical conductivity, starch must be modified or other substances added. Addition of water increases conductivity, reaching 10^{-6} S/cm at 40% moisture (Finkenstadt and Willett 2005). As shown in Table 14.2, addition of inorganic salts such as NaCl increases conductivity from 10^{-9}–10^{-11} to 10^{-5}–10^{-6} S/cm while keeping moisture constant at 20% (Finkenstadt 2005; Finkenstadt and

Table 14.2 Electrical Conductivity of Starch Composites

Starch	Relative Humidity or Water	Plasticizer (%)	Conductor (%)	Conductivity (S/cm)	Reference
Corn	20% water	None	None	10^{-9}–10^{-11}	Finkenstadt and Willett (2005)
Corn	20% water	None	NaCl, NaI, LiCl, LiI, 12	10^{-5}–10^{-6}	Finkenstadt and Willett (2005)
Unknown	0	None	None	10^{-9}	Khiar and Arof (2010)
Unknown	0	None	NH_3NO_3, 20	$10^{-4.5}$	Khiar and Arof (2010)
Corn	50% r.h.	None	BMIM Cl, 23	$10^{-4.6}$	Sankri et al. (2010)
Corn	50% r.h.	None	AMIM Cl, 23	$10^{-1.6}$	Ning et al. (2009a)
Corn	50% r.h.	DMAC, 23	LiCl, 10.8	$10^{-0.5}$	Ning et al. (2009b)
Corn	50% r.h.	Glycerol, 20	None	10^{-8}	Ma et al. (2008)
Corn	50% r.h.	Glycerol, 20	Carbon black, 5	10^{1}	Ma et al. (2008)
Corn	50% r.h.	None	Carbon nanotubes, 1.5	10^{-4}	Zhanjun et al. (2011)

Abbreviations: DMAC, dimethylacetamide; BMIM Cl, 1-butyl-3-methylimidazolium chloride; AMIM Cl, 1-allyl-3-methylimidazolium chloride.

Willett 2005). The effects of other inorganic and organic (ionic liquid) salts on starch conductivity have been examined (Table 14.2), and conductivities up to $10^{-1.6}$ S/cm were measured. Added salts act as coplasticizers with water and usually result in lower starch film strengths and increased elongations to break. Much higher conductivities (10 S/cm) were attained with addition of 5% carbon black to glycerol-plasticized starch films (Ma et al. 2008). It was thought that the high conductivity was due to percolating networks of linked carbon black nanoparticles. Addition of carbon nanotubes at the 1.5% level also increased conductivity to 10^{-4} S/cm (Zhanjun et al. 2011). There has been little published work on the effect of chemical modification on the electrical conductivity of starch, so this might be an area of future research.

Some possible applications for electrically conductive starches include solid polymer electrolytes for batteries, antistatic packaging for electronics, anticorrosive coatings for metals, engineered tissues, and sensors (Finkenstadt 2005). Tatarka (1996) surveyed the surface resistivity of commercial starch foams and found that one (Eco-Foam) made from hydroxypropyl high-amylose starch had sufficiently low surface electrical resistivity to be considered antistatic. Thus, it could be used where static cling is undesirable or where static discharge could be dangerous. Some reports have shown that native and modified starches have potential as anticorrosive coatings (Patru et al. 2006; Bello et al. 2010; Rosliza and Wan Nik 2010). Further work is needed in these areas to improve conductivity and to improve mechanical property stability to changes in humidity.

14.3.4 Optically Active Systems

Starch has no absorption in the visible range and hence is colorless. Starch does, however, rotate polarized light due to asymmetric carbons of glucose residues as well as the helical tertiary structures of A, B, C, and V starch crystals. An interesting application of this inherent asymmetry of starch is the ability of amylose and some of its derivatives to separate optically active isomers. For example, amylose phenylcarbamates are made commercially as chiral stationary phases to separate and purify small molecule pharmaceutical enantiomers by HPLC (Okamoto et al. 1988; Ikai et al. 2005). The molecular basis of this selective adsorption is only beginning to be understood (Ye et al. 2007), so there may be expansion of this technology as understanding grows.

Although most chemically modified starches are colorless, some starch derivatives can organize into ordered, supramolecular structures which reflect certain wavelengths of light. Jeong and Ma (2007) prepared high-DS amyloses bearing cholesterol groups, and these displayed different reflection colors due to thermotropic cholesteric liquid crystal formation. The direction of orientation of the derivatized amylose molecules is thought to rotate, forming a helicoidal array with pitches on the order of wavelength of visible light. Shogren et al. (2010) prepared starch stearates of moderate DS (0.6) using a simple heating step in ionic liquids and showed that these reflected blue light when dispersed in toluene. The color disappeared upon heating to over 60°C and reappeared on cooling. It is thought that the amylose component of starch is responsible for liquid crystalline behavior (Zhao et al. 1998; Jeong and Ma 2007). Such liquid crystalline modified starches may have potential in applications such as liquid crystalline displays, high-strength fibers, or temperature-responsive materials.

Wondraczek et al. (2011) recently published a review of photoactive polysaccharides. Their focus was chemical modification of primarily nonstarch polysaccharides with optically active moieties. However, some of the approaches they summarized could probably be applied to starch as well. For example, dextran nanoparticles derivatized with fluorescein can be used to monitor intracellular pH (Hornig et al. 2008). Functionalized celluloses can serve as organic light-emitting diodes via electrical stimulation (Karakawa et al. 2007). Photovoltaic cells using cellulose–porphyrin films showed higher quantum yields than those using only porphyrin (Sakakibara et al. 2007). Dye-sensitized solar cells containing an amylose layer were found to be more efficient due to retardation of charge recombination losses (Handa et al. 2007). Starch films containing a small amount of photosensitizer (sodium benzoate) were cross-linked by exposure to UV light (Delville et al. 2002). It was hypothesized that light decomposed the benzoate into free radicals which led to starch radical formation and cross-linking. Very reactive, hazardous chemicals such as phosphorus oxychloride are often used to cross-link starch, so this represents a safer alternative.

Certain molecules can, upon forming helical inclusion complexes with amylose, undergo conformational or structural changes which lead to color formation. The most familiar of these is the intense blue color formed by iodine in the presence of starch or amylose. The gross structure of the amylose–iodine complex is a left-handed amylose helix, six glucose residues per turn, 0.8 nm pitch with polyiodide embedded inside the helix, and average I-I distance of 0.29 nm (Bluhm and Zugenmaier 1981). The x-ray diffraction data were, however, of rather low resolution since only 16 reflections could be identified (Bluhm and Zugenmaier 1981). There have been many other studies of the structure of the amylose–iodine complex (Cesaro et al. 1986; Yu et al. 1996; Konishi et al. 2001) and cycloamylose-iodine analogs (Nimz et al. 2003), but the structure of the iodine chain and source of the blue color are still debated. Various species such as I_2, I_3^-, and I_5^- are thought to contribute, depending on experimental conditions.

Despite the volume of work on starch–iodine complexes, there is little mention of applications for these in the literature besides as a colorimetric assay for the presence of starch. Rietman (1984) reported that amylose-iodine films had electrical conductivities typical of semiconductors but did not give units for measurements. One might speculate that amylose–iodine complexes having well-defined architecture might have interesting and useful properties and applications as dyes for solar cells or organic semiconductors.

Amylose is known to form helical inclusion complexes with other polymers (Shogren et al. 1991; Bengang and Liming 2010), and some of these have displayed some interesting optical properties. Coiling of amylose around a ladder polysilane induced the formation of a twisted conformation which displayed optical activity (Kato et al. 2009). Frampton and Anderson (2007) have reviewed the literature on insulated molecular wires and describe how amylose complexing of conjugated polymers enhances their chemical stability, electrical transport, and luminescence. Frampton et al. (2008) prepared amylose inclusion complexes with conjugated polymers poly(p-phenylene) and poly(4,4′-diphenylene-vinylene). Photoluminescence efficiencies of the complexes were higher than the naked polymers, demonstrating that encapsulation by amylose was effective in reducing quenching processes. Electroluminescence was also enhanced by complexing, indicating that complexes can function as light-emitting diodes. Similar results for amylose-poly(phenylenevinylene) complexes were obtained by Nishino et al. (2007) and Kato et al. (2009). Conjugated polymers are gaining commercial importance as light-emitting diodes, thin-film field-effect transistors, photovoltaic cells, and sensors (Frampton and Anderson 2007), so enhancement of properties by complexing with amylose could lead to expansion in these applications.

14.3.5 Barrier/Selective Permeability

The permeability of starch to water, common gases, and organics has been studied by a number of authors. As expected, the permeability of amylose films to water is very high (Table 14.3) and increases with humidity due to strong interactions between starch hydroxyl groups and water (Rankin et al. 1958; Rindlav-Westling et al. 1998). Permeabilities of amylose or starch films to nonpolar gases such as oxygen are very low under dry or moderate humidity conditions and are similar to commercial barrier films such as poly(ethylene-co-vinyl alcohol) (Forssell et al. 2002). At water contents over 20% (humidities > 80%), permeability of starch to oxygen increases exponentially leading to loss of barrier properties (Forssell et al. 2002). This is likely due to the onset of large-scale molecular motions as the glass transition temperature of starch decreases to room temperature at >20% moisture content (Shogren 1992). Amylose films when dry are also good barriers to

Table 14.3 Permeability of Films to Water Vapor, Oxygen, and Organic Vapors

Film	Thickness (μm)	r.h. (%)	WVTR	OTR	ETR	EATR	Reference
Amylose	30	0	3100	~0	7.1	18	Rankin et al. (1958)
Starch	30	50		2.9			Forssell et al. (2002)
EVOH	30	50		2.9			Forssell et al. (2002)
PE/starch + 20% glycerol/PE	65	50		10			Dole et al. (2005)
LDPE	30	50	0.5	5000			Combellick (1985)
PLA	25	50	80–170	330			Shogren (1997), Zenkiewicz et al. (2009)

WVTR, water vapor transmission rate in g/m²/day at a gradient of 0%–100% r.h.; OTR, oxygen transmission rate in cm³/m²/day/atm; ETR, ethanol transmission rate in g/m²/day; EATR, ethyl acetate transmission rate in g/m²/day; EVOH, poly(ethylene-*co*-vinyl alcohol); PE, polyethylene; PLA, polylactic acid.

organics such as ethanol, butanol, acetone, ethyl acetate, and benzene (Rankin et al. 1958). It could be presumed that permeability of amylose to such compounds would also increase with humidity, and this has been demonstrated for ethanol and other aroma compounds (Sereno et al. 2009).

Permeabilities of starch films are also influenced by additions of plasticizers, fillers, and other polymers. Permeability generally increases with addition of plasticizers such as glycerol or sorbitol, but at low levels (~10%), permeability can actually decrease slightly (Adhikari et al. 2010). This is thought to be due to antiplasticization which reduces free volume between starch chains and hence reduces starch chain mobility (Shogren et al. 1992; Adhikari et al. 2010). Addition of particles such as nanoclays to starch decreases permeability due to the tortuous path molecules must travel to diffuse around platelet-shaped particles (Zhao et al. 2008; Ivanova et al. 2009; Zeppa et al. 2009; Arora and Padua 2010). Since starch loses its barrier and mechanical properties at high moisture levels, films having a starch inner layer and moisture-resistant outer layers such as polyethylene and polylactic acid have been prepared (Wang et al. 2000; Dole et al. 2005; Fang et al. 2005), and these have good oxygen barrier properties (Table 14.3).

Such starch laminate films have potential as oxygen barriers for food packaging and as organic barriers for mulch films, chemical packaging, or controlled release. Oxygen is usually deleterious to food stability, so barrier films composed of laminates of various petroleum-based polymers are now used commercially for food packaging (Brody et al. 2008; Mensitieri et al. 2011). Similar laminates are also used as mulch films to keep soil fumigants from evaporating prematurely (Santos et al. 2007). Although methyl bromide is being phased out, replacements will likely be organic molecules with similar permeability characteristics. Starch films and foams can also be used as controlled release matrices for volatile agricultural chemicals such as 2-heptanone (Glenn et al. 2006) and for oral drug dosage forms (Ispas-Szabo et al. 1999).

Instead of amorphous starch/plasticizer films now being studied, more effective amylose-based barriers could be envisioned. For example, short, uniform amylose chains having a stearate bonded at both ends might self-assemble into alternating layers of crystalline amylose and crystalline stearic acid. The crystalline amylose layers would be much more effective barrier layers due to lower molecular mobility and much less sensitivity to moisture. The selective or relative permeabilities of different molecules through amylose films have not been studied, so this may be a fruitful area for future work.

14.3.6 Responsive Systems

Stimuli-responsive polymers undergo a change in structure and properties in response to an environmental change and are an active area of current research (Meng and Hu 2010; Stuart et al.

2010; Zhang and Han 2010). Stimuli may include changes in temperature, humidity, pH, light, magnetic field, and electricity, while responses may include changes in conformation, shape, permeability, swelling, electrical conductivity, etc. Applications can range from drug delivery, tissue engineering, microdevices, self-healing surfaces, sensors, etc.

A number of examples of stimuli-responsive systems have been developed from modified starches. Starch-poly(methyl acrylate) blown films containing urea as plasticizer shrink upon exposure to high humidities (Fanta and Otey 1989; Willett 2008). This is thought to be due to relaxation of partially oriented starch as water absorption softens the starch phase and T_g drops below room temperature. Degree of shrinkage can be controlled by urea content. Shape-memory processes have also been described for starch-water extrudates (Chaunier and Lourdin 2009). In their demonstration, a starch-water rod acquires a helical shape exiting the extruder, the helix is frozen in place by cooling and pulled to a straight rod by pulling at elevated temperature, and then the helical shape is restored on releasing tension. Polyether-polyurethane cross-linked starch nanoparticles showed considerable variation of electrical conductivity with temperature and frequency, suggesting use as temperature sensors (Valodkar and Thakore 2010). A graft copolymer of starch with poly(N-isopropylacrylamide-co-N,N-dimethylacrylamide) and poly(2-acrylamido-2-methyl-1-propanesulfonic acid) showed swelling in saline and drug release when temperature was decreased below 36°C (Fundueanu et al. 2010).

REFERENCES

Adhikari, B., D.S. Chaudhary, and E. Clerfeuille. 2010. Effect of plasticizers on the moisture migration behavior of low-amylose starch films during drying. *Drying Technology* 28(4):468–480.

Arora, A. and G.W. Padua. 2010. Review: Nanocomposites in food packaging. *Journal of Food Science* 75(1):R43–R49.

Bastioli, C. 2001. Global status of the production of biobased packaging materials. *Starch/Stärke* 53:351–355.

Bello, M., N. Ochoa, V. Balsamo, F. Lopez-Carrasquero, S. Coll, A. Monsalve, and G. Gonzalez. 2010. Modified cassava starches as corrosion inhibitors of carbon steel: An electrochemical and morphological approach. *Carbohydrate Polymers* 82(3):561–568.

Bengang, L. and Z. Liming. 2010. Inclusion complexation of amylose. *Progress in Chemistry* 22(6):1161–1168.

Bhosale, R.G. and G.R. Ziegler. 2010. Preparation of spherulites from amylose-palmitic acid complexes. *Carbohydrate Polymers* 80(1):53–64.

Blanshard, J.M.V. 1987. Starch granule structure and function: A physicochemical approach. In *Starch: Properties and Potential*, ed. T. Galliard, pp. 16–54. John Wiley & Sons, New York.

Bluhm, T.L. and Zugenmaur, P. 1981. Detailed structure of the Vh-amylose–iodine complex: A linear polyiodine chain. *Carbohydrate Research* 89(1):1–10.

Brody, A.L., B. Bugusu, J.H. Han, C.K. Sand, and T.H. McHugh. 2008. Innovative food packaging solutions. *Journal of Food Science* 73(8):R107–R116.

Cardoso, M.B., J.-L. Pautaux, Y. Nishiyama, W. Helbert, M. Hytch, N.P. Siveira, and H. Chanzy. 2007. Single crystals of V-amylose complexed with α-naphthol. *Biomacromolecules* 8(4):1319–1326.

Carvalho, A.J.F. 2008. Starch: Major sources, properties and applications as thermoplastic materials. In *Monomers, Polymers and Composites from Renewable Resources*, eds. M.N. Belgacem and A. Gandini, pp. 321–342. Elsevier, Oxford, U.K.

Cesaro, A., J.C. Benegas, and D.R. Ripoll. 1986. Molecular model of the cooperative amylose-iodine-triiodide complex. *Journal of Physical Chemistry* 90(12):2787–2791.

Chaunier, L. and Lourdin, D. 2009. The shape memory of starch. *Starch/Stärte* 61:116–118.

Combellick, W.A. 1985. Barrier polymers. In *Encyclopedia of Polymer Science and Technology*, vol. 2, eds. H.F. Mark, N.M. Bikales, C.G. Overberger, G. Menges, and J.I. Kroschwitz, p. 180. John Wiley & Sons, New York.

Davies, T., D.C. Miller, and A.A. Procter. 1980. Inclusion complexes of free fatty acids with amylose. *Starch/Stärke* 32:149–158.

Delville, J., C. Joly, P. Dole, and C. Bliard. 2002. Solid state photocrosslinked starch based films: A new family of homogeneous modified starches. *Carbohydrate Polymers* 49:71–81.

Dole, P., L. Averous, C. Joly, G. Della Valle, and C. Bliard. 2005. Evaluation of starch-PE multilayers: Processing and properties. *Polymer Engineering and Science* 45(2):217–224.

Donald, A.M. 2001. Plasticization and self assembly in the starch granule. *Cereal Chemistry* 78(3):307–314.

Fang, J.M., P.A. Fowler, C. Escrig, R. Gonzalez, J.A. Costa, and L. Chamudis. 2005. Development of biodegradable laminate films derived from naturally occurring carbohydrate polymers. *Carbohydrate Polymers* 60(1):39–42.

Fanta, G.F., F.C. Felker, R.L. Shogren, and J.H. Salch. 2008. Preparation of spherulites from jet cooked mixtures of high amylose starch and fatty acids. Effect of preparative conditions on spherulite morphology and yield. *Carbohydrate Polymers* 71(2):253–262.

Fanta, G.F. and Otey, F.H. 1989. Moisture-shrinkable films from starch graft copolymers. U.S. Patent 4839450.

Finkenstadt, V.L. 2005. Natural polysaccharides as electroactive polymers. *Applied Microbiology and Biotechnology* 67:735–745.

Finkenstadt, V. and J.L. Willett. 2005. Preparation and characterization of electroactive biopolymers. *Macromolecular Symposia* 227:367–371.

Fitt, L.E. and H.T. McNary. 1992. Absorbable dusting powder derived from starch. U.S. Patent 5126334.

Forssell, P., R. Lahtinen, M. Lahelin, and P. Myllarinen. 2002. Oxygen permeability of amylose and amylopectin films. *Carbohydrate Polymers* 47:125–129.

Frampton, M.J. and H.L. Anderson. 2007. Insulated molecular wires. *Angewandte Chemie International Edition* 46(7):1028–1064.

Frampton, M.J., T.D.W. Claridge, G. Latini, S. Brovelli, F. Cacialli, and H.L. Anderson. 2008. Amylose-wrapped luminescent conjugated polymers. *Chemical Communications* 2797–2799.

Fundueanu, G., M. Constantin, P. Ascenzi, and B.C. Simionescu. 2010. An intelligent multicompartmental system based on thermo-sensitive starch microspheres for temperature-controlled release of drugs. *Biomedical Microdevices* 12(4):693–704.

Glenn, G.M., A.P. Klamczynski, C. Ludvik, J. Shey, S.H. Imam, B.-S. Chiou, T. McHugh, G. DeGrandi-Hoffman, W. Orts, D. Wood, and R. Offeman. 2006. Permeability of starch gel matrices and select films to solvent vapors. *Journal of Agricultural and Food Chemistry* 54(9):3297–3304.

Gordon, G.C., D.W. Cabell, L.N. Mackey, J.G. Michael, and P.D. Trekhan. 2006. Electro-spinning process for making starch filaments for flexible structure. U.S. Patent 7029620.

Halley, P.J., R.W. Truss, M.G. Markotsis, C. Chaleat, M. Russo, A.L. Sargent, I. Tan, and P.A. Sopade. 2008. A review of biodegradable thermoplastic starch polymers. *ACS Symposium Series* 978:287–300.

Handa, S., S.A. Haque, and J.R. Durrant. 2007. Saccharide blocking layers in solid state dye sensitized solar cells. *Advanced Functional Materials* 17(15):2878–2883.

Helbert, W. and H. Chanzy. 1994. Oriented growth of V amylose n-butanol crystals on cellulose. *Carbohydrate Polymers* 24:119–122.

Hornig, S., C. Biskup, A. Grafe, J. Wotschadlo, T. Liebert, and G.J. Mohr. 2008. Biocompatible fluorescent nanoparticles for pH-sensing. *Soft Matter* 4:1169–1172. http://www.ers.usda.gov/briefing/baseline/crops.htm

Ikai, T., C. Yamamoto, M. Kamigaito, and Y. Okamoto. 2005. Enantioseparation by HPLC using phenylcarbonate, benzoylformate, p-toluenesulfonylcarbamate, and benzoylcarbamates of cellulose and amylose as chiral stationary phases. *Chirality* 17:299–304.

Ispas-Szabo, P., F. Ravenelle, I. Hassan, M. Preda, and M.A. Mateescu. 1999. Structure-properties relationship in cross-linked high-amylose starch for use in controlled drug release. *Carbohydrate Research* 323(1–4):163–175.

Ivanova, T., N. Lilichenko, J. Zicans, and R. Maksimov. 2009. Starch based biodegradable nanocomposites: Structure and properties. *Diffusion and Defect Data Part B: Solid State Phenomena* 151:150–153.

Jane, J.-L. 1992. Preparation and food applications of physically modified starches. *Trends in Food Science and Technology* 3C:145–148.

Jeong, S.Y. and Y.D. Ma. 2007. Influence of spacer and degree of esterification on thermotropic liquid crystalline properties of amylases bearing cholesterol group. *Polymer (Korea)* 31(4):356–367.

Karakawa, M., M. Chikamatsu, C. Vakamoto, Y. Maeda, S. Kubota, and K. Yase. 2007. Organic light-emitting diode application of fluorescent cellulose as a natural polymer. *Macromolecular Chemistry and Physics* 208:2000–2006.

Kato, N., T. Sanji, and M. Tanaka. 2009. A luminescent oligo(p-phenylenevinylene) wrapped with amylose. *Chemistry Letters* 38(12):1192.

Khiar, A.S.A. and Arof, A.E. 2010. Conductivity studies of starch-based polymer electrolytes. Jonics 16(2): 123–129.

Konishi, T., W. Tanaka, T. Kawai, and T. Fujikawa. 2001. Iodine l-edge XAFS study of linear polyiodide chains in amylose and α-cyclodextrin. *Journal of Synchrotron Radiation* 8(2):737–739.

Le Corre, D., J. Bras, and A. Dufresne. 2010. Starch nanoparticles: A review. *Biomacromolecules* 11(5):1139–1153.

Lee, K.Y., L. Jeong, Y.O. Kang, S.J. Lee, and W.H. Park. 2009. Electrospinning of polysaccharides for regenerative medicine. *Advanced Drug Delivery Reviews* 61(12):1020–1032.

Lourdin, D., H. Bizot, and P. Colonna. 1997. "Antiplasticization" in starch-glycerol films. *Journal of Applied Polymer Science* 63:1047–1053.

Ma, X., P.R. Chang, J. Yu, and P. Lu. 2008. Electrically conductive carbon black (cb)/glycerol plasticized-starch (gps) composites prepared by microwave radiation. *Starch/Stärke* 60:373–375.

Mauritz, K.A., E. Baer, and A.J. Hopfinger. 1978. Epitaxial crystallization of macromolecules. *Journal of Polymer Science Macromolecular Reviews* 13:1–60.

Meng, H. and J. Hu. 2010. A brief review of stimulus-active polymers responsive to thermal, light, magnetic, electric and water/solvent stimuli. *Journal of Intelligent Material Systems and Structures* 21(9):859–885.

Mensitieri, G., E. DiMaio, G.G. Buonocore, I. Nedi, M. Oliviero, L. Sansone, and S. Iannace. 2011. Processing and shelf life issues of selected food packaging materials and structures from renewable resources. *Trends in Food Science & Technology* 22:72–80.

Nimz, O., K. Gebler, I. Uson, S. Laettig, H. Welfle, G.M. Sheldrick, and W. Sainger. 2003. X-ray structure of the cyclomaltohexaicosaose triiodide inclusion complex provides a model for amylose-iodine at atomic resolution. *Carbohydrate Research* 338:977–986.

Ning, W., Z. Xingxiang, L. Haihui, and H. Benqiao. 2009a. 1-Allyl-3-methylimidazolium chloride plasticized-corn starch as solid biopolymer electrolytes. *Carbohydrate Polymers* 76:482–484.

Ning, W., Z. Xingxiang, L. Haihui, and W. Jianping. 2009b. N,N-dimethylacetamide/lithium chloride plasticized starch as solid biopolymer electrolytes. *Carbohydrate Polymers* 77:607–611.

Nishino, S., T. Mori, S. Tanahara, K. Maeda, M. Ikeda, Y. Furusho, and E. Yashima. 2007. Application of soluble poly(phenylenevinylene) wrapped in amylose to organic light-emitting diodes. *Molecular Crystals and Liquid Crystals* 471:29–38.

Nishiyama, Y., K. Mazeau, M. Morin, M.B. Cardoso, H. Chanzy, and J.-L. Putaux. 2010. Molecular and crystal structure of 7-fold V-amylose complexed with 2-propanol. *Macromolecules* 43(20):8628–8636.

Ohdan, K., K. Fujii, M. Yanase, T. Takaha, and T. Kuriki. 2006. Enzymatic synthesis of amylose. *Biocatalysis and Biotransformation* 24(1–2):77–81.

Okamoto, Y., R. Aburatani, K. Hatano, and K. Hatada. 1988. Optical resolution of racemic drugs by chiral HPLC on cellulose and amylose tris(phenylcarbamate). *Journal of Liquid Chromatography* 11(9–10):2147–2163.

Patru, A., M. Floroiu, B. Tutunaru, and M. Preda. 2006. Inhibition of carbon steel corrosion in HCL solution using poly(O-2-hydroxyethyl)starch. *Revista de Chimie* 57(8):908–909.

Pearlstein, R.W., K.E. Broglie, and C.F. Hines. 2007. Maize starch containing elevated amounts of actual amylose. U.S. Patent 7244839.

Perez, S. and E. Bertoft. 2010. The molecular structures of starch components and their contribution to the architecture of starch granules: A comprehensive review. *Starch/Stärke* 62(8):389–420.

Putseys, J.A., L. Lamberts, and J.A. Delcour. 2010. Amylose-inclusion complexes: Formation, identity and physico-chemical properties. *Journal of Cereal Science* 51(3):238–247.

Rankin, J.C., I.A. Wolff, H.A. Davis, and C.E. Rist. 1958. Permeability of amylose film to moisture vapor, selected organic vapors and the common gases. *Industrial and Engineering Chemistry* 3(1):120–123.

Rietman, E.A. 1984. Electronic properties of the amylose-iodine complex. *Journal of Materials Science Letters* 3:1043–1045.

Rindlav-Westling, A., M. Stading, A.-M. Hermansson, and P. Gatenholm. 1998. Structure, mechanical and barrier properties of amylose and amylopectin films. *Carbohydrate Polymers* 36:217–224.

Rosliza, R. and W.B. Wan Nik. 2010. Improvement of corrosion resistance of AA6061 alloy by tapioca starch in seawater. *Current Applied Physics* 10(1):221–229.

Rutenberg, M.W. and D. Solarek. 1984. Starch derivatives: Production and uses. In *Starch: Chemistry and Technology*, 2nd edn., eds. R.L. Whistler, J.N. BeMiller, and E.F. Paschall, pp. 312–366. Academic Press, New York.

Sakakibara, I.K., Y. Ogawa, and F. Nakatsubo. 2007. First cellulose Langmuir-Blodgett films toward photocurrent generation systems. *Macromolecular Rapid Communications* 28:1270–1275.

Sankri, A., A. Arhaliass, I. Dez, A.C. Gaumont, Y. Grohens, D. Lourdin, I. Pillin, A. Rolland-Sabate, and E. Leroy. 2010. Thermoplastic starch plasticized by an ionic liquid. *Carbohydrate Polymers* 82:256–263.

Santos, B.M., J.P. Gilreath, and M.N. Siham. 2007. Comparing fumigant retention of polyethylene mulches for nutsedge control in Florida spodosols. *HortTechnology* 17(3):308–311.

Scott, A. 2002. Nanoscale starch challenges silica in fuel efficient 'green' tires. *Chemical Week* 164(15):26.

Sereno, N.M., S.E. Hill, A.J. Taylor, J.R. Mitchell, and S.J. Davies. 2009. Aroma permeability of hydroxypropyl maize starch films. *Journal of Agricultural and Food Chemistry* 57(3):985–990.

Shen, L., E. Worrell, and M. Patel. 2010. Present and future development in plastics from biomass. *Biofuels, Bioproducts and Biorefining* 4:25–40.

Shogren, R.L. 1992. Effect of moisture content on the melting and subsequent physical aging of cornstarch. *Carbohydrate Polymers* 19:83–90.

Shogren, R. 1997. Water vapor permeability of biodegradable polymers. *Journal of Polymers and the Environment* 5(2):91–95.

Shogren, R.L. 2007. Effect of orientation on the physical properties of potato amylose and high-amylose corn starch films. *Biomacromolecules* 8:3641–3645.

Shogren, R.L. 2009. Estimation of the intrinsic birefringence of the A, B and V crystalline forms of amylose. *Starch/Stärke* 61:578–581.

Shogren, R.L., A. Biswas, and J.L. Willett. 2010. Preparation and physical properties of maltodextrin stearates of low to high degree of substitution. *Starch/Stärke* 62:333–340.

Shogren, R.L., G.F. Fanta, and F.C. Felker. 2006. X-ray diffraction study of crystal transformations in spherulitic amylose/lipid complexes from jet-cooked starch. *Carbohydrate Polymers* 64:444–451.

Shogren, R.L., C.L. Swanson, and A.R. Thompson. 1992. Extrudates of cornstarch with urea and glycols: Structure/mechanical property relations. *Starch/Stärke* 44:335–338.

Shogren, R.L., A.R. Thompson, R.V. Greene, S.H. Gordon, and G. Cote. 1991. Complexes of starch polysaccharides and poly(ethylene-co-acrylic acid). Structural characterization in the solid state. *Journal of Applied Polymer Science* 42(8):2279–2286.

Srichuwong, S. and J.-L. Jane. 2007. Physicochemical properties of starch affected by molecular composition and structures: A review. *Food Science and Biotechnology* 16(5):663–674.

Stuart, M.A.C., W.T.S. Huck, J. Genzer, M. Muller, C. Ober, M. Stamm, G.B. Sukhorukov, I. Szleifer, V.V. Tsukruk, M. Urban, F. Winnik, S. Zauscher, I. Luzinov, and S. Minko. 2010. Emerging applications of stimuli-responsive polymer materials. *Nature Materials* 9(2):101–113.

Tatarka, P.D. 1996. Electrical resistance characteristics of starch foams. *Journal of Environmental Polymer Degradation* 4(3):149–156.

Tester, R.F., J. Karkalas, and X. Qi. 2004. Starch—Composition, fine structure and architecture. *Journal of Cereal Science* 39(2):151–165.

Thuwall, M., A. Bodizar, and M. Rigdahl. 2006. Extrusion processing of high amylose potato starch materials. *Carbohydrate Polymers* 65:441–446.

Valodkar, M. and S. Thakore. 2010. Isocyanate crosslinked reactive starch nanoparticles for thermo-responsive conducting applications. *Carbohydrate Research* 345(16):2354–2360.

Van Soest, J.J.G. and D.B. Borger. 1997. Structure and properties of compression-molded thermoplastic starch materials from normal and high-amylose maize starches. *Journal of Applied Polymer Science* 64(4):631–644.

Wang, Z.F., L. Fang, K.X. Zhang, and X. Fu. 2008. Applications and research progress of starch in polymer materials. *Journal of Clinical Rehabilitative Tissue Engineering Research* 12(19):3789–3792.

Wang, L., R.L. Shogren, and C. Carriere. 2000. Preparation and properties of thermoplastic starch-polyester laminate sheets by coextrusion. *Polymer Engineering and Science* 40(2):499–506.

Willett, J.L. 2008. Humidity responsive starch-poly(methyl acrylate) films. *Macromolecular Chemistry and Physics* 209:764–772.

Willett, J.L. 2009. Starch in polymer compositions. In *Starch: Chemistry and Technology*, 3rd edn., eds. J.N. BeMiller and R.L. Whistler, pp. 715–744. Academic Press, New York.

Winter, W.T. and A. Sarko. 1974. Crystal and molecular structure of V-anhydrous amylose. *Biopolymers* 13:1447–1460.

Wittmann, J.C. and B. Lotz. 1990. Epitaxial crystallization of polymers on organic and polymeric substrates. *Progress in Polymer Science* 15:909–948.

Wondraczek, H., A. Kotiaho, P. Fardim, and T. Heinze. 2011. Photoactive polysaccharides. *Carbohydrate Polymers* 83:1048–1061.

Yamane, H. 2006. Fiber formation and solution rheology of the enzymatically synthesized amylose. *Sen'i Gakkaishi* 62:P359–P365.

Ye, Y.K., S. Bai, S. Vyas, and M.J. Wirth. 2007. NMR and computational studies of chiral discrimination by amylose tris(3,5-dimethylphenylcarbamate). *Journal of Physical Chemistry B* 111:1189–1198.

Yu, X., C. Houtman, and R.H. Atalla. 1996. The complex of amylose and iodine. *Carbohydrate Research* 292:129–141.

Zenkiewicz, M., J. Rickert, P. Rytlewski, K. Moraczewski, M. Stepczynska, and T. Karasiewicz. 2009. Characterization of multi-extruded poly(lactic acid). *Polymer Testing* 28:412–418.

Zeppa, C., F. Gouanve, and E. Espuche. 2009. Effect of a plasticizer on the structure of biodegradable starch/clay nanocomposites: Thermal, water-sorption and oxygen-barrier properties. *Journal of Applied Polymer Science* 112(4):2044–2056.

Zhang, J. and Y. Han. 2010. Active and responsive polymer surfaces. *Chemical Society Reviews* 39(2):676–693.

Zhanjun, L., Z. Lei, C. Minnan, and Y. Jiugao. 2011. Effect of carboxylate multi-walled carbon nanotubes on the performance of thermoplastic starch nanocomposites. *Carbohydrate Polymers* 83:447–451.

Zhao, W., A. Kloczkowski, and J.A. Mark. 1998. Novel high-performance materials from starch. 1. Factors influencing the lyotropic liquid crystallinity of some starch ethers. *Chemistry of Materials* 10:784–793.

Zhao, R., P. Torley, and P.J. Halley. 2008. Emerging biodegradable materials: Starch- and protein-based bio-nanocomposites. *Journal of Materials Science* 43(9):3058–3071.

Zobel, H.F. and A.M. Stephen. 2006. Starch: Structure, analysis and application. *Food Science and Technology* 160:25–85.

CHAPTER 15

Recent Progress in Starch-Based Biodegradable Hybrids and Nanomaterials

Peter R. Chang, Xiaofei Ma, Debbie P. Anderson, and Jiugao Yu

CONTENTS

15.1 INTRODUCTION

New material development has experienced fast growth in the past two decades, thanks to public concern over the environment, climate change, and the depletion of fossil fuels. Biopolymers (e.g., starch, cellulose, chitin, protein) represent an important alternative to petroleum-based polymers with well-known attributes such as biodegradability, renewability, and low cost. Although starch-based biodegradable materials have exhibited the desirable features of environmental friendliness and sustainability, their functional performance has disappointingly lagged behind. To improve the situation, strategies, such as blending a starch matrix with other biopolymers or fillers, have been explored with an aim to improve the functional performance of the resultant hybrids and composites. Nanomaterials have also attracted much attention, in both academic research and practical application, due to the reinforcing function of incorporated nanoscale fillers (organic and inorganic), as well as the contributions of the nanofillers to electrical, magnetic, and optical properties. Nanometer-sized biofillers from renewable resources show unique advantages over traditional inorganic nanoparticles by virtue of their biodegradability and biocompatibility. Currently, in practical research and applications, the category of biomass-based nanofillers includes mainly the rodlike whiskers of cellulose and chitin and the plateletlike nanocrystals of starch. As a result, over the past decade, the possibility of using nanometer-sized fillers in starch-based materials has received considerable interest. Recent progress in biodegradable starch-based hybrids (where starch serves as a cocontinuous phase with other biodegradable natural polymers) and nanomaterials (where starch serves as a matrix) are therefore focused on in this chapter. Starch/polyester blends are covered in Chapter 4.

15.2 STARCH/NATURAL POLYMER BLENDS

15.2.1 Starch/Chitosan Blends

Chitosan (obtained after *N*-deacetylation of chitin), a linear copolymer of glucosamine and *N*-acetyl glucosamine units linked by β-1,4 glycoside, is nontoxic, edible, biodegradable, biofunctional, and biocompatible. Functionalities of chitosan depend on several factors such as the deacetylation degree, molecular weight, pH of the medium, and temperature. Chitosan-tapioca-starch-based films have been reported to reduce external yeast spoilage caused by *Zygosaccharomyces bailii* [1] in semisolid products. Antimicrobial films have also been made with sweet potato starch and varied levels of chitosan. Chitosan improved the tensile strength and elongation at break of these films but decreased the water solubility and permeabilities to oxygen, water, and vapor [2].

In order to facilitate better interaction between them, either starch or chitosan can be chemically modified. Banana starch was oxidized at three different levels and then acetylated [3]; the modified starch was later blended with chitosan for film preparation. The oxidation level increased the moisture content of the film, but acetylation and the addition of chitosan decreased this characteristic [3]. Incorporation of ferulic acid was found to improve the barrier properties and tensile strength of starch/chitosan blends and significantly inhibited lipid oxidation. Such films can be used to extend the shelf life of food with high lipid content [4]. Chitosan/starch blends will likely exhibit excellent properties because of the combination of hydrogen bonding and opposite charge attraction between the cationic chitosan (in its natural form) and the negatively charged starch (easily transformed by chemical modification). A polysaccharide-based polyelectrolyte complex was prepared by electrostatic interaction between maleic starch half-ester acid (MSA) and chitosan [5]. The carboxyl group content of MSA was a key factor for the self-assembly of MSA-chitosan. MSA containing a higher number of carboxyl groups is more beneficial to the formation and functional properties of MSA-chitosan. MSA-chitosan may be quite useful in controlled drug release or other biomedical applications.

15.2.2 Starch/Alginate Blends

Alginates, the salts of alginic acid, are found in all members of Phaeophyceae, a class of brown seaweed in which alginic acid represents the structural component of the intercellular walls [6], providing both strength and flexibility to the algal tissue. Alginates exist in the form of an insoluble gel of mixed calcium, magnesium, sodium, and potassium salts. The alginic acid is a complex mixture of oligomers, polymannuronic acid, and polyguluronic acid [7]. In the presence of water, guluronic residues in alginate form cross-links with divalent ions. These cross-links are believed to create a stiff egg-box structure that imparts viscoelastic solid behavior to the material. Blends of cornstarch/sodium alginate plasticized with water/glycerol mixtures were reported on by Souza and Andrade [8]. They found a significant increase in Young's modulus for blends with 1% alginate; however, the microstructural characteristics and mechanical properties of the plasticized starch (PS) and its blends were highly dependent on the processing conditions and the plasticizers used. The effect of alginate on the processing of glycerol-plasticized cornstarch, with water excluded from the blends, was investigated using torque rheometry. The addition of alginate to PS had a significant effect on plasticization. The plasticization energy and steady state torque showed a significant decrease when the alginate content was increased from 0% to 15%. Without water, alginate acted synergistically with the plasticizer, and the granular starch was significantly more disrupted than with glycerol alone. The effect of alginate on the processing, mechanical, and microstructural properties of water-excluded PS was so significant that it could be proposed as a secondary plasticizer [9].

15.2.3 Starch/Konjac Glucomannan Blends

Konjac glucomannan (KGM) is a high-molecular-weight, water-soluble, nonionic (neutral) polysaccharide found in roots and tubers of the *Amorphophallus konjac* plant. It is made up of D-mannose (M) and D-glucose (G), in an M/G molar ratio of 1.5–1.6, linked by β-1,4-glycosidic linkages with one acetyl group at the C-6 position in every 17–19 sugar units. KGM has been used widely in processed foods, inks and paints, and biomedical materials. Recently, KGM was reported to interact synergistically with carrageenan, xanthan, gellan, and cornstarch [10]. Chen et al. [11] prepared ST/KGM blend films to study the effect of KGM on the performance of PS-based films. Strong hydrogen bonds formed between the PS and KGM resulting in good miscibility in the blends. Compared with neat PS, the tensile strength of the blend films was significantly enhanced, from 7.4 to 68.1 MPa, as the KGM content increased from 0 to 70 wt.%. The value of the elongation at break of the blend films was higher than that of the PS and reached a maximum value of 59.0% when the KGM content was 70 and 20 wt.% glycerol was used as plasticizer.

15.2.4 Starch/Protein Blends

Blending starch with water-insoluble and/or denatured protein reduces its sensitivity to water. Wheat flour, a starch-rich material, is composed mainly of starch and protein and is quite useful in making thermoplastics [12]. It is well known that wheat gluten is capable of creating a protein network [13] while corn gluten lacks this ability [14]. Zein, which accounts for approximately 45%–50% of the protein present in corn, can form tough, glossy, hydrophobic, and greaseproof coatings with excellent flexibility and compressibility. Some studies have used a starch–zein blend as a model system to understand the behavior of corn flour under extrusion conditions. Batterman-Azcona et al. [15] showed the formation of zein aggregates, which was attributed to the creation of disulfide bonds during thermomechanical treatment. Chanvrier et al. [16] highlighted the influence of blend morphology on mechanical properties. Chanvrier et al. [17] also showed that blends in a glassy state exhibited weak adhesion between the two phases and that dispersed zein particles behaved like solid filler and had a large influence on the mechanical properties of the blend. Starch/protein blends were greatly affected by the processing method used [18]. Habeych et al. [19] designed a shearing device to explore the formation of new types of microstructures in starch–zein blends. The use of simple shear to process blends led to the formation of larger zein flocs than in similar blends obtained through extrusion and mixing. Under shearless conditions, starch–zein formed a cocontinuous blend; conversely, under the influence of shear, zein appeared in the dispersed phase.

15.2.5 Starch/Cellulose Blends

The potential advantages of starch and cellulose, apart from their environmental gains, include their low cost, independence from petroleum sources, availability from renewable resources, and their ability to replace some synthetic polymers. Their inherent disadvantages, such as the relatively low tensile strength and high water absorbency of starch, and the poor solubility and processability of cellulose, however, have limited their wider application [20].

Much attention has been paid to preparing totally biodegradable starch/cellulose derivative blends to overcome the disadvantages of unmodified starch and cellulose. Mater-Bi Y (produced by Novamont SPA, Italy) is a commercial blend based on cellulose derivatives and starch, which has been used as a matrix reinforced by sisal short fibers in injection molding [21]. Acetylated starch and cellulose acetate were pretreated in an acetone solution, and the mixture was extruded as a uniform phase blend [22]. Hydrophilic blend matrices of cross-linked starch and carboxymethyl cellulose (CMC) were also prepared [23]. The cross-linking agent is known to exert a profound effect on the overall properties of the blends. Two cross-linkers, epichlorohydrin and calcium chloride, were used for cross-linking starch and CMC, respectively. The percentage degradation of the blends was increased at first but later decreased [23]. In melt processing, epoxidized soybean oil (a lubricant) and triacetine (a plasticizer) were added to blends of cellulose diacetate and starch. Increasing the amount of starch enhanced the processability of cellulose diacetate [24].

These starch/cellulose derivative blends present good compatibility and retain biodegradability; however, compared to materials based on unmodified natural polymers, these derivatives have increased cost and complicacy. The strong inter- and intramolecular hydrogen bonding in unmodified cellulose can be destroyed using different approaches. Cellulose can be dissolved in an aqueous sodium hydroxide solution after its intramolecular hydrogen bonds have been partially broken down by physical treatment. Miyamoto et al. studied the structure and properties of cellulose/starch blend films regenerated from aqueous sodium hydroxide solution. The blends had porous structures, and the pore size increased as the starch content increased. Because the pores were separate from each other, these films had high water and oil absorption [25]. Recently, room-temperature imidazolium structured ionic liquids (ILs) have been found to be nonderivatizing solvents for cellulose and starch [26]. 1-Allyl-3-methylimidazolium chloride (AmimCl) dissolves cellulose in high concentrations, while 1-butyl-3-methylimidazolium chloride (BmimCl) dissolves starch. A cellulose/starch gel was obtained by storing a homogeneous mixture of cellulose (10% w/w) and starch (5% w/w) in BmimCl solution [27]. Wu et al. [20] prepared cellulose/starch/lignin blend films in AmimCl and coagulated them with water. The cellulose and lignin contents significantly affected the mechanical properties of the films, while starch contributed to the film flexibility.

15.3 BIODEGRADABLE PS-BASED NANOCOMPOSITES

PS has two prevailing disadvantages when compared to most plastics currently in use, i.e., moisture sensitivity and poor mechanical properties; therefore, blending PS with other biodegradable polymers or their derivatives may improve these qualities. Another approach for improving these qualities is to use filler (fiber or mineral, e.g., montmorillonite (MMT)) to reinforce the PS matrix. When these fillers are introduced into the PS matrix, their intrinsic properties may also bring specific functionalities (e.g., electrical and optical properties) to the PS matrix. Such enhancements usually rely on both the geometry and surface structure of the fillers.

15.3.1 Reinforced PS-Based Nanocomposites

15.3.1.1 Nanofillers from Natural Polysaccharides

As the continuous matrix, PS provides good compatibility with numerous natural polysaccharide fillers because of similarities in their chemical structures. Natural cellulose fibers [28], commercial cellulose fibers [29], lignin [30], and granular starch [31] have all been used as micron scale fillers in preparation of full polysaccharide composites by conventional extrusion. Pea starch-based composites, reinforced with citric acid–modified pea or rice starch granules, are complete starch composites in which both the filler and the matrix are from starch components. Due to their abundance, high strength, stiffness, low weight, and biodegradability, nanoscale polysaccharide materials serve as promising candidates for the preparation of bionanocomposites [32]. Recently, much attention has been focused on the use of polysaccharide nanofillers in applications for edible films and/or food packaging. Many natural polysaccharides (such as cellulose, starch, chitosan, chitin, and their derivatives) have been used to generate nanofillers. The size of nanofillers from natural polysaccharides and the properties of the composites are listed in Table 15.1.

15.3.1.1.1 Cellulose

Cellulose constitutes the most abundant renewable biopolymer resource available today. For nearly 150 years, cellulose has been used in the form of fibers or derivatives for a wide spectrum of products and materials [33].

Celluloses can be modified by acid hydrolysis and precipitation to obtain nanofillers. Cellulose nanocrystals are prepared by acid hydrolysis, which removes regions of low lateral order and retains a water-insoluble and highly crystalline residue. The resultant nanocrystals are rodlike particles or whiskers whose dimensions are dependent on the cellulose source [34]. Cellulose nanocrystals deserve greater attention not only because of their unsurpassed quintessential physical and chemical properties but also because of their inherent renewability, sustainability, and abundance [33]. Cellulose nanocrystals have been generated by acid hydrolysis of cellulose from many different sources including bamboo fiber [35], cassava bagasse [36], cottonseed linter [37], flax [38], hemp [39], pea hull [40], ramie fiber [41], wheat straw nanofiber [42], and tunicin (a sea animal) [43–45]. These nanocrystals have been used as reinforcing filler in PS matrices. Improvements to the properties of these nanocomposites may be attributed to the nanometer size effect of the filler, which results in homogeneous dispersion of the nanocrystals in the PS matrix, and to the strong interactions between matrix and filler.

Another efficient method for the production of nanofiller from cellulose is precipitation. Successive addition of a nonsolvent (solvent which is incapable of dissolving cellulose in this case, e.g., ethanol) to a dilute solution of cellulose, or dropwise addition of dissolved cellulose into a nonsolvent, can lead to precipitation of nanoscale material [46,47]. Cellulose acetate, cellulose acetate propionate, cellulose acetate butyrate, and cellulose acetate phthalate, as well as cellulose acetates prepared with varying degrees of substitution, self-assembled into regular nanoparticles, ranging in size from 86 to 368 nm, using nanoprecipitation [46]. Dialysis of cellulose derivatives previously dissolved in *N,N*-dimethylacetamide (DMAc) resulted in the formation of regular nanospheres, while preparation in acetone by successive addition of water led to bean-shaped nanoscale particles [46]. Recently, a new solvent system (i.e., NaOH/urea aqueous solution precooled to −10°C) has been used to easily dissolve cellulose [47]. Cellulose nanoparticles (CENP) were prepared from an NaOH/urea aqueous solution of microcrystalline cellulose using a novel method of dropwise addition of an ethanol/HCl aqueous solution as the precipitant. CENP/PS composites were prepared using a casting process [48].

Table 15.1　The Size of Nanofillers from Natural Polysaccharides and Properties of the Composites

Sources	Dimension of Fillers	Matrix of Composites	Properties of Matrix	Properties of Composites (Wt. Filler Contents)
CEN from bamboo fiber [35]	L=20 μm W=50 nm	Glycerol/pea starch (0.4)	TS=2.5 MPa YM=20.4 MPa	TS=12.8 MPa (8%) YM=210.3 MPa (8%)
CEN from cassava bagasse [36]	L=360–1700 nm W=2–11 nm	Glycerol/cassava starch (0.6) Glycerol/sorbitol/cassava starch (3/3/10)	TS=1.8 MPa YM=16.8 MPa TS=4.1 MPa YM=44.5 MPa	TS=2.5 MPa (5%) YM=23.7 MPa (5%) TS=4.8 MPa (5%) YM=84.3 MPa (5%)
CEN from cottonseed linters [37]	L=350±70 nm d=40±8 nm	Glycerol/wheat starch (3/7)	TS=2.5 MPa YM=36 MPa WDC=2.84×10^{-10} cm^2·s^{-1} (RH 98%)	TS=7.8 MPa (30%) YM=301 MPa (30%) WDC=2.53×10^{-10} cm^2·s^{-1} (30%)
CEN from flax [38]	L=100–500 nm d=10–30 nm	Glycerol/pea starch (0.36)	TS=3.9 MPa YM=31.9 MPa WUE=70% (RH 98%)	TS=11.9 MPa (30%) YM=498.2 MPa (30%) WUE=50% (30%)
CEN from hemp [39]	W=30±10 nm	Glycerol/pea starch (0.36)	TS=3.9 MPa YM=31.9 MPa WUE=70% (RH 98%)	TS=11.5 MPa (30%) YM=823.9 MPa (30%) WUE=50% (30%)
CEN from pea hull [40]	L=400±200 nm d=12±6 nm	Glycerol/pea starch	TS=4.1 MPa YM=40.3 MPa	TS=7.6 MPa (30%) YM=415.2 MPa (30%)
CEN from ramie fibers [41]	L=538.5±125.3 nm d=85.4±25.3 nm	Glycerol/wheat starch (3/7)	TS=2.8 MPa YM=56 MPa WDC=2.8×10^{-10} cm^2·s^{-1} (RH 98%)	TS=6.9 MPa (40%) YM=480 MPa (40%) WDC=2.19×10^{-10} cm^2·s^{-1} (40%)
CEN from wheat straw [42]	L=several μm d=10–80 nm	Glycerol/modified potato starch	TS=4.45±0.30 MPa YM=111±6.3 MPa SM=112 MPa	TS=7.771±0.67 MPa (10%) YM=271±27.4 MPa (10%) SM=308 MPa (10%)
CEN from tunicin whiskers [43–45]	L=500 nm to 1–2 μm W=10 nm	Glycerol/waxy cornstarch (0.33) [43] Sorbitol/waxy cornstarch (0.5) [44] Sorbitol/waxy cornstarch (0.33) [45]	WDC=1.76×10^{-9} cm^2·s^{-1} WDC=10.1×10^{-8} cm^2·s^{-1} TS=8 MPa (RH 31%) YM=350 MPa (RH 31%) TS=3 MPa (RH 75%) YM=48 MPa (RH 75%)	WDC=1.53×10^{-9} cm^2·s^{-1} (3.2%) WDC=1.47×10^{-9} cm^2·s^{-1} (6.2%) WDC=7.0×10^{-8} cm^2·s^{-1} (5%) WDC=6.1×10^{-8} cm^2·s^{-1} (10%) TS=ca. 25 MPa (20%) YM=ca. 950 MPa (20%) TS=ca. 9 MPa (20%) YM=ca. 300 MPa (20%)
CENP from microcrystalline cellulose [47]	d=50–100 nm	Glycerol/wheat starch (0.3)	TS=3.15 MPa WVP=5.75×10^{-10} g m^{-1} s^{-1} Pa^{-1}	TS=10.98 MPa (5%) WVP=3.43×10^{-10} g m^{-1} s^{-1} Pa^{-1} (5%)

Material	Matrix	Dimension	Properties	Properties (loading)
MFC from sulfite softwood cellulose pulp [49]	Glycerol/amylopectin (1)	$W = 20–40\,nm$	$TS = 0.35 \pm 0.05\,MPa$ $YM = 1.6 \pm 0.88\,MPa$	$TS = 5.0 \pm 0.2\,MPa$ (10%) $TS = 31 \pm 2.5\,MPa$ (30%) $TS = 160 \pm 7.9\,MPa$ (70%) $YM = 180 \pm 23\,MPa$ (10%) $YM = 1600 \pm 140\,MPa$ (30%) $YM = 6200 \pm 240\,MPa$ (70%)
Nanofibers from BC [50]	Glycerol/cassava starch (0.3)		$TS = 1.09 \pm 0.39\,MPa$ $YM = 33.4 \pm 4.3\,MPa$	$TS = 4.15 \pm 0.66\,MPa$ (2.5%) $TS = 8.45 \pm 2.35\,MPa$ (2.5% after enzymatic hydrolysis) $YM = 140.6 \pm 40.3\,MPa$ (2.5%) $575.7 \pm 166.7\,MPa$ (2.5% after enzymatic hydrolysis)
STN from waxy cornstarch [51]	Glycerol/waxy cornstarch (1/4) Glycerol/waxy cornstarch (1/3) Glycerol/waxy cornstarch (3/7)	Aggregates $4.4\,\mu m$	$TS = 2.4 \pm 0.5\,MPa$ $YM = 49 \pm 12\,MPa$ $TS = 1.0 \pm 0.1\,MPa$ $YM = 11 \pm 3.6\,MPa$ $TS = 0.26 \pm 0.1\,MPa$ $YM = 0.46 \pm 0.3\,MPa$	$TS = 13.6 \pm 1.6\,MPa$ (10%) $YM = 333 \pm 54\,MPa$ (10%) $TS = 9.8 \pm 1.4\,MPa$ (15%) $YM = 241 \pm 46\,MPa$ (15%) $TS = 3.6 \pm 0.3\,MPa$ (15%) $YM = 44 \pm 5\,MPa$ (15%)
STN from waxy cornstarch [52]	Glycerol/waxy cornstarch (1/2)	Below $50\,nm$; aggregates $1–5\,\mu m$	$WDC = 4.5 \pm 0.6 \times 10^{-10}\,cm^2 \cdot s^{-1}$ (RH58%)	$WDC = 2.7 \pm 0.7 \times 10^{-10}\,cm^2 \cdot s^{-1}$ (2.5%)
CASN from cornstarch [54]	Glycerol/pea starch (0.3)	$d = 50–100\,nm$	$TS = 3.94\,MPa$ $YM = 49.8\,MPa$ $WVP = 5.75 \times 10^{-10}\,g\,m^{-1}\,s^{-1}\,Pa^{-1}$ (RH 75%)	$TS = 8.12\,MPa$ (4%) $YM = 125.1\,MPa$ (4%) $WVP = 2.72 \times 10^{-10}\,g\,m^{-1}\,s^{-1}\,Pa^{-1}$ (4%)
CTNP [56]	Glycerol/potato starch (0.3)	$d = 50–100\,nm$	$TS = 2.84\,MPa$ $WVP = 5.62 \times 10^{-10}\,g\,m^{-1}\,s^{-1}\,Pa^{-1}$ (RH 100%)	$TS = 7.79\,MPa$ (5%) $WVP = 3.41 \times 10^{-10}\,g\,m^{-1}\,s^{-1}\,Pa^{-1}$ (5%)
CSNP [57]	Glycerol/ potato starch (0.3)	$d = 50–100\,nm$	$TS = 2.84\,MPa$ $WVP = 5.8 \times 10^{-10}\,g\,m^{-1}\,s^{-1}\,Pa^{-1}$ (RH 100%)	$TS = 10.8\,MPa$ (6%) $WVP = 3.15 \times 10^{-10}\,g\,m^{-1}\,s^{-1}\,Pa^{-1}$ (6%)

Abbreviation for materials: BC, bacterial cellulose; BC, bacterial cellulose from *Acetobacter xylinum*; CASN, citric acid-modified starch nanoparticles; CEN, cellulose nanoparticles; CENP, cellulose nanoparticles; CTNP, chitin nanoparticles; CSNP, chitosan nanoparticles; MFC, microfibrillated cellulose nanofiber; STN, starch nanocrystals.
Abbreviation for attributes: d, diameter; L, length; SM, storage modulus; TS, tensile strength; W, width; WDC, water diffusion coefficient; WUE, water uptake at equilibrium; WVP, water vapor permeability; YM, Young's modulus.

Microfibrillated cellulose (MFC) nanofiber has been prepared by a method other than acid hydrolysis and precipitation. Sulfite softwood cellulose pulp, consisting of 40% pine and 60% spruce with a hemicellulose content of 13.8% and a lignin content of 1%, was used as the source for the MFC and was treated in four steps: a refining step in order to increase accessibility of the cell wall to the subsequent monocomponent endoglucanase treatment, an enzymatic treatment, a second refining stage, and finally a process in which the pulp slurry was passed through a high-pressure homogenizer. The MFC suspension was added to a 50/50 amylopectin–glycerol matrix to obtain the nanocomposites [49].

Acetobacter xylinum produced an extracellular gel-like material that was a random assembly of cellulose microfibers or nanofibers. Bacterial cellulose (BC) from *Acetobacter xylinum* was used as the reinforcing agent in glycerol-plasticized cassava starch nanocomposites before and after treatment with endoglucanases. As shown in Table 15.1, BC nanofibers showed excellent reinforcement of the starch-based matrix [50].

15.3.1.1.2 Starch Nanofillers

The predominant model for starch is a concentric semicrystalline multiscale structure that allows for the production of new nanoelements: (1) starch nanocrystals, resulting from the disruption of amorphous domains in semicrystalline granules by acid hydrolysis and (2) starch nanoparticles. It is worth emphasizing that another advantage of starch, besides the aforementioned low cost, is that the raw material is relatively pure and does not require an intensive purification procedure [51].

Starch nanocrystals obtained by acid hydrolysis of native waxy cornstarch granules have also been used in a PS matrix [52]. The relative reinforcing effect of starch nanocrystals in PS was most significant when the glycerol content was high. This reinforcing effect was attributed to the establishment of strong interactions, not only between starch nanocrystals but also between the filler and the matrix, and probably also to crystallization that may occur at the filler/matrix interface. Cassava starch was plasticized and reinforced with waxy starch nanocrystals [53]; it showed a 380% increase in the rubbery storage modulus (at 50°C) and a 40% decrease in water vapor permeability. Ma et al. [54] prepared starch nanoparticles by precipitating a starch paste solution with ethanol and fabricating citric acid–modified starch nanoparticles (CASN), which did not swell or gelatinize in hot water as a result of the cross-linking reaction. Compared to starch nanocrystals which may be destroyed at high processing temperatures, CASN did not gelatinize during nanocomposite processing and could therefore be used as reinforcing filler for the PS matrix.

Liu et al. used a high-pressure homogenization method to reduce the size of starch granules from micro to nano, depending on the number of passes through the homogenizer under pressure. The size of starch particles reached 10–20 nm under a pressure of 207 MPa as a result of the bonding interaction between large particles being broken by the strong mechanical shear forces. It is interesting to note that the crystal structure and thermal stability of the starch particles did not change with particle size due to the unchanging nature of starch itself. This method simply produces starch particles of varied size as opposed to acid hydrolysis which is used to generate starch nanocrystals [55].

15.3.1.1.3 Chitin and Chitosan

Chitin is also a biopolymer that is abundant in nature. It is extracted from crab and shrimp shells as a byproduct of the seafood industry. Under hydrolytic conditions of boiling HCl and vigorous stirring, chitin whiskers of slender parallelepiped rods have been successfully prepared from different chitin sources (such as crab shells, shrimp shells, squid pens, and tubes of *Tevnia jerichonana* and *Riftia pachyptila* tubeworms). Unlike the most recognizable slender parallelepiped rod whiskers, unique chitin nanoparticles (CTNP) of about 50–100 nm and lower in crystallinity were obtained from chitin after consecutive acid hydrolysis and mechanical ultrasonication treatments. When CTNP was added to the PS matrix at low levels, it dispersed uniformly, and the tensile strength, storage modulus, glass transition temperature, and water vapor permeability of PS/CTNP composites improved because of good interfacial interaction between the CTNP filler and PS matrix [56].

Chitosan nanoparticles (CSNP) were prepared by a very simple, mild method for physical cross-linking between tripolyphosphate and protonated chitosan. Physical cross-linking by electrostatic interaction, instead of chemical cross-linking, was used to avoid undesirable effects such as the possible toxicity of reagents. The obtained CSNP were incorporated into glycerol-plasticized potato starch matrices [57].

15.3.1.2 Mineral Nanofillers

15.3.1.2.1 Montmorillonite

Intensive research on mineral nanofillers has focused on nanoclays, such as MMT, due to their availability, versatility, and low environmental and health concerns. The chemical formula of MMT is $M_x(Al_{4-x}Mg_x)Si_8O_{20}(OH)_4$, where "M" is a monovalent cation and "x" is the degree of isomorphous substitution (between 0.5 and 1.3) [58]. MMT is a loose-layer silicate made up of two tetrahedral sheets fused to an edge-shared octahedral sheet of aluminum hydroxide. These are flexible anisotropic particles with a high aspect ratio, a thickness of about 1 nm, and a length of several 100 nm [59]. MMT/polymer composites with four different structures, dependent on the matrix, the compatibility between MMT and the polymer matrix, and the processing conditions, are shown in Figure 15.1 [60]. These structures include the following: (a) granular dispersion, where MMT

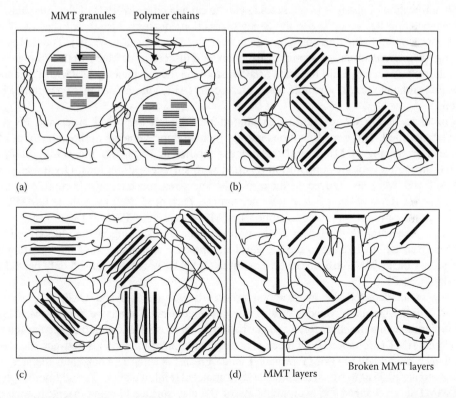

Figure 15.1   Different structures of MMT/polymer composites. (a) granular dispersion; (b) layer aggregated particle dispersion; (c) intercalated dispersion; and (d) exfoliated dispersion. (Reproduced from Polysaccharide/montmorillonite composite: Intercalation and application to superabsorbents, 2004, p. 4 (Chinese), Qiu, H.X., PhD thesis, Tianjin University, Tianjin, China. With permission.)

granules are dispersed in the matrix, the polymer chains have not entered the MMT granules, and compatibility between MMT and the matrix is poor; (b) layer aggregated particle dispersion, where the polymer chains have entered into and destroyed the MMT granules but have not penetrated the clay interlayer, and particles composed of aggregated layers are dispersed in the matrix; (c) intercalated dispersion, where the polymer chains have entered the clay interlayer and the interlayer spacing is enlarged, but the layered structure of MMT is not destroyed; and (d) exfoliated dispersion, where the layered structure of MMT is completely destroyed; the exfoliated layer of MMT is individually dispersed in the matrix. Some MMT layers (long black bars) are broken and become short bars. Structures (a) and (b) belong to microcomposites in which microscale fillers are dispersed in the matrix; nanoscale fillers are dispersed in the matrix of the nanocomposite structures (c) and (d). Nanocomposites have received more attention than microcomposites because they show much better mechanical properties, thermal stability, and water barrier properties because of a more homogeneous dispersion of the nanofillers in the matrix.

The structural style of PS/MMT nanocomposites can be determined by the layered structure of MMT and by the dispersion of MMT layers in the PS matrix. The structure is usually characterized by wide-angle X-ray diffraction (WXRD) analysis and morphological observation. By monitoring the position, shape, and intensity of basal reflections from the distributed MMT layers, the nanocomposite structure (intercalated or exfoliated) can be identified. In exfoliated nanocomposites, extensive layer separation associated with delamination of the original MMT layers in the PS matrix results in the eventual disappearance of any coherent x-ray diffraction from the distributed MMT layers. The finite layer expansion associated with PS intercalation results in the appearance of a basal reflection which corresponds to the larger spacing in the intercalated nanocomposites [58]. Quantification of changes in the spacing can be calculated by WXRD of the intercalated PS/MMT nanocomposites according to the Bragg diffraction equation: $2d\sin\theta = \lambda$ (where λ, θ, and d are the wavelength, diffraction angle, and interlayer spacing, respectively) [61]. The morphology, including the internal structure, spatial distribution, and dispersion of MMT in the PS matrix, can be qualitatively observed. TEM has been used for studying the morphology of PS/MMT nanocomposites [62], although recently, atomic force microscopy (AFM) has gained more popularity in this area [63].

In order to obtain intercalated or exfoliated MMT, both the MMT filler and PS matrix should be modified. These modifications mainly include improvements to the polarity of MMT, the interlayer spacing of MMT, and the process fluidity of PS. MMT has traditionally been modified through a cationic substitution reaction with surface sodium ions. Organically modified MMT (OMMT) has proven to have greatly improved mechanical, thermal, and barrier properties over those of the pristine polymer matrix. Park et al. [62] used three OMMTs with different ammonium cations and one unmodified MMT to prepare PS/MMT composites by melt processing. According to the Bragg diffraction equation, the spacing of the unmodified MMT expanded from 1.17 to 1.78 nm in the PS/MMT composites, an indication of intercalation of PS in the gallery of the MMT silicate layer. The three OMMTs, however, showed only partial or no intercalation of PS.

Zhang et al. depicted the intercalation process between starch and OMMT as shown in Figure 15.2 [64]. The sum of van der Waals forces and electrostatic attractive forces worked against clay exfoliation, while the shear force and elastic force that arose due to starch molecular intercalation favored MMT exfoliation. Both molecular weight and polar interactions between the clay and PS influence intercalation. OMMT provided much greater spacing (3.4 nm) than the size of the starch monomer (0.55 nm), which facilitated starch molecular intercalation. The addition of glycerol to the MMT layer changed the hydrophobicity of the clay surface to some extent; however, low compatibility between the hydrophobic OMMT and hydrophilic starch reduced diffusion of starch molecules into the galleries and lowered the effect of shear force. Hydrophilic modification of MMT is a promising way to further improve the exfoliation of MMT.

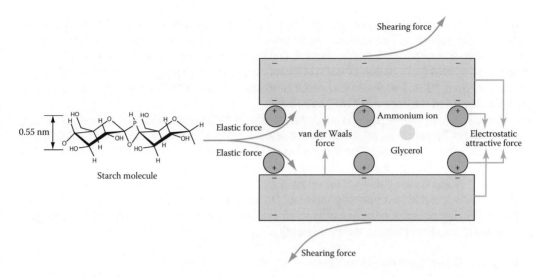

Figure 15.2 Intercalation of starch in the silicate layers of OMMT. (Reproduced from Zhang, Q.X., Yu, Z.Z., Xie, X.L., Naito, K., and Kagawa, Y., Preparation and crystalline morphology of biodegradable starch/clay nanocomposites, *Polymer* 48(24), 7193–7200, Copyright 2007. With permission from Elsevier.)

Huang et al. prepared ethanolamine-activated MMT (EMMT) with the aid of ethanolamine, sulfuric acid, water, and mechanical energy at high temperature (80°C) [65]. Using a similar method, citric acid–activated MMT (CMMT) was also obtained [66]. The spacing was transformed from 1.01 for pristine MMT to 1.25 nm for EMMT and 1.55 nm for CMMT, indicating that the ethanolamine (or citric acid) was intercalated into the MMT layers. The increasing of interlayer spacing and hydrophilicity of MMT facilitated starch molecules to intercalate or exfoliate MMT. Ma et al. [67] adopted a dual-melt extrusion process for the production of PS/MMT nanocomposites. In the first melt extrusion, starch was plasticized to obtain PS, which was then granulated and mixed with EMMT or CMMT for the second melt extrusion. In this way, granular starch was transformed into PS in the first step, and the fluidity of the PS made it easier to enter the MMT layers in the second step. In the matrix of formamide and ethanolamine-plasticized cornstarch [65], the EMMT layers were intercalated and formed intercalated nanocomposites. By increasing the EMMT content from 0 to 5 wt.%, the yield tensile strength changed from 3.6 MPa for PS to 7.0 MPa for PS/EMMT nanocomposites and the Young's modulus went from 47.2 to 145.1 MPa. In addition, thermal stability and water resistance were also improved. In the urea and formamide-PS matrix [66], the CMMT layers were fully exfoliated and formed exfoliated nanocomposites. The tensile strength of PS/CMMT nanocomposites with 10 wt.% CMMT reached 24.9 MPa.

A simple and green processing method was developed to prepare PS/MMT-sorbitol nanocomposites without adding any organic solution or generating any wastewater [67]. In this approach, MMT layers were intercalated by sorbitol, and an MMT–sorbitol complex (with the MMT spacing ranging from 1.01 to 1.80 nm) was obtained in the first melt extrusion. During the second melt extrusion, both sorbitol in the MMT–sorbitol complex and formamide acted as plasticizers for the starch, and the resultant PS was intercalated into the MMT layers.

Chiou et al. [68] used a twin-screw extruder to extrude wheat starch/MMT nanocomposites containing different levels of both glycerol and water. Plasticizer (47 wt.%) was mixed with wheat starch and unmodified MMT to prepare the PS/MMT nanocomposites. The high glycerol and water contents improved the process fluidity of PS, and made it easier for starch molecules to enter the MMT layers. In view of the effect of glycerol on MMT dispersion, composites with 42 wt.% water/5 wt.% glycerol contained the exfoliated MMT, while composites with 37 wt.% water/10 wt.%

glycerol (or 32 wt.% water/15 wt.% glycerol) contained the intercalated MMT. The composites containing higher amounts of glycerol had an increase in starch–glycerol interactions which competed with the interactions between starch, glycerol, and the MMT surface. Tang et al. [69] also found that the degree of MMT exfoliation in PS/MMT nanocomposites increased when the glycerol content decreased from 20 to 5 wt.%. As compared to glycerol as a plasticizer for starch, urea and formamide improved the degree of MMT exfoliation.

Cationic starch (CS) was used as an organomodifier for MMT, and OMMT-CS was subsequently fabricated by Chivrac et al. [70]. XRD did not detect a diffraction peak in the PS/OMMT-CS nanocomposites suggesting an exfoliated morphology; this was confirmed by TEM analysis [70]. The influence of morphology on melt viscosity has also been studied. Because of the presence of a large stack of layers that could not be oriented by shear, the incorporation of pristine MMT into the PS matrix resulted in a viscosity increase. On the contrary, the almost individually dispersed OMMT-CS clay platelets, corresponding to an exfoliated structure, were easily oriented by shear stress and thus did not increase the melt viscosity.

15.3.1.2.2 Other Layered Nanofillers

Similar to MMT, hectorite is among the most common smectite clay minerals. The chemical formula is $M_x(Mg_{6-x}Li_x)Si_8O_{20}(OH)_4$, where "M" is a monovalent cation and "x" is the degree of isomorphous substitution (between 0.5 and 1.3). Composites were made with PS and natural hectorite, with a density of 2600 kg m^{-3} and outer specific surface area of 50 m^2 g^{-1}, and with hectorite modified with 2 methyl, 2 hydrogenated tallow quaternary ammonium chloride (type: Bentone 109, density 1700 kg m^{-3}) by melt processing on a twin roll mill [71]. TEM showed that the untreated hectorite/PS nanocomposites were partially exfoliated while the PS/treated hectorite composites were conventional (not intercalated or exfoliated). Wilhelm et al. [72] also prepared PS/hectorite nanocomposites. Substitution of the PS matrix by a plasticized oxidized starch or a native/oxidized starch blend gave rise to composites with higher interplanar basal distances, indicating that both short oxidized starch chains and glycerol molecules could be intercalated between clay layers. In the absence of glycerol, oxidized starch was preferentially intercalated over native starch chains due to its lower chain size and probable higher diffusion rate.

Kaolinite is a 1:1 dioctahedral clay mineral with the ideal composition of $Al_2Si_2O_5(OH)_4$. It is composed of $AlO_2(OH)_4$ octahedral sheets and SiO_4 tetrahedral sheets. Unlike natural MMT and hectorite, kaolinite is a nonswelling clay. The interlayer space of kaolinite is unsymmetrical, thereby creating large superposed dipoles in the lamellar structure, resulting in a large cohesive energy. The intercalation chemistry of kaolinite is consequently less developed than that of MMT [73]. XRD showed that the (001) peak of kaolinite did not shift after mixing with PS and that the spacing remained at 0.72 nm indicating there was no intercalation and that kaolinite formed conventional composites with PS [71,72]. Kaolinite interacts with small polar molecules, such as dimethyl sulfoxide (DMSO), which are used as precursors for the intercalation of starch. Kaolinite/DMSO/carboxymethyl starch nanocomposites were prepared, and the results showed that DMSO moderately intercalated the interlayer of kaolinite and that the spacing of kaolinite swelled from 0.715 to 1.120 nm [74]. Exfoliated nanocomposites were obtained because the kaolinite nanolayer dispersed into the CMS matrix [73].

The layered double hydroxide (LDH) structure is referred to as natural hydrotalcite and is described by the ideal formula $[M^{II}{}_xM^{III}{}_{1-x}(OH)_2]_{intra}[A^{m-}{}_{x/m}{\cdot}nH_2O]_{inter}$, where "$M^{II}$" and "$M^{III}$" are metal cations, "A" is the anion, and "intra" and "inter" denote the intralayer domain and the interlayer space, respectively. The structure consists of brucitelike layers of edge-sharing $M(OH)_6$ octahedra. Partial "M^{II}" and "M^{III}" substitution induces a positive charge for the layers, balanced with the presence of the interlayered anions [75]. Well-dispersed starch-LDH nanocomposites were prepared by growing LDH crystallites in starch (or acid-modified starch) dispersions under hydrothermal treatment conditions [76]; in this way, LDH could be embedded in a starch matrix. During the reaction, the LDH nuclei were first precipitated in a partially gelatinized starch dispersion, and then gradually aged in the dispersed starch under hydrothermal conditions. The acid modification process reduced the molecular weight and pasting viscosity of the starch, facilitating dispersion of the LDH crystallites in the nanocomposite. The addition of LDH to the acid-modified starch matrix proved to be effective for improving the modulus of the starch–clay nanocomposites. The chemical formulas for LDH in a starch matrix and an acid-modified starch matrix are proposed to be $[M^{2+}_{0.66}M^{3+}_{0.34}(OH)_2]\cdot(NO_3^-)_{0.34}\cdot nH_2O$ and $[M^{2+}_{0.65}M^{3+}_{0.35}(OH)_2]\cdot(NO_3^-)_{0.35}\cdot nH_2O$, respectively.

Synthetic α-zirconium phosphate (α-ZrP), that is, $Zr(HPO_4)_2\cdot H_2O$, exhibits structural characteristics similar to natural MMT clay. The advantages of α-ZrP over MMT clay as a model system include a much higher purity and ion exchange capacity, and the ease of intercalation and exfoliation. In the study by Wu et al. [77], α-ZrP with a large interlayer spacing was synthesized to make PS/ZrP nanocomposites. For this purpose, α-ZrP was exfoliated using commercially available n-butylamine; the resulting material was then incorporated into starch. The objective of this work was to study the effect of exfoliated α-ZrP nanoplatelets on the performance of PS/ZrP nanocomposites. With an increase in α-ZrP content, the maximum thermal decomposition temperature of the nanocomposites decreased. The tensile strength of the nanocomposites increased from 4.01 to 9.44 MPa as the α-ZrP content increased from 0 to 0.3 wt.% with 25 wt.% glycerol content.

15.3.1.3 Other Nanofillers

15.3.1.3.1 Metal, Metal Oxide, and Sulfide

Much effort has been focused on polysaccharides for use as stabilizers of metal and metal oxide nanoparticles. Raveendran et al. [78] selected soluble starch as the template and α-D-glucose as the reductant in an aqueous solution of $AgNO_3$ for the growth of silver nanoparticles. Arabinogalactan was also reported to be a novel protection agent for maintaining metal (platinum, palladium, and silver) nanoparticles in colloidal suspension [79]. He et al. [80] synthesized platinum, palladium, and silver nanoparticles with narrow size distributions using porous cellulose fiber as the stabilizer. He et al. further reported that most metal nanoparticles remaining stable and unchanged in size during carbonization of cellulose and carbon/metal nanoparticle composites were thus prepared [81]. Walsh et al. [82] demonstrated that self-supporting macroporous sponges of silver, gold, and copper oxide, as well as composites of silver/copper oxide or silver/titania, could be routinely prepared by heating metal-salt-containing dextran pastes as soft templates. Nanoparticles of metal oxides and sulfides were also prepared with polysaccharide stabilizers. Zinc oxide (ZnO) nanoparticles were synthesized using water as the solvent and soluble starch as stabilizer [83,84]. CdS nanoparticles were prepared in a sago starch matrix [85] as were silver (Ag) and silver sulfide (Ag_2S) nanoparticles [86]. CMC sodium could be used as the stabilizer for preparation of Sb_2O_3 and ZnO nanoparticles [87,88]. In principle, all of the nanoparticles mentioned earlier are compatible with the PS matrix because of the similar structures of the PS matrix and the polysaccharide encapsulating the nanoparticles. Some of these have, in fact, been used as fillers in the PS matrix.

The encapsulation of cadmium selenide (CdSe) nanoparticles in a plasticized sago starch matrix was investigated [89], and thermogravimetric analysis showed that CdSe nanoparticles significantly reduced the starch degradation rate of the PS matrix. ZnO-soluble starch, ZnO-CMC, and Sb_2O_3-CMC nanoparticles were also incorporated into a PS matrix and resulted in nanocomposites with improved mechanical properties and thermal stability [84,88,90].

15.3.1.3.2 Nonlayered Mineral Nanofillers

Sepiolite (SEP) is a microcrystalline hydrated magnesium silicate with a needlelike structure and a theoretical unit cell formula $Mg_8Si_{12}O_{30}(OH)_4\cdot(H_2O)_4\cdot 8H_2O$. It has alternating blocks and channels that grow in the direction of the fiber. The blocks consist of two layers of tetrahedral silica sandwiching a central octahedral magnesium oxide-hydroxide layer. Some isomorphic substitutions occur inside these central layers creating a negative charge that is naturally counterbalanced by inorganic cations (Na^+, Ca^{2+}, etc.). The SEP channels are filled with both water molecules, which are bonded to the Mg^{2+} ions located at the edges of octahedral sheets, and with zeolitic water, which is associated with the structure by hydrogen bonding. The discontinuity of the silica sheets gives rise to the presence of silanol groups (Si-OH) at the edges of the external surface of the SEP. SEP has been organomodified with CS, and the resultant OSEP-CS was incorporated into a PS matrix [91]. These needle-shaped nanofillers increased the stiffness of the PS matrix without decreasing its stress at break. Because CS acted as a compatibilizer between the SEP and the PS matrix, four different types of organomodified SEP, OSEP-1CS, OSEP-2CS, OSEP-4CS, and OSEP-6CS, corresponding, respectively, to one, two, four, and six charge equivalences between the SEP and the CS, were used. There were fewer SEP aggregates in nanocomposites filled with OSEP with higher charge equivalences.

Hydroxyapatite (HA) is a naturally occurring mineral form of calcium apatite with the formula $Ca_5(PO_4)_3(OH)$, but it is usually written $Ca_{10}(PO_4)_6(OH)_2$ to denote the fact that the crystal unit cell comprises two entities. HA is a fundamental mineral-based biomaterial used for preparing composites for bone repair and regeneration. A porous scaffold of gelatin-starch-HA nanocrystals (nHA) composite was fabricated via microwave vacuum drying and cross-linking using trisodium citrate. This biocompatible composite with enhanced mechanical properties offered enhanced materials for biomedical applications [92].

15.3.1.3.3 Carbon

Much attention has been focused on the unique electrical and mechanical properties of carbon nanotubes (CNT) since a report on them in 1991. Amylose was found to encapsulate single-walled carbon nanotubes (SWNT) when it was in a prearranged helical state produced by complexing with iodine [93]. Kim et al. [94], however, stated that the helical state of amylose was not a prerequisite for amylose encapsulation of SWNT. Their view was that the helical complexation of amylose took place more favorably via a cooperative binding of SWNTs by amylose in a loosely elongated helical state. The best solvent condition for amylose to assume an interrupted loose helix was 10%–20% DMSO in a DMSO-H_2O mixture. Amylose/SWNT composites were prepared in aqueous solutions and made into films [95]. Hydroxypropylated amylose was used as a surfactant to stabilize individual SWNT aqueous dispersions. Ma et al. [96] prepared multiwall CNT (MWCNT) aqueous solution with the aid of sodium dodecyl sulfate. PS/MWCNT composites were obtained by adding glycerol and starch to the solution before casting the films. The introduction of MWCNTs improved the tensile strength, Young's modulus, and electrical conductivity of the PS matrix. Nanocomposites were also prepared using glycerol-plasticized pea starch as the matrix with acid-treated MWCNTs as the reinforcing filler [97]. Besides improving the mechanical properties, the incorporation of MWCNTs into the PS matrix also led to a decrease in water sensitivity. Another form of carbon, 40 nm carbon black (CB), has also been added to a PS matrix in preparation of PS/CB nanocomposites by melt extrusion or microwave radiation [98].

15.3.2 PS-Based Composites with Specific Functionality

15.3.2.1 Electronically Conductive Polymer Composites

Electronically conductive polymer composites (CPC) have attracted much research because they have potential in applications such as electromagnetic shielding materials (e.g., conductive plastic films for the packaging of integrated circuit devices), positive temperature coefficient materials, supercapacitors for charge storage, chemical vapor detection devices, pressure sensors, and electronic noses. CPC is often composed of a polymer matrix, usually petroleum-based polymers (e.g., polyolefin), and electrically conductive fillers, typically metallic powders or carbonaceous fillers including graphite, CNT, and the most widely used CB [99]. The increase in electrical conductivity in CPC is mainly ascribed to the formation of a conductive particle network in the matrix. PS could also serve as the matrix for the formation of a conductive network containing CB.

Electrically conductive glycerol-plasticized starch (GPS)/CB polymer composites were prepared by melt extrusion and microwave radiation [98]. Figure 15.3 shows the electrical conductivity of GPS/CB membranes as a function of CB content at room temperature. GPS/CB membranes prepared by microwave radiation (GPS/CB-MR) had characteristics similar to those of CPC because the electrical conductivity increased sharply at a critical CB concentration and tended to level off above this critical CB level. The critical concentration was interpreted to be the percolation threshold, which was required to form the interconnecting conductive networks in the polymer matrix. The percolation threshold of CB, at which the conductivity increased about 7 orders of magnitude, was 2.398 vol.% for GPS/CB-MR membranes. The conductivity of the GPS/CB-MR membrane containing 4.236 vol.% CB reached 7.08 S cm^{-1}; however, the conductivity of GPS/CB membranes prepared by melt extrusion (GPS/CB-ME) changed only slightly. When the CB content increased from 0 to 4.684 vol.%, the conductivity increased only one order of magnitude. It was discovered that agglomerates of CB particles were isolated and the interconnecting conductive networks could not form in GPS/CB-ME.

The extraordinary mechanical and electrical properties of CNTs make them ideal for blending with polymers in preparation of potentially multifunctional nanocomposites. Good dispersion of CNTs is critical and is usually achieved by dispersing them in solvents or polymers with the aid of a surfactant or a copolymer. CNTs are well dispersed in high-density polyethylene, poly(propylene), poly(ethylene oxide), bisphenol-A polycarbonate (PC), epoxy resin, polyaniline, polyurethane,

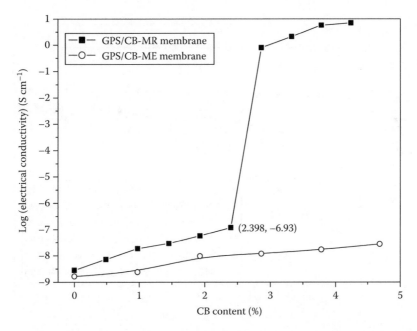

Figure 15.3   Electrical conductivity as a function of CB content for GPS/CB-MR membranes. (Reproduced from *Carbohydr. Polym.*, 74(4), Ma, X.F., Chang, P.R., Yu, J.G., and Lu, P.L., Characterizations of glycerol plasticized-starch (GPS)/carbon black (CB) membranes prepared by melt extrusion and microwave radiation, 895–900, Copyright 2008, with permission from Elsevier.)

poly(ethylene terephthalate) (PET), and so on. MWCNT were added to the GPS matrix to prepare GPS/MWCNT composites by casting in order to improve the mechanical and electrical properties of starch-based materials [96].

Since starch is hydrophilic, water sensitivity is an important criterion for many practical applications of starch-based materials. As shown in Figure 15.4, electrical conductivity of the GPS/MWCNT composite was dependent on water content. GPS/MWCNT composites with different MWCNT contents exhibited a relationship similar to that of conductivity versus water content, which can be described well with a second-order polynomial. The binomial correlation of conductivity (y) and water content (x) was assumed to be $y = B_2 x^2 + B_1 x + B_0$. This model gave good agreement ($R^2 > 0.97$) except for the GPS/MWCNT composite with 4.75 wt.% MWCNTs, which fit the line $y = -0.14 + 0.13x$ and $R = 0.94$ better. The second-order polynomial correlations for the GPS/MWCNT composites are listed in Figure 15.4. Starting at a very low MWCNT content (<2.85 wt.%), conductivity gradually increased as the MWCNT content increased; however, the conductivity of the composite containing 4.75 wt.% MWCNTs increased rapidly to 10^0 S cm^{-1} from that of 10^{-5} S cm^{-1} for the 2.85 wt.% MWCNTs. This change in conductivity was a result of the formation of an interconnected structure of MWCNTs and could be regarded as an electrical percolation threshold. In other words, at a level of about 3.8 wt.% MWCNT, a very high percentage of electrons were permitted to flow through the sample (when an electrical field was applied) due to the creation of interconnecting conductive channels.

As shown in Figure 15.4, the conductivity of GPS without MWCNTs increased about 5 orders of magnitude when the water content increased from 0 to 0.6. The conductivity of the GPS/MWCNT composite with 4.75 wt.% MWCNTs remained almost constant with increasing water content. As the MWCNT content increased, the sensitivity of the conductivity to water was reduced. It was obvious that both the monomial coefficient B_1 and the binomial coefficient B_2 approached zero when the MWCNT content increased. An interconnected structure of MWCNTs was formed when

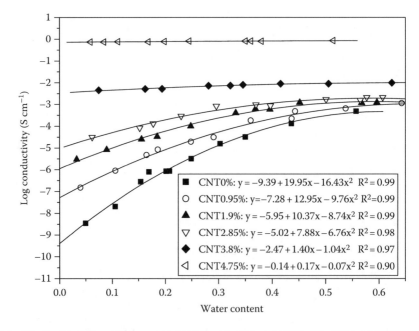

Figure 15.4 Effect of water content on electrical conductivity of GPS with different MWCNT contents. (Reproduced from *Compos. Sci. Technol.*, 68(1), Ma, X.F., Yu, J.G., and Wang, N., Glycerol plasticized-starch/multiwall carbon nanotube composites for electroactive polymers, 268–273, Copyright 2008, with permission from Elsevier.)

greater than 3.8 wt.% of MWCN was incorporated. This structure spatially restrained movement of the starch chain, and the effect of water content on the conductivity was therefore weakened.

15.3.2.2 Solid Polymer Electrolytes

Ionically conducting polymers (ICPs) have attracted a great deal of interest because of their electrochemical application as solid polymer electrolytes (SPEs). They have been extensively studied in the last two decades because of their potential for use in technological areas including solid-state batteries, chemical sensors, and electrochromic devices. Currently, several synthetic polymer matrices have been developed and characterized including poly(ethylene oxide), poly(propylene oxide), poly(acrylonitrile), poly(methyl methacrylate), poly(vinyl chloride), poly(vinylidene fluoride), and poly(vinylidene fluoride-hexafluoropropylene). Many synthetic polymers are fabricated as intractable films, gels, or powders because they are insoluble in most solvents; others are prepared by casting with acidic aqueous or nonaqueous solvents such as ether.

SPEs obtained from natural polymers such as starch, cellulose, chitosan, pectin, hyaluronic acid, agarose, and carrageenan have attracted attention because of their superior mechanical and electrical properties. Among these polymers, starch has many advantages: it is abundant, renewable, low cost, and biodegradable; both starch melt extrusion (thermoplastic processing) and casting processing are available; and its use offers a promising alternative for the development of new SPE materials. Lopes et al. [100] gelatinized amylopectin-rich starch with water on a hotplate. The solution was then combined with glycerol, mixed with LiClO$_4$, cast onto Teflon plates, and allowed to dry. The starch-glycerol-LiClO$_4$ films exhibited a conductance of around 10^{-5} S cm^{-1}. Finkenstadt et al. obtained the accurate moisture content of native starch using a direct-current resistance technique [101] and prepared thermoplastic starch films simply by adding water doped with metal halides (without adding other plasticizers) to produce solid ion-conducting materials [102].

GPS mixed with alkali metal chlorides was prepared as a SPE. Glycerol was added as a plasticizer because the incorporation of plasticizers into the polymer electrolyte reduced its crystallinity, improved movement of the polymeric molecules, and enhanced ionic conductance. Increasing the salt content increased the conductance of GPS, but superfluous salts resulted in the congregation of salt crystals, which definitely affected mechanical properties and conductivity [103]. In another study, two kinds of plasticizers were used to prepare PS with NaCl salt as SPE by melt extrusion. Plasticizers containing amide groups had stronger hydrogen bond–forming abilities with starch than polyols, and formamide-PS had the best conductance over all [104]. Different plasticizers (refer to Chapter 4 for more description) inevitably form different interactions with starch in thermoplastic starch and consequently affect starch recrystallization, movement of starch molecules, and ionic conductance.

Room-temperature ILs have an imidazolium structure and can be used as solvents for starch [27]. AmimCl was found to be a novel plasticizer for cornstarch. AmimCl-PS films have potential in applications as solid biopolymer electrolytes. Hydrophilic AmimCl has a high chloride ion concentration and has strong hydrogen bond–forming abilities with starch, which improves the water absorption and conductance of PS film. The conductance of PS film with 30 wt.% AmimCl was $10^{-1.6}$ S cm^{-1} at a water content of 14.5 wt.%. In another study [105], LiCl was introduced into AmimCl-PS film. An increased concentration of LiCl tended to decrease the proportion of residual starch granules and form homogeneous PS. LiCl also increased the interaction of AmimCl with the PS. The supramolecular structure of starch was destroyed by AmimCl and LiCl. LiCl not only increased the water absorption of PS but also improved its conductivity; the latter achieved values as high as $10^{1.0}$ S cm^{-1} at 20 wt.% water absorption.

DMAc/LiCl is a solvent system that has been utilized to dissolve natural crystalline cellulose. A generally accepted dissolution mechanism is that Li$^+$ is tightly linked with the carbonyl group of DMAc while Cl$^-$ is left unencumbered and thereby highly active as a nucleophilic base. The Li$^+$

therefore plays a major role by breaking up the inter- and intrahydrogen bonds that exist in natural polymers. DMAc and DMAc/LiCl mixtures with different LiCl contents were used as plasticizers in preparation of ionic conductive PS by melt extrusion [106]. The conductance of PS with 18 wt.% LiCl was $10^{-0.5}$ S cm^{-1} with a water content of 18 wt.%.

15.3.2.3 Ultraviolet Shielding Materials

It is well known that nano-ZnO can absorb ultraviolet (UV) radiation. Carboxymethylcellulose (CMC) was used as the stabilizer for production of ZnO-CMC nanoparticles [88]. Figure 15.5 shows the UV absorbance of GPS/ZnO-CMC composites with different ZnO-CMC contents. The addition of ZnO-CMC particles remarkably enhanced the UV absorbance as compared to that of pure GPS. In the UV range (290–400 nm), pure GPS had a very low absorbance; however, by increasing the ZnO-CMC content, the UV absorbance and the intensity of the peaks increased. This optical change was caused by a quantum effect of the nanoparticles. GPS/ZnO-CMC composites could potentially be applied as UV-shielding materials [88]. Similarly, ZnO-starch nanoparticles were synthesized using soluble starch as the stabilizing agent, and the PS/ZnO-starch nanoparticles composites also exhibited absorbance of UV [84].

Antimony trioxide (Sb$_2$O$_3$) is also an excellent UV filter. Since UV absorbance of the PS/Sb$_2$O$_3$-CMC composites increased with increasing Sb$_2$O$_3$-CMC content, these composites may have a potential application on a UV light–emitting device [87].

15.3.2.4 Shape-Memory Materials

Shape-memory polymers (SMP) are a new class of smart materials. They possess the ability to recover a "programmed" permanent shape from a temporary one when subjected to an external

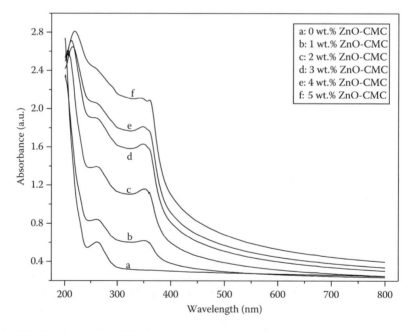

Figure 15.5 UV absorbance of GPS/ZnO-CMC composites with different ZnO-CMC contents. (Reproduced from *Bioresour. Technol.*, 100(11), Yu, J.G., Yang, J.W., Liu, B.X., and Ma, X.F., Preparation and characterization of glycerol plasticized-pea starch/ZnO-carboxymethylcellulose sodium nanocomposites, 2832–2841, Copyright 2009, with permission from Elsevier.)

stimulus. Many recent studies have focused on biopolymer-based materials such as polylactide-block-poly(glycolide-*co*-caprolactone) for the production of thermally responsive materials to replace those that are generally sourced from petroleum, for example, polyurethane and biodegradable multiblock oligo(ε-caprolactone)diol/oligo(*p*-dioxanone)diol copolymers. Melt extruded amorphous starch-based materials were evaluated for their shape-memory properties, such as shape fixity and recovery ratio, for single and cyclic recovery processing [107]. The initial permanent shape was formed by extrusion below the glass transition temperature (T_g) as it is possible to fix amorphous starch in an unstressed state. When the second shape was formed by thermomolding above T_g, residual stress appeared and remained stable when cooled below T_g. Recovery of the initial shape occurred when the residual stresses were relaxed above T_g, making it possible to return to the unstressed equilibrium. Shape recovery was triggered by water sorption; a high recovery ratio ($R_r > 90\%$) was obtained at high relative humidity at deformation ratios of up to 200%. In the case of plasticized blends, a decrease in the glass transition temperature diminished the shape recovery capacity (R_r about 40% for a fixed deformation of 200%) as a result of crystallization at ambient temperature.

15.4 CONCLUSIONS

The poor water resistance and mechanical properties of starch-based materials severely hinder their applications. Many strategies have been adopted to try to overcome these drawbacks. Blending with other materials seems straightforward, though interfacial compatibility is pivotal. Undoubtedly, the intrinsic rigidity of nanofillers, the strong interfacial interactions on the nanofiller surface, and the percolation network organized by nanofillers all contribute to the ultimate properties of nanomaterials. To achieve the best performing starch-based nanomaterials, the key is to balance the interfacial interactions between nanofiller and starch matrix and the self-association of nanofillers. To improve interfacial adhesion between the nanofiller and starch matrix, chemical modification and grafting of nanofillers is helpful. In the future, other plasticizers and different starch sources should also be considered. Further development of biodegradable starch-based hybrids and nanomaterials should include, but not be limited to, biodegradation, the use of new functional materials, processing technology, and cost reduction.

REFERENCES

1. Vasconez MB, Flores SK, Campos CA, Alvarada J, and Gerschenson LN. 2009. Antimicrobial activity and physical properties of chitosan-tapioca starch based edible films and coatings. *Food Research International* 42(7): 762–769.
2. Shen XL, Wu JM, Chen YH, and Zhao GH. 2010. Antimicrobial and physical properties of sweet potato starch films incorporated with potassium sorbate or chitosan. *Food Hydrocolloids* 24(4): 285–290.
3. Zamudio-Flores PB, Torres AV, Salgado-Delgado R, and Bello-Pérez LA. 2010. Influence of the oxidation and acetylation of banana starch on the mechanical and water barrier properties of modified starch and modified starch/chitosan blend films. *Journal of Applied Polymer Science* 115(2): 991–998.
4. Mathew S and Abraham TE. 2008. Characterisation of ferulic acid incorporated starch-chitosan blend films. *Food Hydrocolloids* 22(5): 826–835.
5. Xiao CM and Fang F. 2009. Ionic self-assembly and characterization of a polysaccharide-based polyelectrolyte complex of maleic starch half-ester acid with chitosan. *Journal of Applied Polymer Science* 112(4): 2255–2260.
6. Immirzi B, Santagata G, Vox G, and Schettini E. 2009. Preparation, characterisation and field-testing of a biodegradable sodium alginate-based spray mulch. *Biosystems Engineering* 102(4): 461–472.

7. Avella M, Di Pace E, Immirzi B, Impallomeni G, Malinconico M, and Santagata G. 2007. Addition of glycerol plasticizer to seaweed derived alginates: Influence of microstructure on chemical–physical properties. *Carbohydrate Polymers* 69(3): 503–511.

8. Souza RCR and Andrade CT. 2001. Processing and properties of thermoplastic starch and its blends with sodium alginate. *Journal of Applied Polymer Science* 81(2): 412–420.

9. Cordoba A, Cuellar N, Gonzalez M, and Medina J. 2008. The plasticizing effect of alginate on the thermoplastic starch/glycerin blends. *Carbohydrate Polymers* 73(3): 409–416.

10. Yoshimura M, Takaya T, and Nishinari K. 1998. Rheological studies on mixtures of corn starch and konjac-glucomannan. *Carbohydrate Polymers* 35(1–2): 71–79.

11. Chen JG, Liu CH, Chen YQ, Chen Y, and Chang PR. 2008. Structural characterization and properties of starch/konjac glucomannan blend films. *Carbohydrate Polymers* 74(4): 946–952.

12. Ma XF, Yu JG, and Ma YB. 2005. Urea and formamide as a mixed plasticizer for thermoplastic wheat flour. *Carbohydrate Polymers* 60(1): 111–116.

13. Bache IC and Donald AM. 1998. The structure of the gluten network in dough: A study using environmental scanning electron microscopy. *Journal of Cereal Science* 28(2): 127–133.

14. di Gioia L, Cuq B, and Guilbert S. 2000. Mechanical and water barrier properties of corn-protein-based biodegradable plastics. *Journal of Materials Research* 15(12): 2612–2619.

15. Batterman-Azcona SJ, Lawton JW, and Hamaker BR. 1999. Microstructural changes in zein proteins during extrusion. *Scanning* 21(3): 212–216.

16. Chanvrier H, Colonna P, Della Valle G, and Lourdin D. 2005. Structure and mechanical behaviour of corn flour and starch-zein based materials in the glassy state. *Carbohydrate Polymers* 59(1): 109–119.

17. Chanvrier H, Della Valle G, and Lourdin D. 2006. Mechanical behaviour of corn flour and starch–zein based materials in the glassy state: A matrix–particle interpretation. *Carbohydrate Polymers* 65(3): 346–356.

18. Gonzalez-Gutierrez J, Partal P, Garcia-Morales M, and Gallegos C. 2010. Development of highly-transparent protein/starch-based bioplastics. *Bioresource Technology* 101(6): 2007–2013.

19. Habeych E, Dekkers B, van der Goot AJ, and Boom R. 2008. Starch-zein blends formed by shear flow. *Chemical Engineering Science* 63(21): 5229–5238.

20. Wu RL, Wang XL, Li F, Li HZ, and Wang YZ. 2009. Green composite films prepared from cellulose, starch and lignin in room-temperature ionic liquid. *Bioresource Technology* 100(9): 2569–2574.

21. Alvarez VA and Vazquez A. 2004. Thermal degradation of cellulose derivatives/starch blends and sisal fibre biocomposites. *Polymer Degradation and Stability* 84(1): 13–21.

22. Chen Y, Ishikawa Y, Zhang ZY, and Maekawa T. 2003. Properties of extruded acetylated starch plastic filled with cellulose acetate. *Transactions of the ASAE* 46(4): 1167–1173.

23. Bajpai AK and Shrivastava J. 2005. In vitro enzymatic degradation kinetics of polymeric blends of cross-linked starch and carboxymethyl cellulose. *Polymer International* 54(11): 1524–1536.

24. Lee SY, Cho MS, Nam JD, and Lee Y. 2006. Melting processing of biodegradable cellulose diacetate/starch composites. *Macromolecular Symposia* 242(1): 126–130.

25. Miyamoto H, Yamane C, Seguchi M, and Okajima K. 2009. Structure and properties of cellulose-starch blend films regenerated from aqueous sodium hydroxide solution. *Food Science and Technology Research* 15(4): 403–412.

26. El Seoud OA, Koschella A, Fidale LC, Dorn S, and Heinze T. 2007. Applications of ionic liquids in carbohydrate chemistry: A window of opportunities. *Biomacromolecules* 8(9): 2629–2647.

27. Kadokawa JI, Murakami MA, Takegawa A, and Kaneko Y. 2009. Preparation of cellulose-starch composite gel and fibrous material from a mixture of the polysaccharides in ionic liquid. *Carbohydrate Polymers* 75(1): 180–183.

28. Ma XF, Yu JG, and Kennedy JF. 2005. Studies on the properties of natural fibers-reinforced thermoplastic starch composites. *Carbohydrate Polymers* 62(1): 19–24.

29. Funke U, Bergthaller W, and Lindhauer MG. 1998. Processing and characterization of biodegradable products based on starch. *Polymer Degradation and Stability* 59(1–3): 293–296.

30. Baumberger S, Lapierre C, Monties B, and Della Valle G. 1998. Use of kraft lignin as filler for starch films. *Polymer Degradation and Stability* 59(1–3): 273–277.

31. Ma XF, Chang PR, Yu JG, and Stumborg M. 2009. Properties of biodegradable citric acid-modified granular starch/thermoplastic pea starch composites. *Carbohydrate Polymers* 75(1): 1–8.

32. Dufresne A. 2010. Processing of polymer nanocomposites reinforced with polysaccharide nanocrystals. *Molecules* 15(6): 4111–4128.

33. Habibi Y, Lucia LA, and Rojas OJ. 2010. Cellulose nanocrystals: Chemistry, self-assembly, and applications. *Chemical Reviews* 110(6): 3479–3500.

34. Dufresne A. 2008. Polysaccharide nano crystal reinforced nanocomposites. *Canadian Journal of Chemistry-Revue Canadienne De Chimie* 86(6): 484–494.

35. Liu DG, Zhong TH, Chang PR, Li KF, and Wu QL. 2010. Starch composites reinforced by bamboo cellulosic crystals. *Bioresource Technology* 101(7): 2529–2536.

36. Teixeira EDM, Pasquini D, Curvelo AAS, Corradini E, Belgacem MN, and Dufresne A. 2009. Cassava bagasse cellulose nanofibrils reinforced thermoplastic cassava starch. *Carbohydrate Polymers* 78(3): 422–431.

37. Lu YS, Weng LH, and Cao XD. 2005. Biocomposites of plasticized starch reinforced with cellulose crystallites from cottonseed linter. *Macromolecular Bioscience* 5(11): 1101–1107.

38. Cao XD, Chen Y, Chang PR, Muir AD, and Falk G. 2008. Starch-based nanocomposites reinforced with flax cellulose nanocrystals. *Express Polymer Letters* 2(7): 502–510.

39. Cao XD, Chen Y, Chang PR, Stumborg M, and Huneault MA. 2008. Green composites reinforced with hemp nanocrystals in plasticized starch. *Journal of Applied Polymer Science* 109(6): 3804–3810.

40. Chen Y, Liu CH, Chang PR, Anderson DP, and Huneault MA. 2009. Pea starch-based composite films with pea hull fibers and pea hull fiber-derived nanowhiskers. *Polymer Engineering and Science* 49(2): 369–378.

41. Lu YS, Weng LH, and Cao XD. 2006. Morphological, thermal and mechanical properties of ramie crystallites-reinforced plasticized starch biocomposites. *Carbohydrate Polymers* 63(2): 198–204.

42. Alemdar A and Sain M. 2008. Biocomposites from wheat straw nanofibers: Morphology, thermal and mechanical properties. *Composites Science and Technology* 68(2): 557–565.

43. Angles MN and Dufresne A. 2000. Plasticized starch/tunicin whiskers nanocomposites. 1. Structural analysis. *Macromolecules* 33(22): 8344–8353.

44. Mathew AP and Dufresne A. 2002. Morphological investigation of nanocomposites from sorbitol plasticized starch and tunicin whiskers. *Biomacromolecules* 3(3): 609–617.

45. Mathew AP, Thielemans W, and Dufresne A. 2008. Mechanical properties of nanocomposites from sorbitol plasticized starch and tunicin whiskers. *Journal of Applied Polymer Science* 109(6): 4065–4074.

46. Hornig S and Heinze T. 2008. Efficient approach to design stable water-dispersible nanoparticles of hydrophobic cellulose esters. *Biomacromolecules* 9(5): 1487–1492.

47. Cai J, Zhang L, Zhou J, Qi H, Chen H, Kondo T, Chen X, and Chu, B. 2007. Multifilament fibers based on dissolution of cellulose in NaOH/urea aqueous solution: Structure and properties. *Advanced Materials* 19(6): 821–825.

48. Chang PR, Jian RJ, Zheng PW, Yu JG, and Ma XF. 2010. Preparation and properties of glycerol plasticized-starch (GPS)/cellulose nanoparticle (CN) composites. *Carbohydrate Polymers* 79(2): 301–305.

49. Svagan AJ, Azizi Samir MAS, and Berglund LA. 2007. Biomimetic polysaccharide nanocomposites of high cellulose content and high toughness. *Biomacromolecules* 8(8): 2556–2563.

50. Woehl MA, Canestraro CD, Mikowski A, Sierakowski MR, Ramos LP, and Wypych F. 2010. Bionanocomposites of thermoplastic starch reinforced with bacterial cellulose nanofibres: Effect of enzymatic treatment on mechanical properties. *Carbohydrate Polymers* 80(3): 866–873.

51. Le Corre D, Bras J, and Dufresne A. 2010. Starch nanoparticles: A review. *Biomacromolecules* 11(5): 1139–1153.

52. Angellier H, Molina-Boisseau S, Dole P, and Dufresne A. 2006. Thermoplastic starch-waxy maize starch nanocrystals nanocomposites. *Biomacromolecules* 7(2): 531–539.

53. Garcia NL, Ribba L, Dufresne A, Aranguren MI, and Goyanes S. 2009. Physico-mechanical properties of biodegradable starch nanocomposite. *Macromolecular Materials and Engineering* 294(3): 169–177.

54. Ma XF, Jian RJ, Chang PR, and Yu JG. 2008. Fabrication and characterization of citric acid-modified starch nanoparticles/plasticized-starch composites. *Biomacromolecules* 9(11): 3314–3320.

55. Liu D, Wu Q, Chen H, and Chang PR. 2009. Transitional properties of starch colloid with particle size reduction from micro- to nanometer. *Journal of Colloid and Interface Science* 339(1): 117–124.

56. Chang PR, Jian RJ, Yu JG, and Ma XF. 2010. Starch-based composites reinforced with novel chitin nanoparticles. *Carbohydrate Polymers* 80(2): 420–425.

57. Chang PR, Jian RJ, Yu JG, and Ma XF. 2010. Fabrication and characterisation of chitosan nanoparticles/plasticized-starch composites. *Food Chemistry* 120(3): 736–740.
58. Ray SS and Bousmina M. 2005. Biodegradable polymers and their layered silicate nanocomposites: In greening the 21st century materials world. *Progress in Materials Science* 50(8): 962–1079.
59. Qiu HX and Yu JG. 2008. Polyacrylate/(carboxymethylcellulose modified montmorillonite) superabsorbent nanocomposite: Preparation and water absorbency. *Journal of Applied Polymer Science* 107(1): 118–123.
60. Qiu HX. 2004. Polysaccharide/montmorillonite composite: Intercalation and application to superabsorbents. PhD thesis, Tianjin University, Tianjin, China, p. 4 (Chinese).
61. Huang MF, Yu JG, and Ma XF. 2004. Studies on the properties of Montmorillonite-reinforced thermoplastic starch composites. *Polymer* 45(20): 7017–7023.
62. Park HM, Li XC, Jin CZ, Park CY, Cho WJ, and Ha CS. 2002. Preparation and properties of biodegradable thermoplastic starch/clay hybrids. *Macromolecular Materials and Engineering* 287(8): 553–558.
63. Dai HG, Chang PR, Geng FY, Yu JG, and Ma XF. 2009. Preparation and properties of thermoplastic starch/montmorillonite nanocomposite using *N*-(2-Hydroxyethyl) formamide as a new additive. *Journal of Polymers and the Environment* 17(4): 225–232.
64. Zhang QX, Yu ZZ, Xie XL, Naito K, and Kagawa Y. 2007. Preparation and crystalline morphology of biodegradable starch/clay nanocomposites. *Polymer* 48(24): 7193–7200.
65. Huang MF, Yu JG, Ma XF, and Jin P. 2005. High performance biodegradable thermoplastic starch—EMMT nanoplastics. *Polymer* 46(9): 3157–3162.
66. Huang MF, Yu JG, and Ma XF. 2006. High mechanical performance MMT-urea and formamide-plasticized thermoplastic cornstarch biodegradable nanocomposites. *Carbohydrate Polymers* 63(3): 393–399.
67. Ma XF, Yu JG, and Wang N. 2007. Production of thermoplastic starch/MMT-sorbitol nanocomposites by dual-melt extrusion processing. *Macromolecular Materials and Engineering* 292(6): 723–728.
68. Chiou BS, Wood D, Yee E, Imam SH, Glenn, GM, and Orts WJ. 2007. Extruded starch-nanoclay nanocomposites: Effects of glycerol and nanoclay concentration. *Polymer Engineering and Science* 47(11): 1898–1904.
69. Tang XZ, Alavi S, and Herald TJ. 2008. Effects of plasticizers on the structure and properties of starch-clay nanocomposites films. *Carbohydrate Polymers* 74(3): 552–558.
70. Chivrac F, Pollet E, Schmutz M, and Avérous L. 2008. New approach to elaborate exfoliated starch-based nanobiocomposites. *Biomacromolecules* 9(3): 896–900.
71. Chen BQ and Evans JRG. 2005. Thermoplastic starch-clay nanocomposites and their characteristics. *Carbohydrate Polymers* 61(4): 455–463.
72. Wilhelm HM, Sierakowski MR, Souza GP, and Wypych F. 2003. The influence of layered compounds on the properties of starch/layered compound composites. *Polymer Intentional* 52(6): 1035–1044.
73. Zhao XP, Wang BX, and Li J. 2008. Synthesis and electrorheological activity of a modified kaolinite/carboxymethyl starch hybrid nanocomposite. *Journal of Applied Polymer Science* 108(5): 2833–2839.
74. Wang BX and Zhao XP. 2006. The influence of intercalation rate and degree of substitution on the electrorheological activity of a novel ternary intercalated nanocomposite. *Journal of Solid State Chemistry* 179(3): 949–954.
75. Leroux F and Besse JP. 2001. Polymer interleaved layered double hydroxide: A new emerging class of nanocomposites. *Chemistry of Materials* 13(10): 3507–3515.
76. Chung YL and Lai HM. 2010. Preparation and properties of biodegradable starch-layered double hydroxide nanocomposites. *Carbohydrate Polymers* 80(2): 525–532.
77. Wu HX, Liu CH, Chen JG, Chang PR, Chen Y, and Anderson DP. 2009. Structure and properties of starch/alpha-zirconium phosphate nanocomposite films. *Carbohydrate Polymers* 77(2): 358–364.
78. Raveendran P, Fu J, and Wallen SL. 2003. Completely "green" synthesis and stabilization of metal nanoparticles. *Journal of the American Chemical Society* 125(46): 13940–13941.
79. Mucalo MR, Bullen CR, Manley-Harris M, and McIntire TM. 2002. Arabinogalactan from the Western larch tree: A new, purified and highly water-soluble polysaccharide-based protecting agent for maintaining precious metal nanoparticles in colloidal suspension. *Journal of Materials Science* 37: 493–504.
80. He JH, Kunitake T, and Nakao A. 2003. Facile in situ synthesis of noble metal nanoparticles in porous cellulose fibers. *Chemistry of Materials* 15(23): 4401–4406.

81. He JH, Kunitake T, and Nakao A. 2004. Facile fabrication of composites of platinum nanoparticles and amorphous carbon films by catalyzed carbonization of cellulose fibers. *Chemical Communications* 4: 410–411.

82. Walsh D, Arcelli L, Ikoma T, Tanaka J, and Mann S. 2003. Dextran templating for the synthesis of metallic and metal oxide sponges. *Nature Materials* 2: 386–390.

83. Vigneshwaran N, Kumar S, Kathe AA, Varadarajan PV, and Prasad V. 2006. Functional finishing of cotton fabrics using zinc oxide–soluble starch nanocomposites. *Nanotechnology* 17(20): 5087–5095.

84. Ma XF, Chang PR, Yang JW, and Yu JG. 2009. Preparation and properties of glycerol plasticized-pea starch/zinc oxide–starch bionanocomposites. *Carbohydrate Polymers* 75(3): 472–478.

85. Radhakrishnan T, Georges MK, Nair PS, Luyt AS, and Djokovic V. 2007. Study of sago starch-CdS nanocomposite films: Fabrication, structure, optical and thermal properties. *Journal of Nanoscience and Nanotechnology* 7(3): 986–993.

86. Bozanic DK, Djokovic V, Blanusa J, Sreekumari Nair P, Georges MK, and Radhakrishnan T. 2007. Preparation and properties of nano-sized Ag and Ag_2S particles in biopolymer matrix. *The European Physical Journal E: Soft Matter and Biological Physics* 22(1): 51–59.

87. Chang PR, Yu JG, and Ma XF. 2009. Fabrication and characterization of Sb_2O_3/carboxymethyl cellulose sodium and the properties of plasticized starch composite films. *Macromolecular Materials and Engineering* 294(11): 762–767.

88. Yu JG, Yang JW, Liu BX, and Ma XF. 2009. Preparation and characterization of glycerol plasticized-pea starch/ZnO-carboxymethylcellulose sodium nanocomposites. *Bioresource Technology* 100(11): 2832–2841.

89. Bozanic DK, Djokovic V, Bibic N, Sreekumari Nair P, Georges MK, and Radhakrishnan T. 2009. Biopolymer-protected CdSe nanoparticles. *Carbohydrate Research* 344(17): 2383–2387.

90. Zheng PW, Chang PR, Yu JG, and Ma XF. 2009. Preparation of Sb_2O_3-carboxymethyl cellulose sodium nanoparticles and their reinforcing action on plasticized starch. *Starch-Starke* 61(11): 665–668.

91. Chivrac F, Pollet E, Schmutz M, and Avérous L. 2010. Starch nano-biocomposites based on needlelike sepiolite clays. *Carbohydrate Polymers* 80(1): 145–153.

92. Sundaram J, Durance TD, and Wang RZ. 2008. Porous scaffold of gelatin-starch with nanohydroxyapatite composite processed via novel microwave vacuum drying. *Acta Biomaterialia* 4(4): 932–942.

93. Star A, Steuerman DW, Heath JR, and Stoddart JF. 2002. Starched carbon nanotubes. *Angewandte Chemie International Edition* 41(14): 2508–2512.

94. Kim O-K, Je J, Baldwin JW, Kooi S, Pehrsson PE, and Buckley LJ. 2003. Solubilization of single-wall carbon nanotubes by supramolecular encapsulation of helical amylose. *Journal of the American Chemical Society* 125(15): 4426–4427.

95. Bonnet P, Albertini D, Bizot H, Bernard A, and Chauvet O. 2007. Amylose/SWNT composites: From solution to film—Synthesis, characterization and properties. *Composites Science and Technology* 67(5): 817–821.

96. Ma XF, Yu JG, and Wang N. 2008. Glycerol plasticized-starch/multiwall carbon nanotube composites for electroactive polymers. *Composites Science and Technology* 68(1): 268–273.

97. Cao XD, Chen Y, Chang PR, and Huneault MA. 2007. Preparation and properties of plasticized starch/ multiwalled carbon nanotubes composites. *Journal of Applied Polymer Science* 106(2): 1431–1437.

98. Ma XF, Chang PR, Yu JG, and Lu PL. 2008. Characterizations of glycerol plasticized-starch (GPS)/carbon black (CB) membranes prepared by melt extrusion and microwave radiation. *Carbohydrate Polymers* 74(4): 895–900.

99. Yu J, Zhang LQ, Rogunova M, Summers J, Hiltner A, and Baer E. 2005. Conductivity of polyolefins filled with high-structure carbon black. *Journal of Applied Polymer Science* 98(4): 1799–1805.

100. Lopes LVS, Dragunski DC, Pawlicka A, and Donoso JP. 2003. Nuclear magnetic resonance and conductivity study of starch based polymer electrolytes. *Electrochimica Acta* 48(14–16): 2021–2027.

101. Finkenstadt VL and Willett JL. 2004. A direct-current resistance technique for determining moisture content in native starches and starch-based plasticized materials. *Carbohydrate Polymers* 55(2): 149–154.

102. Finkenstadt VL and Willett JL. 2004. Electroactive materials composed of starch. *Journal of Polymers and the Environment* 12(2): 43–46.

103. Ma XF, Yu JG, and He K. 2006. Thermoplastic starch plasticized by glycerol as solid polymer electrolytes. *Macromolecular Materials and Engineering* 291(11): 1407–1413.

104. Ma XF, Yu JG, He K, and Wang N. The effects of different plasticizers on the properties of thermoplastic starch as solid polymer electrolytes. *Macromolecular Materials and Engineering* 292(4): 503–510.
105. Wang N, Zhang XX, Liu HH, and Han N. 2010. Ionically conducting polymers based on ionic liquid-plasticized starch containing lithium chloride. *Polymers and Polymer Composites* 18(1): 53–58.
106. Wang N, Zhang XX, Liu HH, and Wang JP. 2009. *N,N*-dimethylacetamide/lithium chloride plasticized starch as solid biopolymer electrolytes. *Carbohydrate Polymers* 77(3): 607–611.
107. Véchambre C, Chaunier L, and Lourdin D. 2010. Novel shape-memory materials based on potato starch. *Macromolecular Materials and Engineering* 295(2): 115–122.

CHAPTER 16

Mechanical, Rheological, and Thermal Properties of Starch-Based Nanocomposites

Jasim Ahmed

CONTENTS

16.1 INTRODUCTION

Starch is one of the most versatile macromolecules and can be found in many plants. It typically has two major components and appears as a mixture of two glucosidic macromolecules very dissimilar in structure and properties: largely linear amylose of molecular weight between one thousand and one million, consisting of α-(1 → 4)-linked D-glucose, and amylopectin, having the same backbone as amylose but with a myriad of α-(1 → 6)-linked branch points (Dufresne and Vignon 1998). Virgin starch is brittle and difficult to be processed into particles due to its relatively high glass-transition temperature (T_g, ≈230°C) (Yu et al. 1996), which is even above the thermal degradation temperature (Wang et al. 2003). The other limitations of native starch are lower thermal resistance, shear resistance, and proneness to retrogradation.

Starch has been used in different forms: neat starch, high-amylose starch, modified starch, plasticized starch (PS), etc. Starch can be modified to obtain materials which melt below the decomposition temperature and, therefore, are processed by conventional polymer processing techniques such as injection, extrusion, and blow molding (Yu et al. 1996; Martin et al. 2001). Furthermore, the starch can be fluidized through reactive extrusion in the presence of a plasticizer, which is well known as thermoplastic starch (TPS) and behaves similar to conventional polymers. The modification involves breakdown of the starch granular structure by the use of plasticizers at high temperatures (90°C–180°C) and shear, which will result in a continuous phase in the form of a viscous melt (Rodriguez-Gonzalez et al. 2003). The main disadvantages of TPS are its pronounced hydrophilic character, the fast degradation rate, and, in some cases, unsatisfactory mechanical properties (Wang et al. 2008). PS has been described well in Chapter 4.

Currently, there is a considerable interest in biopolymer–clay nanocomposites (NCs) due to significant property enhancements by incorporating of only a few weight percent of rigid particles (nanofillers) in the nanometer range (≈100 nm). Secondly, NCs have significant industrial impact related to the possibility to design and create new materials and structures with unprecedented flexibility and physical properties. The NC properties depend upon the state of nanoparticle dispersion in a polymer matrix (Mackay et al. 2006) and the quality of nanoclay dispersion in the matrix. Interestingly, it has been observed in the literature that when different clays are mixed under identical conditions, different results are observed for dispersion (Bashir 2008). Properties of starch and TPS can further be improved by incorporating nanoclay into starch formulations (McGlashan and Halley 2003; Park et al. 2003; Chiou et al. 2005). Montmorillonites (MMTs) in nanoscale (usually <5 wt.%) can increase the mechanical properties, thermal stability, and water resistance of TPS dramatically (Dean et al. 2007; Wang et al. 2009; Majdzadeh-Ardakani et al. 2010). However, the relationship between mechanical behavior and nanoparticle dispersion still remains unresolved (Akcora et al. 2010).

Rheological and mechanical properties of NCs provide insight of the nanoparticle–biopolymer interaction and mechanical spectra of the blend. Furthermore, the particle–particle and the particle–polymer interactions play in the suspensions an important role. In particular, the nanoparticles tend to form clusters in the absence of external forces or if the external forces are weak (Eslami et al. 2010). The clusters are then responsible for certain characteristic to NCs' rheological behavior. Incorporating only 5 wt.% nanoclay into potato starch improved mechanical strength (elastic moduli and tensile strengths) and reduced water vapor transmission rates by nearly one-half

(Park et al. 2003). The reason for these improvements is that the nanoclays became intercalated, resulting in better dispersed platelets. In addition, the dispersed platelets provide more surface area for starch–nanoclay interactions, resulting in better reinforcement and improved mechanical properties.

The glass-transition temperature (T_g) is perhaps the most important thermal property in determining the suitability of a biopolymer for an engineering application. At low temperatures, polymers are glassy, hard, and brittle. As the temperature is increased, polymers undergo a phase transition, known as the glass–rubber transition, which is accompanied by drastic changes in their properties. At the T_g, the polymer softens due to the onset of long-range coordinated molecular mobility accompanied by a change in the free volume properties in the polymer. It has been known for more than a decade that nanoconfinement can lead to significant changes in the T_g behavior of polymer films. In nanoparticle-filled polymers, the surface area increased substantially, and a wide variety of polymer NCs have shown interesting changes in the bulk glass-transition behavior. Most researchers report an increase in the T_g as a function of filler content; however, decreases in the T_g or even no changes of T_g have also been reported (Ash et al. 2004; Rittigstein and Torkelson 2006; Siqueira et al. 2009; Ahmed et al. 2010a,b; Barick and Tripathy 2010).

The focus of this chapter is to examine the rheological/mechanical and thermal properties of starch when nanoparticles are dispersed through the starch polymer matrix. A brief description of the nanoclay, NC preparation, and characterization has been discussed in the beginning of the chapter for better understanding.

16.2 NANOCOMPOSITES

NCs are materials that incorporate nanosized particles into a matrix of standard material. The properties of NC materials depend not only on the properties of their individual parents but also on their morphology and interfacial characteristics. Polymer-layered silicate NCs were reported as early as 1950 in the patent literature (Carter et al. 1950). The revolution in NCs has started after the report from the Toyota research group of a nylon-6 (N6)/MMT NC (Okada et al. 1990), for which very small amounts of layered silicate loadings resulted in pronounced improvements of thermal and mechanical properties.

The NCs of interest are mostly made from polymers and layered silicate clay minerals. The organoclay is based on MMT which has high surface area, about $750 \, m^2/g$; a large aspect ratio, greater than 50; and a platelet thickness of $10 \, Å$. Nanoclays have a stacked platelet structure with each platelet having a thickness of approximately $1 \, nm$ and lateral dimensions on the order of micrometers. These nanofillers have been found to improve the mechanical, barrier, electrical, gas permeability, and thermal properties of base polymer significantly at very low filler concentration with respect to their conventional microcomposite counterparts (Ray et al. 2002; Angellier et al. 2005a,b; Joshi et al. 2006; Ginzburg et al. 2009; Ahmed et al. 2010a,b). These remarkable property enhancements make NCs superior candidates for materials application in the food packaging, electronics, and automotive industries (Treece and Oberhauser 2007; Ahmed and Varshney 2011).

Bio-NCs are also another emerging group of nanostructured hybrid materials (Siqueira et al. 2010). Bio-NCs can be defined as NC materials made from renewable nanoparticles [e.g., cellulose whiskers (CWs) and microfibrillated cellulose (MFC)] and also derived from biopolymers (e.g., PLA, PHA, and starch) and synthetic or inorganic nanofillers (e.g., carbon nanotubes and nanoclay). The term "whiskers" is used to designate elongated crystalline rodlike nanoparticles, whereas the "microfibrils" should be used to designate long flexible nanoparticles consisting of alternating crystalline and amorphous strings (Siqueira et al. 2009). The obtaining of the former involves a specific step for the digestion of amorphous cellulosic domains, generally acid hydrolysis, whereas the latter is obtained from a mechanical treatment. Cellulose nanoparticles are extracted from lignocellulosic fibers and are available all around the world (Sreekumar et al. 2007).

16.2.1 Structure and Properties of Layered Silicates

Three groups of clay minerals are used for polymer NCs (Zeng et al. 2005) and are presented in Table 16.1. They are 2:1 type, 1:1 type, and layered silicic acids, as shown in Figure 16.1 (Zeng et al. 2005). The commonly used layered silicates for the preparation of polymer-layered silicate (PLS) NCs belong to the same general family of 2:1 layered or phyllosilicates. Among clay minerals, MMT, hectorite, and saponite are the most commonly used layered silicates. Layered silicates have two types of structure: tetrahedral substituted and octahedral substituted. In the case of tetrahedrally substituted layered silicates, the negative charge is located on the surface of silicate layers, and hence, the polymer matrices can react and interact more readily with these than with octahedrally substituted material.

In 1:1 type, the clays consist of layers made up of one aluminum octahedron sheet and one silicon tetrahedron sheet. Each layer bears no charge due to the absence of isomorphic substitution in either octahedron or tetrahedron sheet. In layered silicic acids, the clays consist mainly of silicon tetrahedron sheets with different layer thickness. Their basic structures are composed of layered silicate networks and interlayer hydrated alkali metal cations.

The crystal structure of MMT consists of layers made up of two tetrahedrally coordinated silicon atoms fused to an edge-shared octahedral sheet of either aluminum or magnesium hydroxide. The layer thickness is around 1 nm, and the lateral dimensions of these layers may vary from 30 nm to several microns or larger, depending on the particular layered silicate. Stacking of the layers leads to a regular van der Waals gap between the layers called the interlayer or gallery (Ray and Bousmina 2005).

Table 16.1   Clay Minerals Used for Polymer NCs

Clay Type	Formula	Origin	Substitution	Layer Charge
2:1 Type				
MMT	$M_x(Al_2–xMg_x)Si_4O_{10}(OH)_2·nH_2O$	N	Octahedral	–ve
Hectorite	$Mx(Mg_3–xLix\_Si_4O_{10}(OH)_2·nH_2O$	N	Octahedral	–ve
Saponite	$MxMg_3(Si_4–xAlx\_O_{10}(OH)_2·nH_2O$	N	Tetrahedral	–ve
Fluorohectorite	$Mx(Mg_3–xLix\_Si_4O_{10}F_2·nH_2O$	S	Octahedral	–ve
Laponite	$Mx(Mg_3–xLix\_Si_4O_{10}(OH)_2·nH_2O$	S	Octahedral	–ve
Fluoromica (Somasif)	$NaMg_2\_5Si_4O_{10}F_2$	S	Octahedral	–ve
1:1 Type				
Kaolinite	$Al_2Si_2O_5(OH)_4$	N	—	Neutral
Halloysite	$Al_2Si_2O_5(OH)_4·2H_2O$	N	—	Neutral
Layered silicic acid				
Kanemite	$Na_2Si_4O_9·5H_2O$	N/S	Tetrahedral	–ve
Makatite	$NaHSi_2O_5·7H_2O$	N/S	Tetrahedral	–ve
Octasilicate	$Na_2Si_8O_{17}·9H_2O$	S	Tetrahedral	–ve
Magadiite	$Na_2Si_{14}O_{29}·10H_2O$	N/S	Tetrahedral	–ve
Kenyaite	$Na_2Si_{20}O_4·10H_2O$	S	Tetrahedral	–ve

Source: Adapted from Zeng, O.H. et al., *J. Nanosci. Nanotechnol.*, 5, 1574, 2005.
Symbols: N (nature) and S (synthetic).

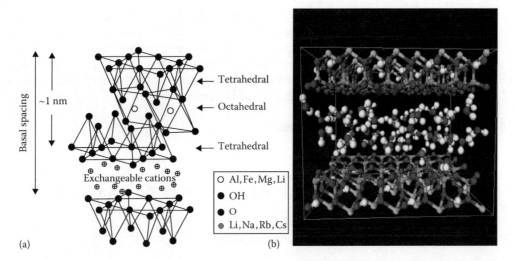

Figure 16.1 Structure of clay minerals. (a) Structure of 2:1 phyllosilicates and (b) 3D crystal image of MMTs. (Adapted from Ray, S.S. and Okamoto, M., *Macromol. Mater. Eng.*, 288, 936, 2003.)

Some excellent reviews on the structure and chemistry for layered silicates are available in the literature (Mittal 2009; Zeng et al. 2005); interested readers can find more details there.

16.2.2 Types of Nanocomposites

The formation of an NC is dependent upon the matrix and its ability to penetrate the silicate layers (Figure 16.2). Depending on the process conditions and on the polymer/nanofiller affinity, different morphologies can be obtained. The most important step in any NC is to achieve the nanolevel dispersion of nanomaterials which could produce the right morphology, structure, and

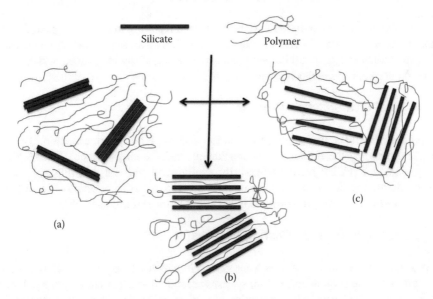

Figure 16.2 Steps showing interaction of silicates and polymer during composite structure formation: (a) phase separated, (b) intercalated, and (c) exfoliated. (Adapted from Schubel, P.J. et al., *Composite: Part A*, 37, 1757, 2006.)

properties. The morphology of the layered silicate–based NC is very complex and may include microparticles, tactoids, and individual layers. Many of the desirable properties of these biopolymer NCs are related to the quality of the dispersion, including polymer intercalation into the clay galleries and/or exfoliation (delamination) into individual clay platelets. Depending on the strength of interfacial interactions between the polymer matrix and layered silicate, three different types of PLS NCs are thermodynamically achievable (Ray and Okamoto 2003):

16.2.2.1 Intercalated Nanocomposites

Intercalated NCs occur when a small amount of polymer moves into the gallery spacing between the silicate platelets leading to a d_{001} increase. The intercalation occurs in a crystallographically regular fashion, regardless of the clay to polymer ratio. Intercalated NCs are normally interlayered by a few molecular layers of polymer.

16.2.2.2 Flocculated or Tactoid Nanocomposites

In a tactoid, the polymer is unable to intercalate between the silicate sheets, and the properties of the composites stay in the same range as the traditional microcomposites. However, silicate layers are sometimes flocculated due to hydroxylated edge–edge interaction of the silicate layers.

16.2.2.3 Exfoliated Nanocomposites

When the silicate layers are completely and uniformly dispersed in a continuous polymer matrix, an exfoliated structure occurred, and the gallery structures are completely destroyed. In an exfoliated NC, the individual clay layers are separated in a continuous polymer matrix by an average distance that depends on clay loading. Usually, the clay content of an exfoliated NC is much lower than that of an intercalated NC.

16.2.3 Preparation of Biopolymer Nanocomposites

Several methods of NC preparation have been utilized resulting in both intercalated and exfoliated nanostructures. Although several successful approaches have been demonstrated, a major barrier to NC preparation is that not all biopolymers can be successfully introduced into clay galleries. Some common preparatory methods are given as follows:

16.2.3.1 In Situ Polymerization Process

This process involves monomer intercalation followed by polymerization. By tailoring the interaction between the monomer, the surfactant, and the clay surface, the exfoliated NCs [poly(ε-caprolactone) (PCL)] were successfully synthesized via ring-opening polymerization. In situ intercalation method is a common method to prepare polymer/phyllosilicate (clay) NCs (Kojima et al. 1993). However, the process is not suitable for starch-based NCs.

16.2.3.2 Solvent Intercalation/Casting Process

This process is based on a suitable solvent system in which the polymer is soluble and the silicate layers are swellable (Figure 16.3). Both systems are mixed together leading to a polymer chain intercalation. The entropy gained by desorption of solvent molecules allows the polymer chains to diffuse between the clay layers, compensating for their decrease in conformational entropy. Then, the solvent is evaporated to obtain NC materials. The process is expensive since solvents are lost

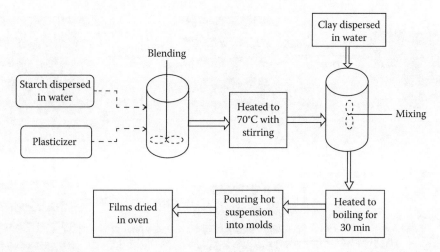

Figure 16.3 Starch-based NC preparation by solution casting.

during evaporation. Furthermore, a small amount of solvent may remain in the final product at the polymer/clay interface creating lower interfacial interaction between the polymer and the clay surfaces. The process is applied to samples which cannot be melt processed due to high thermal or thermomechanical degradations.

16.2.3.3 Direct Melt Intercalation Process

Both the biopolymer and the clay are introduced simultaneously into a melt mixing device (extruder, internal mixer, etc.), and the biopolymer is heated above its T_g (Zanetti et al. 2000). This process has extensively been used to prepare starch nanobiocomposite materials. Manufacturing of nanoclay-incorporated PS through extrusion is the best example for this method (Figure 16.4).

Among various methods which have been used to synthesize NCs, melt intercalation is found as the most appealing approach because of its versatility and compatibility with polymer processing equipment and because it is an environmentally friendly process that requires no solvent and is suitable for industrial uses (Choi et al. 2003; Li and Ha 2003). It involves annealing, statically or under shear, a mixture of the polymer and silicates above the T_g of the polymer (Vaia and Giannelis 1997). During annealing, the polymer chains diffuse from the bulk polymer melt into the galleries

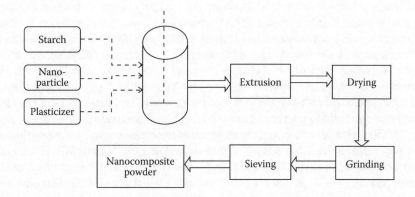

Figure 16.4 Starch-based NC preparation by extrusion.

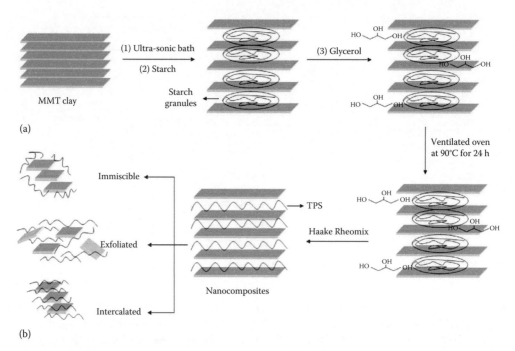

Figure 16.5 Mechanisms of different states of clay dispersion during combinations of (a) intercalation from solution and (b) melt-processing methods utilized to TPS–MMT NCs. (Adapted from Aouada, F.A. et al., *Ind. Crops Prod.*, 2011.)

between the silicate layers. A range of NCs with structures from intercalated to exfoliated can be obtained, depending on the degree of penetration of the polymer chains into the silicate galleries. Experimental results indicate that the outcome of polymer intercalation depends critically on silicate functionalization and constituent interactions. Recently, a simple method based on the combination of the intercalation from solution and melt-processing preparation methods has been used to prepare highly exfoliated and compatible TPS and MMT (Cloisite Na⁺) NCs (Aouada et al. 2011) (Figure 16.5).

16.3 TECHNIQUES USED FOR CHARACTERIZATION OF NANOCOMPOSITES

The state of dispersion of nanoparticles into biopolymer matrix and exfoliation of nanoparticles are desired characteristics to achieve well-defined NCs. Mostly, X-ray diffraction (XRD) analysis and transmission electron micrographic (TEM) observation have been employed for the state of dispersion and exfoliation of nanoparticles (Morgan and Gilman 2003). For its availability and easiness, XRD remains the most frequently used technique to evaluate the NC structure, which allows the interlayer D-spacing (the distance between the basal layers of the nanoclay, or of any layered material) to be measured. The D-spacing observed by XRD for PLS (clay) NCs has been used to describe the nanoscale dispersion of the clay into the polymer. Two main types of scattering techniques (small angle and wide angle) have been used for clay dispersion in the NC. Small-angle X-ray scattering (SAXS) probes structure in the nanometer to micrometer range by measuring scattering intensity at scattering angles 2θ close to 0°. Wide-angle X-ray diffraction (WAXD) is a technique concentrating on scattering angles 2θ larger than 5°. The type of scattering technique used depends on the typical properties of clays that are being studied. Most clay under consideration had typical peaks anywhere between 2θ equal to 3 and 2θ equal to 5. Hence, most of the studies conducted for the dispersion of these nanoclays consisted of SAXS (Bashir 2008).

TEM is also an excellent qualitative method in the characterization of NCs, many of which may have a mixed morphology (regions of both exfoliated and intercalated nanostructures). By monitoring the position, shape, and intensity of the basal reflections from the distributed silicate layers, the NC structure (intercalated or exfoliated) may be identified (Ray and Bousmina 2005). The TEM is time intensive, and only gives qualitative information on the sample as a whole, while wide-angle peaks in XRD allow quantification of changes in layer spacing. TEM and XRD become complimentary techniques, filling in gaps of information that other technique cannot obtain. These techniques are not subject of discussion of this chapter.

Thermomechanical properties (elastic modulus, tensile stress, viscoelasticity, T_g) of starch NCs improve significantly with addition of nanoclay. Further, the processing technique that produces complete exfoliation and good dispersion of clay particles is a prerequisite for the improvement of properties of NCs. A number of measurement techniques, namely, rheometry, differential scanning calorimetry (DSC), dynamic mechanical analysis (DMA), thermomechanical analysis (TMA), dielectric analysis (DEA), and thermogravimetric analysis (TGA), have been employed to characterize NCs. Of the different methods for viscoelastic property characterization, DMA techniques are the most popular, since they are readily adapted for studies of both polymeric solids and liquids. Rheological and thermomechanical behaviors of starch-based NCs are discussed in the following sections.

16.4 RHEOLOGICAL PROPERTIES OF STARCH-BASED NANOCOMPOSITES

Rheological properties (mostly melt rheology) have been considered to provide fundamental insights on the structural organization of bio-NCs and the processability of these materials (Ray and Bousmina 2005). Furthermore, rheological behaviors of NCs are strongly influenced by their nanoscale structure and interfacial characteristics. Rheological measurements of starch NCs have emphasized on steady flow and oscillatory shear measurements as a function of nanoparticle concentration. The steady flow measurement provides information on the effect of shear on the orientation of nanoparticles in NCs, whereas the oscillatory shear measurements provide mechanical characteristic (solid-like or liquid-like) of polymer-based NCs as function of incorporated nanoparticles.

The rheology of polymer melts depends strongly on the temperature at which the measurement is done. For polymer samples, it is expected that at the temperatures and frequencies (ω) at which the rheological measurements were carried out, they should exhibit characteristic homopolymer-like terminal flow behavior, expressed by the power laws $G'\infty\omega^2$ and $''\infty\omega$.

Starch NCs exhibit improved material properties in comparison with the virgin starches or conventional composites. In order to understand the processability of these materials, one must understand the detailed rheological behavior of these materials in the gelatinized state. Understanding the rheological properties of NC melts is not only important in gaining a fundamental knowledge of the processibility but is also helpful in understanding the structure–property relationships in these materials (Ray and Okamoto 2003).

A number of studies have been carried out on starch-based NCs. Interactions of starch, plasticizers, and nanoclay during blend preparation (either solvent casting or melt extrusion) have significant effects on the end composites and functionality. Most studies have focused on the final material properties of the samples. There have been limited studies carried out on measuring the evolving dynamic rheological properties of starch–nanoclay composites during and after gelatinization. These studies can provide insight into starch–nanoclay interactions as well as material property changes during sample processing.

Chiou et al. (2005) extensively studied the oscillatory rheological properties of NCs based on various MMT nanoclays and starches (wheat, potato, corn, and waxy corn). Both hydrophobic (Cloisite 30B, 10A, and 15A) and hydrophilic (Cloisite Na^+) nanoclays have been tested for the NC.

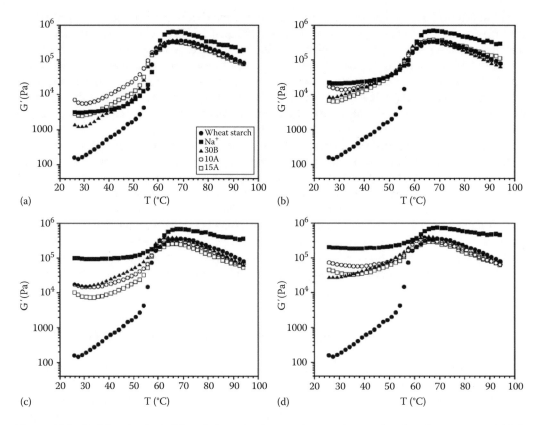

Figure 16.6 Evolving elastic modulus of wheat starch–nanoclay samples as a function of temperature: (a) 2.5, (b) 5, (c) 7.5, and (d) 10 wt.%. (Adapted from Chiou, B.S. et al., *Carbohydr. Polym.*, 59, 467, 2005.)

Pregelatinized wheat starch dispersion (51% moisture) and Cloisite Na⁺ blend exhibited a more gel-like behavior at higher clay concentrations than the other hydrophobic clays. The magnitude of G′ increased almost double, while the Cloisite Na⁺ concentration was increased from 2.5 to 10 wt.% (Figure 16.6). The more hydrophobic nanoclay samples exhibited an increase in G′ of only one order of magnitude over the same concentration range. Ahmed (2011a) observed similar results while working on lentil starch/Cloisite 30B blend NC. The starch/clay blends showed higher magnitude of the G′ (3–4 times) than the G″ and, therefore, showed a predominating solid-like behavior. The complex viscosity (η^*) decreased systematically with frequency, and clay-enriched sample showed higher values throughout the frequency range. The phase angle (δ) decreased in the lower frequency range (0.1–1 Hz) and increased, thereafter. However, mostly, the δ values remained in small domain (9°–12°) and behaved like a true viscoelastic fluid.

The mechanical behavior of the NCs during heating is mostly controlled by the nature of nanoclay whether it is hydrophobic or hydrophilic (Chiou et al. 2005). Further, the evolving G′ of wheat starch–Cloisite Na⁺ samples exhibited a larger value than the other nanoclay samples at higher temperatures. After obtaining the peak temperature, all samples had a modulus that dropped in value, whereas the Cloisite Na⁺ samples had modulus values that decreased at a slower rate. The more hydrophobic nanoclay samples had comparable modulus values to the neat wheat starch sample which indicated that these nanoclays did not have much of a reinforcement effect at elevated temperatures, even though they exhibited a large reinforcement effect at room temperature. Similar observation was reported by Ahmed (2011a) for the evolving complex viscosity (η^*) of lentil starch/clay (Cloisite 30B) blend during linear heating (2.5°C/min) from 25°C to 95°C, as shown in

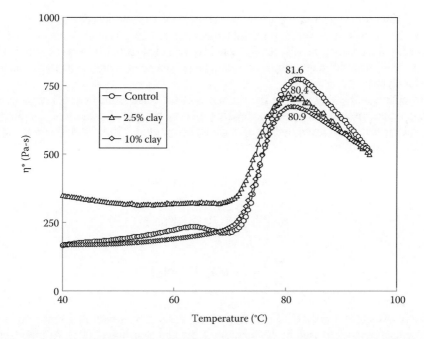

Figure 16.7 Effect of clay loading on linear heating of lentil starch dispersion. (Adapted from Ahmed, 2011a.)

Figure 16.7. The complex viscosity (η^*) development of lentil starch/clay blend dispersion was much slower in the beginning, attained a peak value during nonisothermal heating (function of time and temperature) leading to sol–gel conversion, and decreased, thereafter. No distinct gelatinization temperature was detected while clay was incorporated into starch dispersions. The maximum value of η^* corresponding to temperature T during heating ramp can be considered as the peak gelation temperature. The peak gelation temperature value was marginally dropped from 81.6°C to 80.4°C with increasing the clay concentration.

The gel rigidity of the starch NCs containing Cloisite Na$^+$ increased dramatically on heating (Chiou et al. 2005). As the crystallites melt and the amylose leaches out, the granules swell in size. The swelling leads to an increase in the granule volume fraction, resulting in a rapid rise in the G$'$ between 55°C and 65°C. The leached amylose and smaller amount of amylopectin both contain hydroxy groups that can further interact with the hydrophilic Cloisite Na$^+$ nanoclay. The amylose and amylopectin penetrate into the interlayer between the nanoclay platelets and push them further apart. The clay becomes intercalated, resulting in more interactions between clay and starch molecules. The nanoclays, therefore, serve as a reinforcement agent, leading to increased gel rigidity. On the other hand, the amylose and amylopectin cannot readily penetrate into the hydrophobic nanoclays, and the nanoclay platelets remain stacked, with less surface area available for interaction with starch molecules. This produces less of a reinforcement effect and resulted lowering the gel rigidity.

16.4.1 Time–Temperature Superposition of Nanocomposites

In order to broaden the observation horizon of viscoelastic properties, the method of superposition of time (frequency) and temperature (TTS) was applied (Ferry 1980). This is a basic method for data analysis originating from rheological measurements of synthetic polymers. TTS has been used to obtain the master curves for several properties such as creep, creep compliance, stress compliance against time [or log (time)] or dynamic modulus against frequency, etc. (Ferry 1980). When

TTS applies response, functions are easily determined over many decades of frequency. This is done by combining measurements at different temperatures. This procedure works even if only one or two decades of frequency are directly accessible (Olsen et al. 2001). TTS has applied effectively to demonstrate the dispersion of nanoparticles into biopolymer composites and often compared with the neat biopolymer.

In this technique, log–log plots of the isotherms of the elastic modulus (G′), viscous modulus (G″), and complex viscosity (η*) can be superimposed by horizontal shifts log (a_T), along the log (ω) axis, and vertical shifts given by log (b_T) such that (Ray et al. 2002, 2003; Ptaszek and Grzesik 2007; Ahmed et al. 2010b)

$$b_T G'\left(a_T \omega, T_{ref}\right) = b_T G'(\omega, T) \tag{16.1}$$

$$b_T G''\left(a_T \omega, T_{ref}\right) = b_T G''(\omega, T) \tag{16.2}$$

$$\left(\frac{b_T}{a_T}\right) |\eta*|\left(a_T \omega, T_{ref}\right) = |\eta*|(\omega, T) \tag{16.3}$$

The master curves for G′ and G″ of biopolymer and various NCs with different weight percentages of clay loading are reported in the literature (Ray and Bousmina 2005). At high frequencies ($a_T \omega > 10$), the viscoelastic behaviors of all NCs were alike. On the contrary, at low frequencies ($a_T \omega < 10$), both moduli exhibited weak frequency dependence with increasing clay content, that there are gradual changes of behavior from liquid-like to solid-like with increasing clay content.

The slopes of log G′(ω) and G″(ω) vs. log a_T(ω) indicate cross-linking of biopolymer melts and effect of nanoclay into composites. Slope values of 2 and 1 are expected for non-cross-linked polymer melts, but large deviations occur, especially in the presence of a very small amount of layered silicate loading, which may be due to the formation of a network structure in the molten state. However, NCs based on the in situ polymerization technique show a fairly broad molar mass distribution in the polymer matrix, which hides the relevant structural information and impedes the interpretation of the results (Ray and Okamoto 2003).

Ahmed (2011a) studied effect of nanoclay incorporation into starch blend by using the TTS principle and shifted to a reference temperature (T_{ref}) of 80°C with both the horizontal shift actor a_T and the vertical shift factor b_T. Addition of 2.5%–5% nanoclay to starch blend showed marginal improvement in the superposition (η*-ω data) compared to neat starch dispersion. Further increase in clay concentration showed better superposition (Figure 16.8), and an excellent superposition was observed at 10% clay incorporation. Calculated values of a_T coefficient for 10% clay in the blend are almost close to unity, and therefore, scaling has less effect on the frequency window (Table 16.2). Samples experiencing a short thermal history (corresponding to high frequencies: 1–10 Hz) produce a satisfactory overlapping of the curves. However, longer thermal exposure (0.1–1 Hz) produces significant departures from TTS (especially G″) due to ongoing chemical degradation (Palade et al. 2001).

16.4.2 Clay Miscibility into Nanocomposites as Assessed by Rheology

The miscibility of starch and clay into a blend can be analyzed by the Cole–Cole plot of the rheological data representing the relationship between real (η′) and imaginary (η″) parts of complex viscosity (Cho et al. 1998; Kim et al. 2000; Treece and Oberhauser 2007; Ahmed et al. 2010a,b). A smooth, semicircular shape of the plotted curve indicates good compatibility, i.e., phase homogeneity in the melt, and any deviation from the shape indicates nonhomogenous dispersion or

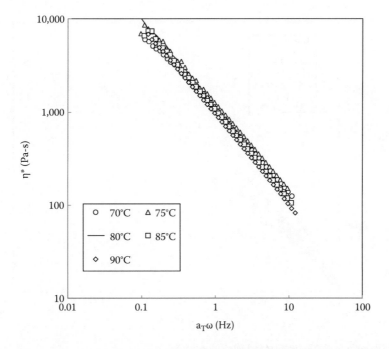

Figure 16.8 Complex viscosity master curve for 7.5% clay-enriched lentil starch dispersion.

Table 16.2   Shift Factor for η-ω TTS Superposition of Lentil Starch/Clay Composite

Sample Type	Temperatures (°C)				
	70	75	80	85	90
Neat starch	1.14	1.03	1.00	1.02	1.12
2.5% clay-blended starch	1.58	1.05	1.00	1.04	1.10
5.0% clay-blended starch	1.49	1.09	1.00	1.03	1.10
7.5% clay-blended starch	1.11	0.97	1.00	1.09	1.22
10% clay-blended starch	1.08	1.00	1.00	1.00	1.02

immiscibility. The η', the real part, and η', the imaginary part, of complex viscosity are calculated according to the following equations (Equations 16.4 and 16.5):

$$\eta' = \frac{G''}{\omega} \qquad (16.4)$$

$$\eta'' = \frac{G'}{\omega} \qquad (16.5)$$

Figure 16.9 illustrates a Cole–Cole plot for starch/clay blend at 80°C for clay/lentil starch blend, as reported by Ahmed (2011a). A well-smooth semicircular shape of the NCs was observed when clay was loaded to starch dispersion with well distribution between starch and nanoclay. This observation is in contrast to polylactide (PLA)–nanoclay miscibility as observed earlier by Ahmed et al. (2010b), where a threshold clay concentration was recorded at 6%, and there was no effect afterward. It is worth to mention that the Cole–Cole plot is not an absolute technique to

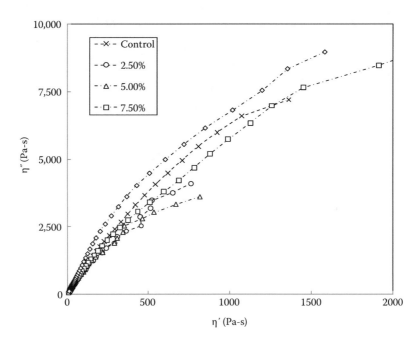

Figure 16.9 Cole–Cole plot for lentil starch/clay blend at 80°C.

determine the miscibility of blends. Other supporting rheological data are required to strengthen the hypothesis.

16.4.3 Steady Shear Measurements

The time-dependent steady shear viscosity of a series of intercalated bio-NCs has been reported by Ray and Okamoto (2003). The shear viscosity of NCs is improved considerably at all shear rates with time and, at a constant shear rate, increases monotonically with increasing clay content. All intercalated NCs exhibited strong rheopexy behavior especially at low shear rate ($0.001\,\mathrm{s}^{-1}$), while pure biopolymer exhibited a time-independent viscosity at all shear rates. With increasing shear rates, the shear viscosity attains a plateau after a certain time, and the time requires to attain this plateau decreases with increasing shear rates. The possible reason for this type of behavior may be due to the planer alignment of the silicate particles toward the flow direction under shear. When shear rate is very slow ($0.001\,\mathrm{s}^{-1}$), silicate particles take a longer time to attain complete planer alignment along the flow direction, and this measurement time ($1000\,\mathrm{s}$) was too short to attain such alignment. For this reason, NCs showed strong rheopexy behavior. On the other hand, under high shear rates (0.005 or $0.01\,\mathrm{s}^{-1}$), this measurement time was considerable enough to attain such alignment, and therefore, NCs showed time-independent shear viscosity after a certain time.

Rheological behavior of reactively extruded starch/PCL NC blends was evaluated by Kalambur and Rizvi (2006) in an off-line capillary rheometer. Starch, PCL, plasticizer, and modified MMT were extruded in a twin-screw microextruder at 120°C and injection molded at 150°C. Power law models for reactive blends (RBs) with different nanoclay volume fractions were developed using appropriate correction factors. Consistency coefficient (K) values for RBs were lower than that of blends containing starch (STPCL), but were significantly higher than that of 100% PCL. Lower values of K for RBs compared with STPCL can result from starch depolymerization occurring during the reactive extrusion process. Starch/PCL NC blends showed shear-thinning behavior with higher pseudoplasticity than 100% PCL. Viscosities of NC blends were significantly lower than

100% PCL, and nonreactive starch/PCL composites synthesized from simple extrusion mixing. An increase in volume fraction or loading levels of nanoclay also increased the K or shear viscosity at low shear rates. The highest degree of shear thinning (lowest n values) was observed in the RBs followed by nonreactive STPCL composite. One hundred percent PCL showed a relatively more Newtonian behavior with a high flow behavior index (n) value of 0.76. Comparing 100% PCL with STPCL, STPCL had a much lower n value than 100% PCL and, thus, was more non-Newtonian and shear thinning.

16.5 MECHANICAL PROPERTIES

There are a number of fundamental techniques used to characterize the mechanical properties of biopolymer-based NCs, including tensile, flexural, tear strength, fatigue, impact, and hardness tests. Some of these techniques are discussed in this section. A tensile test is probably the most fundamental type of test one can measure the mechanical properties of materials. Two methods (ASTM D882-02 and ASTM D638-03) have been prescribed for tensile testing of a biopolymer-based NC film, and the test depends on the thickness of the film. ASTM D882-02 is used to measure tensile testing of materials in the form of thin sheets of less than 1 mm in thickness, whereas ASTM D638-03 can be used to test materials of thickness up to 14 mm. During tensile testing, a rectangular specimen is placed in the grips of movable and stationary fixtures in a testing machine capable of moving the movable fixture at a constant velocity away from the stationary fixture.

The major thrust of incorporation of nanoparticles into biopolymer is to enhance the mechanical properties. However, improvement through addition of nanoscale filler particles would have to rely on a synergetic effect between the rigid-particle filler and the biopolymer matrix. Decreasing the diameter of the filler particles leads to an increase in volume fraction and a consequent improvement in materials' properties.

16.5.1 Clay-Based Nanocomposites

Improved mechanical properties of starch/clay and TPS/clay NCs have been reported in the literature (Park et al. 2002, 2003; Müller et al. 2011). The addition of an organoclay significantly improved both the processing and tensile properties of starch/polyester NC film over the original starch blends (McGlashan and Halley 2003). An improvement in tensile mechanical properties was observed by various researchers (Huang et al. 2004; Müller et al. 2011) while working with TPS/clay NCs and a reduced level of plasticizer. The mechanical strength of glycerol–PS/clay NC films increased of five- and sixfold in the modulus when 3 and 5 wt.% of MMT was added to the starch, respectively (Cyras et al. 2008). The maximum stress also showed an increment of 57% for the composite with 5 wt.% of MMT compared to the neat starch.

The state of clay (natural or modified) and hydrophilic behavior play significant effects on mechanical properties of NCs. Park et al. (2003) compared NCs prepared from TPS/clay through melt intercalation method using natural MMT (Cloisite Na^+) and one organically modified MMT (Cloisite 30B). When 2.5 wt.% of clay was added to TPS, no significant difference in relative elastic modulus (E_c/E_p) was observed, indicating that intercalated NCs do not strongly influence the elastic properties of the matrix. However, for both the hybrids containing 5–10 wt.% Cloisite Na^+, the elastic modulus increased by 20%–50% with increasing clay contents. The tensile properties of the TPS/clay hybrids were generally increased with increasing clay content. However, the TPS/Cloisite Na^+ NCs showed higher tensile properties compared to the TPS/Cloisite 30B hybrids at the same contents and that the TPS/Cloisite Na^+ NCs with 5 wt.% Cloisite Na^+ showed highest tensile properties among all the hybrids. The result is due to the stronger interaction between TPS and Cloisite Na^+ in comparison to that between TPS and Cloisite 30B. In a similar study of TPS/Cloisite Na^+ NC,

Table 16.3   Effect of Clay Concentration on Tensile Properties of TPS and TPS/MMT NCs

Sample Type	Tensile Strength (MPa)	Young's Modulus (MPa)	Elongation at Break (%)
TPS	1.5	8.0	33.5
TPS–1% MMT	2.0	6.7	31.2
TPS–2% MMT	2.3	11.4	40.1
TPS–3% MMT	2.4	15.3	36.1
TPS–5% MMT	2.8	23.8	37.8

Source: Adapted from Aouada, F.A. et al., *Ind. Crops Prod.*, doi:10.1016/j.indcrop.2011.05.003, 2011.

Aouada et al. (2011) found an increase in the tensile strength and Young's modulus values. The Young's modulus values were increased from 8 to 23.8 MPa when the MMT concentration increased from 0 to 5 wt.% (Table 16.3). A possible explanation for such an improvement could be the formation of a three-dimensional network of interconnected long silicate layers, which strengthen the material through mechanical percolation (Park et al. 2003). The presence of MMT did not change the elongation at break values significantly (Table 16.3). The values were around 40% which indicates that TPS and TPS NCs had practically the same flexibility (Aouada et al. 2011).

Chen and Evans (2005) examined NCs of glycerol/PS, with untreated MMT and hectorite. Further, treated hectorite and kaolinite were incorporated to produce conventional composites within the same clay volume fraction range for comparison. In all composites, the presence of clay increased both Young's and shear modulus, although the MMT and unmodified hectorite provided significantly higher increase than kaolinite which formed conventional composites with TPS. Hectorite particles have a lower aspect ratio and larger surface area than MMT, the former factor tending to give a smaller modulus increase and the latter giving a greater increase.

Dean et al. (2007) developed a series of gelatinized high-amylose-enriched starch/clay NCs using a range of plasticizer levels in conjunction with different unmodified nanoclays [sodium MMT (Na-MMT) and sodium fluorohectorite (Na-FHT)] having different cationic exchange capacities. The most significant improvement in modulus for both the Na-MMT and Na-FHT was observed in the systems containing higher levels of clay (2.6–3.2 wt.%). At higher clay concentrations, conventional mixing was found to be a little less effective than dry blending predominantly due to viscosity issues of the clay/water blend. The elongation at break for the Na-MMT TPS samples was dropped compared to the neat TPS samples (Dean et al. 2007). The peak elongation at break was noticed at around 2 wt.% clay. Retrogradation of starches after extrusion is considered as the reason for embrittlement and reduction of the elongation break. Furthermore, the addition of nanoclays and their strong interaction with water molecules (through ion–dipole interactions) may help retain moisture in the samples, leading to more plasticized material with a greater elongation at break. Unlike the Na-MMT NCs, the yield elongation did decrease with increasing clay content, indicating the Na-FHT was not as capable of disrupting recrystallization or did not bind the water molecules as tightly into the matrix. This observation has been explained by the cationic exchange capacity (CEC) of the clay. The CEC is defined by Reichert et al. (1999) as

$$CEC = \frac{\underline{\xi}}{\overline{M}} \times 1000 \tag{16.6}$$

where
$\underline{\xi}$ is the mean layer charge of the clay
$\overline{M}$ is the mean molecular weight for one formula unit

The CEC values of Na-MMT and Na-FHT are 92 mequiv./100 g and 70–80 mequiv./100 g, respectively. The larger CEC relates to the concentration of Na^+ ions and potentially the water retention properties. This difference may account for the variations in elongation to break trends observed where the higher CEC of the Na-MMT allows for a stronger retention of water in the system, leading to a more stable plasticized material. The Na-MMT NC samples showed generally higher break elongations compared to the Na-FHT NC samples.

Acid-treated nanoclay produces higher mechanical strength compared to pristine MMT and organoclay. Citric acid–treated MMT/TPS-blended NC showed the highest Young's modulus than the neat clay (Majdzadeh-Ardakani et al. 2010). Furthermore, a combination of mechanical and ultrasonic mixing led to an extensive dispersion of silicate layers and produced the highest Young's modulus. The maximum stress strength, the stress at break, and Young's modulus were achieved upon addition of 6% clay content. A higher loading showed a weakening effect. The improvement is usually continued with increasing the clay content up to a percent at which the silicate layers cannot be exfoliated further. Addition of more amount of clay into the matrix leads to appearance of clay stacks and even aggregates that deteriorate the mechanical properties. The elongation at break dropped with the addition of clay content up to 6%. This can be attributed to the fact that the layered silicates might provide some new nucleation sites and, thus, contribute in growth of crystallites (Cyras et al. 2008). The crystallization process causes the brittleness of NCs and reduces their strain at break (Magalhães and Andrade 2009).

Majdzadeh-Ardakani and Nazari (2010) further studied effects of clay cation on mechanical properties of extruded starch NCs. MMT with three types of cation or modifier (Na^+, alkyl ammonium ion, and citric acid) was tested. The NCs with modified MMT showed an improvement in mechanical properties in comparison with pristine MMT. However, the organoclay (with ammonium ion) results in weak improvements in tensile properties were compared with the original MMT (with Na^+ cation) and modified MMT with citric acid (CMMT). In starch/PVOH/MMT and starch/PVOH/CMMT NCs, the sodium cation or citric acid molecule provides interfacial interactions with starch/PVOH matrix and, therefore, improves the exfoliation of the layered host into its nanoscale elements, and this, in return, improves the mechanical properties of the sample. The tensile strength and modulus for NCs filled by CMMT were significantly higher ($\sigma = 65.4\,MPa$, $E = 6856\,MPa$). The H-bonding interaction between hydrated sodium cation in MMT and the OH groups of citric acid modifier in CMMT, with the free OH groups residing in the starch and PVOH chains, facilitated the dispersion of clay platelets throughout the starch matrix. The intensive interaction between citric acid and starch/PVOH matrix, as well as the wide gallery spacing of CMMT that facilitates the penetration of starch/PVOH chains, is probably responsible for the highest tensile strength and modulus of starch/PVOH/CMMT NCs.

Effect of clay (Cloisite 30B) incorporation on radial expansion ratio (RER), unit density, bulk compressibility, and bulk spring index (BSI) of the NCs containing tapioca starch (TS) and PLA was investigated by Lee and Hanna (2009). Pure TS foam had a BSI of 0.964, which was significantly different from the BSI of TS/PLA foam. BSI was significantly influenced by the addition of different clay contents into the TS/PLA matrix. The 1 wt.% NC had the highest BSI of 0.958, and the highest clay content produced the lowest BSI of 0.947. These data indicate that a small amount of nanoparticles can act as reinforcing filler to improve the mechanical properties due to the strong interaction between TS/PLA and Cloisite 30B.

Mondragón et al. (2009) prepared TPS/natural rubber/MMT (TPS/NR/Na^+-MMT) NCs by extrusion. After being dried, the NCs were injection molded to produce test specimens. Tensile strength (σ) and elastic modulus (E) of TPS/untreated NR (uNR) blend conditioned at 30% RH were 0.03 and 1.5 MPa, respectively. The use of modified NR (mNR) resulted in a significant increment of both σ and E: σ increased by 167% and E by 30% when compared to TPS/uNR. This significant increase has been attributed to the combination of enhanced compatibility between the mNR and starch and the cross-linking of polyalkenylene chains (Lamb et al. 2001; Rouilly et al. 2004). However, at 30% RH, TPS/uNR/MMT NCs fail in fragile manner regardless of clay content. In the same way, TPS/mNR/MMT NCs remained fragile independently of clay content and RH.

Table 16.4   Tensile Properties of Reactively Extruded Starch/PCL/MMT NCs

Sample Type	Highest Strength (MPa)	Young's Modulus (MPa)	Elongation at Break (%)
Neat PCL	35.7	13.0	1101.9
STPCL[a]	12.6	19.4	246.6
STPCLNC0	3.2	7.1	199.6
STPCLNC1	8.6	11.6	568.1
STPCLNC3	10.2	15.6	926.0
STPCLNC6	10.0	18.6	877.0
STPCLNC9	9.5	21.1	672.0

Source: Adapted from Kalambur, S. and Rizvi, S.S.H., *J. Appl. Polym. Sci.*, 96, 1072, 2005.
[a]   All composites contain about 40/40/20 starch/PCL/glycerol; NC samples substituted clay with calculated amounts of other ingredients.

Low elongation of the TPS/mNR/MMT NCs could be attributed to partial cross-linking of polyalkenylene chains. Conversely, elongation at break of TPS/uNR/MMT NCs conditioned at 60% RH was 65% for TPS/uNR blend having 0 wt.% clay, and this value tended to decrease as clay content increased. Ductility of TPS/uNR/MMT at 60% RH is typical of a plasticized polymer system, charged with a rigid particulate filler (Wilhelm et al. 2003).

Kalambur and Rizvi (2005) developed a reactive extrusion process in which high amounts of starch (approximately 40 wt.%) was blended with a biodegradable polyester (PCL), plasticizer, MMT organoclay, oxidizing/cross-linking agent, and catalysts resulting in tough NC blends. The tensile properties of the starch–PCL composites and NC blends are shown in Table 16.4. The incorporation of starch at a ratio of 1:1 with PCL reduced the strength and elongation by more than 50% in the STPCL composites compared to neat PCL, whereas Young's modulus increased slightly over that of 100% polyester. The elongation was significantly improved in the RB NCs. At 3 and 6 wt.% organoclay concentrations, the elongation approached that of 100% PCL. The elongation at break reached a maximum of 3% and 6% organoclay concentrations and dropped at a 9% organoclay concentration. The increase in the elongation could be attributed by superior interfacial adhesion between the starch and PCL. During the RB extrusion process, a small amount of cross-linked species was developed including starch–starch, PCL–PCL, and starch–PCL. However, only starch–PCL cross-links were expected to improve starch–PCL interfacial adhesion, whereas intramolecular cross-links would reduce it. The tensile results indicated that starch–PCL cross-links might have predominated instead of the other types of cross-links. The glycerol might have also played an important role in cross-linking.

16.5.2 Metal-Based Nanocomposites

In addition to conventional layered silicates, some metal oxides [zinc oxide, graphene oxide (GO)] and phosphates [α-zirconium phosphate (α-ZrP)] have been tested as nanofillers during preparation of NCs. These metallic nanofillers are very interesting multifunctional materials because of its promising application in electronic industries. Incorporation of metallic oxide has significantly improved the mechanical properties of the NCs. Ma et al. (2009) have developed a method to synthesize a bio-NC using ZnO nanoparticles stabilized by soluble starch (nano-ZnO) as filler in a glycerol–plasticized-pea starch (GPS) matrix. The elastic modulus was improved by loading a low level of nano-ZnO particles in NCs (Figure 16.10). The tensile yield strength and Young's modulus increased from 3.94 to 10.80 MPa and from 49.80 to 137.00 MPa, respectively, when the nano-ZnO content varied from 0 to 4 wt.%. The improvement in these properties may be attributed to the interaction between the nano-ZnO filler and GPS matrix. In a similar study, Wu et al. (2009) described mechanical properties of GPS/α-ZrP (PS/ZrP) NC films with different loading levels of α-ZrP. Both the tensile strength (σ) and the elongation at break (ε) of the NC films are remarkably higher than

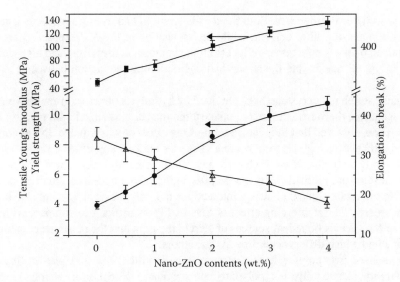

Figure 16.10 Effect of nano-ZnO weight fraction on mechanical properties of GPS/nano-ZnO composites. (Adapted from Ma, X. et al. *Carbohydr. Polym.*, 75, 472, 2009.)

those of the neat starch film (PS/ZrP-0). The maximum values of σ and ε reached 9.44 MPa and 47.5%, respectively, when 0.3% α-ZrP and 25% glycerol as plasticizer were incorporated. The NCs exhibit a remarkable improvement in mechanical properties. The main reason for the improvement in the mechanical properties is the stronger interfacial interaction between the matrix and α-ZrP due to the vast surface of the α-ZrP layers exposed.

Recently, graphene, with one-atom-thick two-dimensional individual sheet structure composed of sp^2-hybridized carbon, has attracted a great deal of attention due to its unique structure and properties (Geim and Novoselov 2007). It combines the lower price and the layered structure of clays with the superior thermal and electrical properties of carbon nanotubes. Li et al. (2011) prepared GPS/GO (PS/GO-n) biocomposite films by solution-casting method. As GO loading increased, both the tensile strength and Young's modulus increased; however, the elongation at break of the biocomposites dropped. When GO loading varied from 0 to 2.0 wt.%, the tensile strength and E increased from 4.56 MPa and 0.11 GPa to 13.79 MPa and 1.05 GPa, respectively, while elongation at break dropped from 36.06% to 12.11%. These results indicated that GO has improved the strength and stiffness of starch films at the expense of flexibility. The improvement in the mechanical properties was due to the good dispersion of GO within the starch matrix and the strong interfacial interactions between GO and PS matrix. However, the good dispersion of GO in PS matrix also restrained the slippage movement among starch molecules, so adding GO significantly decreased elongation at break of PS film.

16.5.3 Starch and Cellulose Nanocrystal–Based Nanocomposites

Recently, cellulose microfibrils and their aggregated nanofibers have received much attention as a new class of cellulose for various applications because they are extremely fine natural polymeric nanofibers that are not achievable using synthetic polymers. Cao et al. (2008) studied the mechanical properties of the neat PS matrix films as well as the NC films reinforced with various contents of flax cellulose nanocrystals (FCNs) by tensile testing at room temperature. Generally, the NC films exhibited two characteristic regions of deformation behavior. At low strains (<10%), the stress increases rapidly with an increase in strain. At higher strain, the stress regularly increases with the strain increasing up to the break of the films. Results indicated that the FCNs content has a profound effect on the mechanical properties. Both the tensile strength and Young's modulus increased significantly with increasing filler content, while the elongation at break decreased. This can probably be explained by the reinforcement effect from the homogeneously dispersed high-performance FCN fillers in the PS matrix and the strong hydrogen bonding interaction between FCNs and PS molecules.

Waxy maize starch nanocrystals (SNs) obtained by hydrolysis of native granules were used as a reinforcing agent in a thermoplastic waxy maize starch matrix plasticized with glycerol (Angellier et al. 2006). It was observed that the relative reinforcing effect of waxy maize SNs in nonaged TPS is the most significant for high glycerol content. The mechanical properties of thermoplastic waxy maize starch plasticized with 25 wt.% glycerol and reinforced with waxy maize SNs are distinctly higher than those of the unfilled matrix, even after aging of the material. This reinforcing effect is attributed to the establishment of strong interactions not only between SNs but also between the filler and the matrix. The reinforcing effect of SNs in TPS is difficult to compare with other fillers because of the various botanical sources of starch, the type and the plasticizer content, and the processing method which differs according to the authors.

Incorporation of two polysaccharide nanocrystals—rodlike CW and platelet-like SN—into an aliphatic thermoplastic polyester [poly(butylene succinate) (PBS)] on mechanical properties of PBS/nanocrystals bio-NCs indicated that at loading levels of 2 wt.% CW or 5 wt.% SN, it enhanced the tensile strength and elongation at break compared to the neat PBS (Lin et al. 2011). However, Young's modulus of the NCs sharply increased at higher clay loading. The major improvements observed in the PNs-filled PBS system may be attributed to the strong interfacial adhesion between fillers and matrix and to the unique mechanical percolation phenomenon.

16.6 CREEP TEST

Creep performance is generally represented by creep compliance, J(t), which is defined as the ratio between creep strain $\varepsilon(t)$ and the applied stress (σ), as shown in the following:

$$J(t) = \frac{\varepsilon(t)}{\sigma} \tag{16.7}$$

Compliance curves generated at different stress levels overlap when data are collected in the range of linear viscoelastic behavior. For a perfectly elastic solid, $J = 1/G$, the reciprocal of the shear modulus; however, due to different time pattern, $J(t) \neq 1/G(t)$.

Generally, several mechanical models have been proposed to explain creep, recovery, and stress relaxation aspects of biopolymeric materials of which the spring-dashpot model (e.g., Maxwell–Voigt model and Burger's fit approach) and models based on a Weibull-type distribution function like the Kohlrausch–Williams–Watts (KWW) model are commonly used (Ganß et al. 2007). Furthermore, Eyring's model based on a potential energy barrier concept has also been very successful in explaining the creep response of materials especially when creep is taken as a thermally activated process.

Among various models, Burger's model has been extensively used for describing viscoelasticity of biopolymers. The model is a series combination of the Maxwell and Kelvin–Voigt model; the total strain can be calculated from the general equation:

$$\varepsilon = \frac{\sigma}{G_0} + \frac{\sigma}{G_1}\left\{1 - \exp\left[\frac{-t}{\lambda_{ret}}\right]\right\} + \frac{\sigma t}{\eta_0} \tag{16.8}$$

where

G$_0$ and G$_1$ are elastic modulus

λ_{ret} is the retardation time ($\lambda_{ret} = \eta/G$) of the Kelvin component

η_0 is the Newtonian viscosity of the free dashpot

Burger's model can also be written in terms of creep compliance as

$$\frac{\varepsilon}{\sigma} = \frac{1}{G_0} + \frac{1}{G_1}\left\{1 - \exp\left[\frac{-t}{\lambda_{ret}}\right]\right\} + \frac{t}{\eta_0} \tag{16.9}$$

Again, the equation for creep compliance can be written as

$$J = J_0 + J_1\left\{1 - \exp\left[\frac{-t}{\lambda_{ret}}\right]\right\} + \frac{t}{\eta_0} \tag{16.10}$$

where

J$_0$ is the instantaneous compliance

J$_1$ is the retarded compliance

The sum of the J$_0$ and J$_1$ is called the steady-state compliance

It is known that creep modulus decreases as a function of time and temperature. Williams et al. (1955) have developed the WLF theory showing that, in the linear viscoelastic range, a viscoelastic curve determined at an arbitrary temperature T$_1$ can be carried to another one T$_2$ making an appropriate translation on the time axis by applying a shift factor a$_T$ at a reference temperature T$_{ref}$. When reference temperature is far from the glass-transition temperature (more than 50 K), the Arrhenius law is preferred. Based on it, the shift factor can be described as a function of temperature as follows:

$$\log a_T = -\frac{E}{2.303R}\left(\frac{1}{T} - \frac{1}{T_{ref}}\right) \tag{16.11}$$

where

E is the activation energy

R is the universal gas constant

Creep behavior is a very important property of a thermoplastic composite that controls its dimensional stability and especially in applications where the material has to support loads for long periods of time (Park and Balatinecz 1998; Cyras et al. 2002). Literature concerning the creep response of NCs is limited and is mainly focused to polymer NCs filled with nanoparticles. Most researchers advocated that nanoparticles improved the creep resistance of the neat matrix and drive to higher shear viscosity (Galgali et al. 2004; Pegoretti et al. 2001). The tensile creep resistance of polymer with different types of TiO$_2$ nanoparticles under different stress levels showed an increase of the creep resistance by the addition of nanofillers and a dependence of the creep behavior on filler size (Yang et al. 2006a,b). Ranade et al. (2005) found an improvement of the creep recovery by adding layered silica particles in polymer, which was correlated to the dispersion and minor crystallinity effects induced by rigid MMT-layered silicate. Starkova et al. (2007) have reported the greater efficiency of TiO$_2$ particles with 21 nm over 300 nm in improving creep resistance, and their observation was further supported by creep modeling using the Boltzmann–Volterra hereditary theory.

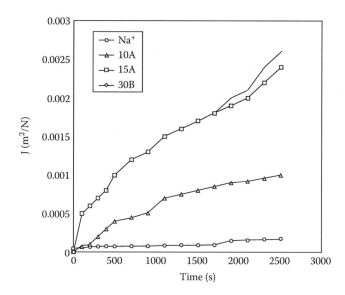

Figure 16.11   Creep compliance of 5 wt.% wheat starch–clay samples as a function of time (51% mc). (Adapted from Chiou, B.S. et al. *Carbohydr. Polym.*, 59, 467, 2005.)

They also reported that the degree of nonlinearity in creep response is significantly reduced by nanoparticles.

Chiou et al. (2005) evaluated the creep behavior of starch/clay NCs by using various clays. A plot of the creep compliance for the clay samples as a function of time is shown in Figure 16.11. The creep results revealed that starch/Cloisite Na$^+$ sample had the lowest creep compliance before gelatinization, and confirmed a more solid-like response. On the other hand, the more hydrophobic nanoclay/starch samples had larger creep compliances, and thus, liquid-like responses were predominated. The decrease on the creep compliance was obviously related with the polymer–clay interaction which is being higher in the case of higher compatibility. The creep compliance of starch/PCL/organo-modified nanoclay commercial blends was decreased by the incorporation of nanofillers due to the enhancement of the modulus, and it was related with the mechanical properties (Pérez et al. 2008). Furthermore, the reduction in the creep strain due to clay addition provides to a material with higher dimensional stability. The parameters of Burger's model also confirmed this trend: the instantaneous and retardant modulus increased, and the permanent viscosity and relaxation time decreased as a function of clay content. The efficiency of nanofillers to get better creep performance of neat matrix was confirmed by the parameters of Findley's power law (Equation 16.5) which properly represent the real strain–time curves of matrix and NCs in the selected range of temperatures. The master curves were built by means of the TTS principle. The master curves displayed a tolerable superposition, and the activation energy for the shift factor was also an evidence for the creep resistance development. Nanofillers contributed to an improvement on the creep resistance which is an important result for the application point of view:

$$\varepsilon_r = A't^n$$

(16.12)

where
  ε_r is the relative strain creep in %
  A' is the slope of the power law

A' and n are parameters that can be obtained by plotting log ε_r vs. log t; log A' is obtained from the intersection with log ε_r and n from the slope of the curve.

16.7 THERMAL ANALYSIS OF STARCH-BASED NANOCOMPOSITES

The analysis of thermal properties of materials is important to determine their processing temperature range and use. The most important techniques in thermal analysis are DSC, TGA, TMA, DMA, DEA, and micro-/nanothermal analysis. It is possible to determine the main characteristics of polymeric systems such as glass–rubber transition temperature (T_g), melting point (T_m), and thermal stability through DSC experiments. Dynamical mechanical analyses can also be used to evaluate the T_g of polymers. Thermal transitions such as melting and crystallization are also important factors to consider in NC processing.

16.7.1 Differential Scanning Calorimetry

Plasticizer and nanoparticles have significant effects on thermal properties of starch-based NCs especially how they interact with each other during composite formation. Two opposite actions are expected in a plasticized NC system. Addition of plasticizer enhances the mobility of the polymer chain resulting decrease in thermal properties (T_g, T_m, and T_c) of the blend. On the other hand, incorporation of nanoparticles into neat polymer restricts the chain mobility. Due to the strong polar–polar interactions between starch, glycerol, and clay surface, a competition mechanism should exist among them. Furthermore, the degree of dispersion of nanoparticles into composite (exfoliated or intercalated) controls the thermal properties of NCs.

16.7.1.1 Clay-Based Nanocomposite

The influence of plasticizers (glycerol and sorbitol) on the thermal properties of amylose/MMT-Na$^+$-based NCs has been studied by Liu et al. (2011a,b). Neat extruded sample exhibited the highest T_g of 49.5°C which reduced to 45.94°C and 25.75°C for 10% and 20% sorbitol concentration (Table 16.5). The corresponding T_g values for glycerol/amylose blends were 40.7°C and 40.3°C with increasing the glycerol concentration to 10% and 20%. The significant drop in T_g values has been

Table 16.5 Effect of Sorbitol and MMT Nanoclay on Thermal Properties of NCs

	T_g (°C)	T_m (°C)	% Crystallinity
PS	49.5	128.5	7.5
S010	45.7	125.8	5.3
S020	25.8	126.0	3.8
S105	49.6	123.4	4.5
S115	47.5	132.0	3.8
S200	60.2	124.4	8.2
S210	41.7	115.3	4.0
S220	31.3	124.9	6.1
S305	53.4	126.3	7.8
S315	50.4	133.9	4.3
S400	68.9	120.6	12.5
S410	55.5	123.8	8.3
S420	33.8	133.3	6.6

Source: Adapted from Liu, H et al., *Carbohydr. Polym.*, 85, 97, 2011b.

S refers to sorbitol; first digit number indicates % MMT used; 2nd and 3rd refer to the amount of glycerol within the samples.

attributed to the plasticizing effect that improves the mobility of starch polymer chain. Incorporation of 4% clay dramatically increased the T_g from 49.5°C to 68.95°C for amylose/clay blend. However, a blend of both clay and plasticizer significantly dropped the T_g of the NC to about 34°C. While studying thermal properties of PS/glycerol/MMT NC, Aouada et al. (2011), however, find it is difficult to detect a T_g of TPS or TPS–MMT NCs by the DSC technique. The possible reason might be caused by the formation of starch-rich or glycerol-rich regions.

Contradictory results are reported in the literature on the melting temperature of the NC when MMT-Na$^+$ has been incorporated into TPS. Liu et al. (2011a) observed a decrease in T_m with addition of MMT (Table 16.5), and authors advocated that the observation was attributed to the reduced crystalline size and presence of crystal imperfections due to compatibility of the MMT with starch which suppressed the crystallization, where the silicate platelet prevented the amylose chains to reorganize. In a similar study, it was found that the T_m of the NC peak increased from 134.9°C to 157.4°C when the MMT Cloisite Na$^+$ content increased from 0 to 5 wt.% (Aouada et al. 2011). The change in the position of this peak indicates that MMT contents favor the formation of larger crystal domains and lowers the mobility of the polymer chains. Also, ΔH_m decreased when the amount of MMT present in NCs increased, showing that the formation of an amorphous phase is favored. These results could be affected by processing techniques, and more studies are required to confirm these results.

Lee and Hanna (2009) reported thermal properties of NC foams based on TS/PLA/Cloisite 30B by a melt intercalation method. The neat TS foam and PLA had a T_m of 162.4°C and 173.9°C, respectively. TS/PLA foam exhibited a T_m of 166.8°C. It was observed that 1, 5, and 7 wt.% clay NCs had T_m at a higher temperature range of 163.8°C–169.8°C as compared to 3 wt.% clay NCs at 152.0°C. The decrease in T_m with addition of Cloisite 30B into TS/PLA matrix was attributed to the compatibility of this organoclay with the starch and PLA mixture. The influence of nanoclay on the reduction of crystallization and melting behavior became distinct when the concentration of clay was around 3 wt.% because of the intercalated nanostructure (Hu and Lesser 2003; Pulta 2004).

Incorporation of nanoclay into pristine polymer has influenced both the T_m and T_c of the NC. The T_m of neat PCL decreased from 57.6°C to 54.2°C when 1% clay was incorporated, and increased, thereafter, to 55°C at higher clay loading (2%–9%) (Kalambur and Rizvi 2005). The starch/polyester blends (without nanoparticles) show a minor reduction in T_m with increasing starch-based plastic content, whereas the effects of organoclay on the T_m for any given starch/polyester blend is not significant (McGlashan and Halley 2003). The T_c of the NC dropped to lower temperature range (21°C–24°C) from the neat PCL (27.4°C) and PCL/starch composite (26.7°C) (Kalambur and Rizvi 2005). The crystallization temperature of the NC blends is significantly lower than the base blend of starch/polyester blend (McGlashan and Halley 2003). This is probably due to the platelets inhibiting order, and hence crystallization, of the starch and polyester. The clay platelets and polymer interact in both intercalated and exfoliated NCs. As the blend is cooled, the platelets can act as nucleation sites, and if they are well dispersed, lead to an increased number of sites for nucleation therefore enhancing the crystallization rate and geometry of crystal growth. However, restrictions in polymer chain mobility through the association of exfoliated platelets can also significantly reduce the degree of crystallinity, further altering the crystallization process. The reduction in T_c with increasing organoclay content indicates that the organoclay is inhibiting crystallization, which, in turn, indicates that an increase in chain mobility restriction is overwhelming any increase in nucleation sites.

Effect of lentil SN and nanoclay on thermal properties of PLA (L-form) and polyethylene-glycol-based NCs (70% PLA + 20% PEG and 10% clay/starch) was compared (Figure 16.12) by Ahmed (2011b). It was observed that nanoclay incorporation showed distinct melting (T_m) and crystallization peak temperature (T_c) for both PEG and PLA over SN. The nanocrystal incorporation reduced the T_m of PEG and PLA by 2°C and 6°C than nanoclay; the corresponding T_c differed by 2°C. The T_g (53°C) remained unaffected with type of nanoparticle incorporation. Furthermore,

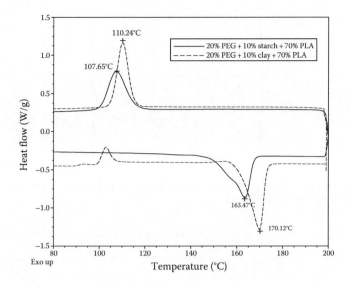

Figure 16.12   Comparison between clay and starch on thermal properties of PLA/PEG/clay or starch NC. (Ahmed, J., Effect of lentil starch nanocrystal and clay on thermal properties of polylactide nanocomposites, 2011b.)

the enthalpy for fusion (ΔH_m) and crystallization (ΔH_c) dropped significantly when starch nanoclay was added to the NC. These results indicate that nanoclay has a slight edge over SN in thermal stability of NCs.

Incorporation of clay into starch (2.5%–10% w/w) affects the peak gelatinization temperature, T_p, significantly (Ahmed 2011a). The T_p increased from 65.7°C to about 70°C although thermograms became flat at higher clay concentrations. An increase in T_p in the blend as function of clay concentration was expected since there was a competition between starch and nanoclay toward water, and furthermore, availability of water was minimized at higher clay concentration. It is also believed that a significant portion of polymer chains become immobilized in the clay NCs (Rao and Pochan 2007).

16.7.1.2 Starch and Cellulose Nanocrystal–Based Nanocomposites

Angellier et al. (2005a,b) investigated the thermal properties of NC films [prepared from NR as the matrix and an aqueous suspension of waxy maize SNs (up to 50%) as the reinforcing phase]. The T_g of unfilled NR occurred at an onset temperature T_{g1} of −66.6°C. The evolutions of the different characteristic temperatures (T_{g1}, T_{g2}, and T_{g3}) are illustrated in Figure 16.13. The magnitude of the specific heat increment at T_g decreased when SNs were added (Figure 16.14). The T_{g1} decreased steadily when increasing the starch content, while T_{g3} remained constant for all compositions. It indicated that the T_g occurred on a broader temperature range for composites. The temperature associated with the maximum differential heat flow, T_{g2}, remained first constant around −62°C for composite materials filled up to 25 wt.% SNs and then decreased down to −64.6°C for the 50 wt.% filled material. Further, it was observed that chemically modified particles did not affect the long-range cooperative motions of the polymeric chains.

Mathew and Dufresne (2002) observed a single T_g in an NC prepared from waxy maize starch plasticized with sorbitol as the matrix and a stable aqueous suspension of tunicin whiskers as the reinforcing phase. No evidence of transcrystallization of amylopectin on CW surfaces was observed; however, antiplasticizing effects were noticed. The T_g of the plasticized amylopectin matrix initially

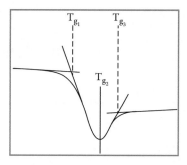

Figure 16.13 Manifestation of the glass–rubber transition in starch/NR NCs. (Angellier, H. et al., *Macromolecules*, 38, 3783, 2005b.)

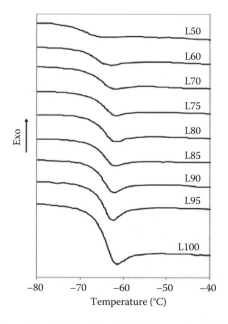

Figure 16.14 DSC thermograms of SNs/NR NC films. (Angellier, H. et al., *Macromolecules*, 38, 3783, 2005b.)

increased up a whisker content of 10–15 wt.% and, thereafter, decreased. The crystallinity was increased significantly in the composites by increasing whisker content.

Thermal properties of nanowhiskers and MFC-reinforced PCL films were described by Siqueira et al. (2009). The T_g of NCs reinforced with either unmodified or modified cellulosic nanoparticles is systematically higher (−59.6°C to −53°C) than the neat matrix (−62°C). This effect is more significant with unmodified sisal whiskers, which indicates that the molecular mobility of amorphous PCL chains is restricted by the presence of the filler. The filler/matrix compatibilization by surface grafting limits this phenomenon. However, it is worth noting that the value of T_g can also be affected by the degree of crystallinity of the matrix. The T_m of the PCL matrix marginally increases when sisal whiskers are added, from about 63°C to 64°C–66°C. It ascertains that the presence of the CWs does not interfere significantly with the crystal growth, regardless of their surface modification state. On the contrary, the melting point tends to decrease upon modified MFC addition due to steric hindrance effects restricting the growth of crystalline PCL regions, probably resulting from entanglements between microfibrils.

For all NCs, the crystallization temperature (22°C) is significantly increased (about 10°C–12°C) compared to the neat matrix. For unmodified cellulosic whiskers, no evolution upon filler loading is observed, whereas it is found to slightly increase for modified nanoparticles. Moreover, the T_c increase is more significant for a given whisker content. The filler most probably acts as a nucleating agent for the crystallization of PCL. The degree of crystallinity of PCL has not been influenced by the presence of chemically modified sisal MFC. However, the presence of sisal whiskers to the PCL matrix seems to significantly increase the degree of crystallinity of the host matrix, regardless of their modification state. It is increased from 51 wt.% for the neat matrix to values close to 60 wt.% for NC films regardless of the filler content, except for the composite reinforced with 3 wt.% of unmodified whiskers. This enhancement of the crystallinity of the PCL matrix probably results, at least partially, in the improvement of the stiffness for these NCs.

Two polysaccharide nanocrystals—rodlike CW and platelet-like SN-incorporated PBS NC—showed no systematic trends of T_m and degree of crystallinity (χ_c) with nanocrystal loading (Lin et al. 2011). With the addition of only 2 wt.% CW or 5 wt.% SN filler, the maximum T_m and χ_c of the NCs were recorded. It was observed that the percolating network did not form, and the interactions between CW or SN nanoparticles and PBS polymer molecules were stronger than the interactions between the nanofillers themselves, preserving the intact and ordered structure of the PBS matrix. Furthermore, the presence of rigid particles and the increase in crystallinity of the NCs inhibited the free motion of PBS polymer chains and resulted in the demand for more energy for the thermal transformation, leading to the higher T_m of PBS/CW-2 and PBS/SN-5 NCs in comparison to the neat PBS.

16.7.2 Dynamic Mechanical Analyzer

DMA has been used to study molecular relaxation processes in NCs and to determine inherent mechanical or flow properties as a function of time and temperature. A number of reports are available on thermomechanical properties of starch-based NCs (Park et al. 2003; Angellier et al. 2005a,b; Garcia et al. 2009; Chivrac et al. 2010; Lin et al. 2011; Liu et al. 2011). Starch has mainly been used in the plasticized state in NC formation, and glycerol is used as a common plasticizer.

16.7.2.1 Clay-Based Nanocomposites

During TMA of starch NCs, two distinct relaxation peaks are detected: the major relaxation peak (called α) is associated to the PS T_g, and the second relaxation peak (termed as β), observed at lower temperature, is consistent with the plasticizer glass transition. The temperature of β-relaxation peak of PS/MMT-Na⁺ nanobiocomposites has been shifted toward higher temperature indicating that the layered clays strongly restricted the starch chain mobility (Park et al. 2003; Avérous and Halley 2009). The tan δ peaks of the TPS/Cloisite Na⁺ 10 wt.% hybrid shifted toward much higher temperatures (−64°C to −32°C) (Park et al. 2003). The result may be attributed to interaction between starch granules and clay at the low temperature ranges. Chivrac et al. (2010), however, reported that clay modification (native sodium or organo modified) has no effect on Tβ (Table 16.6), and therefore, the nanoclays had no influence on the T_g of the domains rich in plasticizer. Similar two relaxation peaks, around −60°C and 0°C for thermoplastic waxy maize starch plasticized with 25 wt.% glycerol, have been reported by Angellier et al. (2006).

Park et al. (2003) observed that the α-relaxation peak of the TPS/Cloisite Na⁺ 10 wt.% hybrid increased from 7.0°C to 10°C–20°C. These authors opined that the α-relaxation peak is influenced by neither the glycerol content nor the filler content. The Tα increased with the clay content irrespective of type of clays and the decrease in glycerol content (Angellier et al. 2006; Chivrac et al. 2010). This result demonstrated that the clay platelets were preferentially located in the domains rich in carbohydrate chains (Chivrac et al. 2010). The reduction in the molecular mobility of

Table 16.6 T_α and T_β Values for Native Wheat Starch-Based Bio-NCs

Filler Type	Filler Content (wt.%)	T_α (°C)	T_β (°C)
WS/MMT-Na	0	11.7	−54.6
	3	15.7	−54.9
	6	23.9	−50.6
WS/OMMT-CS	3	14.9	−53.6
	6	21.7	−51.6

Source: Adapted from Chivrac, F. et al., *Carbohydr. Polym.*, 80, 145, 2010.

starch/amylopectin chains at room temperature for filled materials has been explained by the establishment of hydrogen bonding forces between SNs and TPS (Angellier et al. 2006). For higher clay content, the Tα values observed for MMT-Na were slightly higher than those obtained for the materials elaborated with OMMT-CS (Chivrac et al. 2010). Since the better the clay dispersion, the higher the clay/matrix interface area, a higher Tα was expected for the WS/OMMT-CS samples. This unexpected result could be explained by the morphology of the WS/MMT-Na films. Indeed, it was previously demonstrated that such nanobiocomposites displayed an intercalated structure with preferential plasticizer intercalation.

16.7.2.2 Starch and Cellulose Nanocrystal–Based Nanocomposites

DMA results of NCs of cassava starch reinforced with waxy SNs showed similar two peaks at −60°C (Figure 16.15) attributed by relaxation of glycerol and the second one, wide and of low intensity, between −20°C and 60°C (Garcia et al. 2009). A better-defined second peak is rich in starch. It is believed that the relaxation associated with the glycerol-rich phase shifts slightly to higher temperatures, widens and diminishes its intensity, and is more asymmetric with the nanocrystal addition. Furthermore, the presence of nanocrystals contributed to the establishment of hydrogen bonding between glycerol and SNs as supported by TGA, FTIR, and FE-SEM data.

Angellier et al. (2005a,b) developed NC materials from a latex of NR as matrix and an aqueous suspension of waxy maize SNs as filler and characterized through structural properties. The plot of the logarithm of the elastic modulus, log (E′), and the tangent of the loss angle, tan δ, vs. temperature at 1 Hz are illustrated in Figure 16.16. The evolution of tan δ with temperature displays an α peak located in the temperature range of the glass transition of the NR matrix which is associated

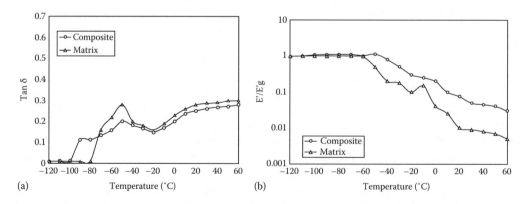

Figure 16.15 Evolution of the dynamic mechanical properties of the neat film and the composite with temperature. (a) Tan δ versus temperature and (b) E′/E′g versus temperature. (Adapted from Garcia, N.L., et al., *Macromol. Mater. Eng.*, 294, 169, 2009.)

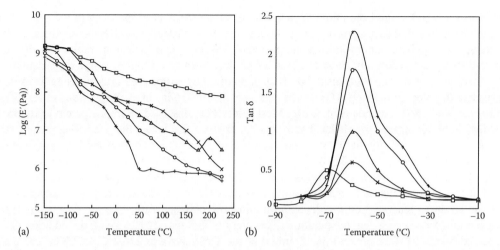

Figure 16.16 (a) Logarithm of the elastic tensile modulus and (b) tangent of the loss angle vs. temperature at 1 Hz for waxy maize SNs/NR NC films (+-L100, ○-L95, ∆-L90, ×-L80, and □-L70). (Adapted from Angellier, H. et al., *Macromolecules*, 38, 9161, 2005a.)

with the anelastic manifestation of the glass–rubber transition of the polymer. This mechanism involves cooperative motions of long-chain sequences. The temperature position (Tα) and magnitude of the peak (Iα) decreased when SNs were added. The decrease in Tα becomes significant for the 20 wt.% starch reinforced material for which Tα decreases from −56°C for the unfilled NR down to −61°C. This is attributed to (1) the broadening of the glass-rubber transition zone toward lower temperatures reported from DSC measurements and (2) a classical mechanical coupling effect. (The shift of the tan δ peak results from the strong decrease of the modulus drop upon filler addition.) The reduction of the magnitude of Iα when increasing the SNs content results from (1) the decrease of the number of mobile units participating to the relaxation process and (2) the decrease of the magnitude of the modulus drop associated with T_g.

16.7.3 Thermal Stability and Thermogravimetric Analysis

The thermal stability of polymeric materials is studied by TGA. The weight loss due to the formation of volatile products after degradation at high temperature is monitored as a function of temperature or time. TGA can be used to quantify the mass change in a polymer associated with transitions or degradation processes. When the heating takes place under an inert gas flow, a nonoxidative degradation occurs, while the use of air or oxygen allows oxidative degradation of the samples. Nevertheless, it is clear that biopolymer NCs containing few organoclay percents are more thermally stable than virgin biopolymer. The role of the clay in the NC structure may be the main reason for the difference in TGA results of these systems compared with pure polymer. This shift to higher temperature for the beginning of decomposition can be explained by a decrease in the diffusion of oxygen and volatile degradation products throughout composite material because of the homogeneous incorporation of clay sheets (Zaidi et al. 2010). The clay acts as a heat barrier, which enhances the overall thermal stability of the system.

16.7.3.1 *Clay-Based Nanocomposites*

According to Aouada et al. (2011), the thermal decomposition of TPS and TPS/MMT NCs can be divided into four decomposition steps: (1) initial temperature until 100°C, corresponding to the water loss; (2) 100°C until the initial decomposition temperature, corresponding to the plasticizer evaporation; (3) initial decomposition temperature until the final decomposition temperature, corresponding to the polymer starch decomposition; and (4) final decomposition temperature until 600°C, corresponding to the decomposition of starch subproducts formed during the starch degradation.

Thermal stability of TPS/Cloisite Na+ NCs is reported by various researchers using TGA and derivative thermogravimetry (DTG) curves (Chiou et al. 2007; Ning et al. 2009; Wang et al. 2009; Masclaux et al. 2011; Aouada et al. 2010). Most researchers observed an improvement in thermal stability in NC compared to pristine starch or TPS, except a few where there is no significant difference in the thermal stability of starch in presence of glycerol and/or after the incorporation of the nanoclays (Masclaux et al. 2010). The extruded wheat starch/Cloisite Na+ nanoclay composites showed improvement in thermal stability compared to the neat starch sample (Chiou et al. 2007). This improvement in thermal stability could be attributed to the nanoclay acting as a barrier to the volatile products generated during decomposition of the polymer. Aouada et al. (2011) observed that the initial decomposition temperature (T_d initial) of TPS was 223.2°C, and the addition of 1, 2, 3, and 5 wt.% MMT increased the T_d initial of the TPS matrix to 226.2°C, 227.0°C, 232.2°C, and 242.1°C, indicating an improvement in the stability. A similar trend was reported by Ning et al. (2009) where the authors attributed the improvement to the intense interactions existing in TPS–MMT NCs. Furthermore, the presence of MMT also shielded the exposed hydroxyl groups of starch which alleviated polysaccharide thermal depolymerization. Wang et al. (2009) attributed the increase in the thermal stability of melt-extruded TPS NCs to the addition of MMT because in intercalated and exfoliated TPS/MMT NCs; a maze or "tortuous path" structure is formed in the matrix which could block normal gas and liquid channels. Additionally, compared with organic materials, inorganic materials had better thermal stability due to their configuration characteristics. Therefore, the introduction of inorganic particles would greatly improve the thermal stability of organic polymer materials (Daí et al. 2009).

16.7.3.2 Metal-Based Nanocomposites

Thermal stability of GPS/α-ZrP (PS/ZrP) NC films was reported by Wu et al. (2009). Results indicated that the initial weight loss of all samples at approximately 100°C is due to the evaporation of water, while the weight loss in the second range (250°C–350°C) corresponds to a complex process including the dehydration of the saccharide rings and depolymerization (Mathew and Dufresne 2002). The TG curves show that all samples are stable up to 275°C, with a maximum rate of decomposition occurring at about 300°C. The thermostability of the PS-based films decreases with an increase in the amount of α-ZrP incorporated in the NC films. This could be attributed to the increase in acidity of α-ZrP with the increase in temperature, which induces the decomposition of the glycoside bonds. Similar starch decomposition temperature at about 291°C in glycerol–PS/clay NC films was observed by Cyras et al. (2008). The decomposition peak in the NCs has been shifted to higher temperatures, which indicates that the introduction of clay improves the thermal stability of starch by approximately 25°C. The clay acts as a heat barrier, which enhances the overall thermal stability of the composites. Consequently, the incorporation of inorganic particles improved the thermal stability of starch.

16.8 CONCLUSIONS

This chapter covers different types of NC and its preparation techniques that are commonly used to manufacture biopolymer-based NC. Direct melt intercalation process (twin-screw extrusion) is the most suitable process for the starch NC. Mostly, layered silicate clay minerals are incorporated into biopolymer matrix to enhance material properties. Few starch-based nanoparticles are recently synthesized and used for NC preparation. The most important step in any NC is to achieve the nanolevel dispersion of nanomaterials which could produce the right morphology, structure, and properties. The primary focus of the chapter remains the mechanical, rheological, and thermal properties of starch-based NCs. Hydrophilic behavior of the nanoclay plays a significant role in mechanical and thermal properties of the starch-based NC. The peak melting and crystallization temperatures of NCs are greatly influenced by nanofillers. A number of the standard mechanical/rheological tests used to characterize NC are discussed in this chapter, including tensile, creep, oscillatory rheology, and steady flow behavior. They generally exhibit improved mechanical properties both in solid and melt states compared to conventional filler composites because reinforcement in NCs occurs two-dimensionally rather than one-dimensionally, and no special processing is required to laminate the composites. Thermal stability of biodegradable polymers also increases after NC preparation because clay acts as a heat barrier, which enhances the overall thermal stability of the system, as well as assists in the formation of char after thermal decomposition. Finally, for biodegradable NCs to meet wide range of applications, NC formulation must be further researched and modified so that mechanical and other properties can be easily manipulated depending on the end users' requirements. To conclude, these polysaccharides/nanoclay materials are a valid answer to produce low-cost, highly competitive, and pioneering environmentally friendly materials.

REFERENCES

Ahmed, J. 2011a. Oscillatory rheology of lentil starch-nanoclay blend suspension: Applicability of time-temperature superposition. *International Journal of Food Properties*. In Press.

Ahmed, J. 2011b. Effect of lentil starch nanocrystal and clay on thermal properties of polylactide nanocomposites. Unpublished.

Ahmed, J. and Varshney, S.K. 2011. Polylactides-chemistry, properties and green packaging-a review. *International Journal of Food Properties*, 14, 37–58.

Ahmed, J., Varshney, S.K., and Auras, R. 2010a. Rheological and thermal properties of polylactide/silicate NCs films. *Journal of Food Science*, 75, N17–N24.

Ahmed, J., Varshney, S.K., Auras, R., and Hwang, S.W. 2010b. Thermal and rheological properties of L-polylactide/polyethylene glycol/silicate NCs films. *Journal of Food Science*, 75, N97–N108.

Akcora P, Kumar SK, Moll J, Lewis S, Schadler LS, Li Y, Benicewicz BC, Sandy A, Narayanan S, Ilavsky J, Thiyagrajan P, Colby RH, and Douglas J. 2010. Gel-like mechanical enforcement in polymer nanocomposite melts. *Macromol* 43:1003–1010.

Angellier, H., Molina-Boisseau, S., Dole, P., and Dufresne, A. 2006. Thermoplastic starch-waxy maize starch nanocrystals nanocomposites. *Biomacromolecules*, 7, 531–539.

Angellier, H., Molina-Boisseau, S., and Dufresne, A. 2005a. Mechanical properties of waxy maize starch nanocrystal reinforced natural rubber. *Macromolecules*, 38, 9161–9170.

Angellier, H., Molina-Boisseau, S., Dufresne, A., Lebrun, L., and Dufresne, A. 2005b. Processing and structural properties of waxy maize starch nanocrystals reinforced natural rubber. *Macromolecules*, 38, 3783–3792.

Aouada, F.A., Mattoso, L.H.C., and Aouada, E.L. 2011. New strategies in the preparation of exfoliated thermoplastic starch–montmorillonite nanocomposites. *Industrial Crops Production*, doi:10.1016/j.indcrop.2011.05.003.

Ash, B.J., Siegel, R.W., and Schadler, L.S. 2004. Glass-transition temperature behavior of alumina/PMMA nanocomposites. *Journal of Polymer Science Part B: Polymer Physics*, 42, 4371–4383.

ASTM D638-03. Standard Test Method for Tensile Properties of Plastics, ASTM International, West Conshohocken, PA.

ASTM D882-02. Standard Test Method for Tensile Properties of Thin Plastic Sheeting, ASTM International, West Conshohocken, PA.

Avérous, L. and Halley, P.J. 2009. Biocomposites based on plasticized starch. *Biofuels, Bioproducts and Biorefining*, 3, 329–343.

Barick, A.K. and Tripathy, D.K. 2010. Preparation and characterization of thermoplastic polyurethane/organoclay nanocomposites by melt intercalation technique: Effect of nanoclay on morphology, mechanical, thermal, and rheological properties. *Journal of Applied Polymer Science*, 117, 639–654.

Bashir, M.A. 2008. Effect of nanoclay dispersion on the processing of polyester nanocomposites. MSc thesis, McGill University, Montreal, Quebec, Canada.

Cao, X., Chang, P.R., and Huneault, M.A. 2008. Preparation and properties of plasticized starch modified with poly(ε-caprolactone) based waterborne polyurethane. *Carbohydrate Polymers*, 71, 119–125.

Carter, L.W., Hendricks, J.G., and Bolley, D.S. 1950. United States Patent No. 2531396. Assigned to National Lead Company.

Chen, B. and Evans, J.R.G. 2005. Thermoplastic starch–clay nanocomposites and their characteristics. *Carbohydrate Polymers*, 61, 455–463.

Chiou, B., Wood, D., Yee, E., Imam, S.H., Glenn, G.M., and Orts, W.J. 2007. Extruded starch–nanoclay NCs: Effects of glycerol and nanoclay concentration. *Polymer Engineering and Science*, 47, 1898–1904.

Chiou, B.S., Yee, E., Glenn, G.M., and Orts, W.J. 2005. Rheology of starch–clay NCs. *Carbohydrate Polymers*, 59, 467–475.

Chivrac, F., Pollet, E., Schmutz, M., and Avérous, L. 2010. Starch nano-biocomposites based on needle-like sepiolite clays. *Carbohydrate Polymers*, 80, 145–153.

Cho, K., Lee, B.H., Hwang, K.M., Lee, H., and Choe, S. 1998. Rheological and mechanical properties in polyethylene blends. *Polymer Engineering & Science*, 38, 1969–1975.

Choi, W.M., Kim, T.W., Park, O.O., Chang, Y.K., and Lee, J.W. 2003. Preparation and characterization of poly(hydroxybutyrate-co-hydroxyvalerate)-organoclay nanocomposites. *Journal of Applied Polymer Science*, 90, 525–529.

Cyras, V.P., Manfredi, L.B., Ton-That, M.T., and Vazquez, A. 2008. Physical and mechanical properties of thermoplastic starch/montmorillonite nanocomposites films. *Carbohydrate Polymers*, 73, 55–63.

Cyras, V.P., Martucci, J.F., Iannace, S., and Vázquez, A. 2002. Influence of the fiber content and the processing conditions on the flexural creep behavior of sisal-PCL-starch composite. *Journal of Thermoplastic Composite Materials*, 15, 253–266.

Daí, H., Chang, P.R., Geng, F., Yu, J., and Ma, X. 2009. Preparation and properties of thermoplastic starch/montmorillonite NC using *N*-(2-hydroxyethyl)formamide as a new additive. *Journal of Polymer and the Environment*, 17, 225–232.

Dean, K., Yu, L., and Wu, D.Y. 2007. Preparation and characterization of melt-extruded thermoplastic starch/clay nanocomposites. *Thermoplastic Composite Materials*, 67, 413–421.

Dufresne, A. and Vignon, M.R. 1998. Improvement of starch film performances using cellulose microfibrils. *Macromolecules*, 31, 2693.

Eslami, H., Grmelaa, M., and Bousmina, M. 2010. Linear and nonlinear rheology of polymer-layered silicate NCs. *Journal of Rheology*, 54, 539–562.

Ferry, J.D. 1980. *Viscoelastic Properties of Polymers*. John Wiley: New York.

Galgali, G., Ramesh, C., and Lele, A. 2001. Rheological study on the kinetics of hybrid formation in polypropylene NCs. *Macromolecules*, 34, 852–858.

Ganß, M., Satapathy, B.K., Thunga, M., Weidisch, R., and Pötschke, P. 2007. Temperature dependence of creep behavior of PP–MWNT nanocomposites. *Macromolecular Rapid Communications*, 28, 1624–1633.

Garcia, N.L., Ribba, L., Dufresne, A., Aranguren, M.I., and Goyanes, S. 2009. Physico-mechanical properties of biodegradable starch NCs. *Macromolecular Materials Engineering*, 294, 169–177.

Geim, A.K. and Novoselov, K.S. 2007. The rise of graphene. *Nature Materials*, 6,183–191.

Ginzburg, V.V., Weinhold, J.D., Jog, P.K., and Srivastava, R. 2009. Thermodynamics of polymer-clay NCs revisited: Compressible self-consistent field theory modeling of melt-intercalated organoclays. *Macromolecules*, 42, 9089–9095.

Hu, X. and Lesser, A.J. 2003. Effect of a silicate filler on the crystal morphology of poly(trimethylene terephthalate)/clay nanocomposites. *Journal of Polymer Science Part B: Polymer Physics*, 41, 2275–2289.

Huang, S.J., Koenig, M.F., and Huang, M. 1993. In *Biodegradable Polymers and Packaging*. Ching, C., Kaplan, D., and Thomas, E., Eds. Technomic: Lancaster, PA, p. 97.

Huang, M., Yu, J., and Ma, X. 2004. Studies on the properties of montmorillonite-reinforced thermoplastic starch composites. *Polymer*, 45, 7017–7023.

Joshi, M., Butola, B.S., Simon, G., and Kukaleva, N. 2006. Rheological and viscoelastic behavior of HDPE/Octamethyl-POSS NC. *Macromolecules*, 39, 1839–1849.

Kalambur, S. and Rizvi, S.S.H. 2005. Biodegradable and functionally superior starch–polyester NCs from reactive extrusion. *Journal of Applied Polymer Science*, 96, 1072–1082.

Kalambur, S. and Rizvi, S.S.H. 2006. Rheological behavior of starch–polycaprolactone (PCL) NC melts synthesized by reactive extrusion. *Polymer Engineering & Science*, 46, 650–658.

Kim, J.W., Noh, M.H., Choi, H.J., Lee, D.C., and Jhon, M. 2000. Synthesis and electrorheological characteristics of SAN-clay composite suspensions. *Polymer*, 41, 1229–1231.

Kojima, Y., Usuki, A., Kawasumi, M., Okada, A., Fukushima, Y., Kurauchi, T., and Kamigaito, O. 1993. Mechanical properties of nylon 6–clay hybrid. *Journal of Materials Research*, 8, 1185–1189.

Lamb, D.J., Anstey, J.F., Fellows, C.M., Monteiro, M.J., and Gilbert, R.G. 2001. Modification of natural and artificial polymer colloids by "topology-controlled" emulsion polymerization. *Biomacromolecules*, 2, 518–525.

Lee, S.Y. and Hanna, M.A. 2009. Tapioca starch-poly(lactic acid)-Cloisite 30B NC foams. *Polymer Composites*, 30, 665–672.

Li, X.C. and Ha, C.-S. 2003. Nanostructure of EVA/organoclay nanocomposites: Effects of kinds of organoclay and grafting of maleic anhydride onto EVA. *Journal of Applied Polymer Science*, 87, 1901–1909.

Li, R., Liu, C., and Ma, J. 2011. Studies on the properties of graphene oxide-reinforced starch biocomposites. *Carbohydrate Polymers*, 84, 631–637.

Li, Y., Mai, Y.W., and Ye, L. 2000. Sisal fibre and its composites: A review of recent developments, composites. *Science and Technology*, 60, 2037–2055.

Lin, N., Yu, J., Chang, P.R., Li, J., and Huang, J. 2011. Poly(butylene succinate)-based biocomposites filled with polysaccharide nanocrystals: Structure and properties. *Polymer Composites*, 32, 472–484.

Liu, H., Chaudhary, D., Yusab, S., and Tadé, M.O. 2011. Glycerol/starch/Na$^+$-montmorillonite NCs: A XRD, FTIR, DSC and 1H NMR study. *Carbohydrate Polymers*, 83, 1591–1597.

Liu, H., Chaudhary, D., Yusa, S., and Tadé, M. O. 2011a. Glycerol/starch/Na+-montmorillonite nanocomposites: A XRD, FTIR, DSC and 1H NMR study and 1H NMR study. *Carbohydrate Polymers*, 83, 1591–1597.

Liu H, Chaudhary D, Yusa S, and Tadé MO. 2011b. Preparation and characterization of sorbitol modified nanoclay with high amylose bionanocomposites. *Carbohydrate Polymers*, 85 (2011) 97–104.

Ma, X., Chang, P.R., Yang, J., and Yu, J. 2009. Preparation and properties of glycerol plasticized-pea starch/zinc oxide-starch bioNCs. *Carbohydrate Polymers*, 75, 472–478.

Ma, X.F., Yu, J.G., and Wang, N. 2007. Production of thermoplastic starch/MMT-sorbitol NCs by dual-melt extrusion processing. *Macromolecular Materials and Engineering*, 292, 723–728.

Mackay, M.E., Tuteja, A., Duxbury, P.M., Hawker, C.J., Van Horn, B., Guan, Z., Chen, G., and Krishnan, R.S. 2006. General strategies for nanoparticle dispersion. *Science*, 311, 1740–1743.

Magalhães, N.F. and Andrade, C.T. 2009. Thermoplastic corn starch/clay hybrids: Effect of clay type and content on physical properties. *Carbohydrate Polymers*, 75, 712–718.

Majdzadeh-Ardakani, K., Navarchian, A.H., and Sadeghi, F. 2010. Optimization of mechanical properties of thermoplastic starch/clay nanocomposites. *Carbohydrate Polymer*, 19, 547–554.

Majdzadeh-Ardakani, K. and Nazari, B. 2010. Improving the mechanical properties of thermoplastic starch/poly(vinyl alcohol)/clay nanocomposites. *Composites Science and Technology*, 70, 1557–1563.

Mali, S. and Grossmann, M.V.E. 2003. Effects of yam starch films on storability and quality of fresh strawberries. *Journal of Agricultural and Food Chemistry*, 24, 7005–7011.

Mali, S., Grossmann, M.V.E., García, M.A., Martino, M.N., and Zaritzky, N.E. 2006. Effects of controlled storage on thermal, mechanical and barrier properties of plasticized films from different starch sources. *Journal of Food Engineering*, 75, 453–460.

Marais, C. and Villoutreix, G. 1998. Analysis and modeling of the creep behavior of the thermostable PMR-15 polyimide. *Journal of Applied Polymer Science*, 69, 1983–1991.

Martin, O., Schawach, E., Averous, L., and Couturier, Y. 2001. Properties of multilayer biodegradable films based on plasticized wheat starch. *Starch/Starke*, 53, 372–381.

Masclaux, C., Gouanvé, F., and Espuche, E. 2010. Experimental and modelling studies of transport in starch nanocomposite films as affected by relative humidity. *Journal of Membrane Science*, 363, 221–231.

Mathew, A.P. and Dufresne, A. 2002. Morphological investigation of nanocomposites from sorbitol plasticized starch and tunicin whiskers. *Biomacromolecules*, 3, 609–617.

McGlashan, S.A. and Halley, P.J. 2003. Preparation and characterization of biodegradable starch-based NC materials. *Polymer International*, 52, 1767–1773.

Mittal, V. 2009. Polymer layered silicate NCs: A review. *Materials*, 2, 992–1057.

Mohanty, A.K., Misra, M., and Drzal, L.T. 2002. Sustainable bio-composites from renewable resources: Opportunities and challenges in the green materials world. *Journal of Polymer and Environment*, 10, 19–26.

Mondragón, M., Hernández, E.M., Rivera-Armenta, J.L., and Rodríguez-González, F.J. 2009. Injection molded thermoplastic starch/natural rubber/clay NCs: Morphology and mechanical properties. *Carbohydrate Polymers*, 77, 80–86.

Morgan, A. B. and Gilman J. W. 2003. Characterization of polymer-layered silicate (Clay) nanocomposites by transmission electron microscopy and X-Ray diffraction: a comparative study. *Journal of Applied Polymer Science*, 87, 1329–1338.

Müller, C.M.O., Laurindoa, J.B., and Yamashita, F. 2011. Effect of nanoclay incorporation method on mechanical and water vapor barrier properties of starch-based films. *Industrial Crops and Products*, 33, 605–610.

Nejad, M.H., Ganster, J., Bohn, A., Volkert, B., and Lehmann, A. 2011. Nanocomposites of starch mixed esters and MMT: Improved strength, stiffness, and toughness for starch propionate acetate laurate. *Carbohydrate Polymers*, 84, 90–95.

Ning, W., Xingxiang, Z., Na, H., and Shihe, B. 2009. Effect of citric acid and processing on the performance of thermoplastic starch/montmorillonite nanocomposites. *Carbohydrate Polymers*, 76, 68–73.

Okada, A., Kawasumi, M., Usuki, A., Kojima, Y., Kurauchi, T., and Kamigaito, O. 1990. Synthesis and properties of nylon-6/clay hybrids. In *Polymer Based Molecular Composites*. Schaefer, D.W. and Mark, J.E., Eds. MRS Symposium Proceedings, Pittsburgh, PA, Vol. 171, pp. 45–50.

Olsen, N.B., Christensen, T., and Dyre, J.C. 2001. Time-temperature superposition in viscous liquids. *Physical Review Letters*, 86, 7, 1271–1273.

Palade, L.I., Lehermeier, H.J., and Dorgan, J.R. 2001. Melt rheology of high L-content poly(lactic acid). *Macromolecules*, 34, 1384–1390.

Park, B.D. and Balatinecz, J.J. 1998. Short term flexural creep behavior of wood-fiber/polypropylene composites. *Polymer Composer*, 19, 377–382.

Park, H.W., Jim, C.Z., Park, C.Y., Cho, W.J., and Ha, C.S. 2002. Preparation and properties of biodegradable thermoplastic starch/clay hybrids. *Macromolecular Materials and Engineering*, 287, 553–558.

Park, H.-M., Lee, W.-K., Park, C.-Y., Cho, W.-J., and Ha, C.-S. 2003. Environmentally friendly polymer hybrids Part 1. Mechanical, thermal, and barrier properties of thermoplastic starch/clay nanocomposites. *Journal of Materials Science*, 38, 909–915.

Pegoretti, A., Kolarik, J., Peroni, C., and Migliaresi, C. 2004. Morphology and tensile mechanical properties. *Polymer*, 45, 2751–2758.

Pérez, C.J., Alvarez, V.A., and Vázquez, A. 2008. Creep behaviour of layered silicate/starch–polycaprolactone blends NCs. *Materials Science and Engineering: A*, 480, 259–265.

Ptaszek, P. and Grzesik, M. 2007. Viscoelastic properties of maize starch and guar gum gels. *Journal of Food Engineering*, 82, 227–237.

Pulta, M. 2004. Morphology and properties of polylactide modified by thermal treatment, filling with layered silicates and plasticization. *Polymer*, 45, 8239–8251.

Ranade, A., Nayak, K., Fairbrother, D., and D'Souza, N.A. 2005. Maleated and non-maleated polyethylene–montmorillonite layered silicate blown films: Creep, dispersion and crystallinity. *Polymer*, 46, 7323–7333.

Rao, Y.Q. and Pochan, J.M. 2007. Mechanics of polymer-clay nanocomposites. *Macromolecules*, 40, 290–296.

Ray, S., Maiti, P., Okamoto, M., Yamada, K., and Ueda, K. 2002. New polylactide/layered silicate Nanocomposites. 1. Preparation, characterization and properties. *Macromolecules*, 35, 3104–3110.

Ray, S.S. and Bousmina, M. 2005. Biodegradable polymers and their layered silicate nanocomposites: In greening the 21st century materials word. *Progress in Materials Science*, 50, 962–1079.

Ray, S.S. and Okamoto, M. 2003. New polylactide/layered silicate nanocomposites, melt rheology and foam processing. *Macromolecular Material Engineering*, 288, 936–944.

Ray, S.S., Yamada, K., Okamoto, M., and Ueda, K. 2003. New polylactide/layered silicate nanocomposites. 2. Concurrent improvements of material properties, biodegradability and melt rheology. *Polymer*, 44, 857–866.

Reichert, P., Kressler, J., Thomann, R., Müllhaupt, R., and Stöppelmann, G. 1999. Nanocomposites based on a synthetic layer silicate and polymide-12. *Acta Polymerica,* 49, 116–123.

Rittigstein, P. and Torkelson, J.M. 2006. Polymer–nanoparticle interfacial interactions in polymer NCs: Confinement effects on glass transition temperature and suppression of physical aging. *Journal of Polymer Science Part B: Polymer Physics*, 44, 2935–2943.

Rodriguez-Gonzalez, F.J., Ramsay, B.A., and Favis, B.D. 2003. High performance LDPE/thermoplastic starch blends: A sustainable alternative to pure polyethylene. *Polymer*, 44, 1517–1526.

Rouilly, A., Rigal, L., and Gilbert, R.G. 2004. Synthesis and properties of composites of starch and chemically modified natural rubber. *Polymer*, 45, 7813–7820.

Schubel, P.J., Johnson, M.S., Warrior, N.A., and Rudd, C.D. 2006. Characterisation of thermoset laminates for cosmetic automotive applications: Part III—Shrinkage control via nanoscale reinforcement. *Composite: Part A*, 37, 1757.

Siqueira, G., Bras, J., and Dufresne, A. 2009. Cellulose whiskers versus microfibrils: Influence of the nature of the nanoparticle and its surface functionalization on the thermal and mechanical properties of NCs. *Biomacromolecules*, 10, 425–432.

Siqueira, G., Bras, J., and Dufresne, A. 2010. Cellulosic BioNCs: A review of preparation. *Properties and Applications Polymers*, 2, 728–765.

Sreekumar, P.A., Joseph, K., Unnikrishnan, G., and Thomas, S. 2007. A comparative study on mechanical properties of sisal-leaf fibre-reinforced polyester composites prepared by resin transfer and compression moulding techniques. *Composites Science and Technology*, 67, 453–461.

Starkova, O., Yang, J.-L., and Zhang, Z. 2007. Application of time-stress superposition to nonlinear creep of polyamide 66 filled with nanoparticles of various sizes. *Composites Science and Technology*, 67, 2691–2698.

Treece, M.A. and Oberhauser, A. 2007. Soft glassy dynamics in polypropylene-clay NCs. *Macromolecules*, 40, 571–582.

Vaia, R.A., Ishii, H., and Giannelis, E.P. 1993. Synthesis and properties of two-dimensional nanostructures by direct intercalation of polymer melts in layered silicates. *Chemistry of Materials*, 5, 1694–1696.

Vaia, R. A. and Giannelis, E.P. 1997. Polymer melt intercalation in organically-modified layered silicates: Model predictions and experiment. *Macromolecules*, 30, 8000–8009.

Vigneshwaran, N., Kumar, S., and Kathe, A.A. 2006. Functional finishing of cotton fabrics using zinc oxide-soluble starch NCs. *Nanotechnology*, 17, 5087–5095.

Wang, Y., Lue, A., and Zhang, L. 2009. Rheological behavior of waterborne polyurethane/starch aqueous dispersions during cure. *Polymer*, 50, 5474–5481.

Wang, X., Yang, K., and Wang, Y. 2003. Properties of starch blends with biodegradable polymers. *Journal of Macromolecular Science: Part C: Polymer Reviews*, 43, 385–409.

Wang, N., Yu, J.G., Chang, P.R., and Ma, X.F. 2008. Influence of formamide and water on the properties of thermoplastic starch/poly(lactic acid) blends. *Carbohydrate Polymers*, 71, 109–118.

Wilhelm, H.M., Sierakowski, M.R., Souza, G.P., and Wypych, F. 2003. Starch films reinforced with mineral clay. *Carbohydrate Polymers*, 52, 101–110.

Willams, M.L., Landel, R.F., and Ferry, J.D. 1955. The temperature dependence of relaxation mechanisms in amorphous polymers and other glass forming liquids. *Journal of American Chemical Society*, 77, 3701.

Wu, H., Liu, C., Chen, J., Chang, P.R., Chena, Y., and Anderson, D.P. 2009. Structure and properties of starch/α-zirconium phosphate NC films. *Carbohydrate Polymers*, 77, 358–364.

Yang, J.L., Zhang, Z., Schlarb, A.K., and Friedrich, K. 2006a. On the characterization of tensile creep resistance of polyamide 66 nanocomposites. Part I: Modeling and prediction of long-term performance. *Polymer*, 47, 2791–2798.

Yang, J.L., Zhang, Z., Schlarb, A.K., and Friedrich, K. 2006b. On the characterization of tensile creep resistance of polyamide 66 nanocomposites. Part II: Modeling and prediction of long-term performance. *Polymer*, 47, 6745–6758.

Yu, J., Gao, J., and Lin, T. 1996. Biodegradable thermoplastic starch. *Journal of Applied Polymer Science*, 62, 1491–1494.

Yu, J., Wang, N., and Ma, X. 2005. The effects of citric acid on the properties of thermoplastic starch plasticized by glycerol. *Starch/Stärke*, 57, 494–504.

Zaidi, L., Bruzaud, S., Bourmaud, A., Médéric, P., Kaci, M., and Grohens, Y. 2010. Mechanical and thermal properties of polylactide/Cloisite 30B NCs. *Journal of Applied Polymer Science*, 116, 1357–1365.

Zanetti, M., Lomakin, S., and Camino, G. 2000. Polymer layered silicate nanocomposites. *Macromolecule Material Engineering*, 279, 1–9.

Zeng, Q.H., Yu, A.B., Lu, G.Q. (Max), and Paul, D.R. 2005. Clay-based polymer nanocomposites: Research and commercial development. *Journal of Nanoscience and Nanotechnology*, 5, 1574–1592.

CHAPTER **17**

Starch Nanocomposites and Nanoparticles
Biomedical Applications

Anitha R. Dudhani

CONTENTS

17.1 INTRODUCTION TO NANOCOMPOSITES AND NANOPARTICLES

Development of biopolymeric nanocomposites (BNCs) and biopolymeric nanoparticles (BNPs) from biodegradable materials has received significant attention over the past few years (Gref et al., 2002; Panyam and Labhasetwar, 2003; Szymonska et al., 2009). Generally, BNCs and BNPs of synthetic or natural polymers are submicro solid colloidal particles, with size ranging from 1 to 1000 nm. In these BNPs and BNCs, nanofillers can be encapsulated, adsorbed, or covalently attached. BNCs are new generation of nanocomposites (NCs), contributing innovative developments in the material science, life science, and nanotechnology (Darder et al., 2007). Research on BNCs has evolved development of natural polymer-based NCs, which showed improvements in mechanical, thermal, and functional properties (Yu et al., 2009). Other additional benefits of natural-based NCs include low density, transparency, flowability, improved surface properties, and recyclability (Zhao et al., 2008). BNCs are a mixture of polymers with organic or inorganic fillers of particular size (nano), geometry, and surface chemistry (Zhao et al., 2008). Nanofillers include solid-layered clays, synthetic polymer nanofibers, cellulose nanowhiskers (CNW), and carbon nanotubes (Ma et al., 2009). Nanoparticles (NPs) are also recognized as possible additives to improve the properties of BNCs.

BNPs are expected to address some specific issues in the drug delivery system, such as to (1) improve solubility, in vivo absorption, and mucosal permeability of the drug (Dudhani and

Kosaraju, 2010) due to their size and surface characters (Singh and Lillard, 2009); (2) increase stability (Szymonska et al., 2009) and controlled release of drug (Zhang et al., 2009); and (3) prevent gut metabolism/degradation. Depending on the preparation method, the BNPs attain different properties and release characteristics for the encapsulated therapeutic agents (Sahoo and Labhasetwar, 2003).

Biodegradable materials used for BNCs and BNPs are mainly polysaccharides, degraded and gradually absorbed or eliminated from the body via hydrolysis or mediated by metabolic processes (Daniels et al., 1990). Biopolymers attract considerable interest as a possible replacement for the petroleum-based plastics due to environmental safety (Ingvild et al., 2007). Polysaccharides have high amount of hydroxyl groups, which allow them to incorporate ligands to produce colloidal system. Other functions which render BNPs effective in drug delivery include mucoadhesion, specific receptor recognition (Listinsky et al., 1998; Santander-Ortega et al., 2010), and neutral coating to prevent the nonspecific protein absorption (Lemarchand et al., 2004).

17.2 TYPES AND METHODS OF PREPARATION OF STARCH NANOCOMPOSITES AND NANOPARTICLES

Polymeric NCs are made of two types of components: matrix which supports and protects filler substance and fillers which are stronger and stiffer components, reinforcing the matrix. These fillers enhance the mechanical properties and physical properties such as permeability. They are classified based on the particle shape as follows:

1. Particulate NCs, which are generally iso dimensional and show moderate reinforcement due to low aspect ratio, e.g., metallic or carbon black–reinforced composites.
2. Elongated NCs particles, which show better mechanical properties owing to their high aspect ratio, e.g., cellulose nanofibrils (known as whiskers or nanocrystals).
3. Layered NCs particles, which can be further classified as flocculated, intercalated NCs (intercalated polymer chains between layered particles), and exfoliated (layers separated), e.g., nanoclays (Alexandre and Dubois, 2000; Le Corre et al., 2010) (Figure 17.1). Starch nanocrystals are classified as layered particles.

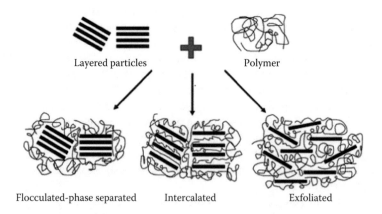

Layered particles   Polymer

Flocculated-phase separated   Intercalated   Exfoliated

Figure 17.1 Different layered particles in polymeric matrix. (Reproduced from *Mat. Sci. Eng. R. Rep.*, 28, Alexandre, M. and Dubois, P., Polymer-layered silicate nanocomposites: Preparation, properties and uses of a new class of materials, 1–63, Copyright 2000, with permission from Elsevier.)

Various techniques have been used to prepare NCs. They usually involve two steps: mixing and processing. Choice of matrix depends on other parameters like application, compatibility between components, process, and cost. The preparation technologies are classified as follows:

1. *Solution intercalation*: In this technique, biopolymer is soluble and inorganic fillers like silicate platelets are swellable.
2. *In situ intercalative polymerization*: In this process, layered silicate (LS) is swollen within liquid monomer or a monomer solution so that polymer formation can occur between the intercalated sheets. Polymerization can be initiated by heat or radiation by diffusing a suitable initiator or organic initiator or catalyst through cation exchange inside the interlayer before swelling.
3. *Melt intercalation*: This process involves annealing a mixture of polymer and LS above softening point of the polymer. During annealing, the polymer chains diffuse between silicate layers. Melt intercalation method has advantage over solution intercalation and in situ intercalative polymerization method as they do not use organic material.

Common methods for preparation of NPs are fluid extraction (inorganic NPs), solvent evaporation (organic NPs), precipitation (Ma et al., 2008; Tan et al., 2009), complex formation and enzymatic hydrolysis, microfluidization (Liu et al., 2009), and emulsion cross-linking. The precipitation method is complex, while emulsion cross-linking method is the simplest of all.

Ma and coworkers (Ma et al., 2008) prepared starch NPs by precipitating starch solution with ethanol as precipitant. Native starch was mixed with water and gelatinized at 90°C for an hour, and ethanol was added dropwise at room temperature with constant stirring for 50 min. The suspension was centrifuged to remove water and dried at 50°C to remove ethanol. Resulted NPs were used to prepare bio-NCs.

An alternative method (complex formation and enzymatic hydrolysis) to prepare starch NPs was studied by Kim and coworkers. Starch (0.5%, w/v) was dissolved in aqueous DMSO solution (90%) with constant heating and stirring in boiling water bath for 1 h and further with magnetic stirring at room temperature for 24 h. This aliquot was passed through 10 μm membrane filter gravimetrically into *n*-butanol and allowed precipitation for 3 days at 70°C. The precipitate was centrifuged (3390 g, 10 min) and washed with *n*-butanol three times. The starch–butanol complex was dispersed in water (20 mL), and then an aliquot (0.2 mL) of alpha-amylase solution (5610 units/mL) was added. The mixture was incubated at 25°C for 60 min with shaking at 170 rpm. Aliquots (10 mL each) of the solution were taken at regular intervals and mixed with ethanol (80 mL to inactivate the enzyme). After centrifugation (11,325 g for 30 min), the starch complex was recovered and washed three times with 80% ethanol. Starch NPs of spherical or oval shape of 10–20 nm were formed (Kim and Lim, 2009).

Starch NPs or nanospheres of starch were prepared by cross-linking starch with different cross-linking agents. Starch restructures its basic configuration by application of alkali, acid, and enzymes. The formation of hydrogen bonds accelerates the formation of NPs (Deetae et al., 2008; Hoover et al., 2010). The starch NPs help to encapsulate and deliver hydrophilic drugs to targeted area without premature degradation of the drug molecule. Encapsulation can reduce the undesired release of the drug (initial burst), improve efficacy of drug delivery, prolong residence time of the drug in the GIT, and reduce drug-related side effects (Kreuter, 2007; Singh and Lillard, 2009).

17.3 STARCH AS A BIOPOLYMER FOR NANOCOMPOSITES AND NANOPARTICLES

Starch is a promising natural, biocompatible, nontoxic, and biodegradable polysaccharide, found mainly in the plants such as corn, potatoes, wheat, and rice (Buleon et al., 1998; Alberta Araújo et al., 2004). In addition, they are renewable and available at low cost (O'Hagan and Illum,

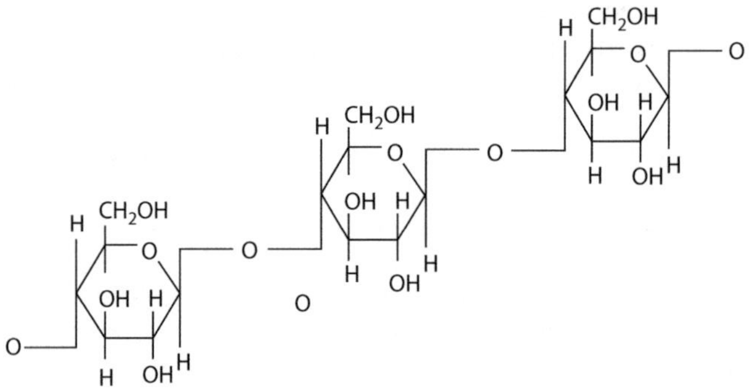

Figure 17.2 Structure of starch. (Reproduced from *Int. J. Biol. Macromol.*, Saboktakin, M.R., Tabatabaie, R.M., Maharramov, A., and Ramazanov, M.A., Synthesis and in vitro evaluation of carboxymethyl starch-chitosan nanoparticles as drug delivery system to the colon, 381–385, In Press, Uncorrected Proof, Copyright 2011, with permission from Elsevier.)

1990), which makes it suitable for biomedical applications and favored over synthetic polymers (Szymonska et al., 2009). Despite of these excellent properties, they are challenged by some limitations such as solubility in cold water, moisture sensitivity, poor processability (highly viscous), and incompatibility with hydrophobic polymers (Simi and Emilia Abraham, 2007). Starch appearance may vary widely depending on the source, but generally, it is white-flour-like, amorphous tasteless powder. Starch powder consists of microscopic granules with diameter 2–100 μm and density of 1.5 (Le Corre et al., 2010). When starch is extracted from plants, they are known as native starch, and if they undergo chemical or physical modifications to attain specific properties, they are termed as modified starch.

Knowledge on structure of starch has been recently advanced by Gidley (Chairam et al., 2009) (Figure 17.2). Starch is made of two glycosidic macromolecules: amylose and amylopectin. Amylose is defined as linear molecule of glucose units linked by (1–4) α-D-glycoside linkages, slightly branched by (1–6) α linkages (Chakraborty et al., 2005) with a molecular weight of 10^5–10^6 g/mol (Wesslén and Wesslén, 2002). Most commercial starches (e.g., potato, corn) contain about 25% amylose and the rest of amylopectin. Amylopectin has molecular weight between 10^6 and 10^7 g/mol, highly branched polymer consisting short branches of α-D (1–4) glycopyranose interlinked by (1–6) α-D linkages (Chakraborty et al., 2005) approximately every 22 units. Starch structure has semicrystalline granules containing concentric growth rings with amylose and amylopectin arranged radically and aligned perpendicularly to the growth rings (Chairam et al., 2009).

Environmentally friendly approaches have been a great interest in using starch and polysaccharides as a template for NP preparation. Wallen and coworkers used concept of green chemistry for the synthesis of silver NPs with solubilized starch using β-D glucose as a nontoxic reducing agent (Raveendran et al., 2006). However, hydrophilic nature of starch is major constraint limiting development of starch NPs. Several physical (Soane et al., 1999) and chemical modifications (Tengamnuay and Mitra, 1990; Callens et al., 2001) are being considered to resolve these issues. Such modified starch NPs were prepared to use in drug delivery system (Simi and Emilia Abraham, 2007; Santander-Ortega et al., 2010). In medical applications, starch is mainly used as a pharmaceutical excipients as a binding agent (Shi et al., 2011).

Hydrogel NCs are a new group of biomaterials recently used for medical and pharmaceutical applications. They are NCs with NPs dispersed in a hydrogel matrix. The hydrogel NCs have been investigated for various biological applications including drug delivery, tissue engineering, antimicrobial materials, and thermal therapy (Satarkar et al., 2009).

17.4 STARCH-BASED NANOCOMPOSITES AND NANOPARTICLES: BIOMEDICAL APPLICATIONS

In the past decade, researchers prioritize considering environment and strongly focus on green chemistry. Utilizing nontoxic chemicals, environmentally benign solvents, and renewable materials are key issues in green strategy. Vigneshwaran and coworkers synthesized first green stable silver NPs of 10–34 nm size with soluble starch. Soluble starch acts as reducing and stabilizing agent in preparations, offering compatibility for pharmaceutical and biomedical applications.

Silver NPs were also fabricated through γ-irradiation reduction of silver ions in aqueous starch solution by Kassaee and coworkers. Smaller-sized silver NPs were produced with higher yields with increased γ-irradiation doses and increased concentration of silver nitrate solution. Economic yield of silver NPs was produced at 5 kGy γ-irradiation with 2×10^{-3} M solution of silver nitrate containing 5% starch with no significant change in particle size. Antibacterial activities of silver NPs against *Escherichia coli* depended on γ-ray doses applied (Kassaee et al., 2008).

Santander-Ortega and coworkers prepared NPs (150–180 nm size) with two different propyl starch derivatives (hydrophobic) using o/w emulsion–diffusion method. NPs showed high entrapment efficiency (EE) for three tested drugs (flufenamic acid, testosterone, and caffeine) and had sustained linear release profile for flufenamic acid and testosterone without burst effect. Hydrophilic drug caffeine showed faster and linear release in first 10 h before reaching a plateau phase. The potential use of flufenamic acid NPs in transdermal drug delivery system was tested and exhibited enhancer effect with heat separated epidermis (Santander-Ortega et al., 2010).

Chang and his group (Chang et al., 2010a) have attempted to use two renewable materials: starch and chitin for the preparation of biocomposites. Chitin NPs of 50–100 nm were obtained after acid hydrolysis and mechanical ultrasonication. Glycerol-plasticized potato starch (GPS) was combined with chitin NPs by casting and evaporation to produce biocomposites. At low loading levels (2 wt%), chitin NPs were uniformly dispersed in GPS matrix, presenting good interaction between the fillers and matrix, with improved tensile strength, storage modulus, glass transition temperature, and water vapor barrier properties compared to higher loading (5%). Higher loading with chitin NPs resulted in aggregation.

Magnetic iron oxide (MIO) NPs have been explored for tumor targeting, but rapid blood clearance of these NPs by reticuloendothelial system limits its applications. Surface modification of MIO NPs with polyethylene glycol (PEG) augmented circulation time and tumor-targeting properties by increasing the hydrophilicity of the magnetic NPs and preventing their protein absorption (Lemarchand et al., 2004). Cole and coworkers prepared PEG-modified cross-linked starch-coated MIO NPs (140–190 nm) of 5 or 20 kDa which exhibited sustained plasma stability in in vitro studies with cell culture medium at 37°C and 7–10-fold less uptake by macrophages compared to unmodified starch NPs. In vivo studies demonstrated increased half life by 7.29 h (5 kDa) and 11.75 h (20 kDa) compared to unmodified MIO NPs and enhanced plasma AUC stability by 100- (D5) and 150 (D20)-fold as measure. Sustained tumor exposure over 24 h was confirmed in rat models, indicating that PEG-modified cross-linked starch MIO NPs show a potential platform for magnetic tumor targeting (Cole et al., 2011).

Development of delivery of drug molecules to the disease site without increase in the tissue level is currently an active area of research. Carboxymethyl starch–iron oxide NPs (CMS/SPIO) were prepared by chemical method of 1:14 drug/polymer ratio yielded spherical magnetic NPs of monodispersed size distribution. The controlled release of mesalamine was studied with CMS–SPIO NPs of 8.4 nm size which demonstrated that diffusion coefficient decreased owing to small particle size. Hydrolysis of mesalamine was observed at pH 7.4 due to increase in hydrophilicity of mesalamine due to both basic and acidic functional groups (Saboktakin et al., 2009).

Modified hydrophobic starch NPs were developed by grafting long fatty acids with potassium per sulfate as catalyst. These NPs were cross-linked with sodium tripolyphosphate for improved stabilization. Oleic acid–grafted NPs produced better yield compared to stearic acid. Grafted starch showed higher swelling properties compared to native starch at neural pH. Controlled release properties of these NPs with indomethacin were studied. Slow release of indomethacin at pH 7.4 showed that starch NPs may be considered as an efficient drug carrier (Simi and Emilia Abraham, 2007).

Systemic delivery of protein bioactive using mucoadhesive NPs for transnasal administration has been widely explored. Factors limiting the drugs like insulin are (1) poor permeability due to hydrophilicity and large size and (2) mucociliary clearance from the absorption site (Soane et al., 1999). Starch NPs were prepared using various cross-linking agents and loaded with insulin. Emulsion cross-linked particles were smaller compared to gel method, and particles produced were smaller in size when epichlorohydrin was used as cross-linking agent. Insulin released from starch NPS followed first-order diffusion-controlled profile with a burst effect (19%–28%). NPs prepared by gel method released insulin 81% in 12 h, whereas emulsion cross-linked particles showed faster release 85%–90% in 12 h. Larger particles (produced by gel method) showed slow release and less burst release compared to smaller particles (produced by emulsion method). The degree of cross-linking was found to have effect on insulin release (Jain et al., 2008). Hypoglycemic effect was found to be prominent with epichlorohydrin cross-linked NP agent compared until 6 h. Release rate, higher surface area, and choice of permeation enhancers make starch NPs efficient mucoadhesive carrier of insulin.

Glycerol-plasticized starch/ZnO BNPs of 10 nm were developed (Ma et al., 2009) using soluble starch as fillers in GPS matrix. Strong interaction between the nano-ZnO filler and GPS contributed enhanced biocomposite properties. Low loading level (below 4%) improved viscosity, glass transition temperature, storage modulus, and water vapor barrier compared to pure GPS. These BNPs have potential applications in medical, agricultural, drug delivery, and packaging. NC was prepared by Yu et al. (2009) using ZnO NPs, stabilized with carboxymethyl cellulose (CMC) as filler in glycerol-plasticized pea starch (GPS) by casting process. ZnO (60 wt%) was encapsulated with CMC (40 wt%) in ZnO–CMC particles of size range 30–40 nm. Low filler loading (5%) imparted improvement in the properties like past viscosity storage modulus, glass transition temperature, and tensile yield strength of GPS/ZnO-CMC NCs. CMC-protected NPs may be capable of readily integrating into system for biomedical and pharmaceutical applications (Ma et al., 2009).

Delivery of drugs to colon was investigated by Saboktakin and coworkers with chitosan (CS) and CMS NPs using 5-aminosalicylic acid (5-ASA) as a model drug. The NPs were prepared by complex conversion under mild conditions. The drug content increased from 10.23 ± 0.5 mg/100 mg to 26.24 ± 0.43 mg/100 mg with increase in amount of drug from 5 to 20 wt%. The percentage of EE increased up to $86.70\% \pm 2.47\%$ with increase in polymer concentration to 4%. Increase in concentration of cross-linking agent exhibited no increase in EE. In vitro release studies of 5-ASA showed $82.2\% \pm 4\%$ and $87.6\% \pm 3.1\%$ release with 15% and 20% drug, respectively, over 5 h. Amount of drug release from the NPs decreased with increase in cross-linking time. There was an initial burst 5-ASA from the NPs that may be due to desorption of loose 5-ASA from the matrix surface (Saboktakin et al., 2011).

Li Yan and coworkers prepared starch-grafted multiwalled carbon nanotube composites (CCNT) by grafting starch onto CCNT. The hydrophilicity of CCNT was improved due to grafted starch which facilitated the dispersion of CCNT–starch in both water and polysaccharide-based films. CCNT–starch promises potential applications in medical, drug release, and as fillers for the matrix of other natural polysaccharides (Yan et al., 2011).

Silver and gold NPs of controlled size and shape were developed with starch vermicelli templates (mung bean). Mung bean vermicelli is of great interest due to high amylose content and transparency, allowing colored particles to the vermicelli. Carbonization of this vermicelli revealed interesting pattern of gold and silver nanorods and nanowired-like assemblies respectively along with carbon nanotubes. The synthesis of these NPs was simple, safe, and easy accessible for large scale production of size- and shape-controlled silver and gold NPs for chemical and biological applications (Chairam et al., 2009).

New synthetic route of producing the starch NPs using high pressure homogenizer combined with water-in-oil mini emulsion cross-linking technique with sodium trimetaphosphate was investigated. Dynamic light scattering and transmission electron microscopy (TEM) showed narrow size distribution, good dispersibility, and uniform spherical shape. Parameters such as surfactant content, w/o ratio, starch concentration, homogenization pressure, and cycles on starch NPs were evaluated. It was reported that optimal surfactant concentration (6%) can produce small NPs with better stability. Apart from the water/oil ratio and starch concentration, homogenization pressure and cycles also significantly affect the starch NP size. Stability analysis of starch NPs in water over 2 days and at temperature of 25°C–45°C showed excellent stability. These NPs are expected to be potential candidates as drug carriers (Shi et al., 2011).

Silver nanowires were prepared by Valodkar's group on waxy cornstarch matrix of 32 nm size using simple and green hydrothermal route. Infrared (IR) spectroscopy and x-ray diffraction (XRD) confirmed silver embedded in waxy starch matrix. Scanning electron microscopy (SEM) and atomic force microscopy (AFM) revealed uniform morphology, demonstrating close association of the metal–polysaccharide composite. These NCs exhibited bactericidal effect which could be of potential biomedical application (Valodkar et al., 2010).

Nanoscale bioactive glasses have been gaining attention due to their reported superior osteoconductivity over conventional (microsized) bioactive glass materials. The combination of bioactive glass NPs or nanofibers with polymeric systems enables the production of NCs with potential to be used in a series of orthopedic applications and regenerative medicine. The recent developments in the preparation methods of nanosized bioactive glasses are covering sol–gel routes, microemulsion techniques, gas-phase synthesis method (flame spray synthesis), laser spinning, and electrospinning. The preparation and properties of NCs based on inorganic bionanomaterials are presented and other polymer matrices such as poly(hydroxybutyrate), poly(lactic acid), and poly(caprolactone) or natural-based polymers such as polysaccharides (starch, chitin, CS) or proteins (silk fibroin, collagen). The physicochemical, mechanical, and biological advantages of incorporating nanoscale bioactive glasses in biodegradable NCs demonstrate possibilities to expand the use of these materials in other nanotechnology concepts aimed to be used in different biomedical applications (Boccaccini et al., 2010).

Dialdehyde starch is prepared by oxidation of starch with periodic acid or periodate, which can cleave the C2–C3 linkage of anhydroglucose units with the formation of dialdehyde. Dialdehyde starch is biocompatible and biodegradable as starch and has more chemical activities. For example, it can react with hydrazine, acid, amine, and imine, and it is widely used in medicine field. Dialdehyde starch NPs of 100 nm (DASNP) were prepared by the redox reaction of $NaIO_4$ and starch in water-in-oil microemulsion. Doxorubicin (DOX) was combined with DASNP at 15:1 w/w, and the product showed efficiency for controlled release of DOX. The cell experiment showed that the drug-carrier particle (DOX-DASNP) can release DOX in controlled manner and strengthened effect of the anticancer drug. DASNP showed good thermal stability, small particle size, low biological toxicity, and slow drug-releasing property proving that it is useful as carrier for anticancer drug (Yu et al., 2007).

Stable silver NPs of 15–500 nm were prepared by treating aqueous solution of $AgNO_3$ with the plant leaf extracts (Pine, Persimmon, Ginkgo, Magnolia, and Platanus) as reducing agent of Ag^+ to Ag. Magnolia leaf broth was the best reducing agent with 90% conversion in 11 min at the reaction temperature of 95°C. The particle size could be controlled by changing the reaction temperature, leaf broth concentration, and AgNO$_3$ concentration. These ecofriendly biological silver NPs can potentially be used in various areas such as cosmetics, foods, and medical applications (Song and Kim, 2009).

Chang and coworkers prepared cellulose NPs (CN) by coagulating from NaOH/urea/H$_2$O solution of microcrystalline cellulose (MC) using an ethanol/HCl aqueous solution as the precipitant. CN ranged in size from about 50 to 100 nm. The glycerol-plasticized wheat starch (GPS)/CN NCs were prepared using CN as filler in GPS matrix by a casting process. At a low loading level, CN was dispersed evenly in the GPS matrix. The tensile strength of resulted matrix increased from 3.15 to 10.98 MPa when CN content went from 0% to 5%. However, water vapor permeability decreased from 5.75×10^{-10} to 3.43×10^{-10} g m^{-1} s^{-1} Pa^{-1}. The improvements in these properties may be attributed to the enhanced interaction between CN filler and GPS matrix because of similar polysaccharide structures of cellulose and starch. These polysaccharide composites can be applied in the medical, agricultural, and drug release (Chang et al., 2010b).

Recently, starch-based biolatex NPs were developed by Jung and coworkers (Oh et al., 2009). The process involved a reactive extrusion process that continuously converted starch into a thermoplastic melt in extruder. Starch polymers were cross-linked with an aldehyde and to produce biopolymer NPs with a diameter of 50–150 nm. Biolatex NPs were developed for replacing synthetic latex particles to offer biocompatibility, high swelling characteristics, and a large surface area with excess functionality for biomedical applications.

Ingvild and coworkers studied characterization of nanostructure and properties of starch-based BNCs with either CNW or LSs (synthetic hectorite) as reinforcements. Starch being hydrophilic and NCs prepared by solution casting and blending of CNW and LS contained well-dispersed reinforcements compared to modified LS. Modified potato starch was used as matrix with water and sorbitol as plasticizers and 5 wt% of both reinforcement contents. TEM examination showed well-distributed reinforcements in the starch matrix. Dynamic mechanical thermal analysis showed that storage modulus was significantly improved at elevated temperatures especially for LS NCs. Both NCs demonstrated a substantial improvement in tensile properties compared to the pure matrix (Ingvild et al., 2007).

17.5 CONCLUSION

Biopolymers are abundant organic compounds in nature with additional benefits being inexpensive, biodegradable, and renewable. Advantages of BNCs over traditional composites are reduced weight, increased flexibility, greater moldability, reduced cost, and renewable nature. Owing to environmental awareness and demand of green technology, nanobiocomposites have potential to replace petroleum-based products. BNCs and BNPs are of immense interest to biomedical technologies such as tissue engineering, bone replacement/repair, dental applications, and controlled drug delivery (Rohan and Darrin, 2007). The high surface energy of BNCS or BNPs makes them to aggregate if the surface is not protected (increasing hydrophilicity). Efforts have been focused on polysaccharides to use as protecting agents of BNCs and BNPs.

ACKNOWLEDGMENT

The author would like to thank Mr. Rajesh Dudhani (Faculty of Pharmacy and Pharmaceutical Sciences, Monash University) for his suggestions and support.

REFERENCES

Alberta Araújo, M., Cunha, A. M., and Mota, M. 2004. Enzymatic degradation of starch-based thermoplastic compounds used in prostheses: Identification of the degradation products in solution. *Biomaterials*, 25, 2687–2693.

Alexandre, M. and Dubois, P. 2000. Polymer-layered silicate nanocomposites: Preparation, properties and uses of a new class of materials. *Materials Science and Engineering: R: Reports*, 28, 1–63.

Boccaccini, A. R., Erol, M., Stark, W. J., Mohn, D., Hong, Z., and Mano, J. F. 2010. Polymer/bioactive glass nanocomposites for biomedical applications: A review. *Composites Science and Technology*, 70, 1764–1776.

Buleon, A., Colonna, P., Planchot, V., and Ball, S. 1998. Starch granules: Structure and biosynthesis. *International Journal of Biological Macromolecules*, 23, 85–112.

Callens, C., Adriaens, E., Dierckens, K., and Remon, J. P. 2001. Toxicological evaluation of a bioadhesive nasal powder containing a starch and Carbopol 974 P on rabbit nasal mucosa and slug mucosa. *Journal of Controlled Release*, 76, 81–91.

Chairam, S., Poolperm, C., and Somsook, E. 2009. Starch vermicelli template-assisted synthesis of size/shape-controlled nanoparticles. *Carbohydrate Polymers*, 75, 694–704.

Chakraborty, S., Sahoo, B., Teraoka, I., and Gross, R. A. 2005. Solution properties of starch nanoparticles in water and DMSO as studied by dynamic light scattering. *Carbohydrate Polymers*, 60, 475–481.

Chang, P. R., Jian, R., Yu, J., and Ma, X. 2010a. Starch-based composites reinforced with novel chitin nanoparticles. *Carbohydrate Polymers*, 80, 420–425.

Chang, P. R., Jian, R., Zheng, P., Yu, J., and Ma, X. 2010b. Preparation and properties of glycerol plasticized-starch (GPS)/cellulose nanoparticle (CN) composites. *Carbohydrate Polymers*, 79, 301–305.

Cole, A. J., David, A. E., Wang, J., Galbán, C. J., Hill, H. L., and Yang, V. C. 2011. Polyethylene glycol modified, cross-linked starch-coated iron oxide nanoparticles for enhanced magnetic tumor targeting. *Biomaterials*, 32, 2183–2193.

Daniels, A. U., Chang, M. K., and Andriano, K. P. 1990. Mechanical properties of biodegradable polymers and composites proposed for internal fixation of bone. *Journal of Applied Biomaterials*, 1, 57–78.

Darder, M., Aranda, P., and Ruiz-Hitzky, E. 2007. Bionanocomposites: A new concept of ecological, bioinspired, and functional hybrid materials. *Advanced Materials*, 19, 1309–1319.

Deetae, P., Shobsngob, S., Varanyanond, W., Chinachoti, P., Naivikul, O., and Varavinit, S. 2008. Preparation, pasting properties and freeze-thaw stability of dual modified crosslink-phosphorylated rice starch. *Carbohydrate Polymers*, 73, 351–358.

Dudhani, A. and Kosaraju, S. L. 2010. Bioadhesuie chitosan nanoparticles: Preparation and characterization. *Carbohydrate Polymers*, 81, 243–251.

Gref, R., Rodrigues, J., and Couvreur, P. 2002. Polysaccharides grafted with polyesters: Novel amphiphilic copolymers for biomedical applications. *Macromolecules*, 35, 9861–9867.

Hoover, R., Hughes, T., Chung, H. J., and Liu, Q. 2010. Composition, molecular structure, properties, and modification of pulse starches: A review. *Food Research International*, 43, 399–413.

Ingvild, K., Junji, S., and Martin, V. 2007. Characterization of starch based nanocomposites. *Journal of Materials Science*, 42, 8163–8171.

Jain, A. K., Khar, R. K., Ahmed, F. J., and Diwan, P. V. 2008. Effective insulin delivery using starch nanoparticles as a potential trans-nasal mucoadhesive carrier. *European Journal of Pharmaceutics and Biopharmaceutics*, 69, 426–435.

Kassaee, M. Z., Akhavan, A., Sheikh, N., and Beteshobabrud, R. 2008. [gamma]-Ray synthesis of starch-stabilized silver nanoparticles with antibacterial activities. *Radiation Physics and Chemistry*, 77, 1074–1078.

Kim, J.-Y. and Lim, S.-T. 2009. Preparation of nano-sized starch particles by complex formation with n-butanol. *Carbohydrate Polymers*, 76, 110–116.

Kreuter, J. 2007. Nanoparticles—A historical perspective. *International Journal of Pharmaceutics*, 331, 1–10.

Le Corre, D., Bras, J., and Dufresne, A. 2010. Starch nanoparticles: A review. *Biomacromolecules*, 11, 1139–1153.

Lemarchand, C., Gref, R., and Couvreur, P. 2004. Polysaccharide-decorated nanoparticles. *European Journal of Pharmaceutics and Biopharmaceutics*, 58, 327–341.

Listinsky, J. J., Siegal, G. P., and Listinsky, C. M. 1998. Alpha-L-fucose: A potentially critical molecule in pathologic processes including neoplasia. *American Journal of Clinical Pathology*, 110, 425–440.

Liu, D., Wu, Q., Chen, H., and Chang, P. R. 2009. Transitional properties of starch colloid with particle size reduction from micro- to nanometer. *Journal of Colloid and Interface Science*, 339, 117–124.

Ma, X., Chang, P. R., Yang, J., and Yu, J. 2009. Preparation and properties of glycerol plasticized-pea starch/zinc oxide-starch bionanocomposites. *Carbohydrate Polymers*, 75, 472–478.

Ma, X., Jian, R., Chang, P. R., and Yu, J. 2008. Fabrication and characterization of citric acid-modified starch nanoparticles/plasticized-starch composites. *Biomacromolecules*, 9, 3314–3320.

O'Hagan, D. T. and Illum, L. 1990. Absorption of peptides and proteins from the respiratory tract and the potential for development of locally administered vaccine. *Critical Reviews in Therapeutic Drug Carrier Systems*, 7, 35–97.

Oh, J. K., Lee, D. I., and Park, J. M. 2009. Biopolymer-based microgels/nanogels for drug delivery applications. *Progress in Polymer Science*, 34, 1261–1282.

Panyam, J. and Labhasetwar, V. 2003. Biodegradable nanoparticles for drug and gene delivery to cells and tissue. *Advance Drug Delivery Review*, 55, 329–347.

Raveendran, P., Fu, J., and Wallen, S. L. 2006. A simple and "green" method for the synthesis of Au, Ag, and Au-Ag alloy nanoparticles. *Green Chemistry*, 8, 34–38.

Rohan, A. H. and Darrin, J. P. 2007. Polymeric nanocomposites for biomedical applications. *MRS Bulletin*, 32, 354–358.

Saboktakin, M. R., Maharramov, A., and Ramazanov, M. A. 2009. Synthesis and characterization of superparamagnetic nanoparticles coated with carboxymethyl starch (CMS) for magnetic resonance imaging technique. *Carbohydrate Polymers*, 78, 292–295.

Saboktakin, M. R., Tabatabaie, R. M., Maharramov, A., and Ramazanov, M. A. 2011. Synthesis and in vitro evaluation of carboxymethyl starch-chitosan nanoparticles as drug delivery system to the colon. *International Journal of Biological Macromolecules*, 48, 381–385.

Sahoo, S. K. and Labhasetwar, V. 2003. Nanotech approaches to drug delivery and imaging. *Drug Discovery Today*, 8, 1112–1120.

Santander-Ortega, M. J., Stauner, T., Loretz, B., Ortega-Vinuesa, J. L., Bastos-González, D., Wenz, G., Schaefer, U. F., and Lehr, C. M. 2010. Nanoparticles made from novel starch derivatives for transdermal drug delivery. *Journal of Controlled Release*, 141, 85–92.

Satarkar, N. S., Hawkins, A. M., and Hilt, J. Z. 2009. Hydrogel nanocomposites in biology and medicine: Applications and interactions. In: Puleo, D. A. and Bizios, R. (eds.), *Biological Interactions on Materials Surfaces*. Springer, New York.

Shi, A. M., Li, D., Wang, L. J., Li, B. Z., and Adhikari, B. 2011. Preparation of starch-based nanoparticles through high-pressure homogenization and miniemulsion cross-linking: Influence of various process parameters on particle size and stability. *Carbohydrate Polymers*, 83, 1604–1610.

Simi, C. K. and Emilia Abraham, T. 2007. Hydrophobic grafted and cross-linked starch nanoparticles for drug delivery. *Bioprocess and Biosystem Engineering*, 30, 173–180.

Singh, R. and Lillard, J. W., Jr. 2009. Nanoparticle-based targeted drug delivery. *Experimental and Molecular Pathology*, 86, 215–223.

Soane, R. J., Frier, M., Perkins, A. C., Jones, N. S., Davis, S. S., and Illum, L. 1999. Evaluation of the clearance characteristics of bioadhesive systems in humans. *International Journal of Pharmaceutics*, 178, 55–65.

Song, J. Y. and Kim, B. S. 2009. Rapid biological synthesis of silver nanoparticles using plant leaf extracts. *Bioprocess and Biosystem Engineering*, 32, 79–84.

Szymonska, J., Targosaz-Korecka, M., and Krok, F. 2009. Characterization of starch nanoparticles. *Journal of Physics: Conference Series*, 146, 012027.

Tan, Y., Xu, K., Li, L., Liu, C., Song, C., and Wang, P. 2009. Fabrication of size-controlled starch-based nanospheres by nanoprecipitation. *ACS Applied Materials and Interfaces*, 1, 956–959.

Tengamnuay, P. and Mitra, A. K. 1990. Bile salt-fatty acid mixed micelles as nasal absorption promoters of peptides. II. In vivo nasal absorption of insulin in rats and effects of mixed micelles on the morphological integrity of the nasal mucosa. *Pharmaceutical Research*, 7, 370–375.

Valodkar, M., Sharma, P., Kanchan, D. K., and Thakore, S. 2010. Conducting and antimicrobial properties of silver nanowire–waxy starch nanocomposites. *International Journal of Green Nanotechnology: Physics and Chemistry*, 2, 10–19.

Wesslén, K. B. and Wesslén, B. 2002. Synthesis of amphiphilic amylose and starch derivatives. *Carbohydrate Polymers*, 47, 303–311.

Yan, L., Chang, P. R., and Zheng, P. 2011. Preparation and characterization of starch-grafted multiwall carbon nanotube composites. *Carbohydrate Polymers*, 84, 1378–1383.

Yu, D., Xiao, S., Tong, C., Chen, L., and Liu, X. 2007. Dialdehyde starch nanoparticles: Preparation and application in drug carrier. *Chinese Science Bulletin*, 52, 2913–2918.

Yu, J., Yang, J., Liu, B., and Ma, X. 2009. Preparation and characterization of glycerol plasticized-pea starch/ZnO-carboxymethylcellulose sodium nanocomposites. *Bioresource Technology*, 100, 2832–2841.

Zhang, L., Dudhani, A., and Kosaraju, S. L. 2009. Macromolecular conjugate based particulates: Preparation, characterisation and evaluation of controlled release properties. *European Polymer Journal*, 45, 1960–1969.

Zhao, R., Torley, P., and Halley, P. J. 2008. Emerging biodegradable materials: Starch- and protein-based bio-nanocomposites. *Journal of Materials Science*, 43, 3058–3071.

CHAPTER 18

Application of Life Cycle Assessment for Starch and Starch Blends

Ali Abas Wani and Preeti Singh

CONTENTS

18.1 INTRODUCTION

Plastic packaging materials derived from fossil fuels have grown at a faster rate than other polymeric materials since past decades. The current production of petrochemical polymers is 12.5 million tons, expected to grow up to 25 million tons by 2020 (Crank et al., 2005). This has been associated with a large number of changes and challenges; many of these changes are beneficial, and many have brought climatic and environmental threats to the present world. The synthetic polymers have been criticized since long primarily due to nonrecyclability, nonrenewability, nonbiodegradability, or incorporation of toxic additives into the packaged materials or surrounding environments (Tang and Alavi, 2011). As a consequence of the societal concerns, legislative pressure, and environmental regulations, the growth of biodegradable materials has steadily increased in order to reduce environmental degradation by selecting more environment-friendly products (Vink et al., 2003; Crank et al., 2005). Since last decade, the development of packaging materials using renewable resources which are naturally biodegradable and the possibility of combining their biodegradability with cost reduction and market needs have been the object of intensive academic and industrial research. Bioplastics will grow at a significant pace over the next few years. In 2010, the global market for bioplastics achieved estimated sale of $2.74 billion, and this value is expected to grow by 32.4% a year from 2011 to 2015, reaching an estimated value of $11.14 billion in 2015 (Crank et al., 2005).

Among these bioplastics, starch-based polymeric materials offer a renewable, economical alternative to existing petroleum-based nonrenewable polymeric materials. These starch-based biopolymers have been considered as one of the most promising because of its availability, biodegradability, and low cost. Starch, a natural biopolymer, is widely used to fabricate biodegradable packaging materials since it is one of the major components of cereal grains and tubers widely used in the food, paper, and textile industries. The integration of naturally occurring materials, such as starch, into commodity plastics has been increasing in recent years.

Although the market share of biobased polymers is considerably small, the continuous growth of this sector is a major challenge to the scientists to replace the conventional petrochemical polymers. The commercialization of starch-based products in the last few years has dominated the biobased polymer market. Starch-based bioplastics represent 85%–90% of the market share which is expected to reduce due to the large-scale production of the polylactic acid (PLA) biopolymeric materials by Nature Works (Bastioli, 2001; Vink et al., 2003). Since 1970, starch has been used as a filler in plastics (Bagheri, 1999) and has recently been plasticized and extruded with traditional plastics (Mohanty et al., 2000), used as a baked foam for thin-walled applications, and used as packaging foams (Tang and Alavi, 2011). Degli-Innocenti and Bastioli (2002) reported that the annual production of 30,000 metric tons of starch-based bioplastics per year constitutes about 75%–80% of the global market for biobased polymers. The major applications of starch polymers include packaging materials for industrial packaging, films for bags and sacks, and loose fill. These materials alone constitute about 75% of the total starch polymer applications. The major producers of starch-based polymeric materials with introduced products in the established market include Novamont, National Starch, Biotec, and Rodenburg. Except Rodenburg, the cost of these materials is relatively higher than the competing petroleum-based polymers.

Native starch has limited applications due to its crystalline nature which affects the polymer melt properties. However, the native starch can be modified to improve melt properties for fabrication of thermoplastic starch (TPS) materials (Sarazin et al., 2008; Tang and Alavi, 2011). Several methods include the use of plasticizers (physical modification), esterification (chemical), and blending with petroleum-based polymers. Plasticized starch is more versatile and can be blended with various polymeric materials for numerous applications. Blending of TPS with other polymers may be a route to overcome these limitations. A number of possibilities have been explored, and a significant progress has been made in this direction to improve the mechanical and barrier properties. Blending of starch polymers with other biomaterials and petroleum-based polymers has also been a matter of intense investigations to combat the drawbacks of the starch-based products. Literature has reported the blending of TPS with PLA, poly vinyl alcohol (PVOH), polycaprolactone (PCL), polybutylene adipate terephthalate (PBAT), and poly ε-caprolactone (PCL) (Crank et al., 2005). Copolymers used for blending or complexing have been reported to constitute up to 50% of the total mass of the starch polymer product (Novamont, 2003). The continuous efforts by the industry and the academia are being geared for the development of pure starch products with versatile properties similar to existing petro polymers. Novamont, a starch-based bioplastic manufacturer, envisages that by 2020, it will be possible to produce a 100% starch-based polymer with similar properties like petrochemical polymers and starch-blended polymers. Expectedly, this major breakthrough will be achieved by the development of technology and efficient chemical and biological starch modification processes (Novamont, 2003). A diverse range of starch-based products currently available in the market are based on the following:

1. Different raw materials
2. Different processing technologies
3. Tailored to various applications
4. Multiple waste management systems for recovery or disposal

This exponential growth of starch-based polymeric materials has generated an interest in the sustainability of starch-based new materials. It has three dimensions: economic, social, and environmental, commonly referred to as "the triple bottom line of sustainability." This reflects that the sustainable development involves the simultaneous pursuit of economic prosperity, environmental protection, and social equity. This means that the starch and its blends with their applications to packaging and allied industries have to expand their responsibility by including the environmental and social dimensions to avert the involved risks. Life cycle assessment (LCA) being a measurement tool is used to assess "products" or "services" for their environmental impact. This shows that LCA can be useful to evaluate and predict all the environmental impacts associated with starch and starch-based products at each stage of the products' life. The stages will start from the extraction of resources to the ultimate disposal of the products being marketed and the waste disposal (Figure 18.1). LCA further allows measurement and reporting on current impacts, alternate scenarios, and improvements achieved in the products under investigation. This chapter focuses on LCA of starch-based polymeric materials for packaging and allied applications. LCA can provide the data on the following listed aspects:

1. To improve the general understanding of the life cycle of starch and its products
2. To validate environmental and economic decisions concerning the process and product improvements, selection of products or services, feedstock, energy carriers, raw materials, production locations, and waste management systems
3. To provide data and information for regulatory and legislative measurements and for corporate environmental and waste management policies
4. To provide methodologies for improvement in product opportunities in the market
5. To the industry and the end-product consumers so as to enable them to more choices

18.2 LIFE CYCLE ASSESSMENT

LCA is defined by the International Organization for Standardization (ISO) as "the compilation and evaluation of the inputs, outputs, and the potential environmental impacts of a product system through its life cycle" (ISO 14040, 2006). The two most commonly used systems chosen in LCA studies are cradle-to-factory gate and cradle-to-grave. A cradle-to-factory gate LCA study includes all steps from the extraction of raw materials and fuels, followed by all conversion steps till the product is delivered at the factory gate. Cradle-to-factory gate analyses are often published by material producers. The system cradle-to-grave covers all steps of the system cradle-to-factory gate and in addition the usage and disposal phase as well. Cradle-to-grave analyses have the advantage of covering all phases of the life cycle. Since waste management differs by country and not all waste treatment options can be taken into account, cradle-to-grave analyses for a given product can lead to

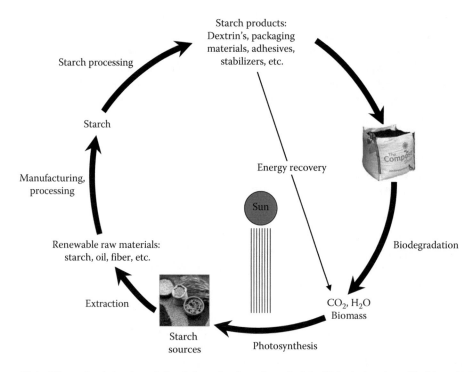

Figure 18.1 Life cycle of starch and starch-based polymeric materials. (Adapted and modified from Guinee, J.B. et al., *Environ. Manag. Health*, 12, 301, 2001; Vink, E.T.H. et al., *Polym. Degrad. Stability*, 80(3), 403, 2003; Murphy, R.J. et al., Life cycle assessment of potato starch based packaging trays, Report prepared (draft) for STI Project Sustainable GB Potato Packaging, Imperial College London, London, U.K., 2004.)

very different results depending on the type of waste management. If comparisons across the various waste management options are not available, cradle-to-factory gate analyses can provide first insight into the environmental impacts.

LCA is primarily divided into four stages, as displayed in Figure 18.2. The first stage, *goal definition*, defines the aim and the scope of the study as well as the function and the functional unit, system boundaries, and quality criteria for inventory data. The second stage, *life cycle inventory (LCI)* deals with the collection and synthesis of information on physical material and energy inputs and outputs in the several stages of the product life cycle. The input and output data are summarized to give pollutant emissions and consumption of resources per functional unit. Third stage, *life cycle impact assessment (LCIA)*, assesses and assigns the environmental impact of the pollutants emitted during the life cycle to different environmental impact categories. This stage also uses characterization models to calculate the contribution of each of these inputs and outputs to category indicators. Lastly, the fourth stage, *interpretation*, allows one to interpret the results and to estimate the uncertainties.

18.2.1 Goal Scope and Definition

This is the first step in the LCA study. Goal starts with the screening of key environmental impacts on the use of starch and its blends for their use in food and packaging industry. In the goal and scope definition, questions or hypotheses should be formulated since the appropriate LCA method depends on the purpose of the individual study (Consoli et al., 1993). The goals for starch and starch-based packaging materials do include the following:

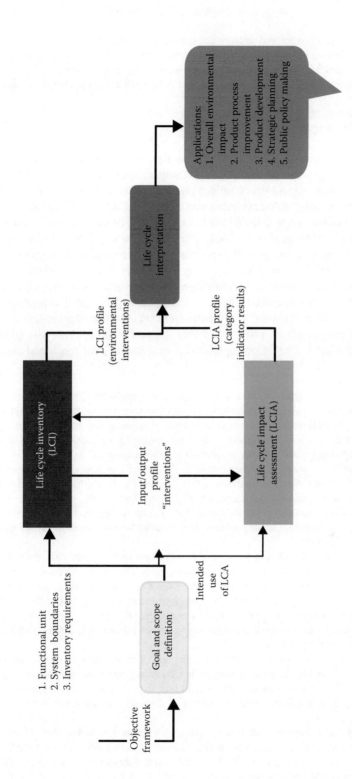

Figure 18.2 Phases of LCA. (Adapted and modified from ISO 14040, International standard, in *Environmental Management–Life Cycle Assessment–Principles and Framework.* International Organisation for Standardisation, Geneva, Switzerland, 2006a; Vink, E.T.H. et al., *Polym. Degrad. Stability,* 80(3), 403, 2003.)

- Identification of product and process development opportunities to improve environmental performance
- To substantiate environmental and economical decisions concerning, e.g., process and product improvements, selection of products or services, selection of feedstock, energy carriers and raw materials, and selection of production locations and waste management system
- For corporate environmental and waste management policies as well as for regulatory and legislative measurements
- Backing up marketing claims with scientific facts
- On how to position products in the market

Furthermore, it also compares environmental performance in terms of energy (emissions), solid waste, and water pollution at different life cycle stages. The goal further extends to assess the starch and its product improvement opportunities on environmental benefits over the life cycle in starch production. This clearly indicates the reasons of doing LCA, its applications, and the target groups. The objective of the scope definition is to identify and define the assessment and to limit it to the critical aspects for the goal of the study. For starch and starch-based materials, the main goal is to carry out a first assessment of the environmental impact of a biodegradable multilayer polymer derived from carbohydrate polymers using LCA. This assessment allows us to find out which life cycle phases and materials of the multilayer polymer should be improved, and to identify the aspects that require further research from an environmental viewpoint. The products developed 100% biodegradable multilayer sheets, films, adhesives, etc., for their application in food packaging obtained from potato, wheat, and cornstarches, presented as an alternative to the existent nonbiodegradable containers. The following critical parameters of *goal scope and definition* may be defined:

1. *Scenarios*—All the three types of scenarios, i.e., predictive, explorative, and normative, may find application for the starch polymeric materials. However, the scenarios must be consistent and capable to fulfill the goal of the study. The continual address of the scenarios will ensure the goal of the study. Different techniques like workshops, time series modeling, and optimized modeling can be used to develop qualitative and quantitative scenarios for biopolymers.
2. *Environmental problem areas*—It is imperative to decide the key environmental aspects of the product life cycle, data collection, analysis, and the impacts to be studied. The LCIA is aimed at understanding and evaluating the magnitude and significance of the potential environmental impacts of the studied system (ISO 14040, 2006). In the interpretation, the results from the previous phases are evaluated in relation to the goal and scope in order to reach conclusions and recommendations (ISO 14040, 2006).
3. *Functional unit*—It is a quantitative measure of the functions that the goods (or services) provide. In other words, "functional unit" is a common denominator upon which all environmental inputs (e.g., energy, water, chemicals) and environmental outputs (e.g., air emissions, solid waste, waste water discharge) are generally measured. The emissions to air, water, or soil in the inventory are determined as the functional unit's proportional share of the full emission from each process. The LCIA, thus, has to operate on mass loads representing a share (often nearly infinitesimal) of the full emission output from the processes.
4. *System and system boundaries*—A system is a collection of connected operations which together perform a defined function. A system boundary is a boundary that separates the internal components of a system from external entities. The system boundary will address the inclusion/exclusion of input and output system. This defines the object of the study and expresses it as the service it provides.
5. *LCA data*—LCA is data intensive, and lack of data can restrict the conclusions drawn from the LCA study. Thus, the authenticity of the results of any LCA study is dependent on the sample and data analysis. This leads to decision making for the inclusion and exclusion of several stages of the study, data collection methodology, evaluation techniques, and result interpretation from impact assessment phase.

18.2.2 Life Cycle Inventory Analysis

This step involves identification and quantification of *environmental inputs and outputs* for a given product throughout its life cycle. It is sometimes also referred to as the "resource and waste inventory," or an "environmental input–output inventory." The LCI thereby involves identification and quantification of inputs and outputs during each life cycle stage (e.g., crop cultivation, storage, transportation, processing, retailing, end-product utilization, and disposal). An exhaustive list of several inputs/outputs and the environmental impacts are available and can be measured for different products. Several considerations are important during the production of an inventory table for a new product, e.g., inventory inputs and outputs of wheat starch are shown in Table 18.1. The other issues may include allocation and validation of data. The key allocation issue concerns the allocation of impacts between the primary systems and according to ISO 14041 standards wherever possible allocation should be avoided. In detailed study, process may be possible to split and exclude the parts of the process that are connected to the other function (Parker, 2008). Materials of low value can be assessed as open loop outputs that leave the system boundaries without contributing to the environmental impacts of the system in question. In the systems where splitting is not possible, expansion of the system boundaries may help in accommodation of the additional input or output boundaries so the allocation is avoided. However, this may add complexity to the LCA, and in this case, ISO 14040 suggests that the system inputs and outputs are partitioned between the products' different functions in such a way that the underlying physical relationships between them are reflected. In case the physical relationship cannot be established, the inputs should be allocated between the products and functions in such a way so that the economic relationships between them are reflected. In case there is uncertainty regarding the data values or there are data gaps, sensitivity analysis can determine the significance of the decision making.

The input–output data may be collected directly from the process stages or through an environmental audit or indirectly through proprietary databases or from LCA practitioners. The national and international databases released include the Swedish SPINE@CPM database (CPM, 2007), the German PROBASE database (UBA, 2007), the Japanese JEMAI database (JEMAI, 2007), the USNREL database (NREL, 2004), the Australian LCI database (RMIT, 2007), the Swiss Ecoinvent database (Ecoinvent, 2007), and the European Reference Life Cycle Database (ELCD) (European Commission, 2007a). Besides these public databases, numerous international business associations have created their own databases as a proactive effort to support the demand for the first-hand industry data (Finnveden et al., 2009). Frequently, a trade-off is made in terms of data coverage and quality depending on the duration of the study and cost of data collection and collation. The quantitative data collected and collated from this step serve as inputs to the LCIA. LCA can be performed using Excel spreadsheets, but in recent years, a number of softwares have been developed and are frequently used to interpret LCA data. Softwares provide guidance on system design, data input, and calculations, and help to undertake the ISO-defined stages and produce results in diverse formats, e.g., graphs, charts, and tables. Currently, the basic LCA tools are available in two variants as follows:

1. Commercial software packages intended for trained practitioners
2. LCA tools for people who want LCA-based results without having to actually develop the LCA data and impact measures

The two principal tools named as GaBi Software and SimaPro developed by PE International and PRe Consultants, respectively, are mainly used by the LCA practitioners. However, for the latter category, different tools operate at different levels. LCA softwares are also available on the World Wide Web for trial basis. The data can be collected from several databases, e.g., Ecoinvent is one

Table 18.1   Summary of Input Materials and Resource Energy Inventory in Environmental Product Chain of Wheat Starch

Stage	Parameters	Inputs per kg of Starch
1.	*Wheat crop cultivation*	
	Gasoline (1)	0.03
	Diesel (1)	0.03
	Electricity (Wh)	140
	Energy used in transportation (kJ)	163
	Nitrogen (g)	50
	Phosphorous (g)	20
	Insecticides (g)	0.7
	Herbicides (g)	5
	Resource energy [14] (MJ_{heat})	4.25
2.	*Wheat crop transportation from farm to grain storage*	
	Diesel (1)	0.00004
	Resource energy [14] (MJ_{heat})	0.02
3.	*Wheat crop storage*	
	Electricity (Wh)	14.7
	Rodenticide (g)	35–85
	PH_3 (fumigating agent) (g)	0.002
	Resource energy (MJ_{heat})	0.175
4.	*Wheat grain transportation from grain storage to wheat flour mill*	
	Diesel (1)	0.023
	Resource energy (MJ_{heat})	0.963
5.	*Wheat flour milling*	
	Electricity (Wh)	92
	Water (1)	0.06
	Resource energy (MJ_{heat})	1.1
6.	*Wheat starch production*	
	Electric consumption (Wh)	450
	Natural gas (MJ)	3.4
	Biogas from effluent treatment plant (normal liters at 298 K and 1 atm)	273.4
	Freshwater to process (1)	10.22
	Resource energy (MJ_{heat})	8.9
7.	*Wheat starch transportation to end use*	
	Diesel (1)	0.05
	Resource energy (MJ_{heat})	1.94
8.	*Emissions from fossil fuel combustion*	
	Carbon dioxide, CO_2 (g)	260
	Carbon monoxide, CO (g)	13
	Nitrogen oxides, NO_x (g)	1.8
	Sulfur dioxide, SO_2 (g)	0.061
	Volatile organic chemicals, VOC (g)	3.7
9.	*Emissions due to electricity use*	
	CO_2 (g)	164
	CO (g)	0.02
	NO_x (g)	0.662
	SO_2 (g)	0.402

Sources: Modified and adapted from Scott, A. et al., *Life Cycle Assessment Case Studies, Centre for Integrated Environmental Protection*, Griffth University, Brisbane, Queensland, Australia, 2000; Narayanaswamy, V. et al., *J. Clean. Prod.*, 11, 375, 2003.

of the largest databases of LCA. Some softwares offer built-in database so that the end user could utilize these databases for their products. The published data of the LCA of starch and its polymeric materials are scanty. Some studies report the use of modified starch particles in tires and their impacts on the environment (Dinkel et al., 1996; Estermann et al., 2000; Würdinger et al., 2001; Patel, 2005).

18.2.3 Life Cycle Impact Assessment

The LCIA aims to examine the product system from an environmental perspective, using impact categories and category indicators connected with the LCI results (ISO 14042). The resource and emission data are assigned to the appropriate impact category and then aggregated within environmental impact category into a single category indicator using characterization factors that reflect the relative contribution of each emission to a given environmental impact category (e.g., using relative global warming potential, relative acidification potential). The category indicators can be based on estimated actual damages on the environment and humans such as the loss of biodiversity, loss of human life, acidification of rivers, release of toxins to the environment, or the release of greenhouse gases (GHGs) to atmosphere. The LCA goal and the expected uses drive the selection of impact categories for the LCIA, whereas the choice of an impact assessment methodology for each impact category is most often set by the availability of data of environmental mechanisms. The material and energy inputs to, and the emissions from, the various stages of the wheat starch were grouped into environmental impact categories using the Society for Environmental Toxicology and Applied Chemistry's (SETAC) environmental impact categorization and imperial chemical industry's environmental burden approach (Anonymous, 1999). Furthermore, the LCA goal and the expected uses drive the selection of impact categories for the LCIA, whereas the choice of an impact assessment methodology for each impact category is most often set by the availability of data of environmental mechanisms. The environmental impact categories can be broadly divided into three categories:

1. Resource depletion and degradation
2. Human health impact
3. Ecosystem health impact

These three main impacts could be further subdivided into more specific impacts. An indicative list of impact categories for starch-based polymeric materials can be exhaustive, but global warming, GHG emissions, energy savings, land use, and fossil fuel consumption are always the key considerations. Loss of energy and biodiversity is related to consumption and depletion resources. The global warming impact is the result of fossil fuel use and deals with adverse impacts on humans and ecosystem health. Human toxicity potential directly addresses human health issues, while other impacts address potential adverse effects on natural ecosystems.

18.2.4 Life Cycle Interpretation

The final element of an LCA deals with structuring the results from the LCI and LCIA phases in order to determine significant issues, in accordance with the goal and scope definition and interactively with an evaluation element. Thus, the LCIA interprets the inventory results into their potential impacts on what is referred to as the "areas of protection" of the LCIA (Consoli et al., 1993). These potential areas of LCA are human health, natural environment, natural resources, and, to some extent, man-made environment (Lindeijer et al., 2002). The impacts on the areas of protection can be simulated based on the relationships between interventions in the form of resource extractions, emissions, land, and water and their impacts to the environment for emissions of substances. The purpose of such interactions is to include the implications of the methods used, assumptions made, etc., in the preceding phases, such as allocation rules, cutoff decisions, selection of impact categories, category indicators, characterization models, etc. (ISO 14043, 2000). According to ISO 14040 and ISO 14044, life cycle interpretation comprises and involves the following elements (ISO, 2006):

1. *Selection of impact categories and classification*—This step involves the identification of the categories with environmental impacts in relevance to the study. Impact categories can be adopted for the existing ones and further be assigned with the emissions from the inventory according to the substances' ability to contribute to different environmental problems (Finnveden et al., 2009).
2. *Selection of characterization methods and characterization*—In this phase, the relevant model is selected, and the impact of each emission is modeled quantitatively according to the environmental mechanism. The results are expressed as an impact score in a unit common to all contributions within the impact category and expressed as an impact score in a unit common to all within the impact category that apply the concept of characterization factors which for each substance expresses its potential impact in terms of the common unit of the category indicator. Global warming potential is frequently used as characterization factor for the time frame of 100 years (GWP_{100}). Characterization sums the contributions from all the emissions and resource extractions within each impact category and translates the inventory data into environmental impact score profile.
3. *Normalization*—This step relates the reference values to the results of characterization. It expresses the relative magnitude of the impact scores on a scale which is common to all the categories of impact in order to facilitate the interpretation of the results.
4. *Grouping or weighting*—This final step of the grouping or weighting of the different environmental impact categories and resource consumptions reflects the relative importance in the study. Normalization expresses the relative magnitudes of the impact scores and resource consumptions, while weighting expresses their relative significance in accordance to the goal of the study. Weighting may be required when trade-off situations exist, e.g., where improvements in one impact score are obtained at the expense on another impact score.

According to ISO standard on LCA, selection of impact categories, classification, and characterization are mandatory steps in LCIA; however, normalization and weighting are optional (ISO 14044, 2006). Although a lot of developments have occurred in the past decade, it is impossible to account all inventory impacts. This requires a combination of impact categories by weighting process. This is the most critical and normative part of the method with its application of preferences and stakeholder values in a ranking, grouping, or quantitative weighting of the impact categories. The ISO does not permit weighting to be performed in studies supporting comparative assertions disclosed to the public (ISO 14040, 2006). Although the ISO standard for LCIA presents the framework and some general principles to adhere to, it refrains from a standardization of more detailed methodological choices. In the last decade, several well-documented LCIA methods have been developed to fill this gap (Bare et al., 2003; Jolliet et al., 2003).

18.3 APPLICATIONS OF LCA TO STARCH AND STARCH-BASED PRODUCTS

Starch, a versatile polymer used in several forms in a wide range of products with many applications, has resulted in a wide-ranging potential life cycle. The technology of starch and its products along with the system boundaries is presented in Figure 18.3. Starch-based products are continuously increasing probably due to the consumer interest in environmentally friendly products and its diverse applications. The other reasons include the regulatory authorities to reduce GHG emissions and the increasing price of fossil fuel stocks. Currently, a number of industries have started large-scale production of starch-based biomaterials. Among these industries, Novamont, National Starch, Biotec, and Rodenburg are the leading manufacturers for these materials. However, the discussion of this chapter is limited to the environmental performance of starch and its products like TPS, PLA, and starch-blended polymeric materials. The performance has to be measured in terms of life cycle impact categories like fossil fuel consumption, carbon emissions, economics, food supply, pollution, energy consumption, and health aspects.

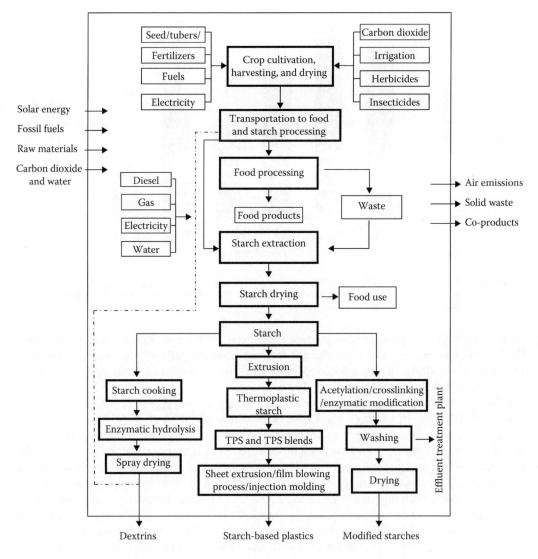

Figure 18.3 Flow diagram and the system boundaries of the starch and starch-based products.

18.3.1 Energy Consumption

The energy requirement for TPS, PLA, and petroleum polymers is presented in Figure 18.4. It is evident from the graph that among all the polymers, TPS consumes the least energy followed by PHB and PLA. The energy savings of starch polymers range from 23 to 52 MJ/kg depending on the type and extent of blended polymer (Boustead, 2003, 2005). TPS has been reported to be the most energy efficient among all the biopolymers used for packaging materials. The TPS requires half the energy to produce the polyethylene (PE) film. It indicates that starch and its products have a potential to reduce the consumption of fossil fuel. The reduction is due to shift from nonrenewable to renewable energy inputs (biomass feedstock), thereby increasing the demand for starch-based bioplastics. Novamont, a commercial starch-packaging material being manufactured by Italian bioplastic manufacturer reported that 1 kg of its starch-based product uses 500 g of petroleum and consumes

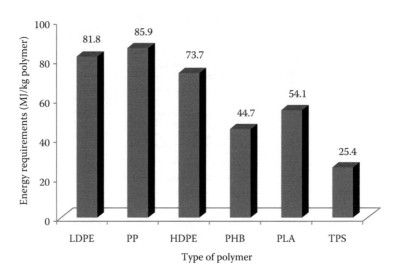

Figure 18.4 Energy requirements of different polymer production. (From Harding, K.G. et al., *J. Biotechnol.*, 130, 57, 2007; Vink, E.T.H. et al., *Polym. Degrad. Stability*, 80(3), 403, 2003; Narayanaswamy, V. et al., *J. Clean. Prod.*, 11, 375, 2003.)

almost 80% of the energy required to produce a traditional PE polymer (Crank et al., 2005). Nature Works LLC, a manufacturer of PLA, claims a 68% reduction of fossil fuel over the traditional plastics (Vink et al., 2003). However, 95% reduction in fossil fuel use for the fabrication of PHB has been claimed by Metabolix (Anonymous, 1999).

The total life cycle energy requirements are quite high for polypropylene (85.9 MJ/kg), high-density PE (HDPE) (73.7 MJ/kg), and LDPE (81.8 MJ/kg) (Harding et al., 2007). On the other hand, TPS has the lowest energy requirements among all the polymeric materials. The energy life cycle requirements for TPS have been reported as 25.4 MJ/kg plastic produced which is far less than the energy requirements of petroleum-based plastics (Narayanaswamy et al., 2003). Starch-based foams vary in energy requirements from 32.4 to 36.5 MJ/kg foam. However, the starch blends have been reported to go up to 52.3 MJ/kg produced (Narayanaswamy et al., 2003). The starch blends are mostly petro plastic in composition. Vink et al. (2003) reported that PLA consumes energy of 54.1 MJ/kg during the entire life cycle. These calculations include the oil feedstock in the energy requirements of the plastics. However, this energy is still less than the petro-based polymers. In the recent past, significant energy savings for starch-based polymers and their products have been achieved by several companies due to innovations in the product process optimization.

18.3.2 Fossil Fuel Consumption

Fossil fuel consumption of starch, starch blends, and the functional units (pellets, bags, films, etc.) versus petroleum-based polymers is shown in Table 18.2. Starch and its polymeric materials have no fossil fuel feedstock, while the petroleum-based polymers are primarily based on fossil fuel stock. However, the production of starch and its polymeric materials may use fossil fuel as energy sources. For energy use during the farming, harvesting, extraction, and processing, the use of fossil fuel energy may be replaced by hydroelectricity, wind energy, nuclear energy, solar energy, etc. Although the current processing of starch into different packaging materials and food-based products is not 100% accurate, the current studies focus on the development of energy analogues for synthesis of biopolymers. Shen and Patel (2008) reported that for the production development of TPS and its blends for packaging use save 25%–75% fossil fuel feedstock. The variation in the

Table 18.2 Comparison of Fossil Fuel Consumption and Environmental Impacts of Starch-Based Materials with Petroleum Polymers

Type of Natural/ Synthetic Polymer	Functional Unit	Cradle-to-Gate Nonrenewable Energy Use[a] (MJ/Functional Unit)	Type of Waste Treatment Assumed for Calculation of Emissions	GHG Emissions (kgCO$_2$eq./ Functional Unit)	Ozone Precursors (g Ethylene eq.)	Acidification (g SO$_2$ eq.)	Eutrophication (g PO$_4$ eq.)	Reference
Petrochemical polymers								
HDPE	1 kg	79.9	Incineration	4.84[b]	n/a	n/a	n/a	Boustead (1999)
LLDPE	1 kg	72.3	Incineration	4.54[b]	n/a	n/a	n/a	
LDPE	1 kg	80.6	Incineration	5.04[b]	n/a	n/a	n/a	Dinkel et al. (1996)
LDPE	1 kg	91.7	80% Incineration + 20% land filling	5.20[b]	13.0	17.4	1.1	
Nylon 6	1 kg	120	Incineration	7.64[b]	n/a	n/a	n/a	Boustead (1999)
PET (bottle grade)	1 kg	77	Incineration	4.93[b]	n/a	n/a	n/a	
PS (general purpose)	1 kg	87	Incineration	5.98[b]	n/a	n/a	n/a	
EPS	1 kg	84	Incineration	5.88[b]	n/a	n/a	n/a	
EPS	1 kg	88	None (cradle-to-factory gate)	2.80	43.0	170.0	5.8	Estermann et al. (2000)
EPS (PS + 2% SBR + Pentan + Butan)	1 kg	87	None (cradle-to-factory gate)	2.72	1.2	18.5	1.5	Würdinger et al. (2001)
Biobased plastics (pellets)								
TPS	1 kg	25.4	Incineration	1.14	n/a	n/a	n/a	Patel et al. (1999)
TPS	1 kg	25.5	80% Incineration + 20% compost	1.20	4.7	10.9	4.7	Dinkel et al. (1996)
TPS	1 kg	25.4	100% Composting	1.14	5.0	10.6	4.7	
TPS (maize starch + 5.4% maize grit + 12.7% PVOH)	1 kg	18.9	None (cradle-to-factory gate)	1.10[c]	0.2	4.6	0.5	Würdinger et al. (2001)
TPS + 15% PVOH	1 kg	24.9	Incineration	1.73	n/a	n/a	n/a	Patel et al. (1999)
TPS + 52.5% PCL	1 kg	48.3	Incineration	3.36	n/a	n/a	n/a	
TPS + 60% PCL	1 kg	52.3	Incineration	3.60	n/a	n/a	n/a	

(continued)

Table 18.2 (continued) Comparison of Fossil Fuel Consumption and Environmental Impacts of Starch-Based Materials with Petroleum Polymers

Type of Natural/Synthetic Polymer	Functional Unit	Cradle-to-Gate Nonrenewable Energy Use[a] (MJ/Functional Unit)	Type of Waste Treatment Assumed for Calculation of Emissions	GHG Emissions (kgCO₂eq./Functional Unit)	Ozone Precursors (g Ethylene eq.)	Acidification (g SO₂ eq.)	Eutrophication (g PO₄ eq.)	Reference
Mater-Bi foam grade	1 kg	32.4	Composting	0.89	5.5	20.8	2.8	Estermann et al. (2000)
Mater-Bi foam grade	1 kg	36.5	Waste water treatment plant	1.43	5.8	20.7	3.1	Estermann and Schwarzwälder (1998)
Mater-Bi film grade	1 kg	53.5	Composting	1.21	5.3	10.4	1.1	Estermann and Schwarzwälder (1998)
PLA	1 kg	54	Incineration	3.45	n/a	n/a	n/a	Vink et al. (2003)
PHA by fermentation	1 kg	81	n/a	n/a	n/a	n/a	n/a	Gerngross and Slater (2000)
PHA, various processes	1 kg	66–573	n/a	n/a	n/a	n/a	n/a	Heyde (1998)
Loose fills								
Starch loose fills	10 kg	492	Waste water treatment plant	21	n/a	n/a	n/a	Estermann et al. (2000)
Starch loose fills	12 kg	277	30% Incineration + 70% landfilling	33.5	n/a	n/a	n/a	Patel et al. (1999)
EPS loose fill	4.5 kg	680	Incineration	56	n/a	n/a	n/a	Estermann et al. (2000)
EPS loose fill	4 kg	453	30% Incineration + 70% landfilling	22.5	n/a	n/a	n/a	Patel et al. (1999)
EPS loose fill (by recycling of PS water)	4 kg	361	30% Incineration + 70% landfilling	18.6	n/a	n/a	n/a	Patel et al. (1999)

Films								
TPS film	100 m²	649	80% Incineration + 20% landfilling	25.3	n/a	n/a	n/a	Dinkel et al. (1996)
PE film	100 m²	1.340	80% Incineration + 20% landfilling	66.7	n/a	n/a	n/a	
Grocery bags								
50% Starch + PBS/A (single use)	3.12 kg	n/a	70.5% Landfill; 10% compost; 0.5% litter; 19% reuse	2.5	n/a	n/a	n/a	James and Grant (2005)
50% Starch + PBAT (single use)	3.12 kg	n/a	70.5% Landfill; 10% compost; 0.5% litter; 19% reuse	2.88	n/a	n/a	n/a	
50% Starch + PCL (single use)	4.21 kg	n/a	70.5% Landfill; 10% compost; 0.5% litter; 19% reuse	4.96	n/a	n/a	n/a	
HDPE (single use)	3.12 kg	n/a	78.5% Landfill; 2% recycle; 0.5% comp.; 19% reuse	6.13	n/a	n/a	n/a	
PP (multiple use)	0.48 kg	n/a	99% Landfill; 0.5% litter	1.95	n/a	n/a	n/a	
LDPE (multiple use)	1.04 kg	n/a	97.5% Landfill; 2% recycle; 0.5% litter	2.76	n./a	n/a	n/a	

[a] Total of process energy and feedstock energy. Nonrenewable energy only, i.e., total of fossil and nuclear energy. In the "cradle-to factory gate" concept, the down-stream system boundary coincides with the output of the polymer or the end product. Hence, no credits are ascribed to valuable by-products from waste management (steam, electricity, secondary materials).

[b] Only CO_2. Embodied carbon: 3.14 kg CO_2/kg PE, 2.34 kg CO_2/kg nylon 6, 2.29 kg CO_2/t PET, 3.38 kg CO_2/t PCL, 2.00 kg CO_2/t PVOH.

[c] No credit for carbon uptake by plants.

use of fossil fuel feedstock primarily depends on the use of blend and the fabrication method. PLA, a product of Nature Works, claims 68% reduction of fossil fuel use over traditional plastics (Vink et al., 2003).

Cradle-to-factory analysis indicated 49 GJ/metric tons of fossil fuel consumption for HDPE, and 39 and 48 GJ/metric tons for the PE terephthalate (PET) and polystyrene, respectively (Narayanaswamy et al., 2003). These substantial amounts of fossil feedstock and the use of starch-based polymeric materials may considerably reduce the fossil fuel demands. It is estimated that if all plastics of the world are replaced by bioplastics like starch-based polymeric materials, the total fossil fuel savings would be approximately 3.49 million barrels a day (British Plastics Federation). This accounts to 4% of the fossil fuel usage. The oil savings would vary from process to process and also from location to location. The oil savings may also vary with transportation costs and technological variations. As discussed earlier, further reduction in the fossil fuel consumption may be achieved by the use of solar, nuclear, hydroelectric, biomass, geothermal, and wind energies for the farm equipment, fertilizer production, and in starch processed and its allied industries.

18.3.3 Starch and Its Coproducts versus Petrochemical Polymers

The use of fossil energy resources is debatable as the fossil fuel feedstock is depleting. This will lead to supply disruptions and limitations within the next few decades (Vink et al., 2003). This section compares the starch and starch blends to traditional petroleum-based polymers, although the starch-based polymeric materials are still in their early stages of development. They are produced on small scale, and optimization is required for transport, conversion, product design, and final disposal which are not being optimized. At this developmental stage, comparison is not wise as it may lead to a biased one as the existing synthetic polymers have been in the market for several decades. However, if future outlook for starch-based materials is given a consideration in LCA studies, the resource and process optimization may result in the novel product development of starch-based polymeric materials for food, packaging, and other applications. Table 18.2 has summarized the performance of the synthetic and biodegradable polymers on the basis of fossil energy use, GHG emissions, and the biological impacts.

The use of petroleum-based polymers for packaging and other applications results in huge carbon emissions along with sulfur, nitrogen oxides, and heavy metals (Vink et al., 2003). These fossil fuels have been reported as global source of anthropogenic GHG rising concentrations of which are considered as driving forces for global warming. GHG is believed to result in unpredictable climatic changes resulting in more frequent and more extreme weather events such as floods, droughts, heat waves, windstorms, ice storms, and cyclones. The other negative effects may include air pollution, increased water- and airborne diseases, fauna migration, decrease in species specific populations, etc. Starch and its polymeric materials provide an alternate to petroleum-based polymers since they are derived from renewable resources, reduce carbon footprint to the atmosphere, reduce risk of discontinuity in supply chain applications, have no GHG emissions, and have ecofriendliness. These advantages indicate starch and starch-based polymeric materials as superior materials with little or no GHG emissions and may therefore enjoy increasing global market.

The environmental assessment of starch polymers, pellets, films, bags, and loose fill packaging materials, indicates that starch polymers offer potential energy and GHG saving relative to PE in the range of 24–52 GJ/ton plastic and 1.2–3.7 ton CO_2/ton plastic. The variation in the energy and GHG depends on the share of petrochemical copolymers (Dinkel et al., 1996; Estermann et al., 2000; Würdinger et al., 2001; Patel, 2005). Besides polymer blending, the variations may be due to differences in waste treatment procedures. A comparison of PE with starch pellets reported 25%–75% energy reduction, while 20%–80% reduction was reported for GHG emissions (Crank et al., 2005). Additionally, except eutrophication, TPS and its blends score better than PE for all other indicators covered by the LCA of starch. Although the environmental impact of starch polymers generally decreases with lower proportion of petrochemical copolymers, the application areas for pure starch polymers and less percentage copolymer blends are limited due to inferior material properties. Hence, blending can extend the applicability of starch polymers and thus lower the overall environmental impact at the macroeconomic level.

18.3.4 Global Climate Change

Global climate change has raised concern among the scientific community and the developed nations to determine the environmental performance of the products manufactured in the world. LCA is considered as a major tool to infer the inputs of carbon footprint of new products introduced to the market. As depicted in Table 18.2, starch-based polymeric materials have reduced carbon footprint, and during their end life cycle, release or return renewable biogenic carbon to the atmosphere. This biogenic carbon is captured from the atmosphere by plants during the growth and converted into the required raw materials as starch, cellulose, lignin, etc. This indicates that at the incineration which is the end of a product's useful life, the biogenic carbon is returned to the atmosphere or recycled in a closed biogenic CO_2 loop. This process is referred to as being carbon neutral. Therefore, the term carbon neutral refers to the biogenic carbon.

In terms of GHG emissions, 1 kg TPS leads to lower GHG emissions than petrochemical plastics. As reported by Patel (2005), the GHG emission savings of TPS are 1.2–3.7 kg CO_2 eq./kg. These variations are based on reference materials of LDPE and LLDPE (Boustead, 2003, 2005). In another study, Narayanaswamy et al. (2003) reported that the life cycle of the TPS produces only 1.14 kg CO_2/kg plastic. The starch foams vary between 0.89 and 1.43 kg CO_2/kg of plastic. Depending on the polymer blend, the CO_2 emissions of the starch blends are as high as 3.60 kg CO_2/kg plastic. Although the carbon emissions of petroleum-based polymers are higher than those of natural biopolymers, the data vary widely between different studies. Vink et al. (2003) reported that the new generation bioplastic from PLA has considerable reduction in GHG emissions (1.8 kg CO_2/kg plastic). The release of CH_4 emissions during landfill disposal process of starch-based biodegradable materials may be a disadvantage in terms of increased GHG emissions (Würdinger et al., 2002). For loose fills of starch polymer films and bags, differences in the film thickness have resulted in variations in GHG emissions. However, the environmental impacts of the starch films/bags are lower with regard to energy, GHG emissions, and ozone precursors. The data on acidification are not available, and the situation is less clear for acidification. Improvements in the production technology and transportation may further reduce the GHG emissions of these biopolymers and their packaging materials in the near future. Concluding this aspect, starch-based polymeric materials along with PLA seem to be the most favorable in terms of GHG emissions.

18.3.5 Market Status and Economics of Starch and Its Polymeric Materials

LCA acts as an important tool to project the future demand and market price based on the available data. The existing market of starch-based polymers and the major producers are reported in Table 18.3. LCA provides an insight in to the environmental performance of the starch and starch-blended polymers along with the comparison with the existing petro polymers, as the existing range of petro-based polymers has developed to maturity, available in a wide range of products in established markets. The comparisons with the starch and its polymeric materials may not be justified at this stage. However, they have a long way to go with product and process optimization for the current and future market. Although major efforts are still required for the replacement of petro-based polymers, the recent economic and technical breakthrough on starch polymers has considerably reduced the cost of the starch-based materials. Starch-based products like compost bags and catering utensils have successfully been commercialized as

Table 18.3 Current and Potential Large Volume Producers of Starch-Based Polymers

Producer and Polymer Type	Region	Trade Name(s)	2002 Production (Kt p.a.) Global	2003 Capacity (Kt p.a.) Global	2010 Capacity (Kt p.a.) Global	2003 Price ($/kg) Global	2010 Price ($/kg) Global
Starch polymers							
Novamont, Italy	EU	Mater-Bi	25	35	>20	2.13–6.38	—
Rodenburg, the Netherlands	EU	Solany	3 (0–7)	40	40	1.42	—
National Starch and Chem., United States	United States	Ecofoam	(20)	(20)	(>20)	—	—
Chinese company	Asia	TPS	—	(100)	(100)	0.85	—
BIOP, Germany	EU	BIOpar	—	10(~2004)	150	—	—
Biotec, Germany	EU	Bioplast TPS	2	2	—	—	—
Japan Corn Starch, Japan	Asia	Cornpol	—	—	—	—	—
Nihon Shokuhin Kako, Japan	Asia	Placorn	—	—	—	—	—
Potatopak, Avebe, Earthshell		Baked starch derivatives	—	—	—	—	—
PLA							
Cargill Dow LLC, United States	United States	Natureworks (Mistui Lacea in Japan)	30	140	208–500	3.12–4.82	1.91
Hycail, the Netherlands	EU	Hycail HM, Hycail LM	—	1	100–250		2.55
Toyota, Japan	Asia	(Toyota Eco-Plastic)	—	50 (in 2004)	150–400		
Project in China	Asia	Conducted by Snamprogetti, Italy	—	2.5 (mid-2003)			

Source: Modified and adapted from Crank, M. et al., Techno-economic Feasibility of Large-scale Production of Bio-based Polymers in Europe (PRO-BIP), Report prepared for the European Commission's Institute for Prospective Technological Studies (IPTS), Sevilla, Spain, edited by O. Wolf, Prepared by the Department of Science, Technology and Society/Copernicus Institute at Utrecht University, Utrecht, the Netherlands and the Fraunhofer Institute for Systems and Innovation Research, Karlsruhe, Germany, 2005.

alternatives to petroleum-based polymers. The poor aesthetics of these materials have limited consumer acceptance, but public awareness on the use of biodegradable materials has promoted this niche market. Further improvements in the surface properties, clarity, color, and tensile strength will improve the market opportunities for these materials. At the same time, the efforts of R&D to improve the design of injection, compression molding equipment, and extruders/ dies for the starch-based biopolymers will bring a major breakthrough for the commercialization of these products. The application of chemical sciences may be used to modify the starch structures to improve its water barrier properties. Market opportunities may increase by educating consumers about the benefits of using starch-based products like nontoxic, reduced GHG emissions, water solubility, etc. The other factors that may be forecasted as driving factors for starch-based materials include the low cost, resource availability, biodegradability, incinerability, renewability, reduction in value-added taxes, waste management practices, energy recovery, and biocomposting.

According to the latest issue of EL Insights, the market of bioplastics has achieved sales of $2.74 billion. This value is expected to grow by 32.4% a year from 2011 to 2015, reaching an estimated value of $11.14 billion in 2015. As mentioned earlier, the present market share of starch polymers will continue to grow with the reduction in the cost and more efficient production methods. Constituting about 80% of the bioplastics market, TPS currently represents the most important and widely used bioplastic. Pure starch possesses the characteristic of being able to absorb humidity and is thus being used for the production of drug capsules in the pharmaceutical sector. In 2002, the global market share of starch bioplastics was about 25,000 ton per year (75%–80%), which is expected to decrease as new biopolymers like PLA are also gaining the market (Degli Innocenti and Bastioli, 2002). Novamont, an Italian company, produces 35,000 ton of bioplastics under different trade names (Novamont, 2003); another German company, Biotec, produces above 2,000 ton/year plasticizer-free thermoplastic processable starch (Biotec, 2003). Other starch-processing companies like BIOP Biopolymer Technologies (Germany), Potatopak (United Kingdom), Japan Corn Starch (Japan), Nihon Shokuhin Kako (Japan), and Rodenburg Biopolymers (the Netherlands) have increased their production due to the existing demand for starch-based materials. Among all these manufacturers, Solanyl, a starch bioplastic from Rodenburg Biopolymers, is price competitive with conventional oil-based plastics (INFORRM, 2003). The price of PLA-based bioplastic made by Nature Works LLC is slightly more expensive than the petroleum-based plastics. Due to the intensive R&D of Cargil Dow LLC, United States, the manufacturer of PLA has recently reduced the production price of PLA from $3.1 to $1.9. Further reduction in the price is expected in the next few years due to R&D focus on production and energy-efficient technologies for biopolymer processing.

18.3.6 Food Supply

Starch and other biopolymeric materials have the potential to affect the world's food supply in a number of ways. Since corn, wheat, rice, potato, yam, cassava, sago, etc., are the major food crops and at the same time raw materials for starch, the use of these crops in starch and allied industries would directly decrease the amount of these crops that would be available for food use. TPS, primarily made from corn, requires 0.971 kg of corn as feedstock. In the United States, production of 77.2 million metric tons of plastic will require 75 million metric tons of corn which accounts to 14.2% of the world's annual production (Dornburg et al., 2003). This reflects that the TPS is not a substitute for petroleum-based polymeric materials. However, in order to maintain stability and to avoid food crisis, the U.S. Conserve Reserve Program (CRP) may be used as bioplastic crop land (Momani, 2009). The use of CRP land would contribute 131 million tons of corn annually, and this would be enough to convert all plastics production to either TPS or PLA. In addition to this, the optimization and improvement potential for starch-based biomaterials is huge, and the strategies are explored for switching to nonfood crops and agricultural waste streams. This will increase the scope of starch biomaterials. Further opportunities will increase by the use of innovative and more efficient processes to increase production and further optimization of conversion technology and product design.

18.3.7 Health

It has been observed that synthetic polymers migrate into the food, thereby raising concerns on their safety for human use. On the other hand, starch-based polymeric materials, when used in packaging food materials, do not leach into food products, or its migration may not be a human health concern, since starch polymers are composed of glucose monomers that are easily metabolized by the human body. The only concern about the use of starch as fillers in packaging materials and TPS is to minimize the use of pesticides and artificial fertilizers. These problems may be averted by organic farming techniques; however, organic farming tends to reduce the crop yield. Genetic modification of crops may be looked into the increase in pest resistance and the enhanced yields under organic farming systems.

18.3.8 Socioeconomic Effects

Apart from the environmental benefits, starch and its polymeric materials offer socioeconomic benefits, particularly in relation to employment in the agricultural sector and the packaging industry. With the increased demand of bioplastics, the income of farm producers is expected to increase with more demand of raw materials. Furthermore, the job avenues in the processing industries are expected to increase in the future as the industry flourishes with the complete replacement of synthetic polymers.

18.3.9 Impact on Land Use

The use of food crops for starch extraction may have a significant impact on the land occupation and the land transformation involved in agriculture and forestry. However, there is currently no agreement on how these impacts should be included in an LCA. Increased land use will affect three of the areas of protection directly: natural environment, natural resources and man-made environment, and, indirectly, human health. This includes loss of biodiversity, loss of soil quality, and loss of biotic production potential. In recent years, excellent approaches for inclusion of land use in LCA studies have been proposed by several researchers (Lindeijer et al., 2002; Pennington et al., 2002; Mila i Canals et al., 2007a; Koellner and Scholz, 2008; Michelson, 2008). The approach based on cumulative energy extraction mentioned earlier and the ecological footprint discussed later is also approaches which include some aspects of land use. The land use aspect will further explore the possibilities of using agricultural by-products as substrate for the synthesis of biopolymers using microbial fermentations. Consideration of this aspect will dramatically decrease the land use requirements for nonfood use of starch-based polymeric materials.

18.4 IMPACT FROM WATER USE

Freshwater as a resource provides fundamental functions for humans and the environment, and the investigation on the gross water use of starch-based polymers should be one of the foci of the LCA study. Till date, there are no published data available on the gross water use of starch or its coproducts. However, studies on PLA indicate that despite of the irrigation water during corn growing, the total amount of water required is competitive with the best performing polymers (Vink et al., 2003). The values of the gross water use may also vary with the type of crop being used for the starch production, as the rice cultivation has high water requirements and may be a serious concern on the water usage aspect. Large-scale production of starch and its coproducts may have significant effect on the freshwater supply and the water quality of the industrial discharges. Water quality parameters have been proposed as additional information to be included in the LCI studies as they can affect the fragile aquatic environment and human health.

18.5 MATERIAL RECYCLING

Starch-based materials can be treated in many different waste management systems in order to recover energy. These treatments include mechanical recycling, composting, anaerobic digestion, and chemical recycling practices. These recovery options have several advantages over traditional products as they are not suitable for composting and at the same instance help to avoid landfill since they represent a loss of useful material and energy. The recovery and final disposal methodology will influence the outcome of an LCA. This requires careful setup and consideration of possible indirect beneficial effects, for instance, the possibility of obtaining homogenous feedstock, waste streams for recycling, and the use of hydroelectricity or wind energy in the case of incineration of renewable starch-based polymeric substances.

18.6 CONCLUSIONS

LCA methodology is a universal tool for the assessment of environmental impacts and the sustainability of existing and new products to the global markets. The current trend shows that the polymers from renewable resources including starch can significantly lower GHG emissions and fossil energy use in comparison with conventional petrochemical-based polymers. The growth of the starch-based bioplastics is further expected in the coming years due to the regulations on lowering GHG emissions and the price hike due to the depletion of fossil fuels. At the same instance, the lower energy consumption and reduced use of fossil fuels will be the driving forces for the industry to reduce the cost of starch polymer product fabrication with innovations in the fabrication process. This bright future is expected to come only with significant investment of time, effort, and money. A final, important benefit of LCA is that it can serve as a tool for monitoring return on these investments over time. To conclude, several options to reduce the environmental impacts related to starch production and use will include the increased energy efficiency and material efficiency (yields) in all processes in the production chain leading to starch polymers and, further, the increased end-use material efficiency, i.e., ensuring the same product service by lower amounts of material (e.g., by the use of thinner starch films) and improved waste management by recycling of materials. Due to the limited published data and the inventory of the starch-based materials, forthcoming LCA studies will give further insight on the starch-based polymers available in the market.

REFERENCES

Anonymous, 1999. SETAC. Guidelines for life cycle assessment. A code of practice. 1993; Imperial Chemical Industries. Environmental Progress Report. 1999.

Bagheri, R. 1999. Effect of processing on the melt degradation of starch-filled polypropylene. *Polymer International* 48: 1257–1263.

Bare, J.C., Norris, G.A., Pennington, D.W., and McKone, T.E. 2003. TRACI, the tool for the reduction and assessment of chemical and other environmental impacts. *Journal of Industrial Ecology* 6: 49–78.

Bastioli, C. 2001. Global status of the production of biobased packaging materials. *Starch/Starke* 53: 351–355.

Biotec. 2003. Biologische Naturverpackungen GmbH & Co. KG, Emmerich, Germany. http://www.biotec. de/engl/index_engl.htm (accessed August 20, 2003).

Boustead, I. 1999. Eco-labels and eco-indices: Do they make any sense? Paper presented at the *Fourth International Ryder Transpak Conference*, Brussels, Belgium, May 18–19, 1999.

Boustead, I. 2003. *Eco-Profiles of Plastics and Related Intermediates* (about 55 products). Prepared for the Association of plastics Manufacturers in Europe (APME). Downloadable from the internet (http://www. apme.org), Brussels, Belgium.

Boustead, I. 2005. *Eco-Profiles of the European Plastics Industry: LDPE Resin*, Data last calculated March 2005. Plastics Europe, Brussels, Belgium. www.plasticseurope.org.

Consoli, F., Allen, D., Boustead, I., Fava, J., Franklin, W., Jensen, A.A., de Oude, N., Parrish, R., Perriman, R., Postlethwaite, D., Quay, B., Sieguin, J., and Vigon, B. (Eds.). 1993. *Guidelines for Life Cycle Assessment. A Code of Practice*. SETAC Press, Pensacola, FL.

CPM. 2007. SPINE@CPM database. Competence center in environmental assessment of product and material systems (CPM), Chalmers University of Technology, Goteborg.

Crank, M., Patel, M., Marscheider-Weidemann, F., Schleich, J., Hüsing, B., and Angerer, G. 2005. Techno-economic feasibility of large-scale production of bio-based polymers in Europe (PRO-BIP). Report prepared for the European Commission's Institute for Prospective Technological Studies (IPTS), Sevilla, Spain, edited by O. Wolf. Prepared by the Department of Science, Technology and Society/Copernicus Institute at Utrecht University, Utrecht, the Netherlands and the Fraunhofer Institute for Systems and Innovation Research, Karlsruhe, Germany.

Degli-Innocenti, F. and Bastioli, B. 2002. Starch based biodegradable polymeric materials and plastics history of a decade of activity. Presentation at UNIDO, Trieste, Italy, September 5–6, 2002. http://www.ics.trieste.it/documents/chemistry/plastics/activities/egmSept2002/DegliInnocenti.pdf.

Dinkel, F., Pohl, C., Ros, M., and Waldeck, B. 1996. Ökobilanz stärkehaltiger Kunststoffe (Nr. 271), 2 volumes. Study prepared by CARBOTECH, Basel, for the Bundesamt für Umwelt und Landschaft (BUWAL), Bern, Switzerland.

Dornburg, V., Lewandowski, I., and Patel, M. 2003. Land requirements and energy savings and greenhouse gas emission reduction by biobased polymers compared to bioenergy—An analysis and system extension of LCA studies. *Journal of Industrial Ecology* 7: 93–116.

Ecoinvent. 2007. Swiss centre for life cycle inventories (Ecoinvent Centre). Ecoinvent Database. Ecoinvent Centre, Dübendorf, 2004 and 2007. http://www.ecoinvent.org.

Estermann, R. and Schwarzwälder, B. 1998. Life cycle assessment of Mater-Bi bags for the collection of compostable waste. Study prepared by COMPOSTO, for Novamont, Novara, Italy. Olten, Uerikon, Switzerland.

Estermann, R., Schwarzwälder, B., and Gysin, B. 2000. Life cycle assessment of Mater-Bi and EPS loose fills. Study prepared by COMPOSTO for Novamont, Novara, Italy. Olten, Switzerland.

European Commission. 2007a. European commission, Directorate General Joint Research Centre (JRC), European Reference Life Cycle Database (ELCD). http://lca.jrc.ec.europa.eu/lcainfohub/.

Finnveden, G., Hauschild, M.Z., Ekvall, T., Guineé, J., Heijungs, R., Hellweg, S., Koehler, K., David Pennington, D., and Suh, S. 2009. Recent developments in life cycle assessment. *Journal of Environmental Management* 91: 1–21.

Gerngross, T.U. and Slater, S. 2000. How green are green polymers? *Scientific American* 282: 37–41.

Guinee, J.B., Huppes, G., and Heijungs, R. 2001. Developing and LCA guide for decision support. *Environmental Management and Health* 12: 301–310.

Harding, K.G., Dennis, J.S., and von Blottniz, S.T.L. 2007. Environmental analysis of plastic production process: Comparing petroleum-based polypropylene and polyethylene with biologically based poly-β-hydroxybutyric acid using life cycle analysis. *Journal of Biotechnology* 130: 57–66.

Heyde, M. 1998. Ecological considerations on the use and production of biosynthetic and synthetic biodegradable polymers. *Polymer Degradation Stability* 59: 3–6.

INFORRM. 2003. Case Study Solyanyl—mouldable bioplastic pellets. Industry Network for Renewable Resources and Materials. http://217.148.32.203/cs1.asp (accessed July 4, 2003).

ISO 14040. 2006a. International standard. In: *Environmental Management—Life Cycle Assessment—Principles and Framework*. International Organisation for Standardisation, Geneva, Switzerland.

ISO 14043. 2000b. International standard. In: *Environmental Management—Life Cycle Assessment—Life Cycle Interpretation*. International Organization for Standardisation, Geneva, Switzerland.

ISO 14044. 2006. International standard. In: *Environmental Management—Life Cycle Assessment—Requirements and Guidelines*. International Organization for Standardisation, Geneva, Switzerland.

James, K. and T. Grant. 2005. LCA of Degradeable plastic bags, Centre for Design at RMIT University, Melbourne.

JEMAI. 2007. Japan environmental management association for Industry (JEMAI). JEMAI database. http://www.jemai.or.jp/english/index.cfm (database available in Japanese language only).

Jolliet, O., Margni, M., Charles, R., Humbert, S., Payet, J., Rebitzer, G., and Rosenbaum, R. 2003. IMPACT 2002þ: A new life cycle impact assessment methodology. *International Journal of Life Cycle Analysis* 8: 324–330.

Koellner, T. and Scholz, R.W. 2008. Assessment of land use impacts on the natural environment—Part 2: Generic characterization factors for local species diversity in Central Europe. *International Journal of Life Cycle Assessment* 13: 32–48

Lindeijer, E., Muller-Wenk, R., and Steen, B. 2002. Impact assessment of resources and land use. In: Udo de Haes, H.A., Finnveden, G., Goedkoop, M., Hauschild, M., Hertwich, E.G., Hofstetter, P., Jolliet, O., Klopffer, W., Krewitt, W., Lindeijer, E.W., Müller-Wenk, R., Olsen, S.I., Pennington, D.W., Potting, J., and Steen, B. (Eds.), *Life-Cycle Impact Assessment: Striving Towards Best Practice*. SETAC Press, Pensacola, FL.

Michelson, O. 2008. Assessment of land use impact on biodiversity. Proposal of a new methodology exemplified with forestry operations in Norway. *International Journal of Life Cycle Assessment* 13: 22–31.

Mila i Canals, L., Bauer, C., Depestele, J., Dubreuil, A., Freiermuth Knuchel, R., Gaillard, G., Michelsen, O., Müller-Wenk, R., and Rydgren, B. 2007a. Key elements in framework for land use impact assessment within LCA. International Journal of Life Cycle Analysis, 12, 5-15.

Mohanty, A.K., Misra, M., and Hinrichsen, G. 2000. Biofibres, biodegradable polymers and biocomposites: An overview. *Macromolecular Materials and Engineering* 276–277: 1–24.

Momani, B. 2009. Assessment of the impacts of bioplastics: Energy usage, fossil fuel usage, pollution, health effects, effects on the food supply, and economic effects compared to petroleum based plastics. A bachelor's thesis, Worcester Polytechnic Institute, Worcester, MA.

Murphy, R.J., Bonin, M., and Hillier, W.R. 2004. Life cycle assessment of potato starch based packaging trays. Report prepared (draft) for STI Project Sustainable GB Potato Packaging, Imperial College London, London, U.K.

Narayanaswamy, V., Scott, J.A., Ness, J.N., and Lochheaad, M. 2003. Resource flow and product chain analysis as practical tools to promote cleaner production initiatives. *Journal of Cleaner Production* 11: 375–387.

Novamont. 2003. Life cycle thinking and life cycle analysis. http://www.novamont.com (accessed May 13, 2003).

NREL. 2004. National renewable energy laboratory (NREL). US Life Cycle Inventory Database. NREL, Golden, CO. http://www.nrel.gov/lci/.

Parker, A.A. 2008. The unintended consequences of trace impurities in polymer product development. Complimentary resource, A.A. Parker Consulting and Product Development, Newtown, PA, pp. 1–10.

Patel, M. 1999. Closing Carbon Cycles: Carbon use for materials in the context of resource efficiency and climate change. PhD thesis (ISBN 90-73958-51-2, http://www.library.uu.nl/digiarchief/dip/diss/1894529/inhoud.htm), Utrecht University, Utrecht, Netherlands.

Patel, M. 2005. Environmental life cycle comparisons of biodegradable plastics. In: Bastioli, C. (Ed.) *Handbook of Biodegradable Polymers*, Chapter 13, Rapra Technology Ltd., Shawbury, U.K., 2005, pp. 431–484.

Pennington, D., Crettaz, P., Tauxe, A., Rhomberg, L., Brand, K., and Jolliet, O. 2002. Assessing human health response in life cycle assessment using ED10s and DALYs: Part 2—Non cancer effects. *Risk Analysis* 22: 947–963.

RMIT. 2007. Centre for design, RMIT University. Australian LCI Database, Available at: http://www.auslci.com/.

Sarazin, P., Li, G., Orts, W.J., and Favis, B.D. 2008. Binary and ternary blends of polylactide, polycaprolactone and thermoplastic starch. *Polymer* 49: 599–609.

Scott, A., Narayanaswamy, V., and Ness, T. 2000. *Life Cycle Assessment Case studies, Centre for Integrated Environmental Protection*, Griffith University, Brisbane, Queensland, Australia.

Shen, L. and Patel, M.K. 2008. Life cycle assessment of polysaccharide materials: A review. *Journal of polymer Environment* 16: 154–167.

Tang, X. and Alavi, S. 2011. Recent advances in starch, polyvinyl alcohol based polymer blends, nanocomposites and their biodegradability. *Carbohydrate Polymers* 85: 7–16.

UBA. 2007. Umweltbundesamt (UBA) (German Environmental Protection Agency). PROBAS Database. http://www.probas.umweltbundesamt.de/php/index.php (available in German language only).

Vink, E.T.H., Rábago, K.R., Glassner, D.A., and Gruber, P.R. 2003. Applications of life cycle assessment to NatureWorks™ polylactide (PLA) production. *Polymer Degradation and Stability* 80(3): 403–419.

Würdinger, E., Roth, U., Wegener, A., Borken, J., Detzel, A., Fehrenbach, H., Giegrich, J., Möhler, S., Patyk, A., Reinhardt, G.A., Vogt, R., Mühlberger, D., and Wante, J. 2001. Kunststoffe aus nachwachsenden Rohstoffen Vergleichende Ökobilanz für Loose-fill-Packmittel aus Stärke bzw. aus Polystyrol (final report, DBU-Az. 04763). Bayrisches Institut für Angewandte Umweltforschung und technik, Augsburg (BIFA; project leader), Institut für Energie und Umweltforschung Heidelberg (IFEU), Flo-Pak GmbH, Germany, March 2002.

Index

For Product Safety Concerns and Information please contact our
EU representative GPSR@taylorandfrancis.com Taylor & Francis
Verlag GmbH, Kaufingerstraße 24, 80331 München, Germany